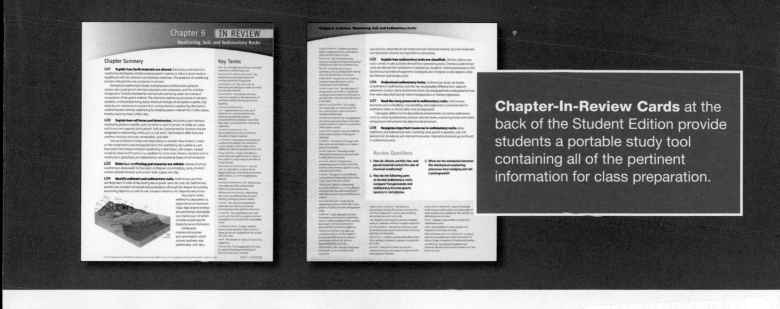

Chapter-In-Review Cards at the back of the Student Edition provide students a portable study tool containing all of the pertinent information for class preparation.

THE SOLUTION

GEOL Are you in?

D0403301

ONLINE RESOURCES INCLUDED!

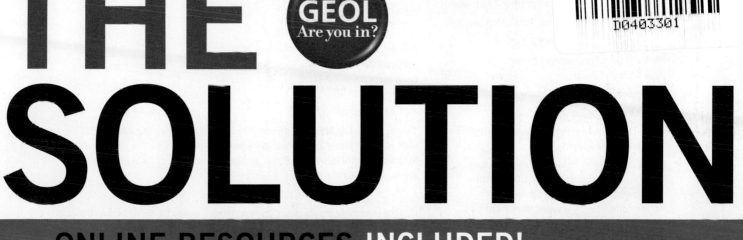

CourseMate — Engaging. Trackable. Affordable.

CourseMate brings course concepts to life with interactive learning, study, and exam preparation tools that support GEOL.

FOR INSTRUCTORS:
- First Day of Class Instructions
- Custom Options through 4LTR+ Program
- Instructor's Manual
- Test Bank
- PowerPoint® Slides
- Dynamic Zoom and Stepped Art
- ExamView® Test Generator
- Instructor Prep Cards
- Engagement Tracker
- Now includes all 15 Virtual Field Trips! See pages x and 1 for more details.
- And everything available to students!

FOR STUDENTS:
- Interactive eBook
- Flashcards
- Chapter Glossary
- Tutorial Quizzes
- Animations
- Videos
- Virtual Field Trips

Students sign in at
login.cengagebrain.com

BROOKS/COLE
CENGAGE Learning®

GEOL2
Reed Wicander, James S. Monroe

Earth Sciences Editor: Aileen Berg

Developmental Editor: Liana Sarkisian

4LTR Press Project Manager: Pierce Denny

Assistant Editor: Kristina Chiapella

Editorial Assistant: Shannon Elderon

Media Editor: Alexandria Brady

Marketing Manager: Jack Cooney

Marketing Coordinator: Julie Stefani

Marketing Communications Manager:
    Darlene Macanan

Content Project Manager: Hal Humphrey

Design Director: Rob Hugel

Art Director: John Walker

Print Buyer: Rebecca Cross

Rights Acquisitions Specialist: Dean Dauphinais

Production Service: Lachina Publishing Services

Text Designer: Diane Beasley

Photo Researcher: Wendy Granger,
    Bill Smith Group

Text Researcher: Ashley Liening

Copy Editor: Maggie Jarpey

Cover Image: Huangshan, China: Merijn van der
    Vliet/Getty Images

Inside Front Cover Images: chair and gears
    © Squared Studios/Photodisc/Getty
    Images; microscope © Siede Preis/
    Photodisc/Getty Images; dartboard and
    binoculars © Photodisc/Getty Images

Back Cover Image: © iStockphoto.com/
    René Mansi

Title Page Image: © iStockphoto.com/CostinT

Compositor: Lachina Publishing Services

For product information and technology assistance, contact us at
Cengage Learning Customer & Sales Support, 1-800-354-9706.

For permission to use material from this text or product,
submit all requests online at **www.cengage.com/permissions**.
Further permissions questions can be e-mailed to
permissionrequest@cengage.com.

Library of Congress Control Number: 2011933969

ISBN-13: 978-1-133-10869-6

ISBN-10: 1-133-10869-5

Brooks/Cole
20 Davis Drive
Belmont, CA 94002-3098
USA

Cengage Learning is a leading provider of customized learning solutions with office locations around the globe, including Singapore, the United Kingdom, Australia, Mexico, Brazil, and Japan. Locate your local office at **www.cengage.com/global**.

Cengage Learning products are represented in Canada by Nelson Education, Ltd.

To learn more about Brooks/Cole visit **www.cengage.com/brookscole**.

Purchase any of our products at your local college store or at our preferred online store **www.CengageBrain.com**.

Printed in the United States of America
4 5 6 7 17 16 15

Brief Contents

Contents

4 IGNEOUS ROCKS AND INTRUSIVE IGNEOUS ACTIVITY   70

5 VOLCANOES AND VOLCANISM   88

6 WEATHERING, SOIL, AND SEDIMENTARY ROCKS   108

7 METAMORPHISM AND METAMORPHIC ROCKS 138

8 EARTHQUAKES AND EARTH'S INTERIOR 154

9 DEFORMATION, MOUNTAIN BUILDING, AND THE CONTINENTS 180

10 MASS WASTING   202

11 RUNNING WATER   224

12 GROUNDWATER   246

13 GLACIERS AND GLACIATION   268

14 THE WORK OF WIND AND DESERTS   288

15 OCEANS, SHORELINES, AND SHORELINE PROCESSES   306

Virtual Field Trips in GEOLogy: Let's Go!

GOALS OF THE TRIPS

1. Improve student understanding of difficult geologic concepts
2. Get students into the field
3. Deliver a more meaningful experience to today's students using today's technology

GET INTO THE FIELD!

With access to the complete set of Virtual Field Trips in Geology, GEOL2 makes it easier than ever to engage students while educating them on the geologic concepts they need to succeed. Virtual Field Trips provide dynamic panoramas, high-definition videos and photos, key concepts, critical thinking questions, and more to serve diverse learning styles. Now available with all 15 Virtual Field Trips, GEOL2 enables students to explore famous national parks throughout the United States while being guided by a virtual instructor!

Instructors: To learn more about Virtual Field Trips, see a video demo, or get access, please visit www.cengage .com/earthscience/community and click on the Virtual Field Trips page.

Students: To access Virtual Field Trips, sign in with your access code at login.cengagebrain.com.

All photographs: Copyright and Photograph by Dr. Parvinder S. Sethi

Mineral Resources → Bingham Canyon Mine

↑ **Plate Tectonics**
Mount St. Helens and Sideling Hill Syncline

Igneous Rocks →
Hawaii Volcanoes and Yosemite National Parks

← **Volcano Types**
Hawaii Volcanoes National Park and Mount St. Helens

← **Metamorphic Rocks** Acadia National Park

↑ **Sedimentary Rocks**
Arches and Capitol Reef National Parks

Roadmap Found in Chapter 6!

Running Water →
Zion National Park

Roadmap
Found in
Chapter 11!

**↑ Earthquakes
and Seismicity**
Los Trancos Open Space
Preserve and Pt. Reyes
National Seashore

Roadmap
Found in
Chapter 16!

↑ Mass Wasting
Arches and Yosemite
National Parks

Roadmap
Found in
Chapter 12!

↑ Groundwater
Mammoth Cave National Park

**↓ Shorelines and
Shoreline Processes**
Bandon Beach

↑ Geologic Time
Grand Canyon and Capitol
Reef National Parks

↑ Hydrothermal Activity
Yellowstone National Park

Roadmap
Found in
Chapter 14!

↑ Glaciers and Glaciation
Kenai Fjords National Park and
Prince William Sound

← Desert Environments
Death Valley National Park

All photographs.
Copyright and
Photograph by Dr.
Parvinder S. Sethi

1 Understanding Earth: A Dynamic and Evolving Planet

Planet Observer/Universal Images Group/Getty Images

True color satellite image of Asia (partly in shadow), the Arctic ice cap, and the Sun. In this book, we examine Earth as a system of interconnected components that interact with each other. The atmosphere, biosphere, hydrosphere, and lithosphere are four of Earth's major subsystems that are visible in this image. The complex interactions among these subsystems, as well as Earth's interior, results in a dynamically changing planet.

Introduction

"**Earth** is a complex, dynamic planet that has **changed continuously** since its origin some 4.6 billion years **ago.**"

A major benefit of the space age has been the ability to look back from space and view our planet in its entirety. Every astronaut has remarked in one way or another on how Earth stands out as an inviting oasis in the otherwise black void of space. We are able to see not only the beauty of our planet, but also its fragility. We can also decipher Earth's long and frequently turbulent history by reading the clues preserved in the geologic record.

A major theme of this book is that Earth is a complex, dynamic planet that has changed continuously since its origin some 4.6 billion years ago. These changes and the present-day features we observe result from the interactions among Earth's internal and external systems, subsystems, and cycles. By viewing Earth as a whole—that is, thinking of it as a *system*—we not only see how its various components are interconnected, but we can also better appreciate its complex and dynamic nature.

> **system** A combination of related parts that interact in an organized fashion; Earth systems include the atmosphere, hydrosphere, biosphere, and solid Earth.

The system concept makes it easier for us to study a complex subject such as Earth, because it divides the whole into smaller components that we can easily understand, without losing sight of how the components fit together as a whole. In the same way, you can think of this book as if it were a large, panoramic landscape painting. Each chapter thus fills in the details of the landscape, thereby enhancing the overall enjoyment and understanding of the entire painting.

A **system** is a combination of related parts that interact in an organized manner. An automobile is a good example of a system. Its various components, or subsystems, such as the engine, transmission, steering, and brakes, are all interconnected in such a way that a change in any one of them affects the others.

We can examine Earth in the same way we view an automobile—that is, as a system of interconnected components that interact and affect each other in many ways. The principal subsystems of Earth are the atmosphere, biosphere, hydrosphere, lithosphere, mantle, and core (Figure 1.1). The complex interactions among these subsystems result in a dynamically changing planet in which matter and energy are continuously recycled into different forms.

We must also not forget that humans are part of the Earth system, and our activities can produce changes with potentially wide-ranging consequences. When people discuss and debate such environmental issues as pollution and global warming, it is important to remember that these are not isolated issues but are part of the larger Earth system. Furthermore, remember that

Learning Outcomes

After reading this unit, you should be able to do the following:

LO1 Define geology

LO2 Understand the impact that the formulation of theories has had on the study of geology

LO3 Explain how geology relates to the human experience

LO4 Explain how geology affects our everyday lives

LO5 Describe global geologic and environmental issues facing humankind

LO6 Describe the origin of the universe and solar system, and Earth's place in them

LO7 Explain why Earth is a dynamic and evolving planet

LO8 Describe the rock cycle

LO9 Define organic evolution and its role in the history of life

LO10 Describe geologic time and uniformitarianism

LO11 Explain how the study of geology benefits us

Atmosphere

Copyright and Photograph by Dr. Parvinder S. Sethi

Atmospheric gases and precipitation contribute to weathering of rocks.

Evaporation, condensation, and precipitation transfer water between atmosphere and hydrosphere, influencing weather and climate and distribution of water.

Plant, animal, and human activity affect composition of atmospheric gases. Atmospheric temperature and precipitation help to determine distribution of Earth's biota.

Copyright and Photograph by Dr. Parvinder S. Sethi

Hydrosphere

Plants absorb and transpire water. Water is used by people for domestic, agricultural, and industrial uses.

Water helps determine abundance, diversity, and distribution of organisms.

Biosphere

Copyright and Photograph by Dr. Parvinder S. Sethi

Plate movement affects size, shape, and distribution of ocean basins. Running water and glaciers erode rock and sculpt landscapes.

Organisms break down rock into soil. People alter the landscape. Plate movement affects evolution and distribution of Earth's biota.

Heat reflected from land surface affects temperature of atmosphere. Distribution of mountains affects weather patterns.

Lithosphere (plates)

Copyright and Photograph by Dr. Parvinder S. Sethi

Convection cells within mantle contribute to movement of plates (lithosphere) and recycling of lithospheric material.

Plate

Mantle

Supplies heat for convection in mantle

Core

© 2013 Cengage Learning

Figure 1.1 Subsystems of Earth

The atmosphere, hydrosphere, biosphere, lithosphere, mantle, and core are all subsystems of Earth. This simplified diagram shows how these subsystems interact, with some examples of how materials and energy are cycled throughout the Earth system. The interactions between these subsystems make Earth a dynamic planet that has evolved and changed since its origin 4.6 billion years ago.

Earth goes through longer time cycles than humans are used to. Although they may have disastrous short-term effects on the human species, global warming and cooling are also part of a longer-term cycle that has resulted in many glacial advances and retreats during the past 2.6 million years.

As you study the various topics covered in this book, keep in mind the themes discussed in this chapter and how, like the parts of a system, they are interrelated. By relating each chapter's topic to its place in the entire Earth system, you will gain a greater appreciation of why geology is so integral to our lives.

LO1 WHAT IS GEOLOGY?

Geology, from the Greek *geo* and *logos*, is defined as the study of Earth, but now must also include the study of the planets and moons in our solar system. It is generally divided into two broad areas—physical geology and historical geology. *Physical geology* is the study of Earth materials, such as **minerals** and **rocks**, as well as the processes operating within Earth and on its surface. *Historical geology* examines the origin and evolution of Earth and its continents, oceans, atmosphere, and life.

Although the discipline of geology is very broad and subdivided into numerous fields, or specialties, nearly every aspect of geology has some economic or environmental relevance. For example, many geologists are involved in exploration for mineral and energy resources, using their specialized knowledge to locate the natural resources on which our industrialized society is based.

Other geologists use their expertise to help solve environmental problems. Finding adequate sources of groundwater for the ever-burgeoning needs of communities and industries is becoming increasingly important, as is the monitoring of surface and underground water pollution and its cleanup. Geologic engineers help find safe locations for dams, waste-disposal sites, and power plants, as well as designing earthquake-resistant buildings.

Geologists are also engaged in making short- and long-range predictions about earthquakes and volcanic eruptions, and the potential destruction that may result.

LO2 GEOLOGY AND THE FORMULATION OF THEORIES

The term **theory** has various meanings. In colloquial usage, it means a speculative or conjectural view of something—hence, the widespread belief that scientific theories are little more than unsubstantiated wild guesses. In scientific usage, however, a theory is a coherent explanation for one or several related natural phenomena supported by a large body of objective evidence. From a theory, scientists derive predictive statements that they test by observations and/or experiments so that their validity can be assessed. The law of universal gravitation is an example of a theory that describes the attraction between masses (an apple and Earth in the popularized account of Newton and his discovery).

Theories are formulated through the process known as the **scientific method**. This method is an orderly, logical approach that involves gathering and analyzing facts or data about the problem under consideration. Tentative explanations, or **hypotheses**, are then formulated to explain the observed phenomena. Next, the hypotheses are tested to see whether what was predicted actually occurs in a given situation. Finally, if one of the hypotheses is found, after repeated tests, to explain the phenomena, then that hypothesis is proposed as a theory. Remember, however, that in science, even a theory is still subject to further testing and refinement as new data become available.

The fact that a scientific theory can be tested and is subject to such testing separates it from other forms of human inquiry. Because scientific theories can be tested, they have the potential for being supported or even proven wrong. Accordingly, science must proceed without any appeal to beliefs or supernatural explanations, not because such beliefs or explanations are necessarily untrue, but because we have no way to investigate them. For this reason, science makes no claim about the existence or nonexistence of a supernatural or spiritual realm.

Each scientific discipline has certain theories that are particularly important. In geology, the formulation

---

geology The science concerned with the study of Earth materials (minerals and rocks), surface and internal processes, and Earth history.

mineral A naturally occurring, inorganic, crystalline solid that has characteristic physical properties and a narrowly defined chemical composition.

rock A solid aggregate of one or more minerals, as in limestone and granite, or a consolidated aggregate of rock fragments, as in conglomerate, or masses of rocklike materials, such as coal and obsidian.

theory An explanation for some natural phenomenon that has a large body of supporting evidence. To be scientific, a theory must be testable (e.g., plate tectonic theory).

scientific method A logical, orderly approach that involves gathering data, formulating and testing hypotheses, and proposing theories.

hypothesis A provisional explanation for observations that is subject to continual testing. If well-supported by evidence, a hypothesis may be called a theory.

of *plate tectonic theory* (discussed later in this chapter) has changed the way geologists view Earth. Geologists now view Earth from a global perspective in which all of its subsystems and cycles are interconnected, and Earth history is seen to be a continuum of interrelated events that are part of a global pattern of change.

Many sketches and paintings depict rocks and landscapes realistically. Leonardo da Vinci's *Virgin of the Rocks* and *Virgin and Child with Saint Anne*, Giovanni Bellini's *Saint Francis in Ecstasy* and *Saint Jerome*, and Asher Brown Durand's *Kindred Spirits* are just a few examples by famous painters.

R. Gino Santa Maria/Shutterstock.com

LO3 HOW DOES GEOLOGY RELATE TO THE HUMAN EXPERIENCE?

You would probably be surprised at the extent to which geology pervades our everyday lives and the numerous references to geology in the arts, music, and literature. Many sketches and paintings by famous painters depict rocks and landscapes realistically. Examples include Leonardo da Vinci's *Virgin of the Rocks*, Giovanni Bellini's *Saint Francis in Ecstasy*, and Asher Brown Durand's *Kindred Spirits*.

In the field of music, Ferde Grofé's *Grand Canyon Suite* was no doubt inspired by the grandeur and timelessness of Arizona's Grand Canyon and its vast rock exposures (Figure 1.2). The rocks on the Island of Staffa in the Inner Hebrides provided the inspiration for Felix Mendelssohn's famous *Hebrides Overture*.

References to geology abound in *The German Legends of the Brothers Grimm*. Jules Verne's novel *Journey to the Center of the Earth* describes an expedition into Earth's interior. There is even a series of mystery books by Sarah Andrews that features the fictional geologist Em Hansen, who uses her knowledge of geology to solve crimes.

Geology has also played an important role in the history and culture of humankind. Empires throughout history have risen and fallen on the distribution and exploitation of natural resources. Wars have been fought to secure such natural resources as oil and gas, and valuable minerals such as gold, silver, and diamonds.

LO4 HOW DOES GEOLOGY AFFECT OUR EVERYDAY LIVES?

The most obvious connection between geology and our everyday lives is when natural disasters such as volcanic

eruptions, earthquakes, landslides, tsunami, and floods strike. Less apparent, but equally significant, are the connections between geology and economic, social, and political issues.

Consider, for example, just how dependent we are on geology in our daily routines (Figure 1.3). Much of the electricity for our appliances comes from the burning of coal, oil, natural gas, or uranium consumed in nuclear-generating plants. Geologists locate the coal, petroleum (oil

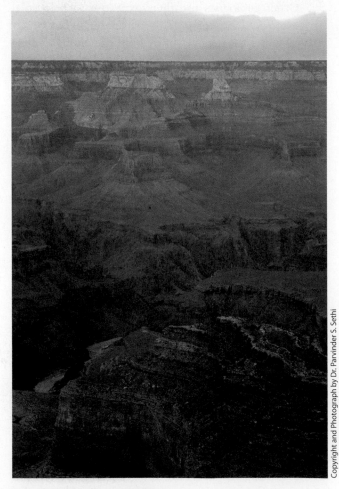

Copyright and Photograph by Dr. Parvinder S. Sethi

Figure 1.2 Geology and Art
Ferde Grofé's Grand Canyon Suite was inspired by the beauty of Arizona's Grand Canyon, where sedimentary rock layers grandly document some of Earth's past history.

Geology and Economic and Political Power

Geology can be closely connected to economic and political power. The configuration of Earth's surface, or its topography, which is shaped by geologic agents, has often played a critical role in military tactics. For example, Napoleon included two geologists in his expeditionary forces when he invaded Egypt in 1798, and the Russians used geologists as advisors in selecting fortification sites during the Russo-Japanese war of 1904–1905. Natural barriers such as mountain ranges and rivers have frequently served as political boundaries, and the shifting of river channels has sparked numerous border disputes. Mineral and energy resources are not equally distributed, and no country is self-sufficient in all of them. Throughout history, people have fought wars to secure these resources. The United States was involved in the 1990–1991 Gulf War largely because it needed to protect its oil interests in that region. Many foreign policies and treaties develop from the need to acquire and maintain adequate supplies of mineral and energy resources.

© John Lund/Blend Images/Jupiterimages

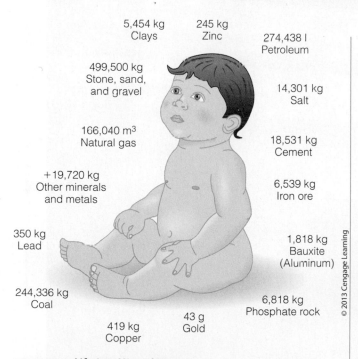

5,454 kg Clays

245 kg Zinc

274,438 l Petroleum

499,500 kg Stone, sand, and gravel

14,301 kg Salt

166,040 m³ Natural gas

18,531 kg Cement

+19,720 kg Other minerals and metals

6,539 kg Iron ore

350 kg Lead

1,818 kg Bauxite (Aluminum)

244,336 kg Coal

6,818 kg Phosphate rock

419 kg Copper

43 g Gold

© 2013 Cengage Learning

Figure 1.3 Lifetime Mineral Usage

According to the Mineral Information Institute in Golden, Colorado, the average American born in 2009 has a life expectancy of 77.7 years and will need 1,305,000 kg of minerals, metals, and fuels to sustain his or her standard of living over a lifetime. This is an average of 16,959 kg of mineral and energy resources per year for every man, woman, and child in the United States.

and natural gas), and uranium. The copper or other metal wires through which electricity travels are manufactured from materials found as the result of mineral exploration. The concrete foundation (concrete is a mixture of clay, sand, or gravel, and limestone), drywall (made largely from the mineral gypsum), and windows (the mineral quartz is the principal ingredient in the manufacture of glass) of the buildings we live and work in owe their very existence to geologic resources.

The car or public transportation we use to go to work is powered and lubricated by some type of petroleum by-product and is constructed of metal alloys and plastics. And the roads or rails we ride over come from geologic materials such as gravel, asphalt, concrete, or steel. All of these items are the result of processing geologic resources.

As individuals and societies, we enjoy a standard of living that is obviously directly dependent on the consumption of geologic materials. We therefore need to be aware of how our use and misuse of geologic resources may affect the environment. We need to support the development of policies that not only encourage management of our natural resources, but also allow for continuing economic development among all the world's nations.

LO5 GLOBAL GEOLOGIC AND ENVIRONMENTAL ISSUES FACING HUMANKIND

Most scientists would argue that overpopulation is the greatest environmental problem facing the world today. The world's population reached 7 billion in 2011, and projections indicate that this number will grow to 9 billion people by 2045. Although this may not seem to be a geologic problem, remember that these people must be fed, housed, and clothed, and all with a minimal impact on the environment. Much of this population growth will be in areas that are already at risk from such hazards as earthquakes, tsunami, volcanic eruptions, and floods. Adequate water supplies must be found and protected from pollution. Additional energy resources will be needed to help fuel the economies of nations with ever-increasing populations (Figure 1.4). New techniques must be developed to reduce the use of our dwindling nonrenewable resource base and to increase our recycling efforts so that we can decrease our dependence on new sources of these materials.

The problems of overpopulation and how it affects the global ecosystem vary from country to country. For many poor and non-industrialized countries, the problem is too many people and not enough food. For the

Overpopulation is the greatest environmental problem facing the world today.

Figure 1.4 Offshore Oil Drilling
With increasing demand for energy, offshore drilling for oil and natural gas has increased in recent years.

a. Short-wavelength radiation from the Sun that is not reflected back into space penetrates the atmosphere and warms Earth's surface.

b. Earth's surface radiates heat in the form of long–wavelength radiation back into the atmosphere, where some of it escapes Into space. The rest is absorbed by greenhouse gases and water vapor and reradiated back toward Earth.

c. Increased concentrations of greenhouse gases trap more heat near Earth's surface, causing a general increase in surface and atmospheric temperatures, which leads to global warming.

Figure 1.5 The Greenhouse Effect and Global Warming

more developed and industrialized countries, it is too many people rapidly depleting both the nonrenewable and renewable natural resource base. And in the most industrially developed countries, it is people producing more pollutants than the environment can safely recycle on a human time scale. The common thread tying these varied situations together is an environmental imbalance created by a human population exceeding Earth's short-term carrying capacity.

Other global issues in the news all the time are the greenhouse effect, global warming, and climate change. The relationship between the greenhouse effect and global warming is an excellent example of how Earth's various subsystems are interrelated. As a by-product of respiration and the burning of organic material, carbon dioxide is a component of the global ecosystem and is constantly being recycled as part of the carbon cycle. The concern in recent years over the increase in atmospheric carbon dioxide levels is related to its role in the greenhouse effect.

The recycling of carbon dioxide between Earth's *crust* (discussed shortly) and atmosphere is an important climate regulator, because carbon dioxide and other gases, such as methane, nitrous oxide, chlorofluorocarbons, and water vapor, allow sunlight to pass through them but trap the heat reflected back from Earth's surface. This retention of heat is called the *greenhouse effect*. It results in an increase in the temperature of Earth's surface and, more importantly, its atmosphere, thus producing global warming (Figure 1.5). The issue is not whether we have a greenhouse effect, because we do, but rather the degree to which human activity, such as the burning of fossil fuels, is increasing the greenhouse effect, and thus contributing to global warming.

Because of the increase in human-produced greenhouse gases during the past 200 years, many scientists are concerned that a global warming trend has already begun and will result in severe global climatic shifts. Presently, most climate researchers use a range of scenarios for greenhouse gas emissions when predicting future warming rates. State-of-the-art climate model simulations published in the *2007 Fourth Intergovernmental Panel on Climate Change* show a predicted increase in global average temperature from 2000 to 2100 of 1°–3°C under the best conditions, to a 2.5°–6.5°C rise under "business as usual" conditions. These predicted increases in temperatures are based on various scenarios that explore different global development pathways.

Regardless of which scenario is followed, the global temperature change will be uneven, with the greatest warming occurring in the higher latitudes of the Northern Hemisphere. As a consequence, rainfall patterns will shift dramatically. That shift will have a major effect on the largest grain-producing areas of the world, such as the American Midwest. Drier and hotter conditions will intensify the severity and frequency of droughts, leading to more crop failures and higher food prices (Figure 1.6). With such shifts in climate, Earth's deserts may expand, with a resulting decrease in valuable crop and grazing land. Higher temperature will also affect regional water supplies, creating potential water crises in the western United States within the next 20 years, as well as other areas such as Peru and western China.

As climates change, diseases such as malaria can easily spread into areas of warmer, wetter climates, regions that heretofore were inhospitable to disease-carrying mosquitoes. Moreover, continued global warming will result in a rise in mean sea level, as icecaps and glaciers melt and contribute their water to the world's oceans. At the current rate of glacial melting, sea level will rise 21 cm by around 2050, thus increasing the number of people at risk from flooding in coastal areas by approximately 20 million!

We would be remiss, however, if we did not point out that many other scientists are not convinced that the global warming trend is the direct result of increased human activity related to industrialization. They indicate that although the level of greenhouse gases has increased, we are still uncertain about their rate of generation and rate of removal, and about whether the rise in global temperatures during the past century resulted from normal climatic variations through time or from human activity. Furthermore, they conclude that even if a general global warming trend occurs during the next hundred years, it is not certain that the dire predictions made by proponents of global warming will come true.

Earth, as we know, is a remarkably complex system, with many feedback mechanisms and interconnections throughout its various subsystems and cycles. It is very difficult to predict all of the consequences that global warming would have for atmospheric and

Figure 1.6 Withered Corn Crop as a Result of Drought Conditions

oceanic circulation patterns and its ultimate effect on Earth's biota. It is, however, important to remember that although everyone is vulnerable to weather-related disasters, large-scale changes brought about by climate change will impact people in poor countries more than those in the more industrialized countries. Whether these climate changes are part of a natural global cycle taking place over thousands or hundreds of thousands of years—that is, on a geologic time scale—or are driven, in part, by human activities, is immaterial. The bottom line is that we already are, or eventually will be, affected by them in some way, be it economic or social.

LO6 ORIGIN OF THE UNIVERSE AND SOLAR SYSTEM, AND EARTH'S PLACE IN THEM

How did the universe begin? What has been its history? What is its eventual fate, or is it infinite? These are just some of the basic questions people have asked and wondered about since they first looked into the nighttime sky and saw the vastness of the universe beyond Earth.

Origin of the Universe: Did It Begin with a Big Bang?

Most scientists think that the universe originated about 14 billion years ago in what is popularly called the **Big Bang**. The Big Bang is a model for the evolution of the universe in which a dense, hot state, was followed by expansion, cooling, and a less-dense state.

According to modern *cosmology* (the study of the origin, evolution, and nature of the universe), the universe has no edge and therefore no center. Thus, when the universe began, all matter and energy were compressed into an infinitely small high-temperature and high-density state in which both time and space were set at zero. Therefore, there is no "before the Big Bang," only what occurred after it. As demonstrated by Einstein's Theory of Relativity, space and time are unalterably linked to form a *space–time continuum*; that is, without space, there can be no time.

How do we know that the Big Bang took place approximately 14 billion years ago? Why couldn't the universe have always existed as we know it today? Two fundamental phenomena indicate that the Big Bang occurred: (1) the universe is expanding, and (2) it is permeated by background radiation.

When astronomers look beyond our own solar system, they observe that everywhere in the universe galaxies are moving away from each other at tremendous speeds. Edwin Hubble first recognized this phenomenon in 1929. By measuring the optical spectra of distant galaxies, Hubble noted that the velocity at which a galaxy moves away from Earth increases proportionally to its distance from Earth. He observed that the *spectral lines* (wavelengths of light) of the galaxies are shifted toward the red end of the spectrum; that is, the lines are shifted toward longer wavelengths. Galaxies receding from each other at tremendous speeds would produce such a redshift. This is an example of the *Doppler effect*, which is a change in the frequency of a sound, light, or other wave caused by movement of the wave's source relative to the observer.

One way to understand the Doppler effect is by analogy to the sound of a passing train's whistle. As the train approaches, the sound waves are slightly compressed, so an individual hears a shorter-wavelength, higher-pitched sound. As the train passes, and recedes from the individual, the sound waves are slightly expanded, and a longer-wavelength, lower-pitched sound is heard.

An easy way to envision how velocity increases with increasing distance is by reference to the popular analogy of a rising loaf of raisin bread, in which the raisins are uniformly distributed throughout the loaf (Figure 1.7). As the dough rises, the raisins are uniformly pushed away from each other at velocities directly proportional to the distance between any two raisins. The farther away a given raisin is to begin with, the farther it must move to maintain the regular spacing during the expansion, and hence the greater its velocity must be.

In the same way that raisins move apart in a rising loaf of bread, galaxies are receding from each other at a rate proportional to the distance between them, which is exactly what astronomers see when they observe the universe. By measuring this expansion rate, astronomers can calculate how long ago the galaxies were all together at a single point, which turns out to be about 14 billion years, the currently accepted age of the universe.

In 1965, Arno Penzias and Robert Wilson of Bell

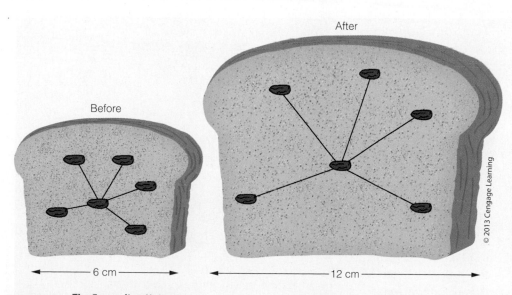

© 2013 Cengage Learning

Figure 1.7 The Expanding Universe
The motion of raisins in a rising loaf of raisin bread illustrates the relationship that exists between distance and speed and is analogous to an expanding universe. In this diagram, adjacent raisins are located 2 cm apart before the loaf rises. After one hour, any raisin is now 4 cm away from its nearest neighbor and 8 cm away from the next raisin over, and so on. Therefore, from the perspective of any raisin, its nearest neighbor has moved away from it at a speed of 2 cm per hour, and the next raisin over has moved away from it at a speed of 4 cm per hour. In the same way that raisins move apart in a rising loaf of bread, galaxies are receding from each other at a rate proportional to the distance between them.

Telephone Laboratories made the second important observation that provided evidence of the Big Bang. They discovered that there is a pervasive background radiation of 2.7 Kelvin (K) above absolute zero (absolute zero equals –273°C; 2.7 K = –270.3°C) everywhere in the universe. This background radiation is thought to be the fading afterglow of the Big Bang.

Currently, cosmologists cannot say what it was like at time zero of the Big Bang, because they do not understand the physics of matter and energy under such extreme conditions. However, it is thought that during the first second following the Big Bang, the four basic forces—(1) *gravity* (the attraction of one body toward another), (2) *electromagnetic force* (combines electricity and magnetism into one force and binds atoms into molecules), (3) *strong nuclear force* (binds protons and neutrons together), and (4) *weak nuclear force* (responsible for the breakdown of an atom's nucleus, producing radioactive decay)—separated, and the universe experienced enormous expansion.

As the universe continued expanding and cooling, stars and galaxies began to form, and the chemical makeup of the universe changed. Initially, the universe was 100% hydrogen and helium, whereas today it is 98% hydrogen and helium and 2% all other elements by weight.

Our Solar System: Its Origin and Evolution

Our solar system, which is part of the Milky Way galaxy, consists of the Sun, eight planets, five dwarf planets (including Pluto), 101 known moons or satellites (although this number keeps changing with the discovery of new moons and satellites surrounding the *Jovian planets*, discussed shortly), a tremendous number of asteroids—most of which orbit the Sun in a zone between Mars and Jupiter—and millions of comets and meteorites, as well as interplanetary dust and gases (Figure 1.8). Any theory formulated to explain the origin and evolution of our solar system must therefore take into account its various features and characteristics.

Many scientific theories for the origin of the solar system have been proposed, modified, and discarded since the French scientist and philosopher René Descartes first proposed, in 1644, that the solar system formed from a gigantic whirlpool within a universal fluid. Today, the **solar nebula theory** for the formation of the solar system not only best explains the features of the solar system, but also provides a logical explanation for its evolutionary history (Figure 1.9).

According to the solar nebula theory, the condensation and subsequent collapse of interstellar material in a spiral arm of the Milky Way galaxy resulted in a counterclockwise-rotating disk of gases and small grains. About 90% of this material was concentrated in

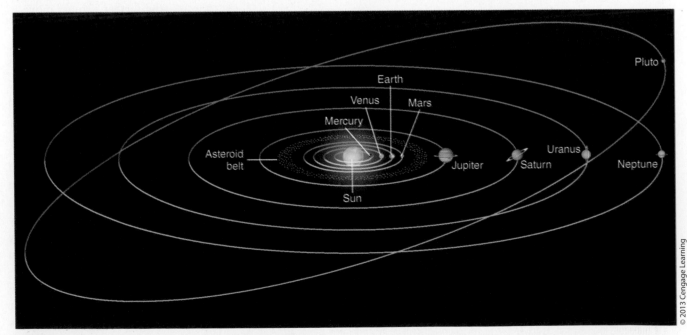

© 2013 Cengage Learning

Figure 1.8 Diagrammatic Representation of the Solar System

This representation of the solar system shows the planets and their orbits around the Sun. On August 24, 2006, the International Astronomical Union downgraded Pluto from a planet to a dwarf planet. A dwarf planet has the same characteristics as a planet, except that it does not clear the neighborhood around its orbit. Pluto orbits among the icy debris of the Kuiper Belt, and therefore does not meet the criteria for a true planet.

the central part of the disk, thus forming an embryonic Sun, around which swirled a rotating cloud of material called a *solar nebula*. Within this solar nebula were localized eddies in which gases and solid particles condensed. During the condensation process, gaseous, liquid, and solid particles began to accrete into ever-larger masses called *planetesimals*, which collided and grew in size and mass until they eventually became planets.

The composition and evolutionary history of the planets are a consequence, in part, of their distance from the Sun. The **terrestrial planets**—Mercury, Venus, Earth, and Mars—so named because they are similar to *terra*, Latin for "earth," are all small and composed of rock and metallic elements that condensed at the high temperatures of the inner nebula. The **Jovian planets**—Jupiter, Saturn, Uranus, and Neptune—so named because they resemble Jupiter (the Roman god was also called Jove), all have small, rocky cores compared to their overall size, and are composed mostly of hydrogen, helium, ammonia, and methane, which condense at low temperatures.

While the planets were accreting, material that had been pulled into the center of the nebula also condensed, collapsed, and was heated to several million degrees by gravitational compression. The result was the birth of a star, our Sun.

During the early accretionary phase of the solar system's history, collisions between various bodies were common, as indicated by the craters on many planets and moons. Asteroids probably formed as planetesimals in a localized eddy between what eventually became Mars and Jupiter in much the same way that other planetesimals formed the terrestrial planets. The tremendous gravitational field of Jupiter, however, prevented this material from ever accreting into a planet. Comets, which are interplanetary bodies composed of loosely bound rocky and icy materials, are thought to have condensed near the orbits of Uranus and Neptune.

Earth: Its Place in Our Solar System

Some 4.6 billion years ago, planetesimals in our solar system gathered enough material together to form Earth and the other planets. Scientists think that this early Earth was probably cool, of generally uniform composition and density throughout, and composed mostly of silicates (compounds consisting of silicon and oxygen), iron and magnesium oxides, and smaller amounts of all the other chemical elements (Figure 1.10a). Subsequently, when the combination of meteorite impacts, gravitational compression, and heat from radioactive decay increased the temperature of Earth enough to melt iron and nickel, this homogeneous composition disappeared (Figure 1.10b) and was replaced by a series of concentric layers of differing composition and density, resulting in a differentiated planet (Figure 1.10c).

This differentiation into a layered planet is probably the most significant event in Earth's history. Not only did it lead to the formation of a crust and eventually continents, but it also was probably responsible for the emission of gases from the interior that eventually led to the origin of the oceans and atmosphere.

a. A huge rotating cloud of gas contracts and flattens

b. to form a disk of gas and dust with the Sun forming in the center

c. and eddies gathering up material to form planets.

Figure 1.9 Solar Nebula Theory
According to the currently accepted theory for the origin of our solar system, the planets and the Sun formed from a rotating cloud of gas.

a. Early Earth probably had a uniform composition and density throughout.

b. The temperature of early Earth reached the melting point of iron and nickel, which, being denser than silicate minerals, settled to Earth's center. At the same time, the lighter silicates flowed upward to form the mantle and the crust.

c. In this way, a differentiated Earth formed, consisting of a dense iron–nickel core, an iron-rich silicate mantle, and a silicate crust with continents and ocean basins.

Crust
Mantle
Liquid outer core
Solid inner core

Figure 1.10 Homogenous Accretion Theory for the Formation of a Differentiated Earth

core The interior part of Earth beginning at a depth of 2,900 km that probably consists mostly of iron and nickel.

mantle The thick layer between Earth's crust and core.

asthenosphere The part of the mantle that lies below the lithosphere; it behaves plastically and flows slowly.

magma Molten rock material generated within Earth.

lithosphere Earth's outer, rigid part, consisting of the upper mantle, oceanic crust, and continental crust.

plate An individual segment of the lithosphere that moves over the asthenosphere.

LO7 WHY EARTH IS A DYNAMIC AND EVOLVING PLANET

Earth is a dynamic planet that has continuously changed during its 4.6-billion-year existence. The size, shape, and geographic distribution of continents and ocean basins have changed throughout time; the composition of the atmosphere has evolved; and life-forms existing today differ from those that lived during the past. Mountains and hills have been worn away by erosion, and the forces of wind, water, and ice have sculpted a diversity of landscapes. Volcanic eruptions and earthquakes reveal an active interior, and folded and fractured rocks are testimony to the tremendous power of Earth's internal forces.

Earth consists of three concentric layers: the core, the mantle, and the crust (Figure 1.11). This orderly division results from density differences between the layers as a function of variations in composition, temperature, and pressure.

The **core** has a calculated density of 10–13 grams per cubic centimeter (g/cm^3) and occupies about 16% of Earth's total volume. Seismic (earthquake) data indicate that the core consists of a small, solid inner region and a larger, apparently liquid, outer portion. Both are thought to consist mostly of iron and a small amount of nickel.

The **mantle** surrounds the core and comprises about 83% of Earth's volume. It is less dense than the core (3.3–5.7 g/cm^3) and is thought to be composed mostly of *peridotite*, a dark, dense, igneous rock containing abundant iron and magnesium. The mantle is divided into three distinct zones based on physical characteristics. The lower mantle is solid and forms most of the volume of Earth's interior. The **asthenosphere** surrounds the lower mantle. It has the same composition as the lower mantle but behaves plastically and flows slowly. Partial melting within the asthenosphere generates **magma** (molten material), some of which rises to the surface, because it is less dense than the rock from which it was derived. The upper mantle surrounds the asthenosphere. The solid upper mantle and the overlying crust constitute the **lithosphere**, which is broken into numerous individual pieces called **plates** that move over the asthenosphere, partially as a result of underlying *convection cells* (Figure 1.12). Interactions

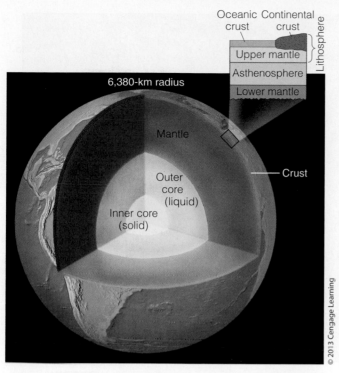

6,380-km radius

Oceanic Continental
crust crust
Upper mantle
Asthenosphere
Lower mantle
Lithosphere

Mantle

Crust

Outer
core
(liquid)

Inner core
(solid)

© 2013 Cengage Learning

Figure 1.11 Cross Section of Earth Illustrating the Core, Mantle, and Crust

The enlarged portion shows the relationship between the lithosphere (composed of the continental crust, oceanic crust, and solid upper mantle) and the underlying asthenosphere and lower mantle.

of these plates are responsible for such phenomena as earthquakes, volcanic eruptions, and the formation of mountain ranges and ocean basins.

The **crust**, Earth's outermost layer, consists of two types. *Continental crust* is thick (20–90 km), has an aver-age density of 2.7 g/cm³, and contains consider-able silicon and alumi-num. *Oceanic crust* is thin (5–10 km), denser than continental crust (3.0 g/cm³), and is com-posed of the dark igne-ous rocks *basalt* and *gabbro*.

Plate Tectonic Theory

The recognition that the lithosphere is divided into rigid plates that move over the asthenosphere (Figure 1.13) forms the foundation of **plate tectonic theory**, a unify-ing theory of geology that will be discussed in greater detail in Chapter 2. Zones of volcanic activity, earthquakes, or both, mark most plate boundaries. Along these boundar-ies, plates separate (diverge), collide (converge), or slide sideways past each other.

The acceptance of plate tectonic theory is recog-nized as a major milestone in the geologic sciences, comparable to the revolution that Darwin's theory of evolution caused in biology. Plate tectonics has pro-vided a framework for interpreting the composition, structure, and internal processes of Earth on a global scale. It has led to the realization that the continents and ocean basins are part of a lithosphere–atmo-sphere–hydrosphere system that evolved together with Earth's interior.

crust Earth's outermost layer, and the upper part of the lithosphere; consists of continental and oceanic crust.

plate tectonic theory The theory holding that large segments of Earth's outer part (lithospheric plates) move relative to one another.

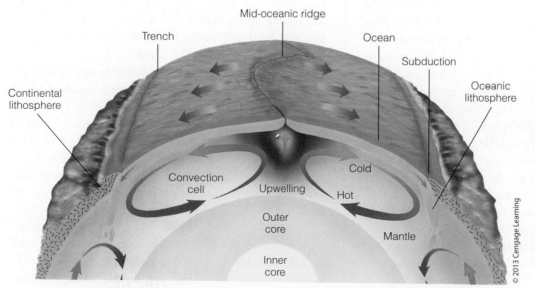

Trench

Mid-oceanic ridge

Ocean

Subduction

Oceanic
lithosphere

Continental
lithosphere

Convection
cell

Upwelling

Cold

Hot

Outer
core

Mantle

Inner
core

© 2013 Cengage Learning

Figure 1.12 Movement of Earth's Plates

Earth's plates are thought to move partially as a result of underlying mantle convection cells in which warm material from deep within Earth rises toward the surface, cools, and then, upon losing heat, descends back into the interior as shown in this diagrammatic cross section.

| Ridge axis | Transform fault | Subduction zone | Zones of extension within continents | Uncertain plate boundary |
| Divergent boundary | Transform boundary | Convergent boundary | | |

Figure 1.13 Earth's Plates

Earth's lithosphere is divided into rigid plates of various sizes that move over the asthenosphere.

rock cycle A group of processes through which Earth materials may pass as they are transformed from one major rock type to another.

igneous rock Any rock formed by cooling and crystallization of magma or lava or by the consolidation of pyroclastic materials.

A revolutionary concept when it was proposed in the 1960s, plate tectonic theory has had far-reaching consequences in all fields of geology, because it provides the basis for relating many seemingly unrelated phenomena, such as the formation and occurrence of Earth's natural resources, as well as the distribution and evolution of the world's biota. Furthermore, the impact of plate tectonic theory has been particularly notable in the interpretation of Earth's history. For example, the Appalachian Mountains in eastern North America and the mountain ranges of Greenland, Scotland, Norway, and Sweden are not the result of unrelated mountain-building episodes but, rather, are part of a larger mountain-building event that involved the closing of an ancient Atlantic Ocean and the formation of the supercontinent Pangaea approximately 251 million years ago.

LO8 THE ROCK CYCLE

A rock is an aggregate of *minerals*, which are naturally occurring, inorganic, crystalline solids that have definite physical and chemical properties. Minerals are composed of elements such as oxygen, silicon, and aluminum, and elements are made up of atoms, the smallest particles of matter that retain the characteristics of an element. More than 3,500 minerals have been identified and described, but only about a dozen make up the bulk of the rocks in Earth's crust.

Geologists recognize three major groups of rocks—igneous, sedimentary, and metamorphic—each of which is characterized by its mode of formation. The **rock cycle** is a pictorial representation of events leading to the origin, destruction and/or changes, and reformation of rocks as a consequence of Earth's internal and surface processes (Figure 1.14). Furthermore, it shows that the three major rock groups are interrelated; that is, any rock type can be derived from the others.

Igneous rocks result when magma or lava crystallizes, or when volcanic ejecta (pyroclastic materials), such as ash, accumulate and consolidate. As magma cools, minerals crystallize, and the resulting rock is characterized

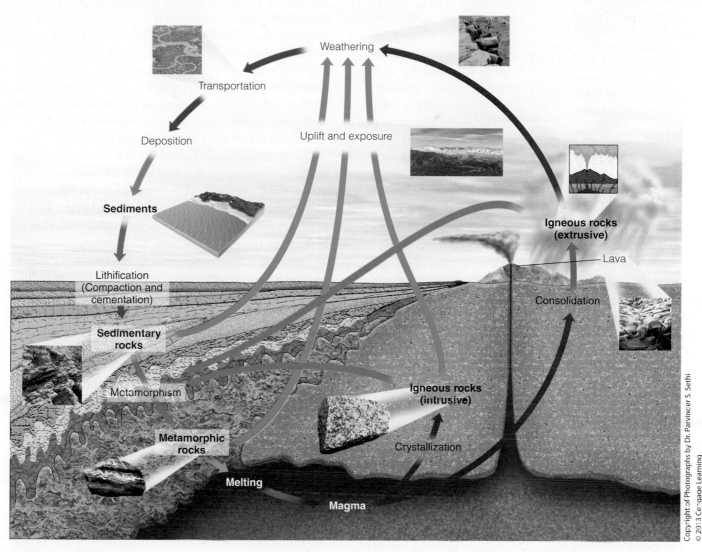

Weathering

Transportation

Deposition

Uplift and exposure

Sediments

Lithification
(Compaction and
cementation)

**Sedimentary
rocks**

Metamorphism

**Metamorphic
rocks**

Melting

Magma

**Igneous rocks
(extrusive)**

Lava

Consolidation

**Igneous rocks
(intrusive)**

Crystallization

Figure 1.14 The Rock Cycle

This cycle shows the interrelationship between Earth's internal and external processes and how the three major rock groups are related. An ideal cycle includes the events on the outer margin of the cycle, but interruptions, indicated by internal arrows, are common.

by interlocking mineral grains. Magma that cools slowly beneath the surface produces *intrusive igneous rocks* (Figure 1.15a); magma that cools at the surface produces *extrusive igneous rocks* (Figure 1.15b).

Rocks exposed at Earth's surface are broken into particles and dissolved by various weathering processes. The particles and dissolved materials may be transported by wind, water, or ice and eventually deposited as *sediment*. This sediment may then be compacted, or cemented (lithified), into sedimentary rock.

Sedimentary rocks form in one of three ways (Figure 1.15c, d): (1) consolidation of mineral or rock fragments, (2) precipitation of mineral matter from solution, or (3) compaction of plant or animal remains. Because sedimentary rocks form at or near Earth's surface, geologists can infer some information about the environment in which they were deposited, the transporting agent, and perhaps even something about the source from which the sediments were derived (see Chapter 6). Accordingly, sedimentary rocks are especially useful for interpreting Earth history.

Metamorphic rocks result from the alteration

**sedimentary
rock** Any rock composed of sediment, such as limestone and sandstone.

**metamorphic
rock** Any rock that has been changed from its original condition by heat, pressure, and the chemical activity of fluids, as in marble and slate.

of other rocks, usually beneath the surface, by heat, pressure, and the chemical activity of fluids. For example, marble—a rock preferred by many sculptors and builders—is a metamorphic rock produced when the agents of metamorphism are applied to the sedimentary rocks limestone or dolostone. Metamorphic rocks are either *foliated* (Figure 1.15e) or *nonfoliated* (Figure 1.15f). Foliation, the parallel alignment of minerals due to pressure, gives the rock a layered, or banded, appearance.

How Are the Rock Cycle and Plate Tectonics Related?

Interactions between plates determine, to some extent, which of the three rock groups will form (Figure 1.16). For example, when plates converge, heat and pressure generated along the plate boundary may lead to igneous activity and metamorphism within the descending oceanic plate, thus producing various igneous and metamorphic rocks.

Some of the sediments and sedimentary rocks on the descending plate are melted, whereas other sediments and sedimentary rocks along the boundary of the nondescending plate are metamorphosed by the heat and pressure generated along the converging plate boundary. Later, the mountain range or chain of volcanic islands formed along the convergent plate boundary will be weathered and eroded, and the new sediments will be transported to the ocean, where they will be deposited.

LO9 ORGANIC EVOLUTION AND THE HISTORY OF LIFE

Plate tectonic theory provides us with a model for understanding the internal workings of Earth and their effect on Earth's surface. The theory of **organic evolution** (whose central thesis is that all present-day organisms are related and that they have descended with modifications from organisms that lived in the past) provides the conceptual framework for understanding the history of life. Together, the theories of plate tectonics and organic evolution have changed the way we view our planet, and we should not be surprised at the intimate association between them. Although the relationship between plate tectonic processes and the evolution of life is incredibly complex, paleontological data provide indisputable evidence of the influence of plate movement on the distribution of organisms.

The publication in 1859 of Darwin's *On the Origin of Species by Means of Natural Selection* revolutionized biology and marked the beginning of modern evolutionary biology. Upon its publication, most naturalists recognized that evolution provided a unifying theory that explained an otherwise encyclopedic collection of biologic facts.

When Darwin proposed his theory of organic evolution, he cited a wealth of supporting evidence, including the way organisms are classified, embryology, comparative anatomy, the geographic distribution of organisms, and, to a limited extent, the fossil record. Furthermore, Darwin proposed that *natural selection*, which results in the survival to reproductive age of those organisms best adapted to their environment, is the mechanism that best accounts for evolution.

Perhaps the most compelling evidence in favor of evolution can be found in the fossil record. Just as the geologic record allows geologists to interpret physical events and conditions in the geologic past, **fossils**, which are the remains or traces of once-living organisms, not only provide evidence that evolution has occurred but also demonstrate that Earth has a history extending beyond that recorded by humans. The succession of fossils in the rock record provides geologists with a means for dating rocks and allowed for a relative geologic time scale to be constructed in the 1800s.

LO10 GEOLOGIC TIME AND UNIFORMITARIANISM

An appreciation of the immensity of geologic time is central to understanding the evolution of Earth and its biota. Indeed, time is one of the main aspects that sets geology apart from the other sciences, except astronomy. Most people have difficulty comprehending geologic time, because they tend to think in terms of the human perspective—seconds, hours, days, and years. Ancient history is what occurred hundreds or even thousands of years ago. When geologists talk of ancient geologic history, however, they are referring to events

Figure 1.15 Hand Specimens of Common Igneous, Sedimentary, and Metamorphic Rocks

a. Granite, an intrusive igneous rock.

b. Basalt, an extrusive igneous rock.

c. Conglomerate, a sedimentary rock formed by the consolidation of rounded rock fragments.

d. Limestone, a sedimentary rock formed by the extraction of mineral matter from seawater by organisms or by the inorganic precipitation of the mineral calcite from seawater.

e. Gneiss, a foliated metamorphic rock.

f. Quartzite, a nonfoliated metamorphic rock.

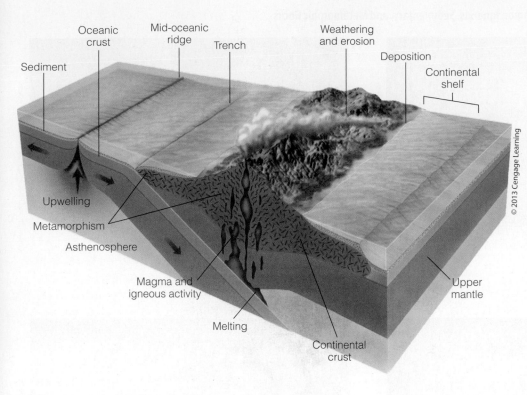

Sediment
Oceanic crust
Mid-oceanic ridge
Trench
Weathering and erosion
Deposition
Continental shelf
Upwelling
Metamorphism
Asthenosphere
Magma and igneous activity
Melting
Continental crust
Upper mantle

© 2013 Cengage Learning

Figure 1.16 Plate Tectonics and the Rock Cycle

Plate movement provides the driving mechanism that recycles Earth materials. This block diagram shows how the three major rock groups—igneous, sedimentary, and metamorphic—are recycled through both the continental and oceanic regions. Subducting plates are partially melted to produce magma, which rises and either crystallizes beneath Earth's surface as intrusive igneous rock or spills out on the surface, solidifying as extrusive igneous rock. Rocks exposed at the surface are weathered and eroded to produce sediments that are transported and eventually lithified into sedimentary rocks. Metamorphic rocks result from pressure generated along converging plates or adjacent to rising magma.

---

geologic time scale
A chart arranged so that the designation for the earliest part of geologic time appears at the bottom, followed upward by progressively younger time designations.

principle of uniformitarianism
A principle holding that we can interpret past events by understanding present-day processes, based on the idea that natural processes have always operated in the same way.

that happened hundreds of millions or even billions of years ago.

It is therefore important to remember that Earth goes through cycles of much longer duration than the human perspective of time. Although they may have disastrous effects on the human species, global warming and cooling are part of a larger cycle that has resulted in numerous glacial advances and retreats during the past 2.6 million years.

The **geologic time scale** (Figure 1.17) subdivides geologic time into a hierarchy of increasingly shorter time intervals, each of which has a specific name. Many 19th-century geologists pieced together information from numerous rock exposures to construct this chronology based on changes in Earth's biota through time. Subsequently, with the discovery of radioactivity

in 1895, and the development of various radiometric dating techniques, geologists have been able to assign numerical ages (also known as *absolute ages*) in years to the subdivisions of the geologic time scale.

One of the cornerstones of geology is the **principle of uniformitarianism**, which is based on the premise that present-day processes have operated throughout geologic time. Therefore, to understand and interpret geologic events from evidence preserved in rocks, we must first understand present-day processes and their results. In fact, uniformitarianism fits in completely with the system approach that we are following for the study of Earth.

Uniformitarianism is a powerful principle that allows us to use present-day processes as the basis for interpreting the past and for predicting potential future events. We should keep in mind, however, that uniformitarianism does not exclude sudden or catastrophic events such as volcanic eruptions, earthquakes, tsunami, landslides, or floods. What uniformitarianism does mean is that even though the rates and intensities of geologic processes have varied during the past, the physical and chemical laws of nature have remained the same. Although Earth is in a dynamic state of change and has been ever since it formed, the processes that shaped it during the past are the same ones operating today.

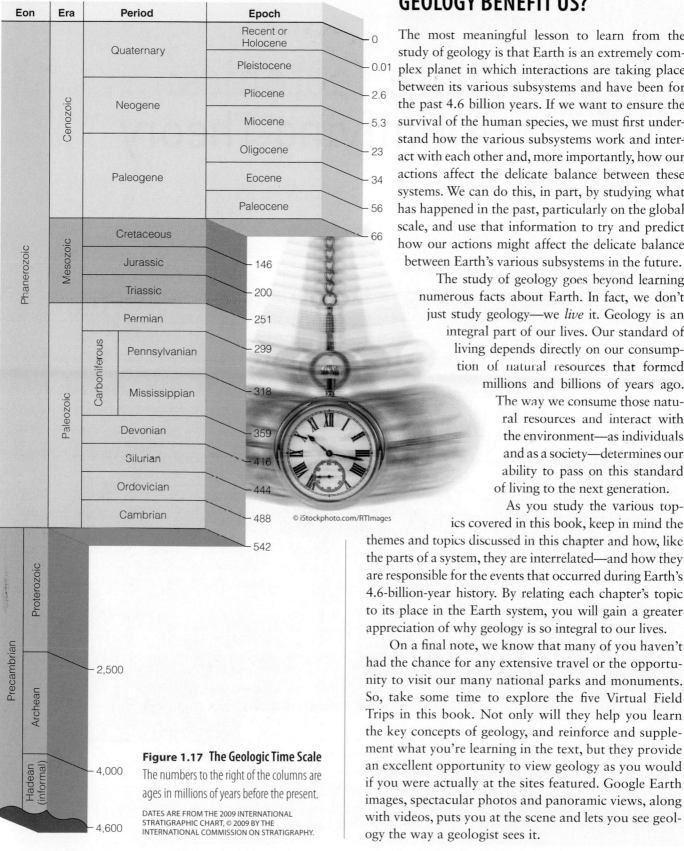

Eon	Era	Period		Epoch	
Phanerozoic	Cenozoic	Quaternary		Recent or Holocene	0
				Pleistocene	0.01
		Neogene		Pliocene	2.6
				Miocene	5.3
		Paleogene		Oligocene	23
				Eocene	34
				Paleocene	56
	Mesozoic	Cretaceous			66
		Jurassic			146
		Triassic			200
	Paleozoic	Permian			251
		Carboniferous	Pennsylvanian		299
			Mississippian		318
		Devonian			359
		Silurian			416
		Ordovician			444
		Cambrian			488
					542
Precambrian	Proterozoic				2,500
	Archean				4,000
	Hadean (informal)				4,600

Figure 1.17 The Geologic Time Scale
The numbers to the right of the columns are ages in millions of years before the present.

DATES ARE FROM THE 2009 INTERNATIONAL STRATIGRAPHIC CHART, © 2009 BY THE INTERNATIONAL COMMISSION ON STRATIGRAPHY.

© iStockphoto.com/RTImages

LO11 HOW DOES THE STUDY OF GEOLOGY BENEFIT US?

The most meaningful lesson to learn from the study of geology is that Earth is an extremely complex planet in which interactions are taking place between its various subsystems and have been for the past 4.6 billion years. If we want to ensure the survival of the human species, we must first understand how the various subsystems work and interact with each other and, more importantly, how our actions affect the delicate balance between these systems. We can do this, in part, by studying what has happened in the past, particularly on the global scale, and use that information to try and predict how our actions might affect the delicate balance between Earth's various subsystems in the future.

The study of geology goes beyond learning numerous facts about Earth. In fact, we don't just study geology—we *live* it. Geology is an integral part of our lives. Our standard of living depends directly on our consumption of natural resources that formed millions and billions of years ago. The way we consume those natural resources and interact with the environment—as individuals and as a society—determines our ability to pass on this standard of living to the next generation.

As you study the various topics covered in this book, keep in mind the themes and topics discussed in this chapter and how, like the parts of a system, they are interrelated—and how they are responsible for the events that occurred during Earth's 4.6-billion-year history. By relating each chapter's topic to its place in the Earth system, you will gain a greater appreciation of why geology is so integral to our lives.

On a final note, we know that many of you haven't had the chance for any extensive travel or the opportunity to visit our many national parks and monuments. So, take some time to explore the five Virtual Field Trips in this book. Not only will they help you learn the key concepts of geology, and reinforce and supplement what you're learning in the text, but they provide an excellent opportunity to view geology as you would if you were actually at the sites featured. Google Earth images, spectacular photos and panoramic views, along with videos, puts you at the scene and lets you see geology the way a geologist sees it.

2 | Plate Tectonics: A Unifying Theory

The Mand River winds through the Zagros Mountains in western Iran. The Zagros Mountains consist of a series of anticlines and synclines, which are rocks that were folded by compressive forces as a result of the collision of the Arabian plate with the Eurasian plate.

USGS EROS Data Center/Landsat/NASA

Introduction

"The idea that Earth's past geography was **different from today** is not new."

Imagine it is the day after Christmas, December 26, 2004, and you are vacationing on a beautiful beach in Thailand. You look up from the book you're reading to see the sea suddenly retreat from the shoreline. Within minutes of this unusual event, a powerful tsunami will sweep over your resort and everything in its path for several kilometers inland. In the next few hours, the coasts of Indonesia, Sri Lanka, India, Thailand, Somalia, Myanmar, Malaysia, and the Maldives will be inundated by the deadliest tsunami in history and more than 220,000 people will die.

Now go forward to October 25, 2010, when Mount Merapi in Central Java, Indonesia, began a series of violent eruptions that lasted into late November. Huge quantities of ash, lava, and ash flows were emitted during these eruptions, and although warnings were issued, at least 350 people were killed by the eruptions, and more than 350,000 were evacuated from areas around the volcano.

Almost five months later, on March 11, 2011, a catastrophic 9.0-magnitude earthquake struck Japan, killing more than 13,000 people and leaving thousands injured and homeless. This devastating earthquake also generated a tsunami, one that sent walls of water crashing into the northeastern shores of the island, causing further damage and causalities.

What do these three recent tragic events have in common? They are part of the dynamic interactions involving Earth's plates. When two plates come together, one plate is pushed or pulled under the other plate, triggering large earthquakes such as the recent ones that shook Haiti in 2010, and New Zealand and Japan in 2011. If conditions are right, earthquakes can also produce a tsunami such as the one in Indonesia in 2004, and Japan in 2011.

As the descending plate moves downward and is assimilated into Earth's interior, magma is generated. Being less dense than the surrounding material, the magma rises toward the surface, where it may erupt as a volcano. It therefore should not be surprising that the distribution of volcanoes and earthquakes closely follows plate boundaries.

If you're like most people, you probably have only a vague notion of what plate tectonic theory is. Yet, plate tectonics affects all of us. Volcanic eruptions, earthquakes, and tsunami are the result of interactions between plates. Global weather patterns

Learning Outcomes

After reading this unit, you should be able to do the following:

LO1 Review early ideas about continental drift

LO2 Explain the evidence for continental drift

LO3 Recognize and describe the features of the seafloor

LO4 Describe Earth's magnetic field

LO5 Explain paleomagnetism and polar wandering

LO6 Explain magnetic reversals and seafloor spreading

LO7 Explain plate tectonic theory as a unifying theory

LO8 Identify and describe the three types of plate boundaries

LO9 Describe hot spots and mantle plumes

LO10 Explain plate movement and motion

LO11 Explain the driving mechanism of plate tectonics

LO12 Recognize the role of plate tectonics in the distribution of natural resources

LO13 Recognize the role of plate tectonics in the distribution of life

© iStockphoto.com/George Clerk

Glossopteris flora A Late Paleozoic association of plants found only on the Southern Hemisphere continents and India; named for its best-known genus, *Glossopteris.*

continental drift The theory that the continents were joined into a single landmass that broke apart, with the various fragments (continents) moving with respect to one another.

Pangaea The name Alfred Wegener proposed for a supercontinent consisting of all Earth's landmasses at the end of the Paleozoic Era.

and oceanic currents are caused, in part, by the configuration of the continents and ocean basins. The formation and distribution of many natural resources are related to plate movement, and thus have an impact on the economic well-being and political decisions of nations. It is therefore important to understand this unifying theory, not only because it affects us as individuals, and as citizens of nation-states, but also because it ties together many aspects of the geology you will be studying.

Figure 2.1 Fossil *Glossopteris* Leaves

Plant fossils, such as these *Glossopteris* leaves from the Upper Permian Dunedoo Formation in Australia, are found on all five of the Gondwana continents. Their presence on continents with widely varying climates today is evidence that the continents were at one time connected. The distribution of the plants at that time was in the same climatic latitudinal belt.

LO1 EARLY IDEAS ABOUT CONTINENTAL DRIFT

The idea that Earth's past geography was different from today is not new. The earliest maps showing the east coast of South America and the west coast of Africa probably provided people with the first evidence that continents may have once been joined, then broken apart and moved to their present positions. As far back as 1620, Sir Francis Bacon commented on the similarity of the shorelines of western Africa and eastern South America. However, he did not make the connection that the Old and New Worlds might once have been joined together.

During the late 19th century, the Austrian geologist Edward Suess noted the similarities between the Late Paleozoic plant fossils of India, Australia, South Africa, and South America, as well as evidence of glaciation in the rock sequences of these continents. The plant fossils comprise a unique flora that occurs in the coal layers just above the glacial deposits of these southern continents. This flora is very different from the contemporaneous coal swamp flora of the northern continents and is collectively known as the *Glossopteris* flora after its most conspicuous genus (Figure 2.1).

Suess also proposed the name Gondwanaland (or *Gondwana* as we will use here) for a supercontinent composed of the aforementioned southern continents. Abundant fossils of the *Glossopteris* flora are found in coal beds in Gondwana, a province in India. Suess thought these southern continents were at one time connected by land bridges over which plants and animals migrated. Thus, in his view, the similarities of fossils on these continents were due to the appearance and disappearance of the connecting land bridges.

Alfred Wegener and the Continental Drift Hypothesis

Alfred Wegener, a German meteorologist, is generally credited with developing the hypothesis of **continental drift**. In his monumental book, *The Origin of Continents and Oceans* (first published in 1915), Wegener proposed that all landmasses were originally united in a single supercontinent that he named **Pangaea**, from the Greek, meaning "all land." Wegener portrayed his grand concept of continental movement in a series of maps showing the breakup of Pangaea and the movement of the various continents to their present-day locations. Wegener amassed a tremendous amount of geologic, paleontologic, and climatologic evidence in support of continental drift; however, initial reaction of scientists to his then-heretical ideas can best be described as mixed.

Nevertheless, the South African geologist Alexander du Toit further developed Wegener's arguments and gathered more geologic and paleontologic evidence in support of continental drift. In 1937, du Toit published *Our Wandering Continents*, in which he contrasted the glacial deposits of Gondwana with coal deposits of the same age found in the continents of the Northern Hemisphere. To resolve this apparent climatologic paradox, du Toit moved the Gondwana continents to the South Pole and brought the northern continents together such that the coal deposits were located at the equator. He named this northern landmass *Laurasia*. It consisted of present-day North America, Greenland, Europe, and Asia (except for India).

LO2 WHAT IS THE EVIDENCE FOR CONTINENTAL DRIFT?

What, then, was the evidence Wegener, du Toit, and others used to support the hypothesis of continental drift? It includes the fit of the shorelines of continents, the appearance of the same rock sequences and mountain ranges of the same age on continents now widely separated, the matching of glacial deposits and paleoclimatic zones, and the similarities of many extinct plant and animal groups whose fossil remains are found today on widely separated continents. Wegener and his supporters argued that this vast amount of evidence from a variety of sources surely indicates that the continents must have been close together in the past.

Continental Fit

Wegener, like some before him, was impressed by the close resemblance between the coastlines of continents on opposite sides of the Atlantic Ocean, particularly South America and Africa. He cited these similarities as partial evidence that the continents were at one time joined together as a supercontinent that subsequently split apart. As his critics pointed out, though, the configuration of coastlines results from erosional and depositional processes and therefore is continuously being modified. So, even if the continents had separated during the Mesozoic Era, as Wegener proposed, it is not likely that the coastlines would fit exactly.

A more realistic approach is to fit the continents together along the continental slope where erosion would be minimal. In 1965, Sir Edward Bullard, an English geophysicist, and two associates showed that the best fit between the continents occurs at a depth of about 2,000 m (Figure 2.2). Since then, other reconstructions, using the latest ocean basin data, have confirmed the close fit between continents when they are reassembled to form Pangaea.

Similarity of Rock Sequences and Mountain Ranges

If the continents were at one time joined, then the rocks and mountain ranges of the same age in adjoining locations on the opposite continents should closely match. Such is the case for the Gondwana continents (Figure 2.3). Marine, nonmarine, and glacial rock sequences of the Pennsylvanian to Jurassic periods are almost identical on all five Gondwana continents, strongly indicating that they were joined at one time.

Furthermore, the trends of several major mountain ranges also support the hypothesis of continental drift. These mountain ranges seemingly end at the coastline of one continent only to apparently continue on another

Figure 2.2 Continental Fit
When continents are placed together based on their outlines, the best fit isn't along their present-day coastlines, but rather along the continental slope at a depth of about 2,000 m, where erosion would be minimal.

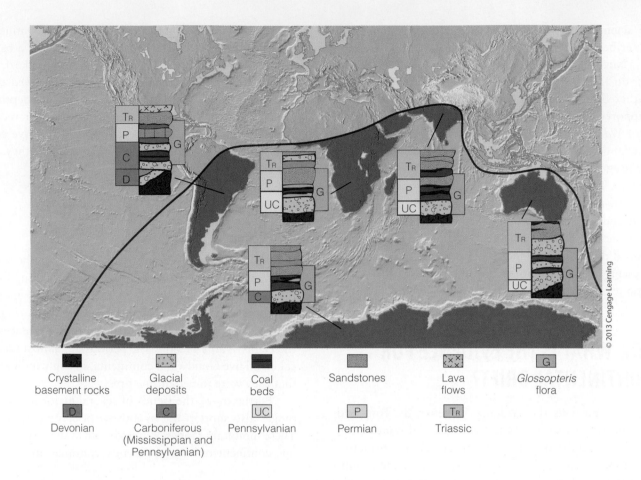

Crystalline basement rocks	Glacial deposits	Coal beds	Sandstones	Lava flows	*Glossopteris* flora

D Devonian	C Carboniferous (Mississippian and Pennsylvanian)	UC Pennsylvanian	P Permian	TR Triassic

Figure 2.3 Similarity of Rock Sequences on the Gondwana Continents

Sequences of marine, nonmarine, and glacial rocks of Pennsylvanian (UC) to Triassic (TR) age are nearly the same on all five Gondwana continents (South America, Africa, India, Australia, and Antarctica). These continents are widely separated today and have different environments and climates ranging from tropical to polar. Thus, the rocks forming on each continent are very different. When the continents were all joined together in the past, however, the environments of adjacent continents were similar, and the rocks forming in those areas were similar. The range indicated by G in each column is the age range of the *Glossopteris* flora.

continent across the ocean. The folded Appalachian Mountains of North America, for example, trend northeastward through the eastern United States and Canada and terminate abruptly at the Newfoundland coastline. Mountain ranges of the same age and deformational style are found in eastern Greenland, Ireland, Great Britain, and Norway. So even though the Appalachian Mountains and their equivalent-age mountain ranges in Great Britain are currently separated by the Atlantic Ocean, they form an essentially continuous mountain range when the continents are positioned next to each other as they were during the Paleozoic Era.

Glacial Evidence

During the Late Paleozoic Era, massive glaciers covered large continental areas of the Southern Hemisphere. Evidence for this glaciation includes layers of *till* (sediments deposited by glaciers) and *striations* (scratch marks) in the bedrock beneath the till (Figure 2.4). Fossils and sedimentary rocks of the same age from the Northern Hemisphere, however, give no indication of glaciation. Fossil plants found in coals indicate that the Northern Hemisphere had a tropical climate during the time that the Southern Hemisphere was glaciated.

Figure 2.4 Glacial Evidence Indicating Continental Drift

a. When the Gondwana continents are placed together so that South Africa is located at the South Pole, the glacial movements indicated by striations (red arrows) found on rock outcrops on each continent make sense. In this situation, the glacier (white area) is located in a polar climate and has moved radially outward from its thick central area toward its periphery.

b. Glacial striations (scratch marks) on an outcrop of Permian-age bedrock exposed at Hallet's Cove, Australia, indicate the general direction of glacial movement more than 200 million years ago. As a glacier moves over a continent's surface, it grinds and scratches the underlying rock. The glacial striations that are preserved on a rock's surface thus provide evidence of the direction (red arrows) the glacier moved at that time.

All of the Gondwana continents except Antarctica are currently located near the equator in subtropical to tropical climates. Mapping of glacial striations in bedrock in Australia, India, and South America indicates that the glaciers moved from the areas of the present-day oceans onto land. Yet, this would be highly unlikely because large continental glaciers (such as occurred on the Gondwana continents during the Late Paleozoic Era) flow outward from their central area of accumulation toward the sea. Therefore, if the continents did not move during the past, one would have to explain how glaciers moved from the oceans onto land and how large-scale continental glaciers formed near the equator. But if the continents are reassembled as a single landmass with South Africa located at the South Pole, the direction of movement of Late Paleozoic continental glaciers makes sense (Figure 2.4). Furthermore, this geographic arrangement places the northern continents nearer the tropics, which is consistent with the fossil and climatologic evidence from Laurasia.

Fossil Evidence

Some of the most compelling evidence for continental drift comes from the fossil record (Figure 2.5). Fossils of the *Glossopteris* flora are found in equivalent Pennsylvanian- and Permian-age coal deposits on all five Gondwana continents. The *Glossopteris* flora is characterized by the seed fern *Glossopteris* (Figure 2.1), as well as by many other distinctive and easily identifiable plants. Pollen and spores of plants can be dispersed over great distances by wind; however, *Glossopteris*-type plants produced seeds that are too large to have been carried by winds. Even if the seeds had floated across the ocean, they probably would not have remained viable for any length of time in saltwater.

The present-day climates of South America, Africa, India, Australia, and Antarctica range from tropical to polar and are much too diverse to support the type of plants in the *Glossopteris* flora. Wegener therefore reasoned that these continents must once have been joined

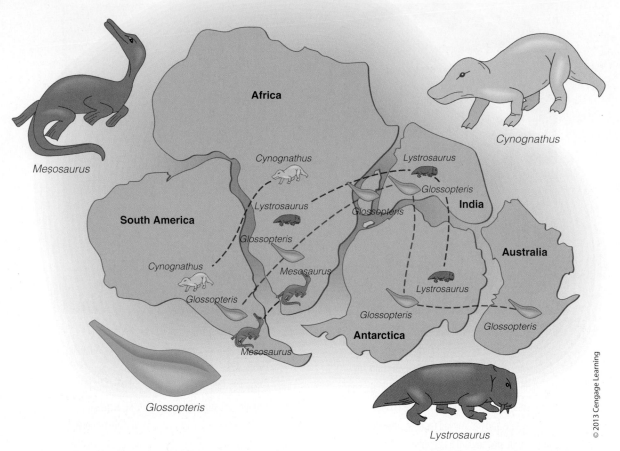

Figure 2.5 Fossil Evidence Supporting Continental Drift

Some of the plants and animals whose fossils are found today on the widely separated continents of South America, Africa, India, Australia, and Antarctica. During the Late Paleozoic Era, these continents were joined together to form Gondwana, the southern landmass of Pangaea. Plants of the *Glossopteris* flora are found on all five continents, which today have widely different climates; however, during the Pennsylvanian and Permian periods, they were all located in the same general climatic belt. *Mesosaurus* is a freshwater reptile whose fossils are found only in similar nonmarine Permian-age rocks in Brazil and South Africa. *Cynognathus* and *Lystrosaurus* are land reptiles that lived during the Early Triassic Period. Fossils of *Cynognathus* are found in South America and Africa, whereas fossils of *Lystrosaurus* have been recovered from Africa, India, and Antarctica. It is hard to imagine how a freshwater reptile and land-dwelling reptiles could have swum across the wide oceans that presently separate these continents. It is more logical to assume that the continents were once connected.

so that these now widely separated localities were at one time all in the same latitudinal climatic belt (Figure 2.5).

The fossil remains of animals also provide strong evidence for continental drift. One of the best examples is *Mesosaurus*, a freshwater reptile whose fossils are found in Permian-age rocks in certain regions of Brazil and South Africa and nowhere else in the world (Figure 2.5). Because the physiologies of freshwater and marine animals are completely different, it is hard to imagine how a freshwater reptile could have swum across the Atlantic Ocean and found a freshwater environment nearly identical to its former habitat. Moreover, if *Mesosaurus* could have swum across the ocean, its fos-

sil remains should be widely dispersed. It is more logical to assume that *Mesosaurus* lived in lakes in what were once adjacent areas of South America and Africa when they were united into a single continent.

Notwithstanding all of the empirical evidence presented by Wegener and later by du Toit and others, most geologists simply refused to entertain the idea that continents might have moved in the past. The geologists were not necessarily being obstinate about accepting new ideas; rather, they found the evidence for continental drift inadequate and unconvincing. In part, this was because no one could provide a suitable mechanism to explain how continents could move over Earth's surface.

LO3 FEATURES OF THE SEAFLOOR

At this point, it is useful to discuss some of the various features of Earth's seafloor. Many of the topographic features found on the seafloor and along the continental margins are the manifestations of Earth's internal processes and activities taking place along plate margins. Thus, it is important to know how these features relate to plate tectonic theory.

Most people think of continents as land areas outlined by the oceans, but the true geologic margin of a continent—where granitic continental crust changes to basalt and gabbro oceanic crust—is below sea level. A **continental margin** is made up of a gently sloping continental shelf, a more steeply inclined continental slope, and, in some cases, a deeper, gently sloping continental rise (Figure 2.6). Thus, the continental margins extend to increasingly greater depths until they merge with the deep seafloor. Continental crust changes to oceanic crust somewhere beneath the continental rise, so part of the continental slope and the continental rise actually rest on oceanic crust.

The Continental Shelf, Slope, and Rise

As one proceeds seaward from the shoreline across the continental margin, the first area encountered is the gently sloping **continental shelf** lying between the shore and the more steeply dipping continental slope (Figure 2.6). The width of the continental shelf varies considerably, ranging from a few tens of meters to more than 1,000 km; the shelf terminates where the inclination of the seafloor increases abruptly from 1 degree or less to several degrees.

The seaward margin of the continental shelf is marked by the *shelf-slope break* (at an average depth of 135 m), where the more steeply inclined **continental slope** begins (Figure 2.6). In most areas around the margins of the Atlantic, the continental slope merges with a more

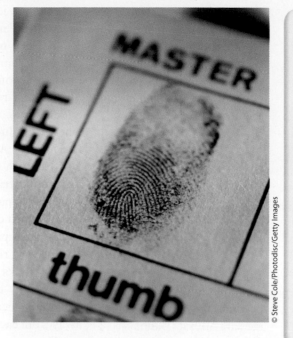

continental margin The area separating the part of a continent above sea level from the deep seafloor.

continental shelf The very gently sloping part of the continental margin between the shoreline and the continental slope.

continental slope The relatively steeply inclined part of the continental margin between the continental shelf and the continental rise or between the continental shelf and an oceanic trench.

Figure 2.6 Features of Continental Margins
A generalized profile showing features of the continental margins. The vertical dimensions of the features in this profile are greatly exaggerated because the vertical and horizontal scales differ.

continental rise The gently sloping part of the continental margin between the continental slope and the abyssal plain.

abyssal plain Vast flat area on the seafloor adjacent to the continental rise of a passive continental margin.

oceanic ridge A mostly submarine mountain system composed of basalt found in all ocean basins.

submarine hydrothermal vent A crack, or fissure, in the seafloor through which superheated water issues.

gently sloping **continental rise**. This rise is absent around the margins of the Pacific, where continental slopes descend directly into an oceanic trench (Figure 2.6).

Abyssal Plains, Oceanic Ridges, Submarine Hydrothermal Vents, and Oceanic Trenches

Beyond the continental rises are **abyssal plains**— flat surfaces covering vast areas of the seafloor. In some areas, they are interrupted by peaks rising more than 1 km, but abyssal plains are nevertheless the flattest, most feature-less areas on Earth (Figure 2.6). Their flatness is a result of sediment deposition covering the rugged topography of the seafloor.

A renewed interest in oceanographic research led to extensive mapping of the ocean basins during the 1960s. Such mapping revealed an **oceanic ridge** system more than 65,000 km long, constituting the most extensive mountain range in the world (Figure 2.6). This system runs from the Arctic Ocean through the middle of the Atlantic and curves around South Africa, where the Indian Ridge continues into the Indian Ocean. The Atlantic–Pacific Ridge extends eastward, and a branch of it, the East Pacific Rise, trends northeast until it reaches the Gulf of California. Perhaps the best-known part of the ridge system is the Mid-Atlantic Ridge, which divides the Atlantic Ocean basin into two nearly equal parts (see Figure 1.13).

Oceanic ridges are composed almost entirely of the igneous rocks basalt and gabbro and possess features produced by tensional forces. Thus, they are the sites where new oceanic crust is generated and plates move away from each other along divergent plate boundaries.

First seen on the seafloor in 1979, **submarine hydrothermal vents** are found at or near spreading ridges. Here, cold seawater seeps through oceanic crust, is heated by the hot rocks at depth, and then rises and discharges into the seawater as plumes of hot water with temperatures as high as 400°C. Many of the plumes are black because dissolved minerals give them the appearance of black smoke—hence the name *black smoker* (Figure 2.7).

Figure 2.7 Submarine Hydrothermal Vents

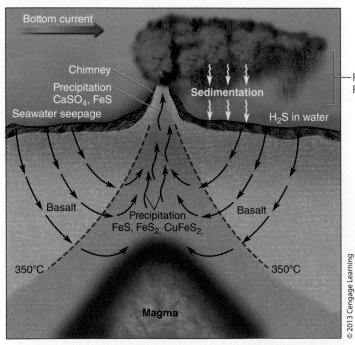

a. Cross section showing the origin of a submarine hydrothermal vent called a black smoker.

b. This black smoker on the East Pacific Rise is at a depth of 2,800 m. The plume of "black smoke" is heated seawater with dissolved minerals.

Submarine hydrothermal vents are interesting from the biologic, geologic, and economic points of view. Near the vents live communities of organisms—such as bacteria, crabs, mussels, starfish, and tube worms—many of which have never been seen before. No sunlight is available, so these organisms depend on bacteria that oxidize sulfur compounds for their ultimate source of nutrients. The vents are also interesting because of their economic potential. The heated seawater reacts with oceanic crust, transforming it into a metal-rich solution that discharges into seawater and cools, precipitating iron, copper, and zinc sulfides and other minerals (Figure 2.7a). A chimney-like vent forms that eventually collapses and forms a mound of sediments rich in the elements just mentioned (Figure 2.7b).

Oceanic trenches are long, steep-sided depressions on the seafloor near convergent plate boundaries (Figure 2.6). They constitute no more than 2 percent of the seafloor. However, it is here that oceanic lithosphere is consumed by subduction; that is, oceanic lithosphere plunges into Earth's interior along convergent plate boundaries. It should also be noted that the greatest oceanic depths are found in trenches; the Challenger Deep of the Mariana Trench in the Pacific is more than 11,000 m deep!

Seamounts, Guyots, and Aseismic Ridges

Except for the abyssal plains, the seafloor is not a flat, featureless expanse. In fact, a large number of volcanic hills, seamounts, and guyots rise above the seafloor in all ocean basins and are particularly abundant in the Pacific. All are of volcanic origin and differ mostly in size. **Seamounts** rise more than 1 km above the seafloor, and if flat-topped, they are called guyots (Figure 2.8). **Guyots** are volcanoes that originally extended above sea level. However, as the plate upon which they were located continued to move, they were carried away from a spreading ridge, and as the oceanic crust cooled, it descended to greater depths. Thus, what was once an island slowly sank beneath the sea, and as it did, wave erosion produced the typical flat-topped appearance (Figure 2.8).

oceanic trench A long, narrow feature restricted to active continental margins and along which subduction occurs.

seamount A submarine volcanic mountain rising at least 1 km above the seafloor.

guyot A flat-topped seamount of volcanic origin rising more than 1 km above the seafloor.

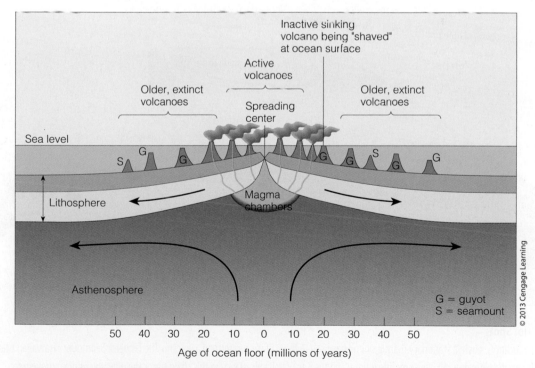

© 2013 Cengage Learning

Figure 2.8 The Origin of Seamounts and Guyots
As the plate on which a volcano rests moves into greater water depths, the submerged volcanic island is called a seamount. Those that are flat-topped are called guyots.

aseismic ridge A ridge, or broad area, rising above the seafloor that lacks seismic activity.

active continental margin A continental margin with volcanism and seismicity at the leading edge of a continental plate where oceanic lithosphere is subducted.

Other common features in the ocean basins are long, narrow ridges and broad, plateau-like features rising as much as 2 to 3 km above the surrounding seafloor. These **aseismic ridges** are so called because they lack seismic (earthquake) activity. A few of these ridges are probably small fragments separated from continents during rifting and are referred to as *microcontinents*. Avalonia is a good example of a Paleozoic microcontinent (see Figure 17.8b).

Most aseismic ridges form as a linear succession of hot-spot volcanoes. These may develop at or near an oceanic ridge, but each volcano so formed is carried laterally with the plate upon which it originated. The net result is a line of seamounts/guyots extending from an oceanic ridge (Figure 2.8). Aseismic ridges also form over hot spots unrelated to ridges—the Hawaiian–Emperor chain in the Pacific, for example (Figure 2.9).

Continental Margins

Although we will be discussing convergent and divergent plate boundaries later in this chapter, here is a good place to cover the two types of continental margins associated with the respective aforementioned plate boundaries.

Active continental margins develop at the leading edge of a continental plate where oceanic lithosphere is subducted. The western margin of South America is a good example of where an oceanic plate is subducted beneath the continent, resulting in seismic activity, a geologically young mountain range (Andes Mountains), and active volcanism (Figure 2.10). In addition, the continental shelf is narrow, and the continental slope descends directly into an oceanic trench, so sediment is dumped into the trench and no continental rise develops (Figure 2.6, left side).

The western margin of North America is also considered an active continental margin, although much of it is now bounded by transform faults (see the section on transform boundaries later in this chapter) rather than a subduction zone. However, plate convergence and subduction continue in the Pacific Northwest along the continental margins of northern California, Oregon, and Washington.

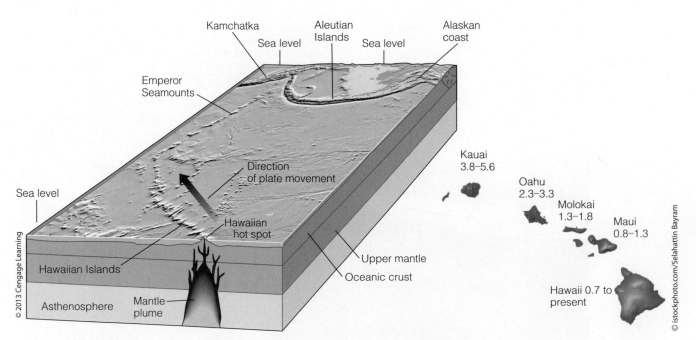

Figure 2.9 Hot Spots

A hot spot is the location where a stationary mantle plume has risen to the surface and formed a volcano. The Emperor Seamount–Hawaiian Island chain formed as a result of the Pacific plate moving over a mantle plume, and the line of volcanic islands in this chain traces the direction of plate movement. The island of Hawaii and the Loihi Seamount are the only current hot spots of this island chain. The numbers indicate the ages of the Hawaiian Islands in millions of years.

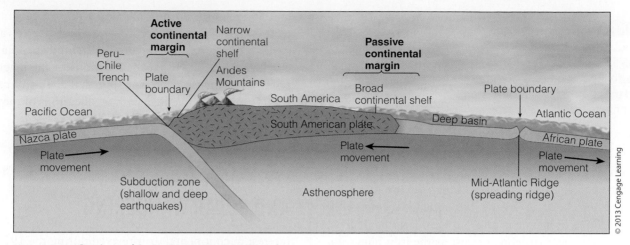

Figure 2.10 Passive and Active Continental Margins
Active and passive continental margins along the west and east coasts of South America. Notice that the passive margins are much wider than active margins. Seafloor sediment is not shown.

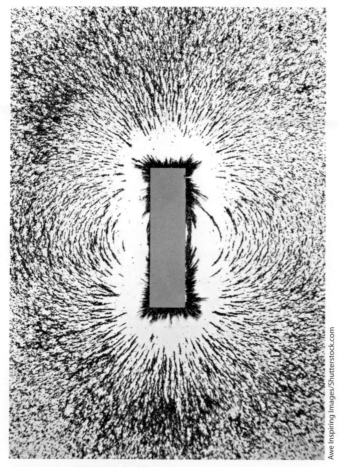

Figure 2.11 Magnetic Field
Iron filings align along the lines of magnetic force radiating from a bar magnet.

The continental margins of eastern North America and South America differ considerably from their western margins. For one thing, they possess broad continental shelves, as well as a continental slope and rise, with abyssal plains adjacent to the rises (Figure 2.6, right side). Furthermore, these **passive continental margins** are within a plate rather than at a plate boundary, and they lack the volcanic and seismic activity found at active continental margins (Figure 2.10). Nevertheless, earthquakes do take place there occasionally.

LO4 EARTH'S MAGNETIC FIELD

Having just discussed those features of the seafloor that are related to plate movement, we now turn our attention to the phenomenon of magnetism and its role in the formulation of plate tectonic theory.

Magnetism is a physical phenomenon resulting from the spin of electrons in some solids—particularly those of iron—and moving electricity. A **magnetic field** is an area in which magnetic substances such as iron are affected by lines of magnetic force emanating from Earth (Figure 2.11). The magnetic field shown in

passive continental margin A continental margin within a tectonic plate, as in the East Coast of North America, where little seismic activity and no volcanism occur; characterized by a broad continental shelf and a continental slope and rise.

magnetism A physical phenomenon resulting from moving electricity and the spin of electrons in some solids in which magnetic substances are attracted toward one another.

magnetic field The area in which magnetic substances are affected by lines of magnetic force emanating from Earth.

Figure 2.12 is *dipolar*, meaning that it possesses two unlike magnetic poles referred to as the north and south poles.

A useful analogy is to think of Earth as a giant dipole magnet in which the magnetic poles are in proximity to the geographic poles (Figure 2.12). This arrangement means that the strength of the magnetic field is not constant, but varies. Notice in Figure 2.12 that the lines of magnetic force around Earth parallel its surface only near the equator, just as the iron filings do around a bar magnet (Figure 2.11). As the lines of force approach the poles, they are oriented at increasingly larger angles with respect to the surface, and the strength of the magnetic field increases; it is strongest at the poles and weakest at the equator.

Another important aspect of the magnetic field is that the magnetic poles, where the lines of force leave and enter Earth, do not coincide with the geographic (rotational) poles. Currently, an 11.5-degree angle exists between the two (Figure 2.12). Studies of Earth's magnetic field show that the locations of the magnetic poles vary slightly over time, but that they still correspond closely, on average, with the locations of the geographic poles.

LO5 PALEOMAGNETISM AND POLAR WANDERING

Interest in continental drift revived during the 1950s as a result of evidence from paleomagnetic studies, a relatively new discipline at the time. **Paleomagnetism** is the remanent magnetism in ancient rocks recording the direction and intensity of Earth's magnetic poles at the time of the rock's formation.

When magma cools, the magnetic iron-bearing minerals align themselves with Earth's magnetic field,

Figure 2.12 Earth's Magnetic Field

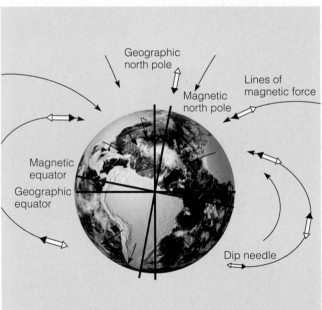

© Istockphoto.com/Jan Rysavy

a. Earth's magnetic field has lines of force like those of a bar magnet.

b. The strength of the magnetic field changes from the magnetic equator to the magnetic poles. This change in strength causes a dip needle (a magnetic needle that is balanced on the tip of a support so that it can freely move vertically) to be parallel to Earth's surface only at the magnetic equator, where the strength of the magnetic north and south pole is equally balanced. Its inclination, or dip, with respect to Earth's surface increases as it moves toward the magnetic poles until it is at 90 degrees, or perpendicular, to Earth's surface at the magnetic poles.

recording both its direction and strength. The temperature at which iron-bearing minerals gain their magnetization is called the **Curie point**. As long as the rock is not subsequently heated above the Curie point, it will preserve that remanent magnetism. Thus, an ancient lava flow provides a record of the orientation and strength of Earth's magnetic field at the time the lava flow cooled.

Remanent magnetism: The permanent magnetism in rocks (resulting from the orientation of the Earth's magnetic field at the time of the rocks' formation).

As paleomagnetic research progressed during the 1950s, some unexpected results emerged. When geologists measured the paleomagnetism of geologically recent rocks, they found that it was generally consistent with Earth's current magnetic field. The paleomagnetism of ancient rocks, though, showed different orientations. For example, paleomagnetic studies of Silurian lava flows in North America indicated that the north magnetic pole was located in the western Pacific Ocean at that time, whereas the paleomagnetic evidence from Permian lava flows pointed to yet another location in Asia. When plotted on a map, the paleomagnetic readings of numerous lava flows from all ages in North America trace the apparent movement of the magnetic pole (called *polar wandering*) through time (Figure 2.13).

Upon analysis, magnetic minerals from European Silurian and Permian lava flows pointed to a different magnetic pole location from those of the same age from North America (Figure 2.13). Furthermore, analysis of lava flows from all continents indicated that each continent seemingly had its own series of magnetic poles. How could this be? Does it really mean there were different north magnetic poles for each continent?

The best explanation for such data is that the magnetic poles have remained near their present locations at the geographic north and south poles, and the continents have moved. When the continental margins are fit together so that the paleomagnetic data point to only one magnetic pole, we find, just as Wegener did, that the rock sequences and glacial deposits match, and that the fossil evidence is consistent with the reconstructed paleogeography.

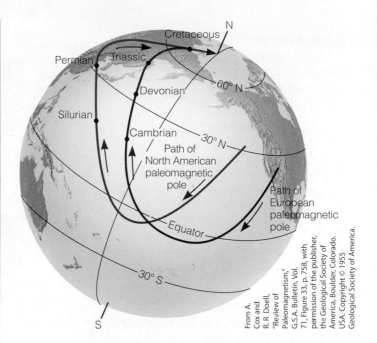

Figure 2.13 Polar Wandering

The apparent paths of polar wandering for North America and Europe. The apparent location of the north magnetic pole is shown for different periods on each continent's polar wandering path. Because Earth has only one magnetic north pole, the paleomagnetic readings taken on different continents for the same time in the past should all point to the same location if the continents have not moved. However, the north magnetic pole has different locations for the same time in the past when measured on different continents, indicating multiple north magnetic poles. The logical explanation for this dilemma is that the magnetic north pole has remained at the same approximate geographic location during the past, and the continents have moved. From A. Cox and R. R. Doell, "Review of Paleomagnetism," *G.S.A. Bulletin*, Vol. 71, Figure 33, p. 758, with permission of the publisher, the Geological Society of America, Boulder, Colorado. USA. Copyright © 1955 Geological Society of America.

LO6 MAGNETIC REVERSALS AND SEAFLOOR SPREADING

Geologists refer to Earth's present magnetic field as being normal—that is, with the north and south magnetic poles located approximately at the north and south geographic poles. At various times in the geologic past, however, Earth's magnetic field has completely reversed. The magnetic north and south

Curie point The temperature at which iron-bearing minerals in cooling magma or lava attain their magnetism.

poles have switched positions so that the magnetic north pole became the magnetic south pole, and the magnetic south pole became the magnetic north pole. During such a reversal, the magnetic field weakens until it temporarily disappears. When the magnetic field returns, the magnetic poles have reversed their position. The existence of such **magnetic reversals** was discovered by dating and determining the orientation of the remanent magnetism in lava flows on land (Figure 2.14). Although the cause of magnetic reversals is still uncertain, their occurrence in the geologic record is well documented.

As a result of oceanographic research conducted during the 1950s, Harry Hess of Princeton University proposed, in a 1962 landmark paper, the theory of **seafloor spreading** to account for continental movement. He suggested that continents do not move through oceanic crust as do ships plowing through sea ice, but rather that the continents and oceanic crust move together as a single unit. Thus, the theory of seafloor spreading answered a major objection of the opponents of continental drift—namely, how could continents move through oceanic crust? In fact, the continents moved with the oceanic crust as part of a lithospheric system.

As a mechanism to drive this system, Hess revived the idea of a

Figure 2.14 Magnetic Reversals

During the time period shown (a–d), volcanic eruptions produced a succession of overlapping lava flows. At the time of these volcanic eruptions, Earth's magnetic field completely reversed—that is, the magnetic north pole moved to the geographic south pole, and the magnetic south pole moved to the geographic north pole. Thus, the end of the needle of a magnetic compass that today would point to the North Pole would point to the South Pole if the magnetic field should again reverse. We know that Earth's magnetic field has reversed numerous times in the past because when lava flows cool below the Curie point, magnetic minerals within the flow orient themselves parallel to the magnetic field at the time. They thus record whether the magnetic field was normal or reversed at that time. The white arrows in this diagram show the direction of the north magnetic pole for each individual lava flow, thus confirming that Earth's magnetic field has reversed in the past.

MODIFIED FROM KIOUS AND TILLING, USGS AND HYNDMAN AND HYNDMAN, *NATURAL HAZARDS AND DISASTERS*, BROOKS/COLE, 2006, P. 15, FIG. 2.6B.

heat transfer system—or **thermal convection cells**—within the mantle to move the plates. According to Hess, hot magma rises from the mantle, intrudes along fractures defining oceanic ridges, and thus forms new crust. Cold crust is subducted back into the mantle at oceanic trenches, where it is heated and recycled, thus completing a thermal convection cell (see Figure 1.12).

How could Hess's hypothesis be confirmed? Magnetic surveys of the oceanic crust revealed a pattern of striped **magnetic anomalies** (deviations from the average strength of Earth's present-day magnetic field) in the rocks that are both parallel to and symmetric around the oceanic ridges (Figure 2.15). A positive magnetic anomaly results when Earth's magnetic field at the time of oceanic crust formation along an oceanic ridge summit was the same as today, thus yielding a stronger than normal (positive) magnetic signal. A negative magnetic anomaly results when Earth's magnetic field at the time of oceanic crust formation was reversed, therefore yielding a weaker than normal (negative) magnetic signal.

Thus, as new oceanic crust forms at oceanic ridge summits and records Earth's magnetic field at the time, the previously formed crust moves laterally away from the ridge. These magnetic stripes therefore represent times of normal and reversed polarity at oceanic ridges (where upwelling magma forms new oceanic crust), and they conclusively confirm Hess's theory of seafloor spreading.

One of the consequences of the seafloor spreading theory is its confirmation that ocean basins are geologically young features whose openings and closings are partially responsible for continental movement. Radiometric dating reveals that the oldest oceanic crust is somewhat younger than 180 million years old, whereas the oldest continental crust is about 4 billion years old. Although geologists do not universally accept the idea of thermal convection cells as the sole driving mechanism for plate movement, most accept that plates are created at oceanic ridges and destroyed at deep-sea trenches, regardless of the driving mechanism involved.

thermal convection cell A type of circulation of material involving only the asthenosphere or the entire mantle during which hot material rises, moves laterally, cools, and sinks, then is reheated and continues the cycle.

magnetic anomaly Any deviation, such as a change in average strength, in Earth's magnetic field.

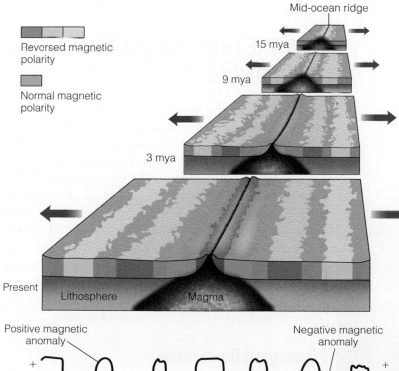

Figure 2.15 Magnetic Anomalies and Seafloor Spreading

The sequence of magnetic anomalies preserved within the oceanic crust is both parallel to and symmetric around oceanic ridges. Basaltic lava intruding into an oceanic ridge today and spreading laterally away from the ridge records Earth's current magnetic field, or polarity (considered by convention to be normal). Basaltic intrusions 3, 9, and 15 million years ago record Earth's reversed magnetic field at those times. This schematic diagram shows how the solidified basalt moves away from the oceanic ridge (or spreading ridge), carrying with it the magnetic anomalies that are preserved in the oceanic crust. Magnetic anomalies are magnetic readings that are either higher (positive magnetic anomalies) or lower (negative magnetic anomalies) than Earth's current magnetic field strength. The magnetic anomalies are recorded by a magnetometer, which measures the strength of the magnetic field.

MODIFIED FROM KIOUS AND TILLING, USGS AND HYNDMAN AND HYNDMAN, *NATURAL HAZARDS AND DISASTERS*, BROOKS/COLE, 2006, P. 15, FIG. 2.6B.

LO7 PLATE TECTONICS: A UNIFYING THEORY

plate tectonic theory The theory holding that large segments of Earth's outer part (lithospheric plates) move relative to one another.

divergent plate boundary The boundary between two plates that are moving apart.

Plate tectonic theory is based on a simple model of Earth. The rigid lithosphere, composed of both oceanic and continental crust, as well as the underlying upper mantle, consists of numerous variable-sized pieces called *plates* (Figure 2.16). There are seven major plates (Eurasian, Indian-Australian, Antarctic, North American, South American, Pacific, and African), and numerous smaller ones ranging from only a few tens to several hundreds of kilometers in width. Plates also vary in thickness; those composed of upper mantle and continental crust are as much as 250 km thick, whereas those of upper mantle and oceanic crust are up to 100 km thick.

The lithosphere overlies the hotter and weaker semiplastic asthenosphere. It is thought that movement resulting from some type of heat-transfer system within the asthenosphere causes the overlying plates to move.

As plates move over the asthenosphere, they separate, mostly at oceanic ridges; in other areas, such as at oceanic trenches, they collide and are subducted back into the mantle.

An easy way to visualize plate movement is to think of a conveyor belt moving luggage from an airplane's cargo hold to a baggage cart. The conveyer belt represents convection currents within the mantle, and the luggage represents Earth's lithospheric plates. The luggage is moved along by the conveyer belt until it is dumped into the baggage cart—in the same way that plates are moved by convection cells until they are subducted into Earth's interior.

Although this analogy allows you to visualize how the mechanism of plate movement takes place, remember that the analogy is limited. The major limitation is that, unlike the luggage, plates consist of continental and oceanic lithosphere, which have different densities, and only oceanic lithosphere is subducted into Earth's interior.

Most geologists accept plate tectonic theory because the evidence for it is overwhelming and because it ties together many seemingly unrelated geologic features and events and shows how they are interrelated. Consequently, geologists now view many geologic processes from the global perspective of plate tectonic theory, in which plate interaction along plate margins is responsible for such phenomena as mountain building, earthquakes, and volcanism.

LO8 THE THREE TYPES OF PLATE BOUNDARIES

Because it appears that plate tectonics has operated since at least the Proterozoic Eon, it is important that we understand how plates move and interact with each other and how ancient plate boundaries are recognized. After all, the movement of plates has profoundly affected the geologic and biologic history of this planet.

Geologists recognize three major types of plate boundaries: *divergent*, *convergent*, and *transform*. Along these boundaries, new plates are formed, consumed, or slide laterally past one another. Interaction of plates at their boundaries accounts for most of Earth's volcanic eruptions and earthquakes, as well as the formation and evolution of its mountain systems.

Divergent Boundaries

Divergent plate boundaries, or *spreading ridges*, occur where plates are separating and new oceanic

7 PLATES

Eurasian

Indian-Australian

Antarctic

North American

South American

Pacific

African

©iStockphoto.com/Nancy Dressel

Ridge axis   Subduction zone   Hot spot   Direction of movement

Figure 2.16 Earth's Plates

A world map showing Earth's plates, their boundaries, their relative motion and rates of movement in centimeters per year, and hot spots.

lithosphere is forming. Divergent boundaries are places where the crust is extended, thinned, and fractured as magma, derived from the partial melting of the mantle, rises to the surface. The magma is almost entirely basaltic and intrudes into vertical fractures to form dikes and pillow lava flows (see Figure 5.4). As successive injections of magma cool and solidify, they form new oceanic crust and record the intensity and orientation of Earth's magnetic field (Figure 2.15).

Divergent boundaries most commonly occur along the crests of oceanic ridges—for example, the Mid-Atlantic Ridge. Oceanic ridges are thus characterized by rugged topography with high relief resulting from displacement of rocks along large fractures, shallow-depth earthquakes, high heat flow, and basaltic flows or pillow lavas.

Divergent boundaries are also present under continents during the early stages of continental breakup. When magma wells up beneath a continent, the crust is initially elevated, stretched, and thinned, producing fractures, faults, rift valleys, and volcanic activity (Figure 2.17a). As magma intrudes into faults and fractures, it solidifies or flows out onto the surface as lava flows; the latter often covering the rift valley floor (Figure 2.17b). The East African Rift Valley is an excellent example of continental breakup at this stage (Figure 2.18).

As spreading proceeds, some rift valleys continue to lengthen and deepen until the continental crust eventually breaks and a narrow linear sea is formed, separating two continental blocks (Figure 2.17c). The Red Sea, separating the Arabian Peninsula from Africa (Figure 2.18), and the Gulf of California, which separates Baja California from mainland Mexico, are good examples of this more advanced stage of rifting.

As a newly created narrow sea continues to enlarge, it may eventually become an expansive ocean basin such as the Atlantic Ocean basin is today, separating North and South America from Europe and Africa by thousands of kilometers (Figure 2.17d). The Mid-Atlantic Ridge is the boundary between these diverging plates; the American plates are moving westward, and the Eurasian and African plates are moving eastward.

It should be noted that extending outward from the eastern coasts of North and South America, as well as the western coasts of Europe and Africa, are broad continental shelves, continental slopes, and rises. These features, discussed earlier, are referred to as passive continental margins (Figure 2.10). And, although they are found within a plate, rather than at a plate boundary, they result from the breakup of a continent and the subsequent movement away from a divergent plate boundary (Figure 2.17).

Figure 2.17 History of a Divergent Plate Boundary

Volcanic activity

Magma

Continental crust

a. Rising magma beneath a continent pushes the crust up, producing numerous fractures, faults, and volcanic activity.

Rift valley

b. As the crust is stretched and thinned, rift valleys develop, and lava flows onto the valley floors, such as seen today in the East African Rift Valley.

Coastal mountain range

Narrow fault-bounded sea

c. Continued spreading further separates the continent until it splits apart and a narrow seaway develops. The Red Sea, which separates the Arabian Peninsula from Africa, is a good example of this stage of development.

Continental "seaboard" (coastal mountains gone)

Wide ocean

d. As spreading continues, an oceanic ridge system forms, and an ocean basin develops and grows. The Mid-Atlantic Ridge illustrates this stage in a divergent plate boundary's history.

© 2013 Cengage Learning

Figure 2.18 East African Rift Valley and the Red Sea—Present-Day Examples of Divergent Plate Boundaries
The East African Rift Valley and the Red Sea represent different stages in the history of a divergent plate boundary. The East African Rift Valley is being formed by the separation of eastern Africa from the rest of the continent along a divergent plate boundary. The Red Sea represents a more advanced stage of rifting, in which two continental blocks (Africa and the Arabian Peninsula) are separated by a narrow sea.

An Example of Ancient Rifting

What features in the geologic record can geologists use to recognize ancient rifting? Associated with regions of continental rifting are faults, dikes (vertical intrusive igneous bodies), sills (horizontal intrusive igneous bod-

ies), lava flows, and thick sedimentary sequences within rift valleys, all features that are preserved in the geologic record. The Triassic fault basins of the eastern United States are a good example of ancient continental rifting (see Figure 17.21). These fault basins mark the zone of rifting that occurred when North America split apart from Africa. The basins contain thousands of meters of continental sediment and are riddled with dikes and sills (see Chapter 17).

Pillow lavas, in association with deep-sea sediment, are also evidence of ancient rifting. The presence of pillow lavas marks the formation of a spreading ridge in a narrow linear sea (Figure 2.17c). Magma, intruding into the sea along this newly formed spreading ridge, solidifies as pillow lavas, which are preserved in the geologic record along with the sediment being deposited on them.

convergent plate boundary The boundary between two plates that move toward each other.

oceanic–oceanic plate boundary A convergent plate boundary along which two oceanic plates collide and one is subducted beneath the other.

Convergent Boundaries

Whereas new crust forms at divergent plate boundaries, older crust must be destroyed and recycled in order for the entire surface area of Earth to remain the same. Otherwise, we would have an expanding Earth. Such plate destruction occurs at **convergent plate boundaries** (Figure 2.19), where two plates collide and the leading edge of one plate is subducted beneath the margin of the other plate and eventually incorporated into the asthenosphere. A dipping plane of earthquake foci, called a *Benioff zone*, defines subduction zones (see Figure 8.5). Most of these planes dip from oceanic trenches beneath adjacent island arcs or continents, marking the surface of slippage between the converging plates.

Deformation, volcanism, mountain building, metamorphism, earthquake activity, and deposits of valuable minerals characterize convergent boundaries. Three types of convergent plate boundaries are recognized: *oceanic–oceanic*, *oceanic–continental*, and *continental–continental*.

Oceanic–Oceanic Boundaries

When two oceanic plates converge, one is subducted beneath the other along an **oceanic–oceanic plate boundary** (Figure 2.19a). The subducting plate bends downward to form

Figure 2.19 Three Types of Convergent Plate Boundaries

Continent · **Marginal sea (back arc)** · **Volcanic island arc** · **Central ocean**

Back arc spreading

Partial melting

Mantle upwelling associated with nearby subduction

Partial melting

Subducting oceanic crust

Forearc

Trench

Upper mantle

Asthenosphere

Japan

Volcanic arc mountain range

Continental interior

Forearc

Subduction complex

Upper mantle

Asthenosphere

Continental crust

Partial melting

Subducting oceanic crust

Panama

Amazon River

N

Pacific Ocean

Andean volcanic arc

South America

Continental crust

0   1000 km

Secondary rifting caused by collision

Mountain range

Continental crust

Oceanic crust

Little or no partial melting

Thrust zone

Continental crust

Upper mantle

Asthenosphere

Himalaya Mts.

India

Indian Ocean

Rift faulting, Tibetan plateau

N

0   1000 km

© 2013 Cengage Learning

a. Oceanic–oceanic plate boundary. An oceanic trench forms where one oceanic plate is subducted beneath another. On the nonsubducted plate, a volcanic island arc forms from the rising magma generated from the subducting plate. The Japanese Islands are a volcanic island arc resulting from the subduction of one oceanic plate beneath another oceanic plate.

b. Oceanic–continental plate boundary. When an oceanic plate is subducted beneath a continental plate, an andesitic volcanic mountain range is formed on the continental plate as a result of rising magma. The Andes Mountains in Peru are one of the best examples of continuing mountain building at an oceanic–continental plate boundary.

c. Continental–continental plate boundary. When two continental plates converge, neither is subducted because of their great thickness and low and equal densities. As the two continental plates collide, a mountain range is formed in the interior of a new and larger continent. The Himalayas in central Asia resulted from the collision between India and Asia approximately 40 to 50 million years ago.

the outer wall of an oceanic trench. A subduction complex, composed of wedge-shaped slices of highly folded and faulted marine sediments and oceanic lithosphere scraped off the descending plate, forms along the inner wall of the oceanic trench. As the subducting plate descends into the mantle, it is heated and partially melted, generating magma that is commonly of andesitic composition (see Chapter 4). This magma is less dense than the surrounding mantle rocks and rises to the surface of the nonsubducted plate to form a curved chain of volcanic islands called a *volcanic island arc* (any plane intersecting a sphere makes an arc). This arc is nearly parallel to the oceanic trench and is separated from it by a distance of up to several hundred kilometers—the distance depends on the angle of dip of the subducting plate (Figure 2.19a).

In those areas where the rate of subduction is faster than the forward movement of the overriding plate, the lithosphere on the landward side of the volcanic island arc may be subjected to tensional stress and be stretched and thinned, resulting in the formation of a *back-arc basin*. This back-arc basin may grow by spreading if magma breaks through the thin crust and forms new oceanic crust (Figure 2.19a). A good example of a back-arc basin associated with an oceanic–oceanic plate boundary is the Sea of Japan between the Asian continent and the islands of Japan.

Most present-day active volcanic island arcs are in the Pacific Ocean basin and include the Aleutian Islands, the Kermadec–Tonga arc, and the Japanese (Figure 2.19a) and Philippine Islands. The Scotia and Antillean (Caribbean) island arcs are in the Atlantic Ocean basin.

Oceanic–Continental Boundaries

When an oceanic and a continental plate converge, the denser oceanic plate is subducted under the continental plate along an **oceanic–continental plate boundary** (Figure 2.19b). Just as at oceanic–oceanic plate boundaries, the descending oceanic plate forms the outer wall of an oceanic trench.

The magma generated by subduction rises beneath the continent and either crystallizes as large intrusive bodies, called *plutons*, before reaching the surface or erupts at the surface to produce a chain of andesitic volcanoes, also called a *volcanic arc*. An excellent example of an oceanic–continental plate boundary is the Pacific Coast of South America where the oceanic Nazca plate is currently being subducted under South America (Figure 2.19b; see also Chapter 9). The Peru–

Chile Trench marks the site of subduction, and the Andes Mountains are the resulting volcanic mountain chain on the nonsubducting plate.

Just as there are passive continental margins, there are also active continental margins (Figure 2.10). The aforementioned oceanic–continental plate boundary between the west coast of the South American plate and the eastern side of the oceanic Nazca plate is an excellent example of an active continental margin. Here, the continental shelf is narrow, and the continental slope descends directly into the Peru–Chile Trench, where sediment is promptly dumped into the trench (Figure 2.10).

Continental–Continental Boundaries

Two continents approaching each other are initially separated by an ocean floor that is being subducted under one continent. The edge of that continent displays the features characteristic of oceanic–continental convergence. As the ocean floor continues to be subducted, the two continents come closer together until they eventually collide. Because continental lithosphere, which consists of continental crust and the upper mantle, is less dense than oceanic lithosphere (oceanic crust and upper mantle), it cannot sink into the asthenosphere. Although one continent may partially slide under the other, it cannot be pulled or pushed down into a subduction zone (Figure 2.19c).

When two continents collide, they are welded together along a zone marking the former site of subduction. At this **continental–continental plate boundary**, an interior mountain belt is formed consisting of deformed sediments and sedimentary rocks, igneous intrusions, metamorphic rocks, and fragments of oceanic crust. In addition, the entire region is subjected to numerous earthquakes. The Himalayas in central Asia, the world's youngest and highest mountain system, resulted from the collision between India and Asia that began 40 to 50 million years ago and is still continuing (Figure 2.19c; see Chapter 9).

oceanic–continental plate boundary A convergent plate boundary along which oceanic lithosphere is subducted beneath continental lithosphere.

continental–continental plate boundary A convergent plate boundary along which two continental lithospheric plates collide.

Recognizing Ancient Convergent Plate Boundaries

How can former subduction zones be recognized in the geologic record? Igneous rocks provide one such clue. The magma that erupted at the surface, forming island arc volcanoes and continental volcanoes, is of andesitic composition. Another clue is the zone of intensely deformed rocks between the deep-sea trench where subduction is taking place and the area of igneous activity. Here, sediments and submarine rocks are folded, faulted, and metamorphosed into a chaotic mixture of rocks termed a *mélange.*

transform plate boundary A plate boundary along which plates slide past one another and crust is neither produced nor destroyed.

transform fault A fault along which one type of motion is transformed into another; commonly displaces oceanic ridges; on land, recognized as a strike-slip fault, such as the San Andreas fault.

During subduction, pieces of oceanic lithosphere are sometimes incorporated into the mélange and accreted onto the edge of the continent. Slices of oceanic crust and upper mantle are called *ophiolites* (Figure 2.20). They consist of a layer of deep-sea sediments that include graywackes (poorly sorted sandstones containing abundant feldspars and rock fragments, usually in a clay-rich matrix), black shales, and cherts (see Chapter 6). These deep-sea sediments are underlain by pillow lavas, a sheeted dike complex, massive gabbro (a dark intrusive igneous rock), and layered gabbro, all of which form the oceanic crust. Beneath the gabbro is peridotite (a dark, intrusive igneous rock composed of the mineral olivine), which probably represents the upper mantle. The presence of ophiolites in an outcrop or drilling core is a key indicator of plate convergence along a subduction zone.

Transform Boundaries

The third type of plate boundary is a **transform plate boundary**. These mostly occur along fractures in the seafloor, known as *transform faults,* where plates slide laterally past one another roughly parallel to the direction of plate movement. Although

Figure 2.20 Ophiolites

Ophiolites are sequences of rock on land consisting of deep-sea sediments, oceanic crust, and upper mantle. Ophiolites are one feature used to recognize ancient convergent plate boundaries.

lithosphere is neither created nor destroyed along a transform boundary, the movement between plates results in a zone of intensely shattered rock and numerous shallow-depth earthquakes.

Transform faults "transform," or change, one type of motion between plates into another type of motion. Most commonly, transform faults connect two oceanic ridge segments (Figure 2.21); however, they can also connect ridges to trenches and trenches to trenches. Although the majority of transform faults are in oceanic crust and are marked by distinct fracture zones, they may also extend into continents.

Figure 2.21 Transform Plate Boundaries

Horizontal movement between plates occurs along transform faults. Extensions of transform faults on the seafloor form fracture zones. Most transform faults connect two oceanic ridge segments.

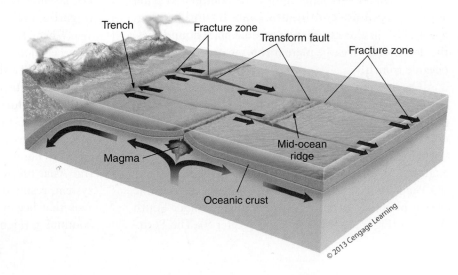

One of the best-known transform faults is the San Andreas fault in California. It separates the Pacific plate from the North American plate and connects spreading ridges in the Gulf of California with the Juan de Fuca and Pacific plates off the coast of northern California (Figure 2.22). Many of the earthquakes that affect California are the result of movement along this fault (see Chapter 8).

Unfortunately, transform faults generally do not leave any characteristic or diagnostic features except for the obvious displacement of the rocks with which they are associated. This displacement is usually large, on the order of tens to hundreds of kilometers. Such large displacements in ancient rocks can sometimes be related to transform fault systems.

LO9 HOT SPOTS AND MANTLE PLUMES

Before leaving the topic of plate boundaries, we should mention an intraplate feature found beneath both oceanic and continental plates. A **hot spot** (Figures 2.9 and 2.16 is the location on Earth's surface where a stationary column of magma, originating deep within the mantle (*mantle plume*), has slowly risen to the surface and formed a volcano. Because the mantle plumes apparently remain stationary (although some evidence suggests that they might not) within the mantle while the plates move over them, the resulting hot spots leave a trail of extinct and progressively older volcanoes (aseismic ridges) that record the movement of the plate.

One of the best examples of aseismic ridges and hot spots is the Emperor Seamount–Hawaiian Island chain (Figure 2.9). This chain of islands and seamounts extends from the island of Hawaii to the Aleutian Trench off Alaska, a distance of some 6,000 km, and consists of more than 80 volcanic structures.

Currently, the only active volcanoes in this island chain are on the island of Hawaii and the Loihi Seamount. The rest of the islands are extinct volcanic structures that become progressively older toward the north and northwest. This means that the Emperor Seamount–Hawaiian Island chain records the direction that the Pacific plate traveled as it moved over an apparently stationary mantle plume.

Mantle plumes and hot spots help geologists explain some of the geologic activity occurring within plates as opposed to activity occurring at or near plate boundaries. In addition, if mantle plumes are essentially fixed with respect to Earth's rotational axis, they can be used to determine not only the direction of plate movement, but also the rate of movement.

> **hot spot** A localized zone of melting below the lithosphere that probably overlies a mantle plume; detected by volcanism at the surface.

LO10 PLATE MOVEMENT AND MOTION

How fast and in what direction are Earth's plates moving? Do they all move at the same rate? Rates of plate movement can be calculated in several ways. The least accurate method is to determine the age of the sediments immediately above any portion of the oceanic crust and then divide the distance from the spreading ridge by that age. Such calculations give an average rate of movement.

A more accurate method of determining both the average rate of movement and relative motion is by dating the magnetic anomalies in the crust of the seafloor. The distance from an oceanic ridge axis to any magnetic anomaly indicates the width of new seafloor that formed during that time interval. For example, if the distance between the present-day Mid-Atlantic Ridge and anomaly 31 is 2,010 km, and anomaly 31 formed 67 million years ago (Figure 2.23), then the average rate of movement during the past 67 million years has been 3 cm per year (2,010 km, which equals 201 million cm divided by 67 million years; 201,000,000 cm/ 67,000,000 years = 3 cm/year). Thus, for a given interval of time, the wider the strip of seafloor, the faster the plate has moved. In this way, not only can the present average rate of movement and relative motion be determined (Figure 2.16),

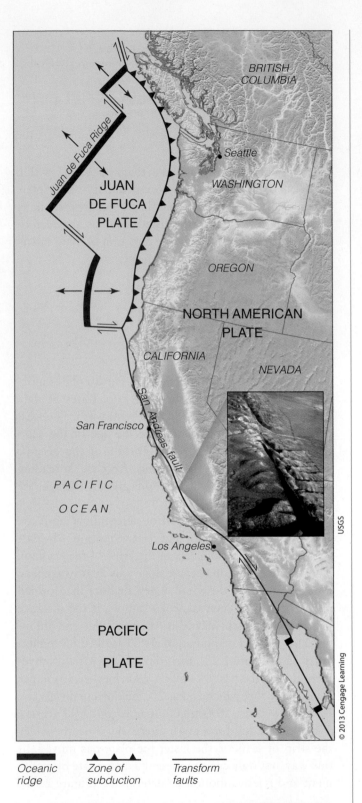

| Oceanic ridge | Zone of subduction | Transform faults |

Figure 2.22 The San Andreas Fault—A Transform Plate Boundary
The San Andreas fault is a transform fault separating the Pacific plate from the North American plate. It connects the spreading ridges in the Gulf of California with the Juan de Fuca and Pacific plates off the coast of northern California. Movement along the San Andreas fault has caused numerous earthquakes. The insert photo shows a segment of the San Andreas fault as it cuts through the Carrizo Plain, California.

USGS

© 2013 Cengage Learning

but the average rate of movement in the past can also be calculated by dividing the distance between anomalies by the amount of time elapsed between anomalies.

Geologists use magnetic anomalies not only to calculate the average rate of plate movement, but also to determine plate positions at various times in the past. Because magnetic anomalies are parallel and symmetric with respect to spreading ridges, all one must do to determine the position of continents when particular anomalies formed is to move the anomalies back to the spreading ridge, which will also move the continents with them (Figure 2.23). Because subduction destroys oceanic crust and the magnetic record that it carries, we have an excellent record of plate movements since the breakup of Pangaea, but not as good an understanding of plate movement before that time.

The average rate of movement, as well as the relative motion between any two plates, can also be determined by satellite-laser ranging techniques. Laser beams

Figure 2.23 Reconstructing Plate Positions Using Magnetic Anomalies

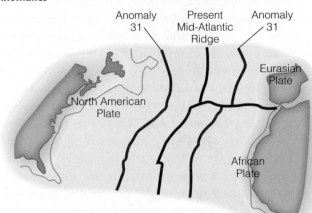

a. The present North Atlantic, showing the Mid-Atlantic Ridge and magnetic anomaly 31, which formed 67 million years ago.

© 2013 Cengage Learning

b. The Atlantic 67 million years ago. Anomaly 31 marks the plate boundary 67 million years ago. By moving the anomalies back together, along with the plates they are on, we can reconstruct the former positions of the continents.

from a station on one plate are bounced off a satellite (in geosynchronous orbit) and returned to a station on a different plate. As the plates move away from each other, the laser beam takes more time to go from the sending station to the stationary satellite and back to the receiving station. This difference in elapsed time is used to calculate the rate of movement and the relative motion between plates.

LO11 THE DRIVING MECHANISM OF PLATE TECTONICS

A major obstacle to the acceptance of the continental drift hypothesis was the lack of a driving mechanism to explain continental movement. When it was shown that continents and ocean floors moved together, not separately, and that new crust is formed at spreading ridges by rising magma, most geologists accepted some type of convective heat system (convection cells) as the basic process responsible for plate motion. However, the question of what exactly drives the plates remains.

Most of the heat from Earth's interior results from the decay of radioactive elements, such as uranium (see Chapter 16), in the core and lower mantle. The most efficient way for this heat to escape Earth's interior is through some type of slow convection. Heat from the core, supplemented by heat generated from radioactive decay, thus drives large mantle convection cells (Figure 2.24). In this manner, hot rock from the interior rises toward the surface, loses heat to the overlying lithosphere, becomes denser as it cools, and then sinks back into the interior, where it is heated and the process repeats itself. This type of convective heat system is analogous to a pot of stew cooking on a stove.

In this mantle convection cell model, spreading ridges mark the ascending limbs of adjacent convection cells, and trenches are present where convection cells descend back into Earth's interior. The convection cells therefore determine the location of spreading ridges and trenches, with the lithosphere lying above the thermal convection cells. Thus, each plate corresponds to a single convection cell that moves as a result of the convective movement of the cell itself (Figure 2.24).

Although most geologists agree that Earth's internal heat plays an important role in plate movement, two other processes, referred to as "slab-pull" and "ridge-push," might also help facilitate the movement of plates (Figure 2.25). Both mechanisms are gravity driven but still depend on thermal differences within Earth.

In slab-pull, the subducting cold slab of lithosphere, being denser than the surrounding warmer asthenosphere, pulls the rest of the plate along as it descends into the asthenosphere (Figure 2.25). As the lithosphere moves downward, there is a corresponding upward flow back into the spreading ridge.

Operating in conjunction with slab-pull is the ridge-push mechanism. As a result of rising magma, the oceanic ridges are higher than the surrounding oceanic crust. It is thought that gravity pushes the oceanic lithosphere away from the higher spreading ridges and

Convection in a Pot of Stew

Heat from the stove is applied to the base of the stew pot, causing the stew to heat up. As heat rises through the stew, pieces of the stew are carried to the surface, where the heat is dissipated, the pieces of stew cool, and then sink back to the bottom of the pot. The bubbling seen at the surface of the stew is the result of convection cells churning the stew. In the same manner, heat from the decay of radioactive elements produces convection cells within Earth's interior.

Figure 2.24 Thermal Convection Cells as the Driving Force of Plate Movement

A cutaway view of Earth shows that the lithosphere glides horizontally across the asthenosphere. Heat from the core, supplemented by heat produced from radioactive decay, drives huge mantle convection cells that move the lithosphere. Spreading ridges mark the location of ascending limbs of the convection cells, and oceanic trenches are the surface expression of subduction, where plates descend into Earth's interior.

toward the trenches, where it is subducted back into Earth's interior (Figure 2.25).

Currently, geologists are fairly certain that some type of convective system is involved in plate movement, but the extent to which other mechanisms, such as slab-pull and ridge-push, are involved is still unresolved. However, the fact that plates have moved in the past and are still moving today has been proven beyond a doubt. And although a comprehensive theory of plate movement has not yet been developed, more and more of the pieces are falling into place as geologists learn more about Earth's interior.

LO12 PLATE TECTONICS AND THE DISTRIBUTION OF NATURAL RESOURCES

In addition to being responsible for the major features of Earth's crust and influencing the distribution and evolution of the world's biota, plate movement also affects the formation and distribution of some natural resources. Consequently, geologists are using plate tectonic theory in their search for petroleum and mineral deposits and in explaining the occurrence of these natural resources. It is becoming increasingly clear that if we are to keep up with the continuing demands of a global industrialized society, the application of plate tectonic theory to the origin and distribution of natural resources is essential.

Petroleum

Although significant concentrations of petroleum occur in many areas of the world, more than 50% of all proven reserves are in the Persian Gulf region. It should therefore not be surprising that many of the conflicts in the Middle East have shared as their underlying cause the desire to control these vast amounts of petroleum. Most people, however, are not aware of *why* there is so much oil in this region of the world. The answer lies in the paleogeography and plate movement of this region during the Mesozoic and Cenozoic eras.

© 2013 Cengage Learning

Figure 2.25 Plate Movement Resulting from Gravity-Driven Mechanisms
Plate movement is also thought to result, at least partially, from gravity-driven "slab-pull" or "ridge-push" mechanisms. In slab-pull, the edge of the subducting plate descends into the interior, and the rest of the plate is pulled downward. In ridge-push, rising magma pushes the oceanic ridges higher than the rest of the oceanic crust. Gravity thus pushes the oceanic lithosphere away from the ridges and toward the trenches.

During the Mesozoic Era, and particularly the Cretaceous Period, when most of the petroleum formed, the Persian Gulf area was a broad marine shelf extending eastward from Africa. This passive continental margin lay near the equator, where countless microorganisms lived in the surface waters. The remains of these organisms accumulated with the bottom sediments and were buried, beginning the complex process of petroleum generation and the formation of source beds in which petroleum forms.

During the Cenozoic Era, convergent plate movement in the Persian Gulf area resulted in the sediments of the passive continental margin being subducted, which generated heat that broke down the organic molecules, leading to the formation of petroleum. The continued collisions in the area folded the rocks, creating traps for petroleum to accumulate, such that this is now a major oil-producing region. Elsewhere in the world, plate tectonics is also responsible for concentrations of petroleum.

Mineral Deposits

Many metallic mineral deposits such as copper, gold, lead, silver, tin, and zinc are related to igneous and associated hydrothermal (hot water) activity. So it is not surprising that a close relationship exists between plate boundaries and the occurrence of these valuable deposits.

The magma generated by partial melting of a subducting plate rises toward the surface, and as it cools, it precipitates and concentrates various metallic ores. Many of the world's major metallic ore deposits are associated with convergent plate boundaries, including those in the Andes of South America, the Coast Ranges and Rockies of North America, Japan, the Philippines, Russia, and a zone extending from the eastern Mediterranean region to Pakistan. In addition, the majority of the world's gold is associated with deposits located at ancient convergent plate boundaries in such areas as Canada, Alaska, California, Venezuela, Brazil, Russia, southern India, and western Australia.

The copper deposits of western North and South America are an excellent example of the relationship between convergent plate boundaries and the distribution, concentration, and exploitation of valuable metallic ores (Figure 2.26a). The world's largest copper deposits are found along this belt. The majority of the copper deposits in the Andes and the southwestern United States were formed less than 60 million years ago when oceanic plates were subducted under the North and South American plates. The rising magma and associated hydrothermal fluids carried minute amounts of copper, which were originally widely

© IStockphoto.com/David G. Freund

disseminated but eventually became concentrated in the cracks and fractures of the surrounding andesites. These low-grade copper deposits contain from 0.2 percent to 2 percent copper and are extracted from large open-pit mines (Figure 2.26b).

Divergent plate boundaries also yield valuable ore resources. For example, the island of Cyprus in the Mediterranean is rich in copper and has been supplying all or part of the world's needs for the past 3,000 years.

LO13 PLATE TECTONICS AND THE DISTRIBUTION OF LIFE

Plate tectonic theory is as revolutionary and far-reaching in its implications for geology as the theory of evolution was for biology. Interestingly, it was the fossil evidence that convinced Wegener, Suess, and du Toit, as well as many other geologists, of the correctness of the theory of continental drift. Together, the theories of plate tectonics and evolution have changed the way we view our planet, and we should not be surprised at the intimate association between them. Although the relationship between plate tectonic processes and the evolution of life is incredibly complex, paleontological data provide convincing evidence of the influence of plate movement on the distribution of organisms.

The present distribution of plants and animals is not random, but is controlled mostly by climate and geographic barriers. The world's biota occupy *biotic provinces*, which are regions characterized by a distinctive assemblage of plants and animals. Organisms within a province have similar ecological requirements, and the boundaries separating provinces are therefore natural ecological breaks. Climatic or geographic barriers are the most common province boundaries, and these are mostly controlled by plate movement.

The complex interaction between wind and ocean currents has a strong influence on the world's climates. Wind and ocean currents, in turn, are strongly influenced by the number, distribution, topography, and orientation of continents. The distribution of continents and ocean basins not only influences wind and ocean currents, but also affects provinciality by creating physical barriers to, or pathways for, the migration of organisms. Intraplate volcanoes, island arcs, mid-oceanic ridges, mountain ranges, and subduction zones all result from the interaction of plates, and their orientation and distribution strongly influence the number of provinces and hence total global diversity.

Figure 2.26 Copper Deposits and Convergent Plate Boundaries

© 2013 Cengage Learning

a. Valuable copper deposits are located along the west coasts of North and South America in association with convergent plate boundaries. Through time, the rising magma and associated hydrothermal activity resulting from subduction carried small amounts of copper that then became trapped and concentrated in the surrounding rocks.

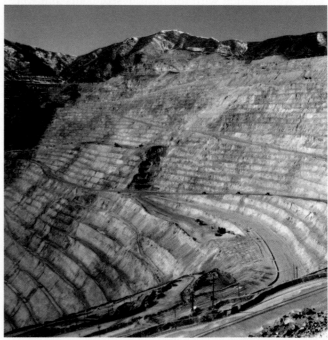

Copyright and photograph by Dr. Parvinder S. Sethi

b. Bingham Copper Mine, near Salt Lake City, Utah, is a huge open-pit copper mine with reserves estimated at 1.7 billion tons. More than 400,000 tons of rock are removed for processing each day. Note the small specks toward the middle of the photograph that are the 12-foot-high dump trucks!

Thus, provinciality and diversity will be highest where numerous small continents are spread across many zones of latitude.

When a geographic barrier separates a once-uniform fauna, species must adapt to the new conditions, migrate, or become extinct. Adaptation to the new environment by various species may involve enough change that new species eventually evolve. The formation of the Isthmus of Panama provides an excellent example of how plate tectonics influences evolution. Prior to the rise of this connection between North and South America, a homogeneous population of bottom-dwelling invertebrates inhabited the shallow seas of the area. After the formation of the Isthmus of Panama by subduction of the Pacific plate approximately 5 million years ago, the original population was divided. In response to the changing environment, new species evolved on opposite sides of the Isthmus.

The formation of the Isthmus of Panama also influenced the evolution of the North and South American mammalian faunas (Figure 2.27). During most of the Cenozoic Era, South America was an island continent, and its mammalian fauna evolved in isolation. When North and South America were connected by the Isthmus, most of the indigenous South American mammals were replaced by migrants from North America. Surprisingly, only a few South American mammal groups migrated northward.

Figure 2.27 Plate Tectonics and the Distribution of Organisms

The Isthmus of Panama forms a barrier that divides a once-uniform fauna of molluscs that inhabited the shallow seas of both the Pacific Ocean and Caribbean Sea. Its creation also formed a land corridor in which migration between the two continents took place. Prior to the formation of this isthmus, South America was isolated from all other landmasses during much of the Cenozoic, and its mammal fauna consisted of marsupials (pouched mammals) and placentals that lived nowhere else. When the Isthmus of Panama formed during the Late Pliocene, many placental mammals migrated south, and many South American mammals became extinct. A few South American mammals migrated north and successfully occupied North America.

© 2013 Cengage Learning

3 | Minerals: The Building Blocks of Rocks

This object is a geode that forms when minerals grow along the margins of a cavity, partially filling it. The outermost layers are thin, sometimes discontinuous layers of agate, a mineral made up of tiny crystals. Next inward toward the cavity's center are quartz crystals that point inward. Many geodes are found in volcanic rocks, but they also occur in other types of rocks.

Sue Monroe

Introduction

> "**We cannot** overstate the importance of minerals in **many** human endeavors."

Liquid water, gases, and trees are not minerals, but ice is because it meets the criteria geologists use to define the term **mineral**: a naturally occurring, inorganic, crystalline solid, with a narrowly defined chemical composition and characteristic physical properties. Of course, ice is naturally occurring and inorganic (it is composed of hydrogen and oxygen (H_2O)), and it has distinguishing properties such as hardness and density. The only part of this definition that may be unfamiliar is "crystalline solid," which means that the atoms in a mineral are arranged in a specific three-dimensional framework, as opposed to the atoms in glass, which have no such orderly arrangement.

We cannot overstate the importance of minerals in many human endeavors. Indeed, ores of iron, copper, and aluminum, as well as minerals and rocks used for glass, fertilizer, and animal feed supplements, are essential to our economic well-being. The United States and Canada owe much of their economic success to abundant mineral and energy resources, but both nations must import some essential commodities, including aluminum ore and manganese. As a matter of fact, the distribution of minerals, rocks, and energy resources has important implications in foreign relations and accounts for economic ties between nations. Even some rather common minerals are valuable: Quartz, for example, is used for sandpaper, and muscovite accounts for the lustrous sheen of paint used on automobiles and appliances.

One important reason to study minerals is their economic importance, but for students it is important to know that minerals are essential constituents of rocks. Granite, for instance, is made up of specific percentages of the minerals quartz, potassium feldspars, and plagioclase feldspars. Some minerals are attractive and eagerly sought out by collectors and for museum displays, whereas others are known as *gemstones*—that is, precious and semiprecious minerals and rocks used for decorative purposes, especially jewelry (Figure 3.1).

From our discussion so far, we have a formal definition of the term *mineral*, and we know that minerals are the basic constituents of rocks. Now let's delve deeper into what minerals are made of by considering matter, atoms, elements, and bonding.

> **mineral** A naturally occurring, inorganic, crystalline solid that has characteristic physical properties and a narrowly defined chemical composition.

Learning Outcomes

After reading this unit, you should be able to do the following:

LO1 Define matter

LO2 Explore the world of minerals

LO3 Identify mineral groups recognized by geologists

LO4 Identify physical properties of minerals

LO5 Recognize rock-forming minerals

LO6 Explain how minerals form

LO7 Recognize natural resources and reserves

Sue Monroe

Figure 3.1 Emerald

These beautiful emeralds are gemstones because they can be cut, polished, and used for decorative purposes. Emerald is actually a variety of the mineral beryl, which is composed of beryllium, aluminum, silica, and oxygen. This specimen is on display at the Natural History Museum in Vienna, Austria.

LO1 MATTER: WHAT IS IT?

Anything that has mass and occupies space is matter and accordingly includes water, plants, animals, the atmosphere, and minerals and rocks. Physicists recognize four states, or phases, of matter: *liquids*, *gases*, *solids*, and *plasma*, an ionized gas, as in the Sun. Here our main concern is with solids because, by definition, minerals are solids.

Atoms and Elements

Matter is made up of chemical **elements**, which, in turn, are composed of **atoms**, the smallest units of matter that retain the characteristics of a particular element (Figure 3.2). That is, elements cannot be changed into different substances except through radioactive decay (discussed in Chapter 16). Thus, an element is made up of atoms, all of which have the same properties. Some of the 92 naturally occurring elements are listed in Figure 3.2. All naturally occurring elements have a name and a symbol—for example, oxygen (O), aluminum (Al), and potassium (K).

At the center of an atom is a **nucleus** made up of one or more **protons**, which have a positive electrical charge, and **neutrons**, which are electrically neutral. The nucleus is only about 1/100,000th of the diameter of an atom, yet it contains virtually all of the atom's mass. **Electrons**, particles with a negative electrical charge, orbit rapidly around the nucleus at specific distances in one or more **electron shells** (Figure 3.2). The electrons determine how an atom interacts with other atoms, but the nucleus determines how many electrons an atom has, because the positively charged protons attract and hold the negatively charged electrons in their orbits.

The number of protons in its nucleus determines an atom's identity and its **atomic number**. Hydrogen (H), for instance, has one proton in its nucleus and thus has an atomic number of 1. The nuclei of helium (He) atoms possess 2 protons, whereas those of carbon (C) have 6, and uranium (U) 92, so their atomic numbers are 2, 6, and 92, respectively. An atom's **atomic mass number** is the sum of protons and neutrons in the nucleus (electrons contribute negligible mass to atoms). However, atoms of the same chemical element might have different atomic mass numbers, because the number of neutrons can vary. All carbon (C) atoms have six protons—otherwise they would not be carbon—but the number of neutrons is 6, 7, or 8. Thus we recognize three types of carbon, or what are known as *isotopes* (Figure 3.3), each with a different atomic mass number.

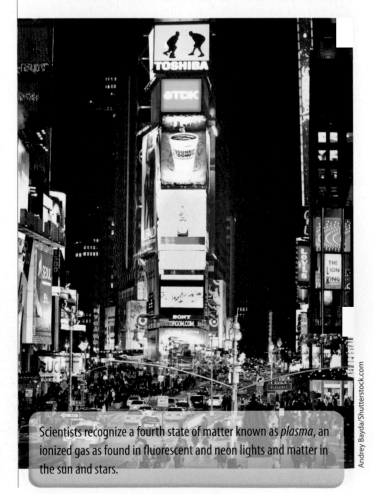

Scientists recognize a fourth state of matter known as *plasma*, an ionized gas as found in fluorescent and neon lights and matter in the sun and stars.

Andrey Bayda/Shutterstock.com

Element	Symbol	Atomic Number	Distribution of Electrons			
			First Shell	Second Shell	Third Shell	Fourth Shell
Hydrogen	H	1	1	—	—	—
Helium	He	2	2	—	—	—
Carbon	C	6	2	4	—	—
Oxygen	O	8	2	6	—	—
Neon	Ne	10	2	8	—	—
Sodium	Na	11	2	8	1	—
Magnesium	Mg	12	2	8	2	—
Aluminum	Al	13	2	8	3	—
Silicon	Si	14	2	8	4	—
Phosphorus	P	15	2	8	5	—
Sulfur	S	16	2	8	6	—
Chlorine	Cl	17	2	8	7	—
Potassium	K	19	2	8	8	1
Calcium	Ca	20	2	8	8	2
Iron	Fe	26	2	8	14	2

© 2013 Cengage Learning

Figure 3.2 Shell Models for Common Atoms

The shell model for several atoms and their electron configurations. A blue circle represents the nucleus of each atom, but remember that atomic nuclei are made up of protons and neutrons, as shown in Figure 3.3.

^{12}C (Carbon 12) ^{13}C (Carbon 13) ^{14}C (Carbon 14)

© 2013 Cengage Learning

Figure 3.3 Carbon Isotopes
Schematic representation of the isotopes of carbon. Carbon has an atomic number of 6 and an atomic mass number of 12, 13, or 14, depending on the number of neutrons in its nucleus.

The isotopes of carbon, or those of any other element, behave the same chemically; carbon 12 and carbon 14 are both present in carbon dioxide (CO_2), for example. However, some isotopes are radioactive, meaning that they spontaneously decay or change to other elements. Carbon 14 is radioactive, whereas both carbon 12 and carbon 13 are not. Radioactive isotopes are important for determining the absolute ages of rocks (see Chapter 16).

Bonding and Compounds

Interactions among electrons around atoms can result in two or more atoms joining together, a process known as **bonding**. If atoms of two or more elements bond, the resulting substance is a **compound**. Gaseous oxygen consists of only oxygen atoms and is thus an element, whereas the mineral quartz, consisting of silicon and oxygen atoms, is a compound. Most minerals are compounds, although gold, platinum, and several others are important exceptions.

To understand bonding, it is necessary to delve deeper into the structure of atoms. Recall that negatively charged electrons orbit the nuclei of atoms in electron shells. With the exception of hydrogen, which has only one proton and one electron, the innermost electron shell of an atom contains only two electrons. The other shells contain various numbers of electrons, but the outermost shell never has more than eight (Figure 3.2). The electrons in the outermost shell are usually involved in chemical bonding.

Ionic and *covalent bonding* are important in minerals, and many minerals contain both types of bonds. Two other types of bonds, *metallic* and *van der Waals*, are much less common, but are important in determining the properties of some useful minerals.

bonding The process whereby atoms join to other atoms.

compound Any substance resulting from the bonding of two or more different elements (e.g., water, H_2O, and quartz, SiO_2).

ion An electrically charged atom produced by adding or removing electrons from the outermost electron shell.

ionic bond A chemical bond resulting from the attraction between positively and negatively charged ions.

Ionic Bonding

Most atoms have fewer than eight electrons in their outermost electron shell, but some, including neon and argon, have complete outer shells with eight electrons (Figure 3.2). Because of this electron configuration, these elements, known as the *noble gases*, do not react readily with other elements to form compounds. Interactions among atoms tend to produce electron configurations similar to those of the noble gases. That is, atoms interact so that their outermost electron shell is filled with eight electrons, unless the first shell (with two electrons) is also the outermost electron shell, as in helium.

One way that the noble gas configuration is attained is by the transfer of one or more electrons from one atom to another. Common salt is composed of the elements sodium (Na) and chlorine (Cl); each element is poisonous, but when combined chemically, they form the mineral halite, a compound of sodium chloride (NaCl). Sodium has 11 protons and 11 electrons; thus the positive electrical charges of the protons are exactly balanced by the negative charges of the electrons, and the atom is electrically neutral (Figure 3.4a). Likewise, chlorine, with 17 protons and 17 electrons, is electrically neutral. However, neither sodium nor chlorine has eight electrons in its outermost electron shell; sodium has only one, whereas chlorine has seven. To attain a stable configuration, sodium loses the electron in its outermost electron shell, leaving its next shell with eight electrons as the outermost one (Figure 3.4a). Sodium now has one fewer electron (negative charge) than it has protons (positive charge), so it is an electrically charged **ion** and is symbolized Na$^+$.

The electron lost by sodium is transferred to the outermost electron shell of chlorine, which had seven electrons to begin with. The addition of one more electron gives chlorine an outermost electron shell of eight electrons, the configuration of a noble gas. But its total number of electrons is now 18, which exceeds by 1 the number of protons. Accordingly, chlorine also becomes an ion, but it is negatively charged (Cl$^-$). An **ionic bond** forms between sodium and chlorine because of the attractive force between the positively charged sodium ion and the negatively charged chlorine ion (Figure 3.4a).

In ionic compounds, such as sodium chloride (the mineral halite), the ions are arranged in a three-dimensional framework that results in overall electrical neutrality. In halite, sodium ions are bonded to chlorine ions on all sides, and chlorine ions are surrounded by sodium ions (Figure 3.4b).

a. Transfer of the electron in the outermost shell of sodium to the outermost shell of chlorine. After electron transfer, the sodium and chlorine atoms are positively and negatively charged ions, respectively.

Sodium atom 11 p$^+$ 11 e$^-$

electron transfer

Chlorine atom 17 p$^+$ 17 e$^-$

Sodium ion 11 p$^+$ 10 e$^-$

Chlorine ion 17 p$^+$ 18 e$^-$

+   **−**

Cl$^-$ Na$^+$

© 2013 Cengage Learning

b. This diagram shows the relative sizes of the sodium and chlorine atoms and their locations in a crystal of halite.

Figure 3.4 Ionic Bond to Form the Mineral Halite (NaCl)

a. The orbits in the outermost electron shell overlap, so electrons are shared in diamond.

b. Covalent bonding of carbon atoms in diamond forms a three-dimensional framework.

c. Covalent bonds in graphite form strong sheets, but the van der Waals bonds between sheets are weak.

Figure 3.5 Covalent Bonds

Covalent Bonding

Covalent bonds form between atoms when their electron shells overlap and they share electrons. For example, atoms of the same element, such as carbon, cannot bond by transferring electrons from one atom to another. Carbon (C), which forms the minerals graphite and diamond, has four electrons in its outermost electron shell (Figure 3.5a). If these four electrons were transferred to another carbon atom, the atom receiving the electrons would have the noble gas configuration of eight electrons in its outermost electron shell, but the atom contributing the electrons would not.

In such situations, adjacent atoms share electrons by overlapping their electron shells. A carbon atom in diamond, for instance, shares all four of its outermost electrons with a neighbor to produce a stable noble gas configuration (Figure 3.5a).

Among the most common minerals, the silicates, the element silicon forms partly covalent and partly ionic bonds with oxygen.

Metallic and van der Waals Bonds

Metallic bonding involves an extreme type of electron sharing. The electrons of the outermost electron shell of metals such as gold, silver, and copper readily move from one atom to another. This electron mobility accounts for the fact that metals have a metallic luster (their appearance in reflected light), provide good electrical and thermal conductivity, and can be easily reshaped. Only a few minerals possess metallic bonds, but those that do are very useful; copper, for example, is used for electrical wiring because of its high electrical conductivity.

Some electrically neutral atoms and molecules* have no electrons available for ionic, covalent, or metallic bonding. They nevertheless have a weak attractive force between them, called a *van der Waals* or *residual bond*, when in proximity. The carbon atoms in the mineral graphite are covalently bonded to form sheets, but the sheets are weakly held together by van der Waals bonds (Figure 3.5c). This type of bonding makes graphite useful for pencil lead; when a pencil point is moved across a piece of paper, small pieces of graphite flake off along the planes held together by van der Waals bonds and adhere to the paper.

> **covalent bond** A chemical bond formed by the sharing of electrons between atoms.

LO2 EXPLORE THE WORLD OF MINERALS

We defined a mineral as a naturally occurring, inorganic crystalline solid with a narrowly defined chemical composition and characteristic physical properties. In the following sections, we examine each part of the definition of the term *mineral*.

---

*A molecule is the smallest unit of a substance that has the properties of that substance. A water molecule (H_2O), for example, possesses two hydrogen atoms and one oxygen atom.

The distinction between minerals and rocks is not easy for beginning students to understand. Minerals are made up of chemical elements, and rocks consist of one or more minerals, but students commonly mistake one for the other. Can you think of analogies that might help you to remember the difference between minerals and rocks?

© Tetra Images/Jupiter Images

Naturally Occurring Inorganic Substances

The criterion *naturally occurring* excludes from minerals all substances manufactured by humans, such as synthetic diamonds and rubies. Some geologists think the term *inorganic* in the mineral definition is unnecessary, but it does remind us that animal matter and vegetable matter are not minerals. Nevertheless, some organisms, including corals, clams, and several other animals and some plants, construct their shells of calcium carbonate ($CaCO_3$), which is either the mineral aragonite or calcite, or their shells are made of silicon dioxide (SiO_2).

crystalline solid A solid in which the constituent atoms are arranged in a regular, three-dimensional framework.

crystal A naturally occurring solid of an element or compound with a specific internal structure that is manifested externally by planar faces, sharp corners, and straight edges.

Mineral Crystals

By definition, minerals are **crystalline solids**, in which the constituent atoms are arranged in a regular, three-dimensional framework (Figure 3.5b). Under ideal conditions, crystalline solids grow and form perfect **crystals** that possess planar surfaces (crystal faces), sharp corners, and straight edges (Figure 3.6). In other words, the regular geometric shape of a mineral crystal is the exterior manifestation of an ordered internal atomic arrangement. Not all rigid substances are crystalline solids; natural and manufactured glass lack the ordered arrangement of atoms and is said to be *amorphous*, meaning "without form." Minerals are crystalline solids, but crystalline solids are not always well-formed crystals. The reason is that when crystals form, they may grow in proximity and form an interlocking mosaic in which individual crystals are not apparent or easily discerned.

So how do we know that minerals with no obvious crystals are actually crystalline? X-ray beams and light transmitted through mineral crystals behave in a predictable manner, which provides compelling evidence for an orderly internal structure. Another way we can determine that minerals are actually crystalline is by their *cleavage*, the property of breaking or splitting repeatedly along smooth, closely spaced planes. Not all minerals have cleavage planes, but many do, and such regularity certainly indicates that cleavage is controlled by internal structure.

As early as 1669, the Danish scientist Nicholas Steno determined that the angles of intersection of equivalent crystal faces on different specimens of quartz are identical. Since then, this *constancy of interfacial angles* has been demonstrated for many other minerals, regardless of their size, shape, age, or geographic occurrence. Steno postulated that mineral crystals are made up of very small, identical building blocks, and that the arrangement of these building blocks determines the external form of mineral crystals, a proposal that has since been verified.

Chemical Composition of Minerals

The chemical composition of minerals is shown by a chemical formula, which is a shorthand way of indicating the numbers of atoms of different elements in a mineral. The mineral quartz consists of one silicon (Si) atom for every two oxygen (O) atoms and thus has the formula SiO_2; the subscript number indicates the number of atoms. Orthoclase is composed of one potassium, one aluminum, three silicon, and eight oxygen atoms,

a. Cubic crystals are typical of the minerals halite and galena.

b. Pyritohedron crystals such as those of pyrite have 12 sides.

c. Diamond has octahedral, or eight-sided, crystals.

d. A prism terminated by a pyramid is found in quartz.

Figure 3.6 A Variety of Mineral Crystal Shapes

so its formula is $KAlSi_3O_8$. Some minerals known as *native elements* consist of a single element and include silver (Ag), platinum (Pt), gold (Au), and graphite and diamond, both of which are composed of carbon (C).

For many minerals, the chemical composition does not vary. Quartz is made up only of silicon and oxygen (SiO_2), and halite contains only sodium and chlorine (NaCl). However, some minerals have a range of compositions because one element can substitute for

another if the atoms of two or more elements are nearly the same size and the same charge. Thus, the definition of the term mineral has the phrase *narrowly defined chemical composition*. Iron and magnesium meet these criteria and can substitute for one another (Figure 3.7), so the chemical formula for olivine is $(Mg,Fe)_2SiO_4$, meaning that in addition to silicon and oxygen, it may contain magnesium, iron, or both. A number of other minerals also have ranges of composition, so they are

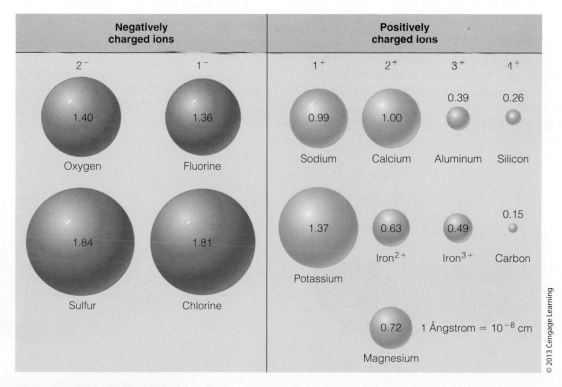

Figure 3.7 Sizes and Charges of Ions
Electrical charges and relative sizes of ions common in minerals. The numbers within the ions are the radii shown in Ångstrom units.

silica A compound of silicon and oxygen.

silicate A mineral that contains silica, such as quartz (SiO_2).

silica tetrahedron The basic building block of all silicate minerals; consists of one silicon atom and four oxygen atoms.

are controlled by composition and structure. The physical properties of minerals are further discussed later in this chapter.

LO3 MINERAL GROUPS RECOGNIZED BY GEOLOGISTS

Geologists have identified and described more than 3,500 minerals, but only a few—perhaps two dozen—are common. One might think that an extremely large number of minerals could form from 92 naturally occurring elements, however, the bulk of Earth's crust is made up of only eight chemical elements, and even among these, silicon and oxygen are by far the most common. In fact, the most common minerals in Earth's crust consist of silicon, oxygen, and one or more of the elements in Figure 3.8.

Geologists recognize mineral classes or groups, each with members that share the same negatively charged ion or ion group (Table 3.1). We have mentioned that ions are atoms that have either a positive actually mineral groups with several members.

Physical Properties of Minerals

The last criterion in our definition of a mineral, *characteristic physical properties*, refers to such properties as hardness, color, and crystal form. These properties

or negative electrical charge resulting from the loss or gain of electrons in their outermost shell. In addition to ions, some minerals contain tightly bonded, complex groups of different atoms known as *radicals* that act as single units. A good example is the carbonate radical, consisting of a carbon atom bonded to three oxygen atoms and thus having the formula CO_3 and a -2 electrical charge. Other common radicals and their charges are sulfate (SO_4, -2), hydroxyl (OH, -1), and silicate (SiO_4, -4) (Figure 3.9).

Silicate Minerals

Because silicon and oxygen are the two most abundant elements in Earth's crust, it is not surprising that many minerals contain these elements. A combination of silicon and oxygen is known as **silica**, and minerals that contain silica are **silicates**. Quartz (SiO_2) is pure silica, because it is composed entirely of silicon and oxygen. But most silicates have one or more additional elements, as in orthoclase ($KAlSi_3O_8$) and olivine [$(Mg,Fe)_2SiO_4$]. Silicate minerals include about one-third of all known minerals, but their abundance is even more impressive when one considers that they make up perhaps 95% of Earth's crust.

The basic building block of all silicate minerals is the **silica tetrahedron**, consisting of one silicon atom and four oxygen atoms (Figure 3.10a). These atoms are arranged so that the four oxygen atoms surround a silicon atom, which occupies the space between the oxygen atoms, thus forming a four-faced pyramidal structure (Figure 3.10b). The silicon atom has a positive charge of 4, and each of the four oxygen atoms has a negative charge of 2, resulting in a radical with a total negative charge of 4 $(SiO_4)^{-4}$.

Earth's crust (by weight)

All others 1.5%
Magnesium 2.1%
Potassium 2.6%
Sodium 2.8%
Calcium 3.6%
Iron 5.0%
Aluminum 8.1%
Oxygen 46.6%
Silicon 27.7%

a. Percentage of crust by weight.

Earth's crust (by atoms)

All others 0.1%
Magnesium 1.8%
Potassium 1.4%
Sodium 2.6%
Calcium 1.0%
Iron 1.9%
Aluminum 6.5%
Oxygen 62.6%
Silicon 21.2%

b. Percentage of crust by atoms.

From G. T. Miller, *Living in the Environment: Principles, Concepts, and Solutions 9/e* (Belmont, CA: Wadsworth Publishing, 1996), Figure 8.3.

Figure 3.8 Common Elements in Earth's Crust

Carbonate CO_3 (-2) Hydroxyl OH (-1) Sulfate SO_4 (-2) Silica SiO_4 (-4)

Figure 3.9 Radicals

Many minerals contain radicals, which are complex groups of atoms tightly bonded together. The silica and carbonate radicals are particularly common in many minerals, such as quartz (SiO_2) and calcite ($CaCO_3$), respectively.

Because the silica tetrahedron has a negative charge, it does not exist in nature as an isolated ion group; rather, it combines with positively charged ions or shares its oxygen atoms with other silica tetrahedra. In the simplest silicate minerals, the silica tetrahedra exist as single units bonded to positively charged ions. In minerals that contain isolated tetrahedra, the silicon-to-oxygen ratio is 1:4, and the negative charge of the silica ion is balanced by positive ions (Figure 3.10c).

Olivine [$(Mg,Fe)_2SiO_4$], for example, has either two magnesium (Mg^{+2}) ions, two iron (Fe^{+2}) ions, or one of each to offset the -4 charge of the silica ion.

Silica tetrahedra may also join together to form chains of indefinite length (Figure 3.10d). Single chains, as in the pyroxene minerals, form when each tetrahedron shares two of its oxygens with an adjacent tetrahedron, resulting in a silicon-to-oxygen ratio of 1:3. Enstatite, a pyroxene-group mineral, reflects this ratio in its chemical formula $MgSiO_3$. Individual chains, however, possess a net -2 electrical charge, so they are balanced by positive ions, such as Mg^{+2}, that link parallel chains together (Figure 3.10d).

The amphibole group of minerals is characterized by a double-chain structure in which alternate tetrahedra in two parallel rows are cross-linked (Figure 3.10d). The formation of double chains results in a silicon-to-oxygen ratio of 4:11, so each double chain possesses a -6 electrical charge. Mg^{+2}, Fe^{+2}, and Al^{+3} are usually involved in linking the double chains together.

In sheet-structure silicates, three oxygens of each tetrahedron are shared by adjacent tetrahedra (Figure 3.10e). Such structures result in continuous sheets of

TABLE 3.1

Mineral Groups Recognized by Geologists

MINERAL GROUP	NEGATIVELY CHARGED ION OR RADICAL	EXAMPLES	COMPOSITION
Carbonate	$(CO_3)^{-2}$	Calcite	$CaCO_3$
		Dolomite	$CaMg(CO_3)_2$
Halide	Cl^{-1}, F^{-1}	Halite	$NaCl$
		Fluorite	CaF_2
Hydroxide	$(OH)^{-1}$	Brucite	$Mg(OH)_2$
Native element	—	Gold	Au
		Silver	$Ag*$
		Diamond	C
Phosphate	$(PO_4)^{-3}$	Apatite	$Ca_5(PO_4)_3(F,Cl)$
Oxide	O^{-2}	Hematite	Fe_2O_3
		Magnetite	Fe_3O_4
Silicate	$(SiO_4)^{-4}$	Quartz	SiO_2
		Potassium feldspar	$KAlSi_3O_8$
		Olivine	$(Mg,Fe)_2SiO_4$
Sulfate	$(SO_4)^{-2}$	Anhydrite	$CaSO_4$
		Gypsum	$CaSO_4 \cdot 2H_2O$
Sulfide	S^{-2}	Galena	PbS
		Pyrite	FeS_2
		Argentite	Ag_2S*

*Note that silver is found as both a native element and a sulfide mineral.

silica tetrahedra with silicon-to-oxygen ratios of 2:5. Continuous sheets also possess a negative electrical charge satisfied by positive ions located between the sheets. This particular structure accounts for the characteristic sheet structure of the *micas*, such as biotite and muscovite, and the *clay minerals*.

Three-dimensional networks of silica tetrahedra form when all four oxygens of the silica tetrahedra are shared by adjacent tetrahedra (Figure 3.10f). Such sharing of oxygen atoms results in a silicon-to-oxygen ratio of 1:2, which is electrically neutral. Quartz is a common framework silicate.

Geologists define two subgroups of silicates: ferromagnesian and nonferromagnesian silicates. The **ferromagnesian silicates** are those that contain iron

Figure 3.10 The Silica Tetrahedron and Silicate Materials

$(SiO_4)^{-4}$

a. Expanded view of the silica tetrahedron (*left*) and how it actually exists with its oxygen atoms touching (*right*).

b. View of the silica tetrahedron from above. Only the oxygen atoms are visible.

			Formula of negatively charged ion group		Example
c.	Isolated tetrahedra		$(SiO_4)^{-4}$	No oxygen atoms shared	Olivine
d.	Continuous chains of tetrahedra	Single chain	$(SiO_3)^{-2}$	Each tetrahedra shares two oxygen atoms with adjacent tetrahedra	Pyroxene group (augite)
		Double chain	$(Si_4O_{11})^{-6}$	Single chains linked by sharing oxygen atoms	Amphibole group (hornblende)
e.	Continuous sheets		$(Si_4O_{10})^{-4}$	Three oxygen atoms shared with adjacent tetrahedra	Micas (muscovite)
f.	Three-dimensional networks	SiO_2	$(SiO_2)^0$	All four oxygen atoms in tetrahedra shared	Quartz Potassium feldspars Plagioclase feldspars

© 2013 Cengage Learning

c–f. Structures of the common silicate minerals shown by various arrangements of the silica tetrahedra.

(Fe), magnesium (Mg), or both. These minerals are commonly dark and more dense than nonferromagnesian silicates. Some of the common ferromagnesian silicate minerals are olivine, the pyroxenes, the amphiboles, and biotite (Figure 3.11a).

The **nonferromagnesian silicates** lack iron and magnesium, are generally light colored, and are less dense than ferromagnesian silicates (Figure 3.11b). The most common minerals in Earth's crust are nonferromagnesian silicates known as *feldspars*. Feldspar is a general name, however, that includes two distinct groups, each of which includes several species. The *potassium feldspars* are represented by microcline and orthoclase ($KAlSi_3O_8$). The second group of feldspars, the *plagioclase feldspars*, range from calcium-rich ($CaAl_2Si_2O_8$) to sodium-rich ($NaAlSi_3O_8$) varieties.

Quartz (SiO_2) is another common nonferromagnesian silicate. It is a framework silicate that can usually be recognized by its glassy appearance and hardness. Another fairly common nonferromagnesian silicate is muscovite, which is a mica (Figure 3.11b).

Carbonate Minerals

Carbonate minerals are those that contain the negatively charged carbonate radical $(CO_3)^{-2}$ and include calcium carbonate ($CaCO_3$) as the minerals *aragonite* or *calcite* (Figure 3.12a). Aragonite is unstable and commonly changes to calcite, the main constituent of the sedimentary rock *limestone*. Several other carbonate minerals are known, but only one need concern us: *dolomite* [$CaMg(CO_3)_2$], formed by the chemical alteration of calcite by the addition of magnesium. Sedimentary rock composed of the mineral dolomite is *dolostone*.

Other Mineral Groups

Even though minerals from these groups are less common than silicates and carbonates, many are found in rocks in small quantities and others are important resources (refer again to Table 3.1). In the oxides, an element combines with oxygen as in hematite (Fe_2O_3) and magnetite (Fe_3O_4). Rocks with high concentrations of these minerals are sources of iron ores for the manufacture of steel. The related hydroxides form mostly by the chemical alteration of other minerals.

We have noted that the *native elements* are minerals composed of a single element, such as diamond and graphite (C) and the precious metals gold (Au), silver (Ag), and platinum (Pt). Some elements, such as silver and copper, are found both as native elements and as compounds and are thus also included in other mineral groups; argentite (Ag_2S), a silver sulfide, is an example.

Several minerals and rocks that contain the phosphate radical $(PO_4)^{-3}$ are important sources of phosphorus for fertilizers. The sulfides, such as galena (PbS), the ore of lead, have a positively charged ion combined with sulfur (S^{-2}) (Figure 3.12b), whereas the sulfates have

> **nonferromagnesian silicate** A silicate mineral that has no iron or magnesium.
>
> **carbonate mineral** A mineral with the carbonate radical $(CO_3)^{-2}$, as in calcite ($CaCO_3$) and dolomite [$CaMg(CO_3)_2$].

a. Ferromagnesian silicates.

b. Nonferromagnesian silicates.

Figure 3.11 Common Silicate Materials

luster The appearance of a mineral in reflected light. Luster is metallic or nonmetallic, although the latter has several subcategories.

cleavage Breakage along internal planes of weakness in mineral crystals.

an element combined with the complex radical $(SO_4)^{-2}$, as in gypsum $(CaSO_4 \cdot 2H_2O)$ (Figure 3.12c). The halides contain the halogen elements, fluorine (F^{-1}) and chlorine (Cl^{-1}); examples are halite (NaCl) (Figure 3.12d) and fluorite (CaF_2).

LO4 PHYSICAL PROPERTIES OF MINERALS

Many physical properties of minerals are remarkably constant for a given mineral species, but some, especially color, may vary. You can identify most common minerals by using the physical properties described next.

Luster and Color

Luster (not to be confused with *color*) is the quality and intensity of light reflected from a mineral's surface. Geologists define luster as *metallic*, having the appearance of a metal, and *nonmetallic*. Of the four minerals shown in Figure 3.12, only galena has a metallic luster. Among the several types of nonmetallic luster are glassy, or vitreous (as in quartz); and dull or earthy, waxy, greasy, and brilliant (as in diamond).

Beginning geology students are distressed by the fact that the color of some minerals varies, making the most obvious physical property of little use for identifying some minerals. In any case, we can make some helpful generalizations about color. Ferromagnesian silicates are typically black, brown, or dark green, although olivine is olive green (Figure 3.11a). Nonferromagnesian silicates vary in color but are rarely very dark. White, cream, colorless, and shades of pink and pale green are typical (Figure 3.11b).

Another helpful generalization is that the color of minerals with a metallic luster is more consistent than is the color of nonmetallic minerals. For example, galena is always lead-gray (Figure 3.12b), and pyrite is invariably brassy yellow. In contrast, quartz, a nonmetallic mineral, may be colorless, smoky brown to almost black, rose, yellow-brown, milky white, blue, or violet to purple.

Figure 3.12 Representative Specimens from Four Mineral Groups

a. Calcite $(CaCO_3)$ is the most common carbonate mineral.

b. The sulfide mineral galena (PbS) is the ore of lead.

c. Gypsum $(CaSO_4 \cdot 2H_2O)$ is a common sulfate mineral.

d. Halite (NaCl) is a good example of a halide mineral.

Crystal Form

As we noted, many mineral specimens do not show the perfect crystal form typical of that mineral species. Nevertheless, some minerals do typically occur as crystals (refer again to Figure 3.6). For example, 12-sided crystals of garnet are common, as are 6- and 12-sided crystals of pyrite. Minerals that grow in cavities or are precipitated from circulating hot water (hydrothermal solutions) in cracks and crevices in rocks also commonly occur as crystals.

Crystal form is useful for mineral identification, but some minerals have the same crystal form. Pyrite (FeS_2), fluorite (CaF_2), and halite (NaCl) all occur as cubic crystals, but they can be easily identified by other properties, such as color, luster, hardness, and density (Figure 3.13).

Cleavage and Fracture

Not all minerals possess cleavage, but those that do break or split along smooth planes of weakness determined by the strength of their chemical bonds. Cleavage is characterized in terms of quality (perfect, good, poor), direction, and angles of intersection of cleavage planes. Biotite, a common ferromagnesian silicate, has perfect cleavage in one direction (Figure 3.14a). Biotite is a sheet silicate, with the sheets of silica tetrahedra weakly bonded to one another by iron and magnesium ions.

Feldspars possess two directions of cleavage that intersect at right angles (Figure 3.14b), and the mineral halite has three directions of cleavage, all of which intersect at right angles (Figure 3.14c). Calcite also possesses three directions of cleavage, but none of the intersection angles is a right angle, so cleavage fragments of calcite are rhombohedrons (Figure 3.14d). Minerals with four directions of cleavage include fluorite and diamond (Figure 3.14e). Ironically, diamond, the hardest mineral, can be easily cleaved. A few minerals, such as sphalerite, an ore of zinc, have six directions of cleavage (Figure 3.14f).

Cleavage is an important diagnostic property of minerals, and recognizing it is essential in distinguishing between some minerals. The pyroxene mineral augite and the amphibole mineral hornblende, for example, look much alike: Both are dark green to black, have the same hardness, and possess two directions of cleavage. But the cleavage planes of augite intersect at about 90 degrees, whereas the cleavage planes of hornblende intersect at angles of 56 degrees and 124 degrees (Figure 3.15).

Figure 3.13 Minerals with the Same Kind of Crystals

a. Pyrite (FeS$_2$).   b. Fluorite (CaF$_2$).   c. Halite (NaCl).

Mineral crystals are found in a variety of shapes (Figure 3.6), but different minerals may have the same kinds of crystals. Differentiating one from the other is easy. Pyrite (a) is brassy yellow (the silvery color results from reflected light), and it is much denser than the other two. You can identify fluorite (b) and halite (c) by their cleavage and taste—halite tastes salty whereas fluorite has no taste.

a. Cleavage in one direction.

Cleavage plane

Micas—biotite and muscovite

b. Cleavage in two directions at right angles.

Potassium feldspars, plagioclase feldspars

c. Cleavage in three directions at right angles.

Halite, galena

d. Cleavage in three directions, not at right angles.

Calcite, dolomite

e. Cleavage in four directions.

Fluorite, diamond

f. Cleavage in six directions.

Sphalerite

© 2013 Cengage Learning

Figure 3.14 Several Types of Mineral Cleavage

Cross section

93°
87°

a. Augite crystal and cross section of crystal showing cleavage.

Cross section

56°
124°

b. Hornblende crystal and cross section of crystal showing cleavage.

© 2013 Cengage Learning

Figure 3.15 Cleavage in Augite and Hornblende

In contrast to cleavage, *fracture* is mineral breakage along irregular surfaces. Any mineral will fracture if enough force is applied, but the fracture surfaces are uneven or conchoidal (curved) rather than smooth.

Hardness

An Austrian geologist, Friedrich Mohs (1773–1839), arbitrarily assigned a hardness value of 10 to diamond, the hardest mineral known, and lesser values to the other minerals. Relative hardness is easily determined by the use of Mohs hardness scale (Table 3.2). Quartz will scratch fluorite but cannot be scratched by fluorite, gypsum can be scratched by a fingernail, and so on. So **hardness** is defined as a mineral's resistance to abrasion and is controlled mostly by internal structure. Both graphite and diamond are composed of carbon, but the former has a hardness of 1 to 2, whereas the latter has a hardness of 10.

Specific Gravity (Density)

A mineral's **specific gravity** is the ratio of its weight to the weight of an equal volume of pure water at 4°C. Thus, a mineral with a specific gravity of 3.0 is three times as heavy as water. **Density**, in contrast, is a mineral's mass (weight) per unit of volume expressed in grams per cubic centimeter. So the specific gravity of galena (refer again to Figure 3.12b) is 7.58, and its density is 7.58 g/cm³. In most instances, we will refer to a mineral's density, and in some of the following chapters, we will mention the density of rocks.

Because ferromagnesian silicates contain iron, magnesium, or both, they tend to be denser than nonferromagnesian silicates. In general, the metallic minerals, such as galena and hematite, are denser than nonmetals. Pure gold, with a density of 19.3 g/cm³, is about 2.5 times as dense as lead. Diamond and graphite, both of which are composed of carbon (C), illustrate how structure controls specific gravity or density. The specific gravity of diamond is 3.5, whereas that of graphite varies from 2.09 to 2.33.

TABLE 3.2

Mohs Hardness Scale

HARDNESS	MINERAL	HARDNESS OF SOME COMMON OBJECTS
10	Diamond	
9	Corundum	
8	Topaz	
7	Quartz	
		Steel file (6½)
6	Orthoclase	
		Glass (5½–6)
5	Apatite	
4	Fluorite	
3	Calcite	Copper penny (3)
		Fingernail (2½)
2	Gypsum	
1	Talc	

© 2013 Cengage Learning

Other Useful Mineral Properties

Talc has a distinctive soapy feel, graphite writes on paper, halite tastes salty, and magnetite is magnetic. Calcite possesses the property of *double refraction*, meaning that an object when viewed through a transparent piece of calcite will have a double image. Some sheet silicates are plastic and, when bent into a new shape, will retain that shape; others are flexible and, if bent, will return to their original position when the forces that bent them are removed.

A simple chemical test to identify the minerals calcite and dolomite involves applying a drop of dilute hydrochloric acid to the mineral specimen. If the mineral is calcite, it will react vigorously with the acid and release carbon dioxide, which causes the acid to bubble or effervesce. Dolomite, in contrast, will not react with hydrochloric acid unless it is powdered.

LO5 ROCK-FORMING MINERALS

A **rock** is a solid aggregate of one or more minerals, but the term also refers to masses of mineral-like matter, as in the natural glass obsidian, and masses of solid organic matter, as in coal. And even though some rocks may contain many minerals, only a few, designated **rock-forming minerals**, are sufficiently common for rock identification and classification (Table 3.3

and Figure 3.16). Others, known as *accessory minerals*, are present in such small quantities that they can be disregarded.

Given that silicate minerals are by far the most common ones in Earth's crust, it follows that most rocks are composed of these minerals. Indeed, feldspar minerals (plagioclase feldspars and potassium feldspars) and quartz make up more than 60% of Earth's crust. So, even though there are hundreds of silicates, only a few are particularly common in rocks.

The most common nonsilicate rock-forming minerals are the carbonates calcite ($CaCO_3$) and dolomite [$CaMg(CO_3)_2$], the main constituents of the sedimentary rocks limestone and dolostone, respectively. Among the sulfates and halides, gypsum ($CaSO_4 \cdot 2H_2O$) in rock gypsum and halite ($NaCl$) in rock salt, are common enough to qualify as rock-forming minerals. Even though these minerals and their corresponding rocks might be common in some areas, their overall abundance is limited compared to the silicate and carbonate rock-forming minerals.

LO6 HOW DO MINERALS FORM?

Thus far, we have discussed the composition, structure, and physical properties of minerals, but we have not addressed how they originate. One phenomenon that accounts for minerals is the cooling of molten rock material known as *magma* (magma that reaches the surface is called *lava*).

As magma or lava cools, minerals crystallize and grow, thereby determining the mineral composition of igneous rocks such as basalt (dominated by ferromagnesian silicates) and granite (dominated by nonferromagnesian silicates). Hot-water solutions derived from magma commonly invade cracks and crevasses in adjacent rocks, and from these solutions, several minerals crystallize, some of which have economic importance. Minerals also originate when water in hot springs cools, and when hot, mineral-rich water discharges onto the seafloor at hot springs known as *hydrothermal vents* (see Figure 2.7).

rock A solid aggregate of one or more minerals, as in limestone and granite, or a consolidated aggregate of rock fragments, as in conglomerate, or masses of rocklike materials, such as coal and obsidian.

rock-forming mineral Any mineral common in rocks that is important in their identification and classification.

TABLE 3.3

Important Rock-Forming Minerals

MINERAL	PRIMARY OCCURRENCE
Ferromagnesian silicates	
Olivine	Igneous and metamorphic rocks
Pyroxene group Augite most common	Igneous and metamorphic rocks
Amphibole group Hornblende most common	Igneous and metamorphic rocks
Biotite	All rock types
Nonferromagnesian silicates	
Quartz	All rock types
Potassium feldspar group Orthoclase, microcline	All rock types
Plagioclase feldspar group	All rock types
Muscovite	All rock types
Clay mineral group	Soils, sedimentary rocks, and some metamorphic rocks
Carbonates	
Calcite	Sedimentary rocks
Dolomite	Sedimentary rocks
Sulfates	
Anhydrite	Sedimentary rocks
Gypsum	Sedimentary rocks
Halides	
Halite	Sedimentary rocks

© 2013 Cengage Learning

Plagioclase feldspar
Quartz
Potassium feldspar
Biotite

Figure 3.16 **Minerals in Granite**

The igneous rock granite (see Chapter 4) is made up of mostly three minerals—quartz, potassium feldspar, and plagioclase feldspar—but it may also contain small amounts of biotite, muscovite, and hornblende.

Some clay minerals form when chemical processes compositionally and structurally alter other minerals, and others originate when rocks are changed during metamorphism. In fact, the agents that cause metamorphism—heat, pressure, and chemically active fluids—are responsible for the origin of many minerals.

LO7 NATURAL RESOURCES AND RESERVES

Geologists at the U.S. Geological Survey define a **resource** as a naturally occurring concentration of solid, liquid, or gaseous material in, or on, Earth's crust in such form and amount that economic extraction of a commodity is currently or potentially feasible. Many natural resources are concentrations of minerals, rocks, or both, but liquid petroleum and natural gas are also included. In fact, we refer to *metallic resources* (copper, tin, iron ore, etc.), *nonmetallic resources* (sand and gravel, crushed stone, sulfur, salt, etc.), and *energy resources* (petroleum, natural gas, coal, and uranium). All are resources, but it is important to make a distinction between a *resource*, the total amount of a commodity whether discovered or not, and a **reserve**, which is only that part of the resource base that is known and can be economically recovered.

In principal, the distinction between a resource and a reserve is simple, but in practice it depends on several factors, not all of which remain constant. For

resource A concentration of naturally occurring solid, liquid, or gaseous material in, or on, Earth's crust in such form and amount that economic extraction of a commodity from the concentration is currently or potentially feasible.

reserve The part of the resource base that can be extracted economically.

Dissolved materials in seawater, more rarely in lake water, combine to form minerals such as halite (NaCl), gypsum ($CaSO_4 \cdot 2H_2O$), and several others when the water evaporates. Aragonite and/or calcite, both varieties of calcium carbonate ($CaCO_3$), might also form from evaporating water, but most originate when organisms such as clams, oysters, corals, and floating microorganisms use this compound to construct their shells. A few plants and animals use silicon dioxide (SiO_2) for their skeletons, which accumulate as mineral matter on the seafloor when the organisms die.

instance, a resource in a remote region might not be mined because transport expenses are too high, and what may be deemed a resource rather than a reserve in a developed nation might be mined in a developing nation where labor costs are low. The value of gold and diamonds makes it profitable to mine them just about anywhere, whereas deposits of sand and gravel for construction must be close to their market areas. Changes in technology are important, too. For example, iron ore mined in the Great Lakes region is not as rich as that mined in decades past, but a technique for separating the iron from its host rock and shaping it into pellets has made mining of low-grade deposits profitable.

In addition to resources such as petroleum, gold, and ores of iron, copper, and aluminum, some common minerals and rocks are important. Pure quartz sand is used to make glass, sandpaper, and optical instruments; clay minerals are needed to make ceramics and paper; feldspars are used for porcelain, ceramics, enamel, and glass; and phosphate-bearing rocks are used for fertilizers. Muscovite is an ingredient in wallboard joint compound, lipstick, glitter, and eye shadow, as well as the lustrous paints on appliances and automobiles. In 2009, the extraction of non-fuel mineral resources in the United States totaled more than $57 billion.

Access to natural resources is essential for industrialization and the high standard of living enjoyed in the developed countries. However, most resources are *nonrenewable*, meaning that there is a limited supply and they cannot be replenished by natural processes as fast as they are depleted. Accordingly, as the amount of a resource diminishes, suitable substitutes, if available, must be found, or we continue to exploit the resource and simply pay more and more as further depletion takes place. For some essential resources, the United States is totally dependent on imports, some from unstable parts of the world. For example, no cobalt was mined in the United States during 2009, so all cobalt needed must be imported for use in gas-turbine aircraft engines, magnets, and for corrosion- and wear-resistant alloys. Manganese is an essential element for making steel, and all used in the United States is imported.

The United States also imports all of the aluminum ore it uses, all or some of many other resources, and as of February 2011, 56.2% of all petroleum needed. Canada, in contrast, is more self-reliant, meeting most of its domestic mineral and energy needs. Nevertheless, it must import aluminum ore, chromium, manganese, and phosphate. Canada produces more crude oil and natural gas than it uses, and it is among the world's leaders in the production and exporting of uranium.

To ensure continued supplies of mineral and energy resources, scientists, government agencies, and executives in business and industry continually assess the status of resources in view of changing economic and political conditions and changes in science and technology. The U.S. Geological Survey keeps detailed statistical records of mine production, imports, and exports, and regularly publishes reports on the status of numerous resources in *Mineral Commodity Summaries*. Similar reports also appear regularly in the *Canadian Minerals Yearbook*.

4 Igneous Rocks and Intrusive Igneous Activity

The Harney Peak Granite in the Black Hills of South Dakota with the images of presidents George Washington, Thomas Jefferson, Theodore Roosevelt, and Abraham Lincoln. This large body of granite, which formed about 1.7 billion years ago, forms the central part of the Black Hills. The 18-m-high images of the presidents were carved between 1927 and 1941 and are now the primary attraction at Mount Rushmore National Memorial.

Sue Monroe

Introduction

"Most magma never reaches Earth's surface, but rather cools and crystallizes underground."

Before we begin a new topic, let's review some terms. You now know that the term *rock* applies to aggregates of minerals, mineral-like matter as in obsidian, and coal, which is altered organic matter. You also know from Chapter 1 that the three groups of rocks are igneous, sedimentary, and metamorphic (see Figure 1.14). Here we are concerned with **igneous rocks**, that is, rocks made up of minerals that crystallized from **magma** (molten rock matter below the surface), and **lava** (molten rock material at the surface), and from particulate matter (pyroclastic materials) ejected from volcanoes during explosive eruptions. Magma and lava are the same material, although the gas content of lava is less because of decreased pressure.

Most people have given little thought to magma but are familiar with volcanoes, which are simply hills or mountains built up where magma reaches the surface. In any case, volcanism is an observable phenomenon, but most magma never reaches the surface. In fact it cools and crystallizes underground, a phenomenon called *intrusive igneous activity*, thus forming several types of igneous rock bodies known as *plutons*, named for

Granite and related rocks are attractive, especially when sawed and polished. They are used for tombstones, mantelpieces, kitchen counters, facing stones on buildings, monuments, pedestals for statues, and for statuary itself.

Pluto, the Roman god of the underworld. Granite, composed of potassium feldspars, plagioclase feldspars, and quartz, is common in many of the larger plutons.

Even though we discuss intrusive igneous activity and the origin of plutons in this chapter and volcanism in Chapter 5, both processes are related. In fact, the same kinds of magmas are involved, but magma varies in its mobility, so only some of it reaches the surface. Furthermore, plutons lie beneath areas of volcanism and are the source of the lava flows and particulate matter erupted at volcanoes or along fissures.

An important reason to study igneous rocks and plutons is that these rocks make up parts of the continents and all of the oceanic crust, which forms continuously at divergent plate boundaries. As a matter of fact, most plutons and volcanoes are found at or near divergent and convergent plate boundaries, so the presence of igneous rocks in the geologic record is one criterion for recognizing ancient plate boundaries. Another reason to study these topics is that when large plutons form, fluids emanating from them follow cracks and crevasses in adjacent rocks where important mineral resources such as copper may form.

igneous rock Any rock formed by cooling and crystallization of magma or lava or the consolidation of pyroclastic materials.

magma Molten rock material generated within Earth.

lava Magma that reaches Earth's surface.

Learning Outcomes

After reading this unit, you should be able to do the following:

LO1 Describe the properties and behavior of magma and lava

LO2 Explain how magma originates and changes

LO3 Identify and classify igneous rocks by their characteristics

LO4 Recognize intrusive igneous bodies, or plutons

LO5 Explain how batholiths intrude into Earth's crust

LO1 THE PROPERTIES AND BEHAVIOR OF MAGMA AND LAVA

lava flow A stream of magma flowing over Earth's surface.

pyroclastic materials Fragmental substances, such as ash, that are explosively ejected from a volcano.

volcanic (extrusive igneous) rock An igneous rock formed when magma is extruded onto Earth's surface where it cools and crystallizes, or when pyroclastic materials become consolidated.

plutonic (intrusive igneous) rock Igneous rock that formed from magma intruded into or formed in place within the crust.

ultramafic magma Magma with a silica content of less than 45%.

mafic magma Magma with between 45% to 52% silica and proportionately more calcium, iron, and magnesium than intermediate and felsic magma.

intermediate magma Magma with a silica content between 53% to 65% and an overall composition intermediate between mafic and felsic magma.

felsic magma Magma with more than 65% silica and considerable sodium, potassium, and aluminum, but little calcium, iron, and magnesium.

Magma is less dense than the rock from which it formed, so it tends to rise toward the surface, but much of it solidifies deep underground, thereby accounting for the origin of plutons. The magma that does reach the surface issues forth as **lava flows**, and some of it is forcefully ejected into the atmosphere as particles known as **pyroclastic materials** (from the Greek *pyro* for "fire" and *klastos* for "broken").

All igneous rocks derive ultimately from magma; however, two separate processes account for them. They form when (1) magma or lava cools and crystallizes to form aggregates of minerals, or (2) pyroclastic materials are consolidated. Those igneous rocks derived from lava flows and pyroclastic materials, both of which are extruded onto the surface, are known as **volcanic rocks** or **extrusive igneous rocks**. When magma cools below the surface it forms **plutonic rocks** or **intrusive igneous rocks**.

Composition of Magma

By far the most abundant minerals in Earth's crust are silicates such as quartz, feldspars, and several ferromagnesian silicates, all made up of silicon and oxygen, and other elements shown in Figure 3.8. As a result, melting of the crust yields mostly silica-rich magmas that also contain considerable aluminum, calcium, sodium, iron, magnesium, and potassium, and several other elements in lesser quantities. Another source of magma is Earth's upper mantle, which is composed of rocks that contain mostly ferromagnesian silicates. Thus, magma from this source contains comparatively less silicon and oxygen (silica) and more iron and magnesium.

With few exceptions, magma is composed mostly of silica, which varies enough to designate magmas as **ultramafic** (less than 45% silica), **mafic** (45-52% silica), **intermediate** (53-65% silica), and **felsic** (more then 65% silica) (Table 4.1). According to Table 4.1, mafic magma is silica-poor but contains more calcium, iron, and magnesium than silica-rich felsic magma, which has more sodium, potassium, and aluminum. And as you would expect, intermediate magma has a composition between mafic and felsic.

How Hot Are Magma and Lava?

Erupting lava has a temperature in the range of 700° to 1,200°C, although a temperature of 1,350°C was recorded above a lava lake in Hawaii where volcanic gases reacted with the atmosphere. Magma must be even hotter than lava, but no direct measurements of magma temperatures have ever been made.

© iStockphoto.com/M-I-S-H-A-

Most lava temperatures are taken at volcanoes that show little or no explosive activity, so our best information comes from mafic lava flows such as those in Hawaii (Figure 4.1). In contrast, eruptions of felsic lava are not as common, and the volcanoes from which these flows issue tend to be explosive and thus cannot be approached safely. Nevertheless, the temperatures of some bulbous masses of felsic lava in lava domes have been measured at a distance with an optical pyrometer. The surfaces of these lava domes are as hot as 900°C, but their interiors must surely be even hotter.

The reason that lava and magma retain heat so well is that rock conducts heat so poorly. Accordingly, the

Table 4.1

The Most Common Types of Magmas and Their Characteristics

TYPE OF MAGMA	SILICA CONTENT (%)	SODIUM, POTASSIUM, AND ALUMINUM	CALCIUM, IRON, AND MAGNESIUM
Ultramafic	< 45		Increase
Mafic	45–52		
Intermediate	53–65		
Felsic	> 65	Increase	

© 2013 Cengage Learning

interiors of thick lava flows and pyroclastic flows may remain hot for months or years. In 1959, lava filled a crater to 85 m deep in Hawaii, but when it was drilled in 1988 it still had not completely solidified near its base. Plutons, depending on their size and depth, may not cool completely for thousands to millions of years.

Viscosity: Resistance to Flow

Viscosity, or resistance to flow, is a property of all liquids. Water's viscosity is very low, so it is highly fluid and flows readily. For other liquids, such as cold motor oil and syrup, viscosity is so high that they flow much more slowly. But when these liquids are heated, their viscosity is much lower and they flow more easily. Accordingly, you might suspect that temperature controls the viscosity of magma and lava, and this inference is partly correct. We can generalize and say that the hotter the magma

or lava, the more readily it moves, but we must qualify this statement by noting that temperature is not the only control of viscosity.

Silica content also strongly controls magma and lava viscosity. With increasing silica content, numerous networks of silica tetrahedra form and retard flow, because for flow to take place, the strong bonds of the networks must be ruptured. Mafic magma and lava with 45% to 52% silica have fewer silica tetrahedra networks and, as a result, are more mobile than felsic magma and lava flows (Figure 4.2). One mafic flow in 1783 in Iceland flowed nearly 80 km. Felsic magma, because of its higher viscosity, does not reach the surface as often as mafic magma. And when felsic lava flows do occur, they tend to be slow moving and thick, and move only short distances.

Temperature and silica content are important controls on the viscosity of magma and lava, but other factors include gases, mostly water vapor and CO_2, as well as the presence of mineral crystals and friction from the surface over which lava flows. Lava with a high content of dissolved gases flows more readily than one with a lesser amount of gases, whereas lava with many crystals or that flows over a rough surface tends to be more viscous.

LO2 HOW DOES MAGMA ORIGINATE AND CHANGE?

Even if you have not witnessed a volcanic eruption, you are probably familiar with lava flows and pyroclastic eruptions. In any case, we know about some aspects of igneous activity, but most people are unaware of how and where magma originates, how it rises, and how it might change. Indeed, there is a misconception that lava comes from a continuous layer of molten rock beneath the crust or that it comes from Earth's molten outer core.

First, let us address how magma originates. We know that the atoms in a solid are in constant motion and that, when a solid becomes hot enough, the energy of motion exceeds the binding forces and the solid melts. We are all familiar with this phenomenon, and we are also aware that not all solids melt at the same temperature. Likewise, if heated, the minerals in rocks begin to melt, but not all at the same time. And once magma forms, it tends to rise, because it is less dense than the rock that melted.

© P. Mouginis-Mark

Figure 4.1
A Geologist Takes the Temperature of an Active Lava Flow in Hawaii

Figure 4.2 Viscosity of Magma and Lava

Temperature is an important control of viscosity, but so is composition. Mafic lava tends to be fluid, whereas felsic lava is much more viscous.

J.D. Griggs/USGS/VHP

G. Iwatsubo/USGS

a. A mafic lava flow in 1984 on Mauna Loa Volcano in Hawaii. These flows move rapidly and form thin layers.

b. The Novarupta lava dome in Katmai National Park in Alaska. The lava is felsic and viscous, so it was extruded as a bulbous mass. This image was taken in 1987.

magma chamber A reservoir of magma within Earth's upper mantle or lower crust.

Bowen's reaction series A series of minerals that form in a specific sequence in cooling magma or lava; originally proposed to explain the origin of intermediate and felsic magma from mafic magma.

Magma may come from 100 to 300 km deep, but most forms at much shallower depths in the upper mantle or lower crust and accumulates in reservoirs known as **magma chambers**. Beneath spreading ridges, where the crust is thin, these chambers lie only a few kilometers deep but along convergent plate boundaries, they are commonly a few tens of kilometers deep. The volume of a magma chamber ranges from a few to many hundreds of cubic kilometers of molten rock within the otherwise solid lithosphere. Some simply cools and crystallizes within Earth's crust, thus accounting for the origin of plutons, whereas some rises to the surface and is erupted as lava flows or pyroclastic materials.

Bowen's Reaction Series

During the early 1900s N. L. Bowen knew that minerals crystallize in a predictable sequence from cooling magma. Based on his observations and laboratory experiments, Bowen proposed a mechanism, now called **Bowen's reaction series**, to account for the derivation of intermediate and felsic magmas from mafic magma. Bowen's reaction series consists of two branches: a *discontinuous branch* and a *continuous branch* (Figure 4.3). As the temperature of magma decreases, minerals crystallize along both branches simultaneously, but for convenience, we will discuss them separately.

In the discontinuous branch, which contains only ferromagnesian silicates, one mineral changes to another over specific temperature ranges (Figure 4.3). As the temperature decreases, a temperature range is reached in which a given mineral begins to crystallize. A previously formed mineral reacts with the remaining liquid magma (the melt) so that it forms the next mineral in the sequence. For instance, olivine [$(Mg,Fe)_2SiO_4$] is the first ferromagnesian silicate to crystallize. As the magma continues to cool, it reaches the temperature range at which pyroxene is stable; a reaction occurs between the olivine and the remaining melt, and pyroxene forms.

With continued cooling, a reaction takes place between pyroxene and the melt, and the pyroxene structure is rearranged to form amphibole. Further cooling causes a reaction between the amphibole and the melt, and its structure is rearranged so that the sheet structure of biotite mica forms. Although the reactions just

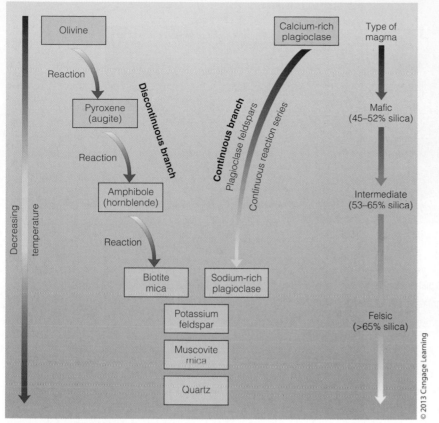

Figure 4.3 Bowen's Reaction Series

Bowen's reaction series consists of a *discontinuous branch*, along which a succession of ferromagnesian silicates crystallize as the magma's temperature decreases, and a *continuous branch*, along which plagioclase feldspars with increasing amounts of sodium crystallize. Notice also that the composition of the initial mafic magma changes as crystallization takes place along the two branches.

described tend to convert one mineral to the next in the series, the reactions are not always complete. Olivine, for example, might have a rim of pyroxene, indicating an incomplete reaction. If magma cools rapidly enough, the early-formed minerals do not have time to react with the melt, and thus all of the ferromagnesian silicates in the discontinuous branch can be in one rock. In any case, by the time biotite has crystallized, essentially all of the magnesium and iron present in the original magma have been used up.

Plagioclase feldspars, which are nonferromagnesian silicates, are the only minerals in the continuous branch of Bowen's reaction series (Figure 4.3). Calcium-rich plagioclase crystallizes first. As the magma continues to cool, calcium-rich plagioclase reacts with the melt, and plagioclase containing proportionately more sodium crystallizes until all of the calcium and sodium are used up. In many cases, cooling is too rapid for a complete transformation from calcium-rich to sodium-rich plagioclase to take place. Plagioclase forming under these

conditions is *zoned*, meaning that it has a calcium-rich core surrounded by zones progressively richer in sodium.

As minerals crystallize simultaneously along the two branches of Bowen's reaction series, iron and magnesium are depleted, because they are used in ferromagnesian silicates, whereas calcium and sodium are used up in plagioclase feldspars. At this point, any leftover magma is enriched in potassium, aluminum, and silicon, which combine to form orthoclase ($KAlSi_3O_8$), a potassium feldspar, and if water pressure is high, the sheet silicate muscovite forms. Any remaining magma is enriched in silicon and oxygen (silica) and forms the mineral quartz (SiO_2). The crystallization of orthoclase and quartz is not a true reaction series, because they form independently rather than by a reaction of orthoclase with the melt.

The Origin of Magma at Spreading Ridges

One fundamental observation regarding the origin of magma is that Earth's temperature, or *geothermal gradient*, increases with depth (see Chapter 8). Accordingly, rocks at depth are hot but remain solid, because their melting temperature rises with increasing pressure. However, beneath spreading ridges, the temperature locally exceeds the melting temperature, at least in part because pressure decreases. That is, plate separation at ridges probably causes a decrease in pressure on the already hot rocks at depth, thus initiating melting (Figure 4.4). In addition, the presence of water decreases the melting temperature beneath spreading ridges, because water aids thermal energy in breaking the chemical bonds in minerals.

Magma formed beneath spreading ridges is invariably mafic (45% to 52% silica). However, the upper-mantle rocks from which this magma is derived are ultramafic (<45% silica), consisting mostly of ferromagnesian silicates and lesser amounts of nonferromagnesian silicates. To explain how mafic magma originates from ultramafic rock, geologists propose that the magma comes from source rock that only partially melts because not all minerals melt at the same temperature.

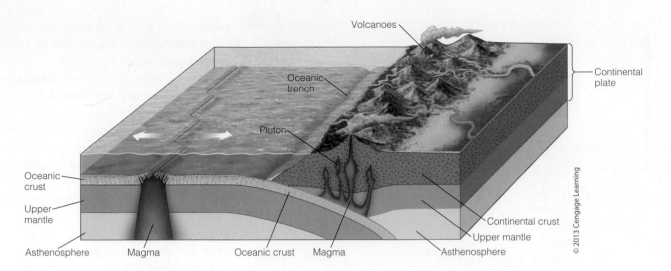

Figure 4.4 The Origin of Magma

Magma forms beneath spreading ridges, because as plates separate, pressure is reduced on the hot rocks and partial melting of the upper mantle begins. Invariably, the magma formed is mafic. Magma also forms at subduction zones, where water from the subducted plate causes partial melting of the upper mantle. This magma is also mafic, but as it rises, melting of the lower crust makes it more felsic.

hot spot Localized zone of melting below the lithosphere that probably overlies a mantle plume.

Recall the sequence of minerals in Bowen's reaction series (Figure 4.3). The order in which these minerals melt is the opposite of their order of crystallization. Accordingly, rocks made up of quartz, potassium feldspar, and sodium-rich plagioclase begin melting at lower temperatures than those composed of ferromagnesian silicates and the calcic varieties of plagioclase. So when ultramafic rock starts to melt, the minerals richest in silica melt first, followed by those containing less silica. Therefore, if melting is not complete, mafic magma containing proportionately more silica than the source rock results.

Subduction Zones and the Origin of Magma

Another fundamental observation regarding magma is that where an oceanic plate is subducted beneath either a continental plate or another oceanic plate, a belt of volcanoes and plutons is found near the leading edge of the overriding plate (Figure 4.4). It would seem, then, that subduction and the origin of magma must be related in some way, and indeed they are. Furthermore, magma at these convergent plate boundaries is mostly intermediate (53% to 65% silica) or felsic (>65% silica).

Once again, geologists invoke the phenomenon of partial melting to explain the origin and composition of magma at subduction zones. As a subducted

plate descends toward the asthenosphere, it eventually reaches the depth where the temperature is high enough to initiate partial melting. In addition, the oceanic crust descends to a depth at which dewatering of hydrous minerals takes place, and as the water rises into the overlying mantle, it enhances melting and magma forms.

Recall that partial melting of ultramafic rock at spreading ridges yields mafic magma. Similarly, partial melting of mafic rocks of the oceanic crust yields intermediate (53% to 65% silica) and felsic (>65% silica) magmas, both of which are richer in silica than the source rock. Moreover, some of the silica-rich sediments and sedimentary rocks of continental margins are probably carried downward with the subducted plate and contribute their silica to the magma. Also, mafic magma rising through the lower continental crust must be contaminated with silica-rich materials, which changes its composition.

Hot Spots and the Origin of Magma

Most volcanism occurs at divergent and convergent plate boundaries, but there are some chains of volcanic outpourings in the ocean basins and on continents that are not near either of these boundaries. The Emperor Seamount–Hawaiian Islands, for instance, form a chain of volcanic islands 6,000 km long, and the volcanic rocks become progressively older toward the northwest (see Figure 2.22). In 1963, Canadian geologist J. Tuzo Wilson proposed that the Hawaiian Islands and other areas showing similar trends lay above a **hot spot** over

Figure 4.5 Mantle Plume and Hot Spot

Direction of Plate Motion

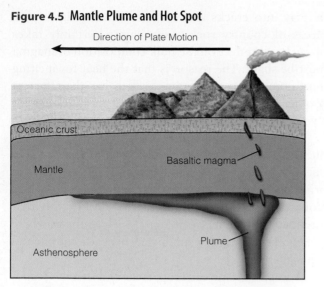

a. A mantle plume beneath oceanic crust with a hot spot. Rising magma forms a series of volcanoes that become younger in the direction of plate movement.

b. A mantle plume with an overlying hot spot yields flood basalts, and some of the continental crust melts to form felsic magma.

© 2013 Cengage Learning

which a plate moves, thereby yielding a succession of volcanoes (Figure 4.5a).

Many geologists now think that hot-spot volcanism results from a rising **mantle plume**, a cylindrical plume of hot mantle rock that rises from perhaps near the core-mantle boundary. As it rises toward the surface, the pressure decreases on the hot rock and melting begins, thus yielding magma. Hot-spot volcanism may also account for vast flat-lying areas of overlapping lava flows on the continents or what geologists call *flood basalts* (Figure 4.5b). Figure 2.15 shows the locations of many hot spots.

Compositional Changes in Magma

Obviously the composition of any magma depends on what was melted in the first place. Should mafic rock melt completely, the resulting magma would also be mafic. However, as we discussed previously, partial melting yields magma that differs from its parent rock. In any case, once magma forms, it may change by **crystal settling**, which involves gravitational settling of minerals as they crystallize (Figure 4.6). A good example is olivine, the first ferromagnesian silicate in the discontinuous branch of Bowen's reaction series, which has a density greater than the remaining melt and tends to sink. As a result, the remaining melt becomes richer in silica, sodium, and potassium, because much of the iron and magnesium were removed as olivine and perhaps pyroxene minerals crystallized. In other words, the remaining melt becomes more felsic.

In some thick, sheet-like plutons called *sills*, the first-formed ferromagnesian silicates are in fact concentrated in their lower parts, thus making their upper parts less mafic. But even so, very little felsic magma forms by this process. Calculations show that to yield a given volume of granite (a felsic igneous rock), about 10 times as much mafic magma would have to be present initially for crystal settling to yield the volume of granite in question. If this were so, mafic igneous rocks should be much more common than felsic ones, but just the opposite is true, so it seems that something other than crystal settling must account for the large volume of felsic magma.

Once again, geologists refer to partial melting to solve this apparent dilemma. Remember that partial melting of oceanic crust and silica-rich sediments of continental margins yields magma richer in silica (more felsic) than the source rock. In addition, magma rising through continental crust changes in composition by **assimilation** as it reacts with preexisting rock, called **country rock**, with which it comes in contact (Figure 4.7a). Country rock may be heated to 1,300°C and partially or completely melt, provided that its

mantle plume A cylindrical mass of magma rising from the mantle toward the surface; recognized at the surface by a hot spot, an area such as the Hawaiian Islands where volcanism takes place.

crystal settling The physical separation and concentration of minerals in the lower part of a magma chamber or pluton by crystallization and gravitational settling.

assimilation A process whereby magma changes composition as it reacts with country rock.

country rock Any preexisting rock that has been intruded by a pluton or altered by metamorphism.

a. Early formed ferromagnesian silicates such as olivine crystallize and because of their density settle to the bottom of the magma chamber.

b. Ferromagnesian silicates continue to form and settle.

c. The remaining melt becomes richer in silicon, sodium, and potassium because much of the iron and magnesium originally present is now in the ferromagnesian minerals that settled.

© 2013 Cengage Learning

Figure 4.6 Crystal Settling Is a Process That Changes the Composition of Magma

magma mixing The process whereby magmas of different composition mix together to yield a modified version of the parent magmas.

melting temperature is lower than that of the magma. Because assimilated rocks rarely have the same composition as the magma, the composition of the magma changes.

The fact that assimilation takes place is clearly indicated by *inclusions*, which are incompletely melted pieces of country rock that are fairly common in many igneous rocks (Figure 4.7b). As magma rises, it forces its way into cracks and crevasses and wedges loose pieces of country rock. Assimilation certainly takes place, but its effect on the bulk composition of magma must be slight. The reason is that the heat for melting must come from the magma, which has the effect of cooling the magma, so only a limited amount of country rock can be assimilated.

Neither crystal settling nor assimilation can produce a significant amount of felsic magma from a mafic one. But both processes, if operating concurrently, can bring about greater changes than either process acting alone. Some geologists think that this is one way that intermediate magma (53% to 65% silica) forms where oceanic lithosphere is subducted beneath continental lithosphere.

A single volcano can erupt lavas of different composition, indicating that magmas of differing composition are present. It seems likely that some of these magmas would come into contact and mix with one another. If this is the case, we would expect that the composition of the magma resulting from **magma mixing** would be a modified version of the parent magmas (Figure 4.7a).

LO3 IGNEOUS ROCKS: THEIR CHARACTERISTICS AND CLASSIFICATION

We have already defined *plutonic*, or *intrusive igneous rocks*, and *volcanic*, or *extrusive igneous rocks*. Here we will have considerably more to say about the texture, composition, and classification of these rocks, which constitute one of the three major rock groups depicted in the rock cycle (see Figure 1.14).

Igneous Rock Textures

Texture refers to the size, shape, and arrangement of the minerals that make up igneous rocks. Size is the most important characteristic, because mineral crystal size is related to the cooling history of magma or lava and usually indicates whether an igneous rock is plutonic or volcanic. The atoms in magma and lava are in constant motion, but when cooling begins, some atoms bond to form small nuclei. As other atoms in the liquid chemically bond to these nuclei, they do so in an orderly geometric arrangement, and the nuclei grow into crystalline *mineral grains*, the individual particles that make up igneous rocks.

If cooling takes place rapidly, as in lava flows, the rate at which mineral nuclei form exceeds the rate of growth and an aggregate of many small mineral grains forms. The result is a fine-grained or **aphanitic texture**, in which individual minerals are too small to be seen without magnification (Figure 4.8a). If cooling is slow, the rate of growth exceeds the rate of nuclei formation, and large mineral grains form, thus yielding a coarse-grained, or **phaneritic texture**, in which minerals are clearly visible (Figure 4.8b). Aphanitic textures usually indicate an extrusive origin, whereas rocks with phaneritic textures are usually intrusive. However, shallow plutons might have an aphanitic texture, and the rocks that form in the interiors of thick lava flows might be phaneritic.

Another common texture in igneous rocks is one termed **porphyritic**, in which minerals of markedly different size are present in the same rock. The larger minerals are *phenocrysts* and the smaller ones collectively make up the *groundmass*, which is simply the grains between phenocrysts (Figure 4.8c).

The only requirement for a porphyritic texture is that the phenocrysts be considerably larger than the minerals in the groundmass, which may be either aphanitic or phaneritic. Igneous rocks with porphyritic textures are designated *porphyry*, as in basalt porphyry. These rocks have more complex cooling histories than those with aphanitic or phaneritic textures and might involve, for example, magma partly cooling beneath the surface, followed by its eruption and rapid cooling at the surface.

Lava may cool so rapidly that its constituent atoms do not have time to become arranged in the ordered, three-dimensional frameworks of minerals. As a consequence, a *natural glass*, such as *obsidian* forms (Figure 4.8d). Some magmas contain large amounts of water vapor and other gases. These gases may be trapped in cooling lava, where they form numerous small holes or cavities known as **vesicles**; rocks with many vesicles are termed *vesicular*, as in vesicular basalt (Figure 4.8e).

aphanitic texture A texture in igneous rocks in which individual mineral grains are too small to be seen without magnification; results from rapid cooling of magma and generally indicates an extrusive origin.

phaneritic texture Igneous rock texture in which minerals are easily visible without magnification.

porphyritic texture An igneous texture with minerals of markedly different sizes.

vesicle A small hole or cavity formed by gas trapped in cooling lava.

Figure 4.7 Assimilation and Magma Mixing

Inclusions

Country rock

Felsic magma

Magma mingling

Injection of mafic or intermediate magma into felsic magma. Low viscosity contrast between magmas.

Injection of mafic magma into felsic magma. High viscosity contrast between magmas.

© 2013 Cengage Learning

a. Fragments of rock dislodged by rising magma may melt and become incorporated into the magma, a process called assimilation, or they may remain as inclusions.

Sue Monroe

b. Dark inclusions in granitic rocks in the Sierra Nevada in California.

a.

b.

Phenocrysts

c.

d.

e.

f.

Sue Monroe

Figure 4.8 Textures of Igneous Rocks

a. Rapid cooling as in lava flows results in many small minerals and an aphanitic (fine-grained) texture.

b. Slower cooling in plutons yields a phaneritic (coarse-grained) texture.

c. These porphyritic textures indicate a complex cooling history.

d. Obsidian has a glassy texture because magma cooled too quickly for mineral crystals to form.

e. Gases expand in lava to yield a vesicular texture.

f. Microscopic view of a rock with a fragmental texture. The colorless, angular particles of volcanic glass measure up to 2 mm.

---

pyroclastic (fragmental) texture A fragmental texture characteristic of igneous rocks composed of pyroclastic materials.

A **pyroclastic**, or **fragmental**, **texture** characterizes igneous rocks formed by explosive volcanic activity (Figure 4.8f). For example, ash discharged high into the atmosphere eventually settles to the surface, where it accumulates; if consolidated, it forms pyroclastic igneous rock.

Composition of Igneous Rocks

Most igneous rocks, like the magma from which they originate, are mafic (45% to 52% silica), intermediate (53% to 65% silica), or felsic (<65% silica). A few are *ultramafic* (<45% silica), but these are probably derived from mafic magma by a process we discuss later. The parent magma plays an important role in determining the mineral composition of igneous rocks, yet it is possible for the same magma to yield a variety of igneous rocks, because its composition can change as a result of the sequence in which minerals crystallize, or by crystal settling, assimilation, and magma mixing (Figures 4.3, 4.6, and 4.7).

Classifying Igneous Rocks

Geologists use texture and composition to classify most igneous rocks, although a few are classified mostly by texture. Notice in Figure 4.9 that all rocks except peridotite are in pairs; the members of a pair have the same composition but different textures. Basalt and gabbro, andesite and diorite, and rhyolite and granite are compositional (mineralogical) pairs, but basalt, andesite, and rhyolite are aphanitic and most commonly volcanic, whereas gabbro, diorite, and granite are phaneritic and mostly plutonic. The volcanic and plutonic members of each pair can usually be distinguished by texture, but remember that rocks in some shallow plutons may be aphanitic, and rocks that formed in thick lava flows may be phaneritic. In other words, all of these rocks exist in a textural continuum.

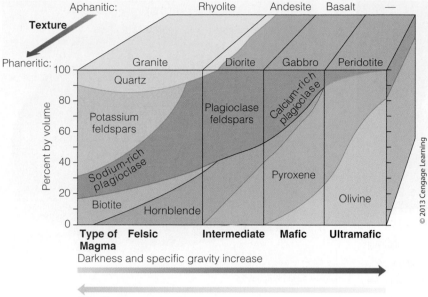

Figure 4.9 Classification of Igneous Rocks
This diagram shows the percentages of minerals, as well as the textures of common igneous rocks. For example, an aphanitic (fine-grained) rock of mostly calcium-rich plagioclase and pyroxene is basalt, whereas a phaneritic (coarse-grained) rock of the same composition is gabbro.

The igneous rocks in Figure 4.9 are also differentiated by their mineral content. Reading across the chart from rhyolite to andesite to basalt, for example, we see that the proportions of nonferromagnesian and ferromagnesian silicates change. The differences in composition, however, are gradual along a compositional continuum. In other words, there are rocks with compositions that correspond to the lines between granite and diorite, basalt and andesite, and so on.

Ultramafic Rocks

Ultramafic rocks (<45% silica) are composed mostly of ferromagnesian silicates. *Peridotite* contains olivine, lesser amounts of pyroxene, and usually a little plagioclase feldspar (Figure 4.10), and *pyroxenite* is composed predominantly of pyroxene. Because these minerals are dark, the rocks are black or green. Ultramafic rocks in Earth's crust probably originate by concentration of the early-formed ferromagnesian minerals that separated from mafic magmas.

Ultramafic lava flows, called *komatiites*, are known in rocks older than 2.5 billion years but are rare or absent in younger ones. The reason is that in order to erupt, ultramafic lava must have a near-surface temperature of about 1,600°C; the surface temperatures of present-day mafic lava flows are rarely more than 1,200°C. During early Earth history, though, more radioactive decay heated the mantle to as much as 300°C hotter than now, and ultramafic lavas formed. Because the amount of radiogenic heat has decreased over time, Earth has cooled, and eruptions of ultramafic lava flows have ceased.

Basalt-Gabbro

Basalt and *gabbro* are the aphanitic and phaneritic rocks that crystallize from mafic magma (45% to 52% silica) (Figure 4.11). Both have the same composition—mostly calcium-rich plagioclase and pyroxene, with smaller amounts of olivine and amphibole (Figure 4.9). Because they contain a large proportion of ferromagnesian silicates, basalt and gabbro are dark; those that are porphyritic contain calcium plagioclase or olivine phenocrysts.

Extensive basalt lava flows cover vast areas in Washington, Oregon, Idaho, and northern California (see Chapter 5). Oceanic islands such as Iceland, the Galápagos, the Azores, and the Hawaiian Islands are composed mostly of basalt, which also makes up the upper part of the oceanic crust.

Gabbro is much less common than basalt, at least in the continental crust, where it can be easily observed. Small intrusive bodies of gabbro are present in the continental crust, but intermediate to felsic intrusive rocks are much more common. However, the lower part of the oceanic crust is composed of gabbro.

Andesite-Diorite

Intermediate-composition magma (53% to 65% silica) crystallizes to form *andesite* and *diorite*, which are compositionally equivalent fine- and coarse-grained igneous rocks (Figure 4.12). Andesite and diorite are composed predominantly of plagioclase feldspar, with the typical ferromagnesian component being amphibole or biotite (Figure 4.9). Andesite is medium to dark gray, but diorite has a

Figure 4.10 Peridotite

This specimen of the ultramafic rock peridotite is made up mostly of olivine. Notice in Figure 4.9 that peridotite is the only phaneritic rock that does not have an aphanitic counterpart. Peridotite is rare at Earth's surface, but is very likely the rock that makes up the mantle.

salt-and-pepper appearance because of its white to light-gray plagioclase and dark ferromagnesian silicates.

Andesite is a common extrusive igneous rock that cooled from lava erupted in volcanic chains at convergent plate boundaries. The volcanoes of the Andes Mountains of South America and the Cascade Range in western North America are composed, in part, of andesite. Intrusive bodies of diorite are fairly common in the continental crust.

Rhyolite-Granite

Rhyolite and *granite* crystallize from felsic magma (>65% silica) and are therefore silica-rich rocks (Figure 4.13). They consist mostly of potassium feldspar, sodium-rich plagioclase, and quartz, with perhaps some biotite and rarely hornblende (Figure 4.9). Because nonferromagnesian silicates predominate, rhyolite and granite are typically light colored. Rhyolite is fine grained, although most often it contains phenocrysts of potassium feldspar or quartz, and granite is coarse grained.

Rhyolite lava flows are much less common than andesite and basalt flows. Recall that one control of magma viscosity is silica content. Thus, if felsic magma rises to the surface, it begins to cool, the pressure on it decreases, and gases are released explosively, usually yielding rhyolitic pyroclastic materials. The rhyolitic lava flows that do occur are thick and highly viscous and move only short distances.

Granite is a coarsely crystalline igneous rock with a composition corresponding to that of the field shown in Figure 4.9. Strictly speaking, not all rocks in this field

a. Basalt is aphanitic.

b. Gabbro is phaneritic. Notice the light reflected from the crystal faces.

Figure 4.11 Mafic Igneous Rocks Basalt and Gabbro

are granites. For example, a rock with a composition close to the line separating granite and diorite is called *granodiorite*. To avoid the confusion that might result from introducing more rock names, we will follow the practice of referring to rocks to the left of the granite-diorite line in Figure 4.9 as *granitic*.

Granitic rocks are by far the most common plutonic igneous rocks, and they are restricted to the continents. Most granitic rocks were intruded at or near convergent plate margins during mountain-building episodes. When these mountainous regions are uplifted and eroded, the vast bodies of granitic rocks forming their cores are exposed.

Figure 4.12 Intermediate Igneous Rocks Andesite and Diorite

a. This specimen of andesite has hornblende phenocrysts that are so numerous the rock is classified as an andesite hornblende porphyry.

b. Diorite has a salt-and-pepper appearance because it contains light-colored nonferromagnesian silicates and dark-colored ferromagnesian silicates.

Pegmatite

The term *pegmatite* refers to a particular texture rather than a specific composition, but most pegmatites are composed primarily of quartz, potassium feldspar, and sodium-rich plagioclase, thus corresponding closely to granite. The most remarkable feature of pegmatites is the size of their minerals, which measure at least 1 cm across, and in some pegmatites they measure tens of centimeters or meters (Figure 4.14). Many pegmatites are adjacent to large granite plutons and are composed of minerals that formed from the water-rich magma that remained after most of the granite crystallized.

When felsic magma cools and forms granite, the remaining water-rich magma has a lower density and viscosity and invades cracks in the nearby rocks where minerals crystallize. This water-rich magma also contains elements that rarely enter into the common minerals that form granite. Pegmatites that are essentially very coarsely crystalline granite are simple pegmatites, whereas those with minerals containing elements such as lithium, beryllium, cesium, boron, and several others are complex pegmatites.

The formation and growth of mineral-crystal nuclei in pegmatites are similar to those processes in other magmas, but with one critical difference: The water-rich magma from which pegmatites crystallize inhibits the formation of nuclei. However, some nuclei do form, and because the appropriate ions in the liquid can move easily and attach themselves to a growing crystal, individual minerals may grow very large.

Other Igneous Rocks

Geologists classify the igneous rocks in Figure 4.9 by texture and composition, but a few others are identified primarily by their textures. Much of the fragmental material erupted by volcanoes is *ash*, a designation for pyroclastic materials measuring less than 2.0 mm, most of which consists of pieces of minerals or shards of volcanic glass (Figure 4.8f). The consolidation of ash forms the pyroclastic rock *tuff* (Figure 4.15a). Most tuff is silica-rich and light colored and is appropriately called *rhyolite tuff*. Some ash flows are so hot that as they come to rest, the ash particles fuse together and form a *welded tuff*. Consolidated deposits of larger pyroclastic materials, such as cinders, blocks, and bombs, are *volcanic breccia*.

Both *obsidian* and *pumice* are varieties of volcanic glass (Figure 4.15b, c). Obsidian may be black, dark gray, red, or brown, depending on the presence of iron. Obsidian breaks with a conchoidal (smoothly curved) fracture typical of glass. Analyses of many samples indicate that most obsidian has a high silica content and is compositionally similar to rhyolite.

Pumice is a variety of volcanic glass containing numerous vesicles that develop when gas escapes through lava and forms a froth (Figure 4.15c). If pumice falls into water, it can be carried great distances, because it is so porous and light that it floats. Another vesicular rock is *scoria*. It is more crystalline and denser than pumice, but it has more vesicles than solid rock (Figure 4.15d).

Figure 4.13 Felsic Igneous Rocks Rhyolite and Granite

These rocks are typically light-colored because they contain mostly nonferromagnesian silicate minerals. The dark spots in the granite speciment are biotite mica. The white and pinkish minerals are feldspar, whereas the glassy-appearing minerals are quartz.

a. Rhyolite.

b. Granite.

pluton An intrusive igneous body that forms when magma cools and crystallizes within the crust, such as a batholith or sill.

concordant pluton Intrusive igneous body whose boundaries parallel the layering in the country rock.

discordant pluton Pluton with boundaries that cut across the layering in the country rock.

dike A tabular, or sheetlike, discordant pluton.

sill A tabular, or sheetlike, concordant pluton.

laccolith A concordant pluton with a mushroom-like geometry.

LO4 INTRUSIVE IGNEOUS BODIES: PLUTONS

Unlike volcanism and the origin of volcanic rocks, we can study intrusive igneous bodies, collectively called **plutons**, only indirectly, because intrusive rocks form when magma cools and crystallizes within Earth's crust (Figure 4.16a). We can observe these rock bodies only when uplift and deep erosion have taken place, thereby exposing them at the surface. Furthermore, geologists cannot duplicate the conditions under which intrusive rocks form except in small laboratory experiments.

Geologists recognize several types of plutons based on their geometry (three-dimensional shape) and relationships to the country rocks. In terms of their geometry, plutons are tabular, cylindrical, or irregular (massive). Furthermore, they may be **concordant**, meaning they have boundaries that parallel the layering in the country rock, or **discordant**, with boundaries that cut across the country rock's layering (Figure 4.16a).

Dikes, Sills, and Laccoliths

Dikes and **sills** are tabular, or sheetlike, igneous bodies that differ only in that dikes are discordant and sills are concordant (Figure 4.16a). Dikes are quite common and range from a few centimeters to more than 100 m thick (Figure 4.16b). Invariably, they are intruded into preexisting fractures or where fluid pressure is great enough for them to form their own fractures as they move upward into country rock.

Sills are tabular, just as dikes are, but they are concordant. Many sills are a meter or less thick, although some are much thicker. Most sills were intruded into sedimentary rocks, but eroded volcanoes also reveal

Figure 4.14 Pegmatite

a. This pegmatite, the light-colored rock, is exposed in the Black Hills of South Dakota.

b. Closeup view of a specimen from a pegmatite with minerals measuring 8 to 10 cm across.

that sills are injected into piles of volcanic rocks. In fact, some of the inflation of a volcano that precedes an eruption may be caused by the injection of sills (see Chapter 5). In contrast to dikes, which follow zones of weakness, sills are intruded between layers in country rock when the fluid pressure is great enough for the magma to actually lift the overlying rocks.

Under some circumstances, a sill inflates and causes the overlying rocks to bow upward, forming an igneous body called a **laccolith** (Figure 4.16a). A laccolith has a flat floor and is domed up in its central part, giving it a mushroom-like geometry. Like sills, laccoliths are rather shallow intrusions that lift the overlying rocks.

Courtesy of David J. Mutty

a. Tuff.

Sue Mcnroe

b. Obsidian.

Sue Monroe

c. Pumice.

Sue Monroe

d. Scoria.

Volcanic Pipes and Necks

Volcanoes have a cylindrical conduit known as a **volcanic pipe** that connects to an underlying magma chamber. Magma rises through this structure; however, when a volcano ceases to erupt, its slopes are attacked by weathering and erosion, but the magma in the pipe is commonly more resistant to erosion and is left as a remnant called a **volcanic neck** (Figure 4.16c). Several volcanic necks are found in the southwestern United States, especially in Arizona and New Mexico, and others are recognized elsewhere.

Batholiths and Stocks

A **batholith**, the largest of all plutons, must have at least 100 km² of surface area, and most are far larger (Figure 4.16a). A **stock**, in contrast, is similar but smaller. Some stocks are simply parts of large plutons that, once exposed by erosion, are parts of batholiths. Both batholiths and stocks are mostly discordant, although locally they may be concordant, and batholiths, especially, consist of multiple intrusions. In other words, a batholith is a large composite body produced by repeated, voluminous intrusions of magma in the same region.

volcanic pipe The conduit connecting the crater of a volcano with an underlying magma chamber.

volcanic neck An erosional remnant of the material that solidified in a volcanic pipe.

batholith An irregularly shaped, discordant pluton with at least 100 km² of surface area.

stock An irregularly shaped, discordant pluton with a surface area smaller than 100 km².

Figure 4.16 Plutons

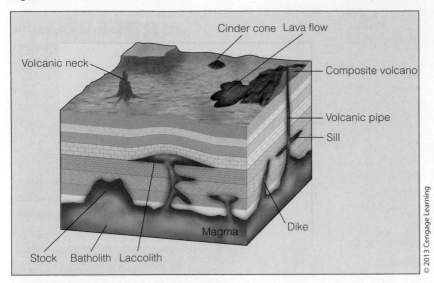

Volcanic neck · Cinder cone · Lava flow · Composite volcano · Volcanic pipe · Sill · Stock · Batholith · Laccolith · Magma · Dike

© 2013 Cengage Learning

a. Block diagram showing various plutons. Some plutons cut across the layering in country rock and are discordant, whereas others parallel the layering and are concordant.

James S. Monroe

c. A volcanic neck in Monument Valley Navajo Tribal Park, Arizona. This landform is 457 m high. Most of the original volcano was eroded, leaving only this remnant.

James S. Monroe

b. The dark material in this image is a dike that cuts through Cenozoic-age sedimentary rocks near Dulce, New Mexico.

The igneous rocks that make up batholiths are mostly granitic, although diorite may also be present. Batholiths and stocks are emplaced mostly near convergent plate boundaries during episodes of mountain building. One example is the Sierra Nevada batholith of California, which formed over millions of years. Other large batholiths in North America include the Idaho batholith, the Boulder batholith in Montana, and the Coast Range batholith in British Columbia, Canada.

Mineral resources are found in rocks of batholiths and stocks, and in the adjacent country rocks. The copper deposits at Butte, Montana, are in rocks near the margins of the granitic rocks of the Boulder batholith. Near Salt Lake City, Utah, copper is mined from the mineralized rocks of the Bingham stock, a composite pluton composed of granite and granite porphyry. Granitic rocks also are the primary source of gold, which forms from mineral-rich solutions moving through cracks and fractures of the igneous body.

LO5 THE ORIGIN OF BATHOLITHS

Geologists realized long ago that the origin of batholiths posed a space problem. What happened to the rock that was once in the space now occupied by a batholith? One proposed solution was that these large igneous bodies melted their way into the crust. In other words, they simply assimilated the country rock as they moved upward (Figure 4.7). The presence of inclusions, especially near the tops of some plutons, indicates that assimilation does occur. Nevertheless, as we noted, assimilation is a limited process, because magma cools as country rock is assimilated. Calculations indicate that far too little heat is available in magma to assimilate the huge quantities of country rock necessary to make room for a batholith.

Geologists now mostly agree that batholiths were emplaced by *forceful injection* as magma moved upward (Figure 4.17a). Recall that granite is derived from viscous felsic magma and therefore rises slowly. It appears that the magma deforms and shoulders aside

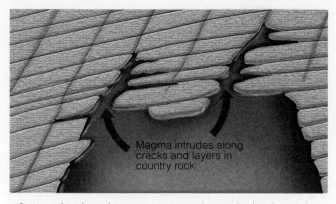

Magma intrudes along cracks and layers in country rock

a. Stoping takes place when magma rises into the crust by detaching and engulfing pieces of country rock.

© 2013 Cengage Learning

b. Some of the detached blocks may be assimilated, and some may remain as inclusions (Figure 4.7b).

Figure 4.17 Emplacement of a Batholith by Stoping

the country rock, and as it rises farther, some of the country rock fills the space beneath the magma.

Some batholiths do show evidence of having been emplaced by forceful injection, but this mechanism probably occurs in the deeper parts of the crust where temperature and pressure are high and the country rocks are easily deformed in the manner described. At shallower depths, the crust is more rigid and tends to deform by fracturing. In this environment, batholiths may move upward by **stoping**, a process in which rising magma detaches and engulfs pieces of country rock (Figure 4.17b). According to this concept, magma moves up along fractures and the planes separating layers of country rock. Eventually, pieces of country rock detach and settle into the magma. No new room is created during stoping; the magma simply fills the space formerly occupied by country rock.

> **stoping** A process in which rising magma detaches and engulfs pieces of the country rock.

5 | Volcanoes and Volcanism

Eyjafjallajökull in Iceland erupting a huge cloud of volcanic ash and steam on April 17, 2010. The eruption, although not very large, rook place beneath a glacier, which added to the cloud over the volcano. Because of the ash that was in the atmosphere, air traffic over the North Atlantic was disrupted for several days.

bt3/ZUMA Press/Newscom

Introduction

"When considered in the context of Earth history, volcanism is actually a constructive process."

About 550 volcanoes on Earth are active—that is, they have erupted during historic time—but only a dozen or so are erupting at any one time. Most eruptions are minor and go unreported in the popular press unless they occur in or near populated areas or have tragic consequences. Recent examples are the April 2010 eruptions of Eyjafjallajökull in Iceland (see the chapter opening photo). The eruptions were not particularly large, but they took place beneath 200 m of glacial ice, resulting in steam explosions that injected huge quantities of volcanic ash into the atmosphere that disrupted air travel over the North Atlantic for several days.

In addition to active volcanoes, Earth has many others that are *dormant*, meaning they have not erupted in historic time but may do so in the future. For instance, in 1991, Mount Pinatubo in the Philippines erupted for the first time in 600 years, and similarly Mount Sinabung in Indonesia lay dormant for 400 years and then began erupting in August 2010. Thousands of volcanoes have not erupted during historic time and show no signs of doing so in the future, so they are *extinct* or *inactive* volcanoes. The Sutter Buttes in northern California last erupted about 1.5 million years ago and now show no indications of any activity.

No other geologic phenomenon has captured the public imagination more than erupting volcanoes, with their fiery streams of lava and particulate matter blasted into the atmosphere in huge clouds. What better subject for a disaster movie, many of which show lava flows as a great danger to humans. In 2002, lava flows in Goma, Zaire (Democratic Republic of the Congo), caused gasoline storage tanks to explode, killing 147 people. And, of course, lava flows may destroy buildings and cover productive land, but overall they are the least dangerous manifestation of volcanic eruptions. Explosive eruptions, in contrast, are quite dangerous, especially if they occur near populated areas.

Erupting volcanoes do cause fatalities and considerable damage, but when considered in the context of Earth history, volcanism is actually a constructive process. Emissions from volcanoes during Earth's early history played an important part in the origin of the atmosphere and surface waters, and oceanic crust is continuously produced by volcanism at spreading ridges. Oceanic islands such as the Hawaiian Islands, Iceland, the Azores, and many others owe their existence to volcanic eruptions, and weathering of lava, pyroclastic materials, and volcanic mudflows in areas such as Indonesia coverts them to productive soils.

An excellent reason to study volcanism is that it illustrates the complex interactions among Earth's systems. The emission of gases and pyroclastic materials has an immediate and profound impact on the atmosphere, hydrosphere, and biosphere, at least in the vicinity of an eruption. And in some cases the effects are worldwide, as they were following the eruptions of Tambora in 1815, Krakatau in 1883, and Pinatubo in 1991.

Learning Outcomes

After reading this unit, you should be able to do the following:

LO1 Understand volcanism and volcanoes

LO2 Identify the types of volcanoes

LO3 Identify other volcanic landforms

LO4 Identify the distribution of volcanoes

LO5 Understand the relationship between plate tectonics, volcanoes, and plutons

LO6 Understand volcanic hazards, volcano monitoring, and forecasting eruptions

Eruptions in the Continental U.S.

Eruptions in the continental United States have occurred only three times since 1914, all in the Cascade Range, which stretches from northern California through Oregon and Washington and into southern British Columbia, Canada.

PANIC

LO1 VOLCANOES AND VOLCANISM

volcano A hill or mountain formed around a vent as a result of the eruption of lava and pyroclastic materials.

volcanism The process whereby magma and its associated gases rise through the crust and are extruded onto the surface or into the atmosphere.

A **volcano** is a hill or mountain that forms around a vent where lava, pyroclastic materials, and gases erupt. Thus, it is a *landform*—a feature on Earth's surface. The term **volcanism** refers to all processes related to the rise and discharge of magma and gases at the surface or into the atmosphere. Thus, volcanism accounts for the origin of all volcanic (extrusive igneous) rocks, such as basalt, tuff, and obsidian, as well as volcanoes, but not all volcanism results in the origin of volcanoes. Some eruptions take place along fissures and build up basalt plateaus (discussed later in this chapter).

All of the other terrestrial planets (Mercury, Venus, and Mars) and Earth's Moon were volcanically active during their early histories, but now the only one with possible active volcanoes is Venus. As recently as April 2010, scientists at the Jet Propulsion Laboratory in Pasadena, California, claimed that three active volcanoes are present on Venus. In addition, volcanoes are known on two other bodies in the solar system. Triton, one of Neptune's moons, probably has active volcanoes, and Jupiter's moon Io is by far the most volcanically active body in the solar system. Many of its one hundred or so volcanoes are erupting at any given time.

Volcanic Gases

Volcanic gases from present-day volcanoes are 50% to 80% water vapor, with lesser amounts of carbon dioxide, nitrogen, sulfur gases—especially sulfur dioxide and hydrogen sulfide—and very small amounts of carbon monoxide, hydrogen, and chlorine. In areas of recent volcanism, emission of gases from fumer-oles (volcanic vents) continues, and one cannot help noticing the rotten-egg odor of hydrogen sulfide gas (Figure 5.1).

As magma rises toward the surface, the pressure is reduced and the contained gases begin to expand. In highly viscous, felsic magma, expansion is inhibited and gas pressure increases. Eventually, the pressure may become great enough to cause an explosion and produce pyroclastic materials. In contrast, low-viscosity mafic magma allows gases to expand and escape easily, so mafic magma usually erupts rather quietly.

Most volcanic gases quickly dissipate in the atmosphere and pose little danger to humans, but on several occasions they have had far-reaching climatic effect or have caused fatalities. In 1783, sulfur gases erupted from Laki fissure in Iceland were responsible for the Blue Haze Famine, crop failures, and the deaths of 24% of Iceland's population. In 1816, a persistent "dry fog" caused unusually cold spring and summer weather in Europe and eastern North America. In North America, the year of 1816 was called "The Year Without a Summer" or "Eighteen Hundred and Froze-to-Death." Killing frost during the summer resulted in crop failures and food shortages. This particularly cold summer was attributed to the huge eruption of Tambora in 1815 in Indonesia, although the eruption of Mayon Volcano in the Philippines during the previous year also probably contributed.

In 1986, in the African nation of Cameroon, 1,746 people died when a cloud of carbon dioxide engulfed them. The gas accumulated in the waters of Lake Nyos, which occupies a volcanic caldera. Scientists disagree about what caused the gas to suddenly burst forth from the lake, but once it did, it flowed downhill along the surface because it was denser than air. In fact, the density and velocity of the gas cloud were great enough to flatten vegetation, including trees, a few kilometers from the lake. Unfortunately, thousands of animals and many people, some as far as 23 km from the lake, were asphyxiated.

Vog

Residents of the island of Hawaii have coined the term *vog* for volcanic smog. Kilauea volcano has been erupting continuously since 1983, releasing small amounts of lava and copious quantities of carbon dioxide and sulfur dioxide. Carbon dioxide is no problem, because it dissipates quickly in the atmosphere, but sulfur dioxide produces a haze and the unpleasant odor of sulfur. Vog probably poses little or no health risk for tourists, but a long-term threat exists for residents of the west side of the island, where vog is most common.

James S. Monroe

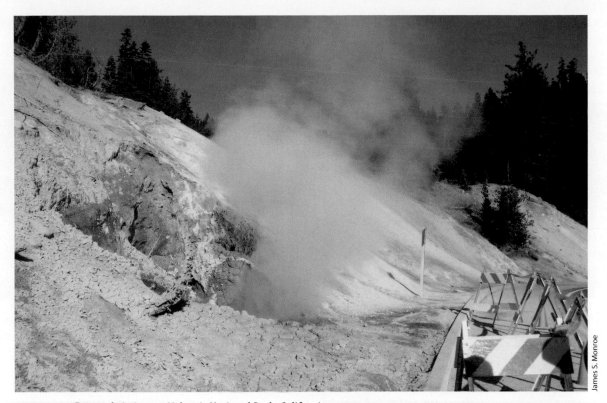

Figure 5.1 Fumerole in Lassen Volcanic National Park, California
This fumerole opened up only a few years ago high on the slope to the left. It has since moved down the slope and now is near the road visible at the lower right.

Lava Flows

Movies and TV shows depict lava flows as a great danger to humans, but only rarely do they cause fatalities. The reason is that most lava flows do not flow very rapidly, and because they are fluid they follow low areas. So once a flow begins moving, determining the path it will take is fairly easy, and anyone in areas likely to be affected can be evacuated. From April 1990 to January 1991, lava flows covered Kalapana, Hawaii, and in the process destroyed 180 homes, highways, and points of archaeological interest. Civil Defense authorities working with geologists at the Hawaiian Volcano Observatory made important decisions regarding evacuations and road closures, and as a result there were no injuries or fatalities.

Even low-viscosity flows, such as those in Hawaii, do not move very rapidly. They can move much more rapidly, though, when their margins cool to form a channel, and especially when insulated on all sides as in a **lava tube**, where geologists have recorded speeds of more than 50 km/hr. A lava tube forms within a flow when its margins and upper surface solidify, thereby forming a conduit through which lava can move rapidly and for great distances. When an eruption ceases, the tube drains, leaving an empty tunnel-like structure (Figure 5.2). Part of the roof of a lava tube may collapse to form a *skylight* through which an active flow can be observed, or access can be gained to an inactive lava tube.

Geologists in Hawaii characterized basalt lava flows as *pahoehoe* or *aa*, although these terms are now used elsewhere, too. **Pahoehoe** (pronouned *pay-hoy-hoy*) has a smooth, ropy surface much like taffy (Figure 5.3a). The surface of an **aa** (pronounced *ah-ah*) flow consists of rough, jagged, angular blocks and fragments (Figure 5.3b). Pahoehoe flows are hotter and thinner than aa flows; indeed, aa flows are viscous enough to break into blocks and move forward as a wall of rubble. One further point: A pahoehoe flow may change along its length to aa as its viscosity increases, partly because it cools, but aa does not change to pahoehoe.

> **lava tube** A tunnel beneath the solidified surface of a lava flow through which lava moves; also, the hollow space left when the lava within a tube drains away.
>
> **pahoehoe** A type of lava flow with a smooth, ropy surface.
>
> **aa** Lava flow with a surface of rough, angular blocks and fragments.

Figure 5.2 Lava Tubes

Lava tubes consisting of hollow spaces beneath the surfaces of lava flows are common in many areas.

Photo by J.B. Judd/USGS

Photo provided by Hawai'i Volcanoes National Park

a. An active lava tube in Hawaii. Part of the tube's roof has collapsed, forming a skylight.

b. A lava tube in Hawaii after the lava has drained out.

Figure 5.3 Pahoehoe and aa Lava Flows

Pahoehoe and aa were named for lava flows in Hawaii, but the same kinds of flows are found in many other areas.

J. D.Griggs/USDS

Robert Tilling/USGS

a. An excellent example of the taffylike appearance of pahoehoe.

b. An aa lava flow advances over an older pahoehoe flow. Notice the rubbly nature of the aa flow.

Much of the igneous rock in the upper oceanic crust is a distinctive type made up of bulbous masses of basalt that resemble pillows, hence the name **pillow lava**. Geologists realized long ago that pillow lava (Figure 5.4) forms when lava is rapidly chilled underwater, but its formation was not observed until 1971. Divers near Hawaii saw pillows form when a blob of lava broke through the crust of an underwater lava flow and cooled very quickly, forming a pillow- or loaf-shaped structure with a glassy exterior. The remaining fluid

Figure 5.4 Pillow Lava

Much of the upper part of the oceanic crust is made up of pillow lava that formed when lava erupted underwater.

a. Pillow lava on the seafloor in the Pacific Ocean about 150 miles west of Oregon that formed about five years before this photo was taken.

b. Ancient pillow lava that formed on the seafloor is now on land in Alaska.

lava inside then broke though the crust of an already-formed pillow and formed another pillow, and so on.

Mafic lava flows and some intermediate ones, as well as some rocks in dikes, sills, and volcanic necks, show a pattern of columns bounded by fractures, or what geologists call **columnar joints**. For columnar joints to form, a lava flow must cease moving, then cool and contract, thereby setting up forces that cause fractures, known as *joints*, to open. On a lava flow's surface the fractures are commonly polygonal (often six-sided) cracks that extend downward, thus outlining columns with their long axes perpendicular to the cooling surface (Figure 5.5).

Pressure on the partly solidified crust of a still-moving lava flow causes the surface to buckle into *pressure ridges* (Figure 5.6a). Gases escaping from a flow hurl globs of lava into the air, which fall back to the surface and adhere to one another, thus forming small, steep-sided *spatter cones*, or spatter ramparts if they

pillow lava Bulbous masses of basalt, resembling pillows, formed when lava is rapidly chilled under water.

columnar jointing The phenomenon of forming columns bounded by fractures in some igneous rocks as they cooled and contracted.

Figure 5.5 Columnar Jointing

Columnar jointing is seen mostly in mafic lava flows and related plutonic rocks.

a. Columnar joints in a basalt lava flow at Devil's Postpile National Monument in California. The rubble in the foreground is collapsed columns.

b. Surface view of the columns from (a). The straight lines and polish resulted from abrasion by a glacier that moved over this surface.

Figure 5.6 Pressure Ridges and Spatter Cones

a. The buckled surface of this lava flow on Medicine Lake Volcano in California is a pressure ridge. It formed when the solidified surface of a lava flow was bent upward by a still-moving flow beneath.

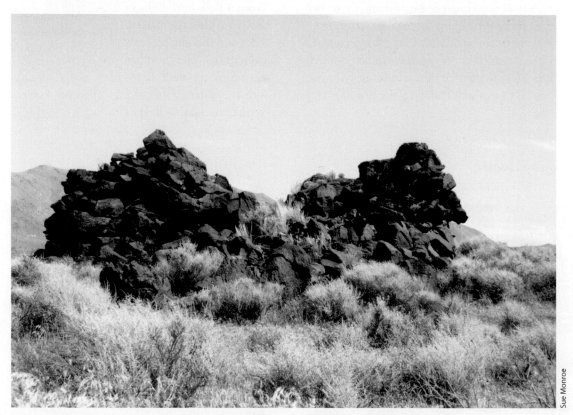

b. These two chimney-like pillars of rocks are spatter cones on the surface of a lava flow in the Coso volcanic field of California.

are elongated (Figure 5.6b). Spatter cones a few meters high are common on the lava flows in Hawaii, and you can see ancient ones in many areas.

Pyroclastic Materials

In addition to lava flows, erupting volcanoes eject pyroclastic materials, especially **volcanic ash**, a designation for pyroclastic particles that measure less than 2.0 mm (Figure 5.7). In some cases, ash is ejected into the atmosphere and settles to the surface as an *ash fall*. In contrast to an ash fall, an *ash flow* is a cloud of ash and gas that flows along or close to the land surface. Ash flows can move faster than 100 km per hour, and some cover vast areas.

In populated areas adjacent to volcanoes, ash falls and ash flows pose serious problems, and volcanic ash in the atmosphere is a hazard to aviation. In 1989, ash from Redoubt volcano in Alaska caused all four jet engines to fail on KLM Flight 867. The plane, carrying 231 passengers, nearly crashed when it fell more than 3 km before the crew could restart the engines. The plane landed safely in Anchorage, Alaska, but it required $80 million in repairs. And as already mentioned, the April 2010 eruptions of Eyjafjallajökull in Iceland disrupted air traffic over the North Atlantic.

In addition to volcanic ash, volcanoes erupt *lapilli*, pyroclastic materials that measure from 2 mm to 64 mm,

and *blocks* and *bombs*, both larger than 64 mm (Figure 5.7). Bombs have a twisted, streamlined shape, which indicates that they were erupted as globs of magma that cooled and solidified during their flight through the air. Blocks, in contrast, are angular pieces of rock ripped from a volcanic conduit or pieces of a solidified crust of a lava flow. Because of their size, lapilli, bombs, and blocks are confined to the immediate area of an eruption.

Christopher Parypa/Shutterstock.com

volcanic ash Pyroclastic materials that measure less than 2 mm.

crater An oval to circular depression at the summit of a volcano resulting from the eruption of lava, pyroclastic materials, and gases.

caldera A large, steep-sided, oval to circular depression, usually formed when a volcano's summit collapses into a partially drained underlying magma chamber.

LO2 WHAT ARE THE TYPES OF VOLCANOES?

Simply put, a volcano is a hill or mountain that forms around a vent where lava, pyroclastic materials, and gases erupt. Although volcanoes vary in size and shape, all have a conduit or conduits leading to a magma chamber beneath the surface. Vulcan, the Roman deity of fire, was the inspiration for calling these mountains volcanoes, and because of their danger and obvious connection to Earth's interior, they have been held in awe by many cultures.

Most volcanoes have a circular depression known as a **crater** at their summit, or on their flanks, that forms by explosions or collapse. Most craters are less than 1 km across, whereas much larger rimmed depressions on volcanoes are **calderas**. In fact, some volcanoes have a summit crater within a caldera. Calderas are huge structures that form following voluminous eruptions, during which part of a magma chamber drains and the mountain's summit collapses into the vacated space below. An excellent example is misnamed Crater Lake in Oregon (Figure 5.8). Crater Lake is actually a steep-rimmed caldera

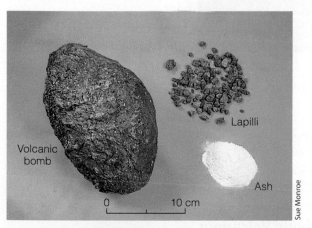

Volcanic bomb

Lapilli

Ash

0   10 cm

Sue Monroe

Figure 5.7 Pyroclastic Materials

Pyroclastic materials are all particles ejected from volcanoes, especially during explosive eruptions. The volcanic bomb is elongate, because it was molten when it descended through the air. The lapilli was collected at a small volcano in Oregon, whereas the ash came from the 1980 eruption of Mount St. Helens in Washington.

a. Eruption begins as huge quantities of ash are ejected from the volcano.

Mt. Mazama volcano (elevation about 3,500 meters)

Magma chamber

c. The eruption continues as more ash and pumice are ejected into the air, and pyroclastic flows move down the flanks of the mountain.

b. The collapse of the summit into the partially drained magma chamber forms a huge caldera.

d. Postcaldera eruptions partly cover the caldera floor and the small cinder cone called Wizard Island forms.

Wizard island

Rim of Crater Lake caldera (elevation 2,000 meters)

FROM HOWELL WILLIAMS, *CRATER LAKE: THE STORY OF ITS ORIGIN* (BERKELEY, CALIF.: UNIVERSITY OF CALIFORNIA PRESS): ILLUSTRATIONS FROM P. 84. © 1941 REGENTS OF THE UNIVERSITY OF CALIFORNIA, © RENEWED 1969, HOWELL WILLIAMS.

e. View from the rim of Crater Lake showing Wizard Island. The lake is 594 m deep, making it the second deepest in North America.

James S. Monroe

Figure 5.8 The Origin of Crater Lake, Oregon
Remember, Crater Lake is actually a caldera that formed when a volcano's summit collapsed into a partly drained magma chamber.

shield volcano A dome-shaped volcano with a low, rounded profile built up mostly by overlapping basalt lava flows.

that formed about 7,700 years ago in the manner just described. As impressive as Crater Lake is, it is not nearly as large as some other calderas, such as the Toba caldera in Sumatra, which is 100 km long and 30 km wide.

Geologists recognize several major types of volcanoes, but one must realize that each volcano is unique in its history of eruptions and development. For instance, the frequency of eruptions varies considerably; the Hawaiian volcanoes and Mount Etna on Sicily have

erupted repeatedly, whereas Pinatubo in the Philippines erupted in 1991 for the first time in 600 years. Some volcanoes are complex mountains that have the characteristics of more than one type of volcano.

Shield Volcanoes

Volcanoes that look much like the outer surface of a shield lying on the ground with the convex side up are **shield volcanoes** (Figure 5.9). They are composed almost entirely of mafic lava flows that had low viscosity, so the flows spread out and formed thin layers that slope only 2 to 10 degrees. Erupting shield volcanoes, sometimes called *Hawaiian-type eruptions*, are rather quiet

Figure 5.9 Shield Volcanoes

a. Shield volcanoes consist of numerous thin basalt lava flows that build up mountains with slopes rarely exceeding 10 degrees.

b. Profile of Mauna Loa in Hawaii. Mauna Loa is one of five huge shield volcanoes that make up the island of Hawaii.

compared to eruptions of many other volcanoes, particularly those at convergent plate boundaries. Magma that rises to the surface and issues as lava flows pose little danger to humans, although incandescent lava may be forcefully ejected at lava fountains, some up to 400 m high, when magma reaches the surface and gases expand.

Although eruptions of shield volcanoes tend to be rather quiet, some of the Hawaiian volcanoes have, on occasion, produced sizable explosions when groundwater instantly vaporized as it came into contact with magma. One such explosion in 1790 killed about 80 warriors in a party headed by Chief Keoua, who was leading them across the summit of Kilauea volcano.

Kilauea volcano is impressive, because it has been erupting continuously since January 3, 1983, making it the longest recorded eruption. During these 28 years, more than 2.5 km³ of molten rock has flowed out at the surface, much of it reaching the sea and forming 2.2 km² of new land on the island of Hawaii. Unfortunately, lava flows from Kilauea have also destroyed many homes and caused some $61 million in damages.

Shield volcanoes are most common in the ocean basins, but some are also present on the continents—in East Africa, for instance. The island of Hawaii is made up of five huge shield volcanoes, two of which, Kilauea and Mauna Loa, are active much of the time. Mauna Loa is nearly 100 km across its base and stands more than 9.5 km above the surrounding seafloor; it has a volume estimated at 50,000 km³, making it the world's largest volcano.

Cinder Cones

Small, steep-sided **cinder cones** made up of particles resembling cinders form when pyroclastic materials accumulate around a vent from which they erupted (Figure 5.10). Cinder cones are small, rarely exceeding 400 m high, with slope angles up to 33 degrees, depending on

the angle that can be maintained by the angular pyroclastic materials. Many have a large, bowl-shaped crater, and if they issue any lava flows, they usually break through the

cinder cone A small, steep-sided volcano made up of pyroclastic materials resembling cinders that accumulate around a vent.

Figure 5.10 Cinder Cones
This 400-m-high cinder cone named Paricutín formed in a short time in Mexico in 1943, when pyroclastic materials began to erupt in a farmer's field. Lava flows from the volcano covered two nearby villages, but all activity ceased by 1952.

composite volcano (stratovolcano) A volcano composed of lava flows and pyroclastic layers, typically of intermediate composition, and mudflows.

lahar A mudflow composed of pyroclastic materials such as ash.

base or lower flanks of the mountains. Although all cinder cones are conical, their symmetry varies from almost perfectly symmetrical to those that formed when prevailing winds caused pyroclastic materials to build up higher on the downwind side of the vent.

Many cinder cones form on the flanks or within the calderas of larger volcanoes and represent the final stages of activity, particularly in areas of basaltic volcanism. Wizard Island in Crater Lake, Oregon, is a small cinder cone that formed after the summit of Mount Mazama collapsed to form a caldera (Figure 5.8). Cinder cones are common in the southern Rocky Mountain states, particularly New Mexico and Arizona, and many others are in California, Oregon, Washington, and Hawaii.

steep-sided near their summits, perhaps as much as 30 degrees, but the slope decreases toward the base, where it may be no more than 5 degrees. Mayon volcano in the Philippines is one of the most nearly symmetrical composite volcanoes anywhere. It erupted in 1999 for the 13th time during the 1900s.

When most people think of volcanoes, they picture the graceful profiles of composite volcanoes, which are the typical large volcanoes found on the continents and island arcs. And some of these volcanoes are indeed large: Mount Shasta in northern California is made up of about 350 km^3 of material and measures 20 km across its base. Mount Pinatubo in the Philippines erupted violently on June 15, 1991. Huge quantities of gas and an estimated 3 to 5 km^3 of ash were discharged into the atmosphere, making this the world's largest eruption since 1912. Fortunately, warnings of an impending eruption were heeded, and 200,000 people were evacuated from around the volcano. Nevertheless, the eruption was responsible for 722 deaths.

Composite Volcanoes (Stratovolcanoes)

Pyroclastic layers, as well as lava flows, both of intermediate composition, are found in **composite volcanoes**, which are also called **stratovolcanoes** (Figure 5.11a). As the lava flows cool, they typically form andesite; recall that intermediate lava flows are more viscous than mafic ones. Geologists use the term **lahar** for volcanic mudflows, which are also common on composite volcanoes. A lahar may form when rain falls on unconsolidated pyroclastic materials and creates a muddy slurry that moves downslope. On November 13, 1985, a minor eruption of Nevado del Ruiz in Colombia melted snow and ice on the volcano, causing lahars that killed 23,000 people.

Composite volcanoes differ from shield volcanoes and cinder cones in composition and shape. Remember that shield volcanoes have very low slopes, whereas cinder cones are small, steep-sided, conical volcanoes. In contrast, composite volcanoes are

Figure 5.11 Composite Volcanoes

a. Composite volcanoes, also called stratovolcanoes, are made up mostly of lava flows and pyroclastic materials of intermediate composition, although mudflows (lahars) are also common.

Pyroclastic layers
Lava flows

© 2013 Cengage Learning

D. R. Crandel/USGS

b. Mount St. Helens in Washington state as it appeared from the east in 1978.

Lava Domes

Less common volcanoes are **lava domes**, also known as *volcanic domes* and *plug domes*. These are steep-sided, bulbous mountains that form when viscous felsic magma, and occasionally intermediate magma, is forced toward the surface (Figure 5.12). Because felsic magma is so viscous, it moves upward very slowly and only when the pressure from below is great.

Lava dome eruptions are some of the most violent and destructive. In 1902, viscous magma accumulated beneath the summit of Mount Pelée on the island of Martinique. Eventually, the pressure increased until the side of the mountain blew out in a tremendous explosion, ejecting a mobile, dense cloud of pyroclastic materials and a glowing cloud of gases and dust called a **nuée ardente** (French for "glowing cloud"). The pyroclastic flow followed a valley to the sea, but the nuée ardente jumped a ridge and engulfed the city of St. Pierre.

A tremendous blast hit St. Pierre, leveling buildings, and hurling boulders, trees, and pieces of masonry down the streets. Accompanying the blast was a swirling cloud of incandescent ash and gases with an internal temperature of 700°C that incinerated everything in its path. The nuée ardente passed through St. Pierre in two or three minutes, only to be followed by a firestorm as combustible materials

> **lava dome** A bulbous, steep-sided mountain formed by viscous magma moving upward through a volcanic conduit.
>
> **nuée ardente** A fast-moving, dense cloud of hot pyroclastic materials and gases ejected from a volcano.

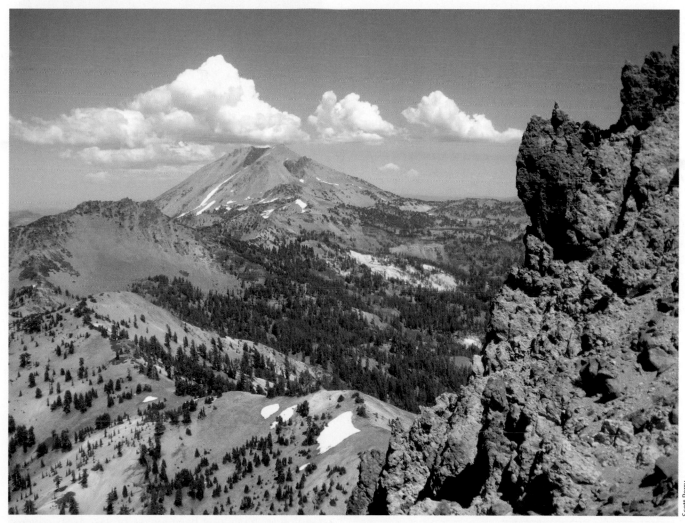

Scott Perry

Figure 5.12 Lava Domes
Lava domes are bulbous masses of viscous magma that form irregularly shaped volcanic mountains flanked by debris shed from the dome. This image shows Lassen Peak in Lassen Volcanic National Park in California. The dark masses of rock cooled from viscous magma. Lassen Peak erupted from 1914 to 1917.

burned and casks of rum exploded. But by then, most of the 28,000 residents of the city were already dead. In fact, in the area covered by the nuée ardente, only two people survived!* One survivor was on the outer edge of the nuée ardente, but even there, he was terribly burned, and his family and neighbors were all killed. The other survivor, a stevedore incarcerated the night before for disorderly conduct, was in a windowless cell partly below ground level. He remained in his cell badly burned for four days after the eruption until rescue workers heard his cries for help.

LO3 OTHER VOLCANIC LANDFORMS

During *fissure eruptions*, fluid lava pours out and simply builds up rather flat-lying areas, whereas huge explosive eruptions might yield *pyroclastic sheet deposits*, which, as their name implies, have a sheetlike geometry. In both cases volcanoes fail to develop.

Fissure Eruptions and Basalt Plateaus

Rather then erupting from central vents, the lava flows making up **basalt plateaus** issue from long cracks or fissures during **fissure eruptions**. The lava has such low viscosity that it spreads out and covers vast areas. A good example is the Columbia River basalt in eastern Washington and parts of Oregon and Idaho. This huge accumulation of 17-million- to 6-million-year-old overlapping lava flows covers about 164,000 km² (Figure 5.13a) and has an aggregate thickness of more than 1,000 m.

Similar accumulations of vast, overlapping lava flows are also found in the Snake River Plain in Idaho (Figure 5.13b). These flows are 5.0 to 1.6 million years old, and they represent a style of eruption between fissure eruptions and those of shield volcanoes. In fact,

---

*Although reports commonly claim that only two people survived the eruption, at least 69 and possibly as many as 111 people survived beyond the extreme margins of the nuée ardente and on ships in the harbor. Many, however, were badly injured.

a. About 20 lava flows of the Columbia River basalt are exposed in the canyon of the Grand Ronde River in Washington.

b. Basalt lava flows of the Snake River Plain near Twin Falls, Idaho.

Figure 5.13 Basalt Plateaus
Basalt plateaus are vast areas of overlapping lava flows that issued from long fissures. Fissure eruptions take place today in Iceland; however, in the past, they formed basalt plateaus in various areas.

there are small, low shields, as well as fissure flows, in the Snake River Plain.

Currently, fissure eruptions occur only in Iceland. Iceland has several volcanoes, but the bulk of the island is composed of basalt lava flows that issued from fissures. Indeed, about half of the lava that erupted during historic time in Iceland came from two fissure eruptions, one in A.D. 930 and the other in 1783. The 1783 eruption from Laki fissure, which is more than 30 km long, accounted for lava that covered 560 km² and, in one place, filled a valley to a depth of about 200 m.

Pyroclastic Sheet Deposits

Geologists have long been aware of vast areas covered by felsic volcanic rocks a few meters to hundreds of meters thick. Based on observations of historic pyroclastic flows, such as the nuée ardente erupted by

Mount Pelée in 1902, it seems that these ancient rocks originated as pyroclastic flows—hence the name **pyroclastic sheet deposits**. They cover far greater areas than any observed during historic time, however, and apparently erupted from long fissures rather than from a central vent. The pyroclastic materials of many of these flows were so hot that they fused together to form *welded tuff*.

Geologists now think that major pyroclastic flows issue from fissures formed during the origin of calderas. The Bishop Tuff of eastern California erupted shortly before the formation of the Long Valley caldera. Interestingly, earthquake activity in the Long Valley caldera and nearby areas beginning in 1978 may indicate that magma is moving upward beneath part of the caldera. Thus, the possibility of future eruptions in that area cannot be discounted.

LO4 DISTRIBUTION OF VOLCANOES

Most of the world's active volcanoes are in well-defined zones or belts rather than being randomly distributed.

The **circum-Pacific belt**, popularly called the Ring of Fire, has more than 60% of all active volcanoes. It includes volcanoes in the Andes of South America; the volcanoes of Central America, Mexico, and the Cascade Range of North America; as well as the Alaskan volcanoes and those in Japan, the Philippines, Indonesia, and New Zealand (Figure 5.14).

The second area of active volcanism is the **Mediterranean belt** (Figure 5.14). About 20% of all active volcanism takes place in this belt, where the famous Italian volcanoes such as Mounts Etna and Vesuvius and the Greek volcano Santorini are found.

pyroclastic sheet deposit A vast, sheetlike deposit of felsic pyroclastic materials erupted from fissures.

circum-Pacific belt A zone of seismic and volcanic activity and mountain building that nearly encircles the Pacific Ocean basin.

Mediterranean belt A zone of seismic and volcanic activity extending through the Mediterranean region of southern Europe and eastward to Indonesia.

Divergent plate boundary (some transform plate boundaries)

Convergent boundary

Volcano

© 2013 Cengage Learning

Figure 5.14 Volcanoes of the World

Most volcanoes are at or near convergent and divergent plate boundaries. The two major volcano belts are the circum-Pacific belt, commonly known as the Ring of Fire, with about 60% of all active volcanoes, and the Mediterranean belt, with 20% of active volcanoes. Most of the rest lie near mid-oceanic ridges.

The **Cascade Range** (Figure 5.15) stretches from Lassen Peak in northern California north through Oregon and Washington into British Columbia, Canada. Most of the large volcanoes in the range are composite volcanoes, but Lassen Peak in California is the world's largest lava dome. It erupted from 1914 to 1917, but has since been quiet except for ongoing hydrothermal activity (Figure 5.15b).

What was once a nearly symmetrical composite volcano changed markedly on May 6, 1980, when Mount St. Helens in Washington erupted explosively, killing 57 people and leveling some 600 km² of forest. A huge lateral blast caused much of the damage and

fatalities, but snow and ice on the volcano melted, and pyroclastic materials displaced water in lakes and rivers, causing lahars and extensive flooding.

Mount St. Helens's renewed activity, beginning in late September 2004, has resulted in dome growth and small steam and ash explosions. Scientists at the Cascades Volcano Observatory in Vancouver, Washington, issued a low-level alert for an eruption and continue to monitor the volcano. However, in January 2008, after 40 months of activity, it became dormant.

LO5 PLATE TECTONICS, VOLCANOES, AND PLUTONS

In Chapter 4, we discussed the origin and evolution of magma and concluded that (1) mafic magma is generated

Figure 5.15 The Cascade Range of the Pacific Northwest

a. Plate tectonic setting for the Pacific Northwest. Subduction of the Juan de Fuca plate accounts for ongoing volcanism in the region.

© 2013 Cengage Learning

Photo by Chester Mullen; courtesy National Park Service/USGS

b. Lassen Peak in California erupted from 1914 to 1917. This eruption took place in 1915.

beneath spreading ridges, and (2) intermediate and felsic magma form where an oceanic plate is subducted beneath another oceanic plate or a continental plate. Accordingly, most volcanism and emplacement of plutons take place at or near divergent and convergent plate boundaries.

Igneous Activity at Divergent Plate Boundaries

Much of the mafic magma that originates at spreading ridges is emplaced as vertical dikes and gabbro plutons, thus composing the lower part of the oceanic crust. However, some rises to the surface and issues forth as submarine lava flows and pillow lava (Figure 5.4), which constitutes the upper part of the oceanic crust. Much of this volcanism goes undetected, but researchers in submersibles have seen the results of recent eruptions.

Mafic lava is very fluid, allowing gases to escape easily, and at great depth in the oceans, the water pressure is so great that explosive volcanism is prevented. In short, pyroclastic materials are rare to absent unless a volcanic center builds up above sea level. Even if this occurs, however, the mafic magma is so fluid that it forms the gently sloping layers found on shield volcanoes.

Excellent examples of divergent plate boundary volcanism are found along the Mid-Atlantic Ridge, particularly where it rises above sea level as in Iceland. The East Pacific Rise and the Indian Ridge are areas of similar volcanism. A divergent plate boundary is also present in Africa as the East African Rift system, which is well known for its volcanoes.

Igneous Activity at Convergent Plate Boundaries

Nearly all of the large active volcanoes in both the circum-Pacific and Mediterranean belts are composite volcanoes near the leading edges of overriding plates at convergent plate boundaries (Figure 5.14). The overriding plate, with its chain of volcanoes, may be oceanic, as in the case of the Aleutian Islands, or it may be continental as is, for instance, the South American plate with its chain of volcanoes along its western margin.

As we noted, these volcanoes at convergent plate boundaries consist mostly of lava flows and pyroclastic materials of intermediate to felsic composition. Remember that when mafic oceanic crust partially melts, some of the magma generated is emplaced near plate boundaries as plutons, and some is erupted to build up composite volcanoes. More viscous magmas, usually of felsic composition, are emplaced as lava domes, thus accounting for the explosive eruptions that typically occur at convergent plate boundaries.

Good examples of volcanism at convergent plate boundaries are the explosive eruptions of Mount Pinatubo and Mayon volcano in the Philippines; both are near a plate boundary beneath which an oceanic plate is subducted. Mount St. Helens, Washington, is similarly situated, but it is on a continental rather than an oceanic plate.

Intraplate Volcanism

Mauna Loa and Kilauea on the island of Hawaii and Loihi just 32 km to the south are within the interior of a rigid plate located far from any divergent or convergent plate boundary (Figure 5.14). The magma is derived from the upper mantle, as it is at spreading ridges, and accordingly is mafic, so it builds up shield volcanoes. Loihi is particularly interesting, because it represents an early stage in the origin of a new Hawaiian island. It is a submarine volcano that rises more than 3,000 m above the adjacent seafloor, but its summit is still about 940 m below sea level.

Even though the Hawaiian volcanoes are not at or near a spreading ridge or a subduction zone, their evolution is nevertheless related to plate movements. Notice in Figure 2.22 that the ages of the rocks that make up the Hawaiian Islands increase toward the northwest. Kauai formed 5.6 to 3.8 million years ago, whereas Hawaii began forming less than 1 million years ago, and Loihi began to form even more recently. The islands have formed in succession as the Pacific plate moves continuously over a hot spot that is now beneath Hawaii and just to the south at Loihi.

LO6 VOLCANIC HAZARDS, VOLCANO MONITORING, AND FORECASTING ERUPTIONS

Undoubtedly you suspect that living near an active volcano poses some risk, and of course this assessment

volcanic explosivity index (VEI) A semi-quantitative scale for determining the size of a volcanic eruption based on evaluation of criteria, such as volume of material explosively erupted and height of eruption cloud.

is correct, but further questions remain: What exactly are volcanic hazards? Is there any way to anticipate eruptions? What can we do to minimize the dangers of eruptions? We have already mentioned that lava flows, with few exceptions, pose little threat to humans, although they may destroy property. Lava flows, nuée ardentes, and volcanic gases are threats during an eruption, but lahars and landslides may take place even when no eruption has occurred for a long time (Figure 5.16).

How Large Is an Eruption, and How Long Do Eruptions Last?

The most widely used indication of the size of a volcanic eruption is the **volcanic explosivity index (VEI)** (Figure 5.17). The VEI ranges from 0 (gentle) to 8 (cataclysmic) and is based on several aspects of an eruption, such as the volume of material explosively ejected and the height of the eruption plume. However, the volume of lava, fatalities, and property damage are not considered. For instance, the 1985 eruption of Nevado del Ruiz in Colombia killed 23,000 people, yet had a VEI value of only 3. In contrast, the huge eruption (VEI = 6) of Novarupta in Alaska in 1912 caused no fatalities or injuries. Since A.D. 1500, only the 1815 eruption of Tambora had a value of 7; it was both large and deadly. Of the several eruptions

Figure 5.16 Volcanic Hazards
A volcanic hazard is any manifestation of volcanism that poses a threat, including lava flows and, more importantly, volcanic gas, ash, and lahars.

MARCO/LONGARI/AFP/Getty Images

a. This 2002 lava flow in Goma, Democratic Republic of Congo, killed 47 people, mostly by causing gasoline storage tanks to explode.

Sue Monroe

b. This sign at Mammoth Mountain volcano in California warns of the potential danger of CO_2 gas, which has killed 170 acres of trees.

Reuters/Corbis

c. When Mount Pinatubo in the Philippines erupted on June 15, 1991, this huge cloud of ash and steam formed over the volcano.

Figure 5.17 The Volcanic Explosivity Index (VEI)
In this example, an eruption with a VEI of 5 has an eruption cloud up to 25 km high and ejects at least 1 km³ of *tephra*, a collective term for all pyroclastic materials. Geologists characterize eruptions as Hawaiian (nonexplosive), Strombolian, Vulcanian, or Plinian.

of Eyjafjallajökull in Iceland during April 2010, the largest one had a VEI of no more than 4, but because of its location and the fact that it erupted beneath glacial ice, it caused disruption of air traffic over the North Atlantic.

The duration of eruptions varies considerably. Fully 42% of about 3,300 historic eruptions lasted less than one month. About 33% erupted for one to six months, but some 16 volcanoes have been active more or less continuously for more than 20 years. Stromboli and Mount Etna in Italy and Erta Ale in Ethiopia are good examples. For some explosive volcanoes, the time from the onset of their eruptions to the climactic event is weeks or months.

A case in point is the explosive eruption of Mount St. Helens on May 18, 1980, which occurred two months after eruptive activity began. Unfortunately, many volcanoes give little or no warning of such large-scale events. Of 252 explosive eruptions, 42% erupted most violently during their first day of activity.

Is It Possible to Forecast Eruptions?

Only a few of Earth's potentially dangerous volcanoes are monitored, including some in Japan, Italy, Russia, New Zealand, and the United States. Volcano monitoring involves recording and analyzing physical and chemical changes at volcanoes (Figure 5.18). Tiltmeters detect changes in the slopes of a volcano as it inflates when magma rises beneath it, and a geodimeter uses a laser beam to measure horizontal distances, which change as a volcano inflates. Geologists also monitor gas emissions, changes in groundwater level and temperature, hot springs activity, and changes in the local magnetic and electrical fields. Even the accumulating snow and ice, if any, are evaluated to anticipate hazards from floods if an eruption takes place.

Of critical importance in volcano monitoring and warning of an imminent eruption is the detection of **volcanic tremor**, continuous ground motion that lasts for minutes to hours as opposed to the sudden, sharp jolts produced by most earthquakes. Volcanic tremor, also known as *harmonic tremor*, indicates that magma is moving beneath the surface.

volcanic tremor Ground motion lasting from minutes to hours, resulting from magma moving beneath the surface, as opposed to the sudden jolts produced by most earthquakes.

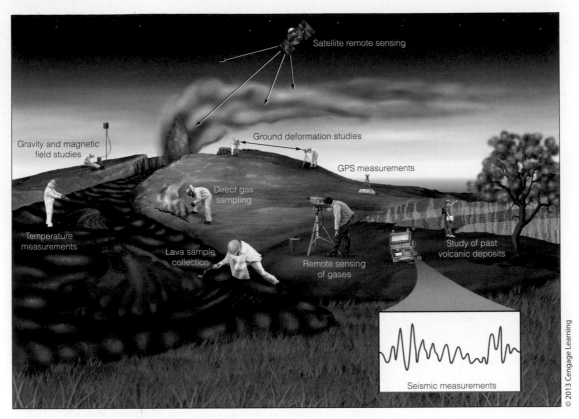

Figure 5.18 Volcanic Monitoring
Some important techniques used to monitor volcanoes.

To more fully anticipate the future activity of a volcano, its eruptive history must be known. Accordingly, geologists study the record of past eruptions preserved in rocks. Detailed studies before 1980 indicated that Mount St. Helens, Washington, had erupted explosively 14 or 15 times during the last 4,500 years, so geologists concluded that it was one of the most likely Cascade Range volcanoes to erupt again.

Geologists successfully gave timely warnings of impending eruptions of Mount St. Helens in Washington and Mount Pinatubo in the Philippines, but in both cases, the climactic eruptions were preceded by eruptive activity of lesser intensity. In some cases, however, the warning signs are much more subtle and difficult to interpret. Numerous small earthquakes and other warning signs indicated to United States Geological Survey (USGS) geologists that magma was moving beneath the surface of the Long Valley caldera in eastern California, so in 1987, they issued a low-level warning, and then nothing happened.

Volcanic activity in the Long Valley caldera occurred as recently as 250 years ago, and there is every reason to think that it will occur again. Unfortunately, the local populace was largely unaware of the geologic history of the region, the USGS did a poor job of communicating its concerns, and premature news releases caused more concern than was justified. In any case, local residents were outraged, because the warnings caused a decrease in tourism (Mammoth Mountain on the margins of the caldera is the second largest ski area in the country) and property values plummeted. Monitoring continues in the Long Valley caldera, and the signs of renewed volcanism, including (1) earthquake swarms, (2) trees being killed by carbon dioxide gas apparently emanating from magma (Figure 5.16b), and (3) hot spring activity, cannot be ignored. In April 2006, three members of a ski patrol were killed by carbon dioxide gas that accumulated in a low area.

WHY CHOOSE?

Every 4LTR Press solution comes complete with a visually engaging textbook in addition to an interactive eBook. Go to CourseMate for **GEOL2** to begin using the eBook. Access at **www.cengagebrain.com**

Complete the Speak Up survey in CourseMate at
www.cengagebrain.com

 Follow us at
www.facebook.com/4ltrpress

6 Weathering, Soil, and Sedimentary Rocks

Differential weathering and erosion in Red Canyon along Utah State Highway 12, which is designated a Scenic Byway. Here the Paleocene to Eocene (40- to 60-million-year-old) Claron Formation has been deeply eroded to form spires and pinnacles because the rock is not structurally and compositionally uniform throughout. The Claron Formation consists of sedimentary rocks made up of gravel, sand, and mud. The rocks are red because they contain small amounts of iron oxide.

Sue Monroe

Introduction

"All rocks are important in deciphering Earth history, but sedimentary rocks have a special place in this endeavor."

All rocks at or near Earth's surface—as well as rocklike substances such as pavement and concrete in sidewalks, bridges, and foundations—decay and crumble with age. In short, they experience **weathering**, defined as the physical breakdown and chemical alteration of Earth materials as they are exposed to the atmosphere, hydrosphere, and biosphere. Actually, weathering is a group of physical and chemical processes that alter Earth materials so that they are more nearly in equilibrium with a new set of environmental conditions. Many rocks form within the crust where little or no oxygen or water is present, but at or near the surface they are exposed to both, as well as to lower temperature and pressure and the activities of organisms.

The rock acted on by weathering, or **parent material**, is disaggregated to form smaller pieces (Figure 6.1), and some of its constituent minerals are altered or dissolved. Some of this weathered material may accumulate and be further modified to form *soil*. Much of it, however, is removed by **erosion**, which is the wearing away of soil and rock by geologic agents such as running water. This eroded material is transported elsewhere by running water, wind, glaciers, and marine currents and is eventually deposited as *sediment*, the raw material for *sedimentary rocks*.

Earth's crust is composed mostly of *crystalline rock*, a term that refers loosely to metamorphic and igneous rocks, except those made up of pyroclastic materials.

Learning Outcomes

After reading this unit, you should be able to do the following:

LO1 Explain how Earth materials are altered

LO2 Explain how soil forms and deteriorates

LO3 Know how weathering and resources are related

LO4 Identify sediment and sedimentary rocks

LO5 Explain how sedimentary rocks are classified

LO6 Understand sedimentary facies

LO7 Read the story preserved in sedimentary rocks

LO8 Recognize important resources in sedimentary rocks

weathering The physical breakdown and chemical alteration of rocks and minerals at or near Earth's surface.

parent material The material that is chemically and mechanically weathered to yield sediment and soil.

erosion The removal of weathered materials from their source area by running water, wind, glaciers, and waves.

Erosion and Natural Resources

Weathering and erosion have yielded many areas of exceptional scenery, including Bryce Canyon and Arches National Parks, both in Utah. In addition to forming interesting landscapes, weathering is also responsible for the origin of some natural resources (e.g., aluminum ore), and it enriches other resources by removing soluble materials. In fact, some sediments and sedimentary rocks are resources in their own right, or they are the host rocks for petroleum and natural gas.

Nevertheless, sediment and sedimentary rocks are by far the most common materials in surface exposures and in the shallow subsurface, even though they make up perhaps only 5.0% of the crust. They cover approximately two-thirds of the continents and most of the seafloor, except spreading ridges. All rocks are important in deciphering Earth history, but sedimentary rocks have a special place in this endeavor, because they preserve evidence of surface processes responsible for them, as well as most fossils, the only evidence of prehistoric life.

LO1 HOW ARE EARTH MATERIALS ALTERED?

Weathering takes place at or near the surface, but the rocks it acts upon are not structurally and compositionally homogeneous throughout, which accounts for **differential weathering**. This is weathering that takes place at different rates even in the same area, so irregular surfaces develop. Differential weathering and *differential erosion*—that is, variable rates of erosion—combine to yield some unusual and even bizarre features, such as *hoodoos*, *spires*, and *arches* (see chapter-opening photo and the Virtual Field Trip in this chapter).

Geologists characterize weathering as *mechanical* and *chemical*, both of which proceed simultaneously on parent material, as well as on materials in transport and those deposited as sediment. In short, all surface or near-surface materials weather, although one type of weathering may predominate depending on such variables as climate and rock type.

Mechanical Weathering

Mechanical weathering takes place when physical forces break earth materials into smaller pieces that retain the composition of the parent material. Granite,

for instance, might yield smaller pieces of granite or individual grains of quartz, potassium feldspars, plagioclase feldspars, and biotite (Figure 6.1).

Frost action involving water repeatedly freezing and thawing in cracks and pores in rocks is particularly effective where temperatures commonly fluctuate above and below freezing. Frost action is so effective because water expands by about 9% when it freezes, thus exerting great force on the walls of a crack, widening and

Figure 6.1 Weathering of Granite

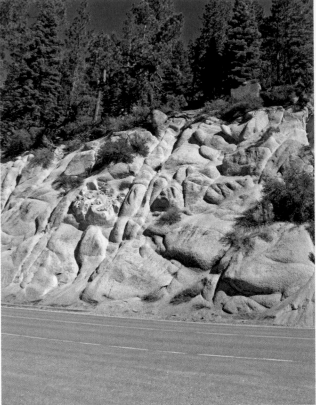

Sue Monroe

a. This exposure of granite has been so thoroughly altered by weathering that only spherical masses of the original rock are visible.

Sue Morroe

b. Close-up view of weathered granite. Mechanical weathering predominates, so the particles are mostly small pieces of granite and minerals including quartz, feldspars, and biotite.

a. Frost wedging takes place when water seeps into cracks and expands as it freezes. Angular pieces of rock are pried loose by repeated freezing.

b. Frost wedging and other mechanical weathering processes produced these talus accumulations in Yukon Territory, Canada.

Talus

Figure 6.2 Frost Wedging

extending it by *frost wedging* (Figure 6.2a). Repeated freezing and thawing dislodge angular pieces of rock from the parent material that tumble downslope and accumulate as **talus** (Figure 6.2b).

Some rocks form at depth and are stable under tremendous pressure. Granite crystallizes far below the surface, so when it is uplifted and eroded, its contained energy is released by outward expansion, a phenomenon known as **pressure release**. The outward expansion results in the origin of fractures called *sheet joints* that more or less parallel the exposed rock surface. Sheet-joint-bounded slabs of rock slip or slide off the parent rock, leaving large, rounded masses known as **exfoliation domes** (Figure 6.3). That solid rock expands and produces fractures might be counterintuitive but is a well-known phenomenon. In deep mines, masses of rock detach from the sides of the excavation, often explosively. These *rock bursts* and less violent *popping* pose a danger to mine workers.

During **thermal expansion and contraction**, the volume of rocks changes as they heat up and then cool down. The temperature may vary as much as 30°C a day in a desert, and because rocks conduct heat poorly, they heat and expand on the outside more than the inside. Even dark minerals absorb heat faster than light-colored ones, so differential expansion takes place between minerals. Surface expansion might generate enough stress to cause fracturing, but experiments in which rocks are heated and cooled repeatedly to simulate years of such activity indicate that thermal expansion and contraction are of minor importance in mechanical weathering.

The formation of salt crystals exerts enough force to widen cracks and dislodge particles in porous, granular rocks such as sandstone. And even in rocks with an interlocking mosaic of crystals, such as granite, **salt crystal growth** pries

talus Accumulation of coarse, angular rock fragments at the base of a slope.

pressure release A mechanical weathering process in which rocks that formed under pressure expand on being exposed at the surface.

exfoliation dome A large, rounded dome of rock resulting when concentric layers of rock are stripped from the surface of a rock mass.

thermal expansion and contraction A type of mechanical weathering in which the volume of rocks changes in response to heating and cooling.

salt crystal growth A mechanical weathering process in which salt crystals growing in cracks and pores disaggregate rocks.

loose individual minerals. It takes place mostly in hot, arid regions, but also probably affects rocks in some coastal areas.

Animals and plants also participate in the mechanical alteration of rocks. Burrowing animals, such as worms, reptiles, rodents, termites, and ants, constantly mix soil and sediment particles and bring material from depth to the surface, where further weathering occurs. The roots of plants, especially large bushes and trees, wedge themselves into cracks in rocks and further widen them (Figure 6.4a).

Chemical Weathering

In contrast to mechanical weathering, **chemical weathering** changes the composition of parent materials. For example, several clay minerals (sheet silicates) form by the chemical and structural alteration of other minerals, such as potassium feldspars and plagioclase feldspars, both of which are framework silicates. Other minerals completely decompose during chemical weathering.

Important agents of chemical weathering include atmospheric gases, especially oxygen, water, and acids. Organisms also play an important role. Rocks with lichens (composite organisms made up of fungi and algae) on their surfaces undergo more rapid chemical alteration than do lichen-free rocks (Figure 6.4b). In addition, plants remove ions from soil water and reduce

the chemical stability of soil minerals, and plant roots release organic acids.

When **solution** takes place, the ions of a substance separate in a liquid, and the solid substance dissolves. Water is a remarkable solvent, because its molecules

Figure 6.3 Sheet Joints and Exfoliation Domes

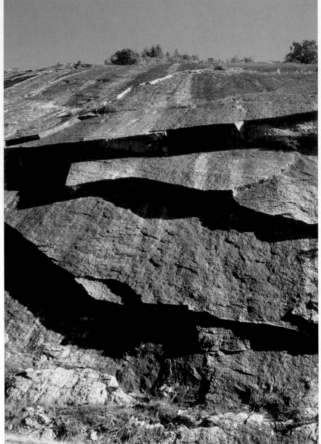

Sue Monroe

a. Slabs of granitic rock bounded by sheet joints in the Sierra Nevada of California. The slabs are inclined downward toward the roadway visible at the lower left.

In South Africa, rock bursts and popping are responsible for approximately 20 deaths per year.

© Stockbyte/Getty Images

Mark Gibson/Visuals Unlimited, Inc.

b. Stone Mountain in Georgia is a large exfoliation dome.

Figure 6.4 Organisms and Weathering

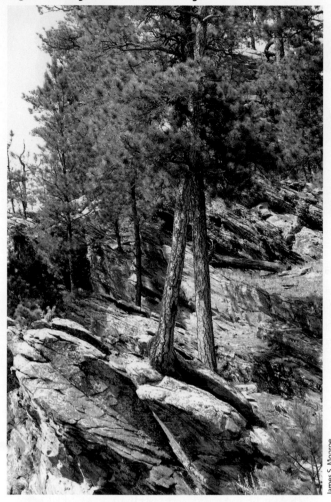

a. These trees in the Black Hills of South Dakota contribute to mechanical weathering as they grow in cracks in the rocks, thereby breaking the parent material into smaller pieces.

b. The orange and gray masses on this rock at Grimes Point Archaeological Site in Nevada are lichens, composite organisms made up of fungi and algae. Lichens derive their nutrients from the rock and thus contribute to chemical weathering.

have an asymmetric shape, consisting of one oxygen atom with two hydrogen atoms arranged so that the angle between the two hydrogen atoms is about 104 degrees (Figure 6.5). Because of this asymmetry, the oxygen end of the molecule retains a slight negative electrical charge, whereas the hydrogen end retains a slight positive charge. When a soluble substance such as the mineral halite (NaCl) comes into contact with a water molecule, the positively charged sodium ions are attracted to the negative end of the water molecule, and the negatively charged chloride ions are attracted to the positive end. Thus, ions are liberated from the crystal structure, and the solid goes into solution; in other words, it dissolves (Figure 6.5).

Most minerals are not very soluble in pure water, because the attractive forces of water molecules are not sufficient to overcome the forces between particles in minerals. The mineral calcite ($CaCO_3$), the major constituent of the sedimentary rock limestone and the metamorphic rock marble, is practically insoluble in pure water, but it rapidly dissolves if a small amount of acid is present. One way to make water acidic is by dissociating the ions of carbonic acid as follows:

$$H_2O \ + \ CO_2 \ \rightleftharpoons \ H_2CO_3 \ \rightleftharpoons \ H^+ \ + \ HCO_3^-$$

water ⁣ carbon dioxide ⁣ carbonic acid ⁣ hydrogen ion ⁣ bicarbonate ion

According to this chemical equation, water and carbon dioxide combine to form *carbonic acid*, a small amount of which dissociates to yield hydrogen and bicarbonate ions. The concentration of hydrogen ions determines the acidity of a solution: the more hydrogen ions present, the stronger the acid.

The atmosphere is mostly nitrogen and oxygen, but approximately 0.03% is carbon dioxide, causing rain to be slightly acidic. Decaying organic matter and the respiration of organisms produce carbon dioxide in soils, so groundwater is also usually slightly acidic. Climate affects the acidity, however, with arid regions tending to have alkaline groundwater (that is, it has a low concentration of hydrogen ions). Whatever the source of carbon dioxide, once an acidic solution is present, calcite rapidly dissolves according to the reaction.

$$CaCO_3 \ + \ H_2O \ + \ CO_2 \ \rightleftharpoons \ Ca^{++} \ + \ 2HCO_3^-$$

calcite ⁣ water ⁣ carbon dioxide ⁣ calcium ion ⁣ bicarbonate ion

oxidation The reaction of oxygen with other atoms to form oxides or, if water is present, hydroxides.

hydrolysis The chemical reaction between hydrogen (H⁺) ions and hydroxyl (OH⁻) ions of water and a mineral's ions.

Figure 6.5 The Solution of Halite
The asymmetric arrangement of hydrogen atoms causes a water molecule to have a slight positive electrical charge at its hydrogen end and a slight negative charge at its oxygen end. The mineral halite (NaCl) goes into solution because sodium atoms are attracted to the oxygen end of a water molecule, whereas chlorine atoms are attracted to the hydrogen end of the molecule.

The term **oxidation** has a variety of meanings for chemists, but in chemical weathering, it refers to reactions with oxygen to form an *oxide* (one or more metallic elements combined with oxygen) or, if water is present, a *hydroxide* (a metallic element or radical combined with OH⁻). For example, iron rusts when it combines with oxygen to form the iron oxide hematite:

$$4Fe \ + \ 3O_2 \ \longrightarrow \ 2Fe_2O_3$$
$$\text{iron} \qquad \text{oxygen} \qquad \text{iron oxide}$$
$$\text{(hematite)}$$

Atmospheric oxygen is abundantly available for oxidation reactions, but oxidation is a slow process unless water is present. Thus, most oxidation is carried out by oxygen dissolved in water.

Oxidation is important in the alteration of ferromagnesian silicates such as olivine, pyroxenes, amphiboles, and biotite. Iron in these minerals combines with oxygen to form the reddish iron oxide hematite (Fe_2O_3) or the yellowish or brown hydroxide limonite [$FeO(OH) \cdot nH_2O$]. The yellow, brown, and red colors of many soils and sedimentary rocks are caused by the presence of small amounts of hematite or limonite.

The chemical reaction between the hydrogen (H⁺) ions and hydroxyl (OH⁻) ions of water and a mineral's ions is known as **hydrolysis**. In hydrolysis, hydrogen ions actually replace positive ions in minerals. Such replacement changes the composition of minerals and liberates iron that then may be oxidized.

The chemical alteration of the potassium feldspar orthoclase provides a good example of hydrolysis. All feldspars are framework silicates, but when altered, they yield soluble salts and clay minerals, such as kaolinite, which are sheet silicates. The chemical weathering of orthoclase by hydrolysis occurs as follows:

$$2KAlSi_3O_8 \ + \ 2H^+ \ + \ 2HCO_3^- \ + \ H_2O \rightarrow$$
$$\text{orthoclase} \qquad \text{hydrogen} \quad \text{bicarbonate} \quad \text{water}$$
$$\text{ion} \qquad \text{ion}$$

$$Al_2Si_2O_5(OH)_4 \ + \ 2K^+ \ + \ 2HCO_3^- \ + \ 4SiO_2$$
$$\text{clay (kaolinite)} \quad \text{potassium} \quad \text{bicarbonate} \quad \text{silica}$$
$$\text{ion} \qquad \text{ion}$$

In this reaction, hydrogen ions attack the ions in the orthoclase structure, and some liberated ions are incorporated into a developing clay mineral. The potassium and bicarbonate ions go into solution and combine to form a soluble salt. On the right side of the equation is excess silica that would not fit into the crystal structure of the clay mineral.

The Rate of Chemical Weathering

Chemical weathering operates on the surfaces of particles, so it alters rocks and minerals from the outside inward, but the rate at which it proceeds depends on several factors. One is simply the presence or absence of fractures, because fluids seep along fractures and thus weathering is more intense here (Figure 6.1a).

Figure 6.6 Particle Size and Chemical Weathering

Surface area = 6 m²

1 m

1 m

Surface area = 12 m²

0.5 m

0.5 m

Surface area = 24 m²

0.25 m

0.25 m

© 2013 Cengage Learning

a. As a rock is divided into smaller particles, its surface area increases, but its volume of 1 m³ remains the same. The surface area is 6 m².

b. The surface area is 12 m².

c. The surface area is 24 m², but the volume remains the same at 1 m³. Small particles have more surface area in relation to their volume than do large particles.

> **spheroidal weathering** A type of chemical weathering in which corners and sharp edges of rocks weather more rapidly than flat surfaces, thus yielding spherical shapes.

Because chemical weathering affects particle surfaces, the greater the surface area, the more effective the weathering. It is important to realize that small particles have larger surface areas compared to their volume than do large particles. Notice in Figure 6.6 that a block measuring 1 m on a side has a total surface area of 6 m², but when the block is broken into particles measuring 0.5 m on a side, the total surface area increases to 12 m². And if these particles are all reduced to 0.25 m on a side, the total surface area increases to 24 m². Note that although the surface area in this example increases, the total volume remains the same at 1 m³.

We can conclude that mechanical weathering contributes to chemical weathering by yielding smaller particles with greater surface area compared to their volume. Actually, your own experiences with particle size verify our contention about surface area and volume. Because of its very small particle size, powdered sugar gives an intense burst of sweetness as the tiny pieces dissolve rapidly, but otherwise it is the same as the granular sugar we use on our cereal or in our coffee.

It is not surprising that chemical weathering is more effective in the tropics than in arid and arctic regions, because temperatures and rainfall are high, evaporation rates are low, and vegetation and animal life are much more abundant. Consequently, the effects of weathering extend to greater depths, perhaps several tens of meters.

Parent material is another control on the rate of chemical weathering, because some rocks are more resistant to chemical alteration than others. The metamorphic rock quartzite is an extremely stable substance that alters slowly compared to most other rocks. In contrast, basalt, which contains large amounts of calcium-rich plagioclase and pyroxene minerals, decomposes rapidly, because these minerals are chemically unstable. In fact, the stability of common minerals is just the opposite of their order of crystallization in Bowen's reaction series: The minerals that form last in this series are more stable, whereas those that form early are easily altered, because they are most out of equilibrium with their conditions of formation (see Figure 4.3).

> If you break open a weathered stone, you will see a rind of weathering at and near the surface, but the stone is completely unaltered inside.

In **spheroidal weathering**, a stone, even one that is rectangular to begin with, weathers to form a more spherical shape, because that is the most stable shape it can assume. The reason is that on a rectangular stone, the corners are attacked by weathering from three sides, and the edges are attacked from two sides, but the flat surfaces weather more or less uniformly (Figure 6.7). Consequently, the corners and edges alter more rapidly, the material sloughs off, and a more spherical shape develops.

Figure 6.7 Spheroidal Weathering

© 2013 Cengage Learning

b. Corners and edges weather most rapidly.

c. When the blocks are weathered so that they are nearly spherical, their surfaces weather evenly and no further change in shape takes place.

a. The rectangular blocks outlined by fractures are attacked by chemical weathering processes.

d. An exposure of granite showing spheroidal weathering in Joshua Tree National Park in California.

Sue Monroe

LO2 HOW DOES SOIL FORM AND DETERIORATE?

regolith The layer of unconsolidated rock and mineral fragments and soil that covers most of the land surface.

soil Weathered materials containing water, air, and humus that can support vegetation.

Much of Earth's land surface is covered by a layer of **regolith** consisting of sediment, pyroclastic materials, and the residue formed in place by weathering. Part of the regolith that contains air, water, and organic matter and supports vegetation is **soil**. Obviously, plants grow in soil, from which they receive most of their nutrients and water, and, of course, many land-dwelling animals depend directly or indirectly on soils.

A good soil for farming and gardening is made up of about 45% solid particles derived by weathering of parent material, and most of the rest of its volume consists of voids filled with air or water or both (Figure 6.8a). Another important constituent of soil is *humus*, which is carbon that forms by bacterial decay of organic matter and is very resistant to further decay. Even fertile soils may have as little as 5% humus, but it is essential as a source of plant nutrients, and it enhances a soil's capacity to retain moisture.

The solid particles in soils may be sand- and silt-sized mineral grains such as quartz, feldspars, and others, which hold soil particles apart and allow oxygen and water to circulate more freely. Clay minerals are also important, because they supply nutrients to plants and aid in soil-water retention. Should excess clay be present, though, a soil drains poorly, and is sticky when wet and hard when dry.

We characterize soils as *residual* or *transported* depending on whether they formed in place or the materials composing them were transported from the weathering site. For example, if a body of granite weathers, and the residue that accumulates over the granite is converted to soil, the soil so formed is residual. In contrast, transported soil forms if this same weathering residue is transported elsewhere, deposited, and then converted to soil.

The Soil Profile

Observed in vertical cross section, soil is made up of distinct layers, or **soil horizons**, that differ in texture, structure, composition, and color (Figure 6.8b). From the surface downward the soil horizons are designed O, A, E, B, and C, although the boundaries between horizons are transitional, and in some cases horizon E is not present. Because soil formation starts at the surface and works downward, horizon A is the most altered from the parent material.

Horizon O is only a few centimeters thick and is composed of organic matter (Figure 6.8b). Plant remains in various states of decomposition are clearly visible in the upper part of this horizon, but its lower part consists of humus. Indeed, the upper and lower parts of horizon O are sometimes referred to as O1 and O2, respectively.

Horizon A, also called *topsoil*, has more organic matter than the horizons below, and is characterized by intense biological activity because plant roots, fungi, bacteria, and worms are abundant (Figure 6.8b). In fact,

The Scent of Soil

Threadlike soil bacteria give freshly plowed soil its earthy aroma.

the earthy aroma of a freshly plowed soil comes from threadlike soil bacteria. In soils developed over a long time, horizon A is composed mostly of clays and chemically stable minerals such as quartz.

Below horizon A in some soils is a pale layer with little carbon from which much of the small particles have been removed. This horizon E, as it is called, is present in older, more mature soils, and results from *eluviation*, a process of leaching of minerals by downward-moving soil water. Some of this material is then deposited in the horizon below.

> **soil horizon** A distinct soil layer that differs from other soil layers in texture, structure, composition, and color.

Figure. 6.8 The Composition of Soil and Soil Horizons

Air 25%

Water 25%

Mineral Particles 45%

Organic Matter 5%

Organisms 10%

Roots 10%

Humus 80%

a. Soils are made up mostly of minerals and rock fragments derived by weathering, air, water, and organic matter. Most of the organic mater is humus.

O — Loose leaves and organic debris
Partly decomposed organic debris

A — Topsoil; dark in color; rich in organic matter

E — Zone of intense leaching or eluviation

B — Subsoil, zone of accumulation

Transition to C

C — Partly weathered parent material

Parent material

b. The soil horizons in a fully developed soil. Horizon O is only a few centimeters thick, but it has been exaggerated here to show some detail.

Horizon B, or *subsoil*, has fewer organisms and less organic matter than horizon A (Figure 6.8b). This horizon is also called the *zone of accumulation* because soluble materials leached from above accumulate as irregular masses. If horizon A is eroded, leaving horizon B exposed, plants do not grow as well, and if it is clayey, it is harder when dry and stickier when wet than the other soil horizons.

Partially altered bedrock grading down into unaltered bedrock with little organic matter characterizes horizon C (Figure 6.8b). In the horizons above C, the parent material has been so thoroughly altered that it is no longer recognizable, but in horizon C minerals and rock fragments of parent material are easy to identify.

It is worth noting that soils are subdivided, classified, and mapped on the basis of the development and composition of the various soil horizons. The Soil Survey Division of the Natural Resources Conservation Service (the NRCS) divides soils into 12 soil orders, and further subdivides them into small groupings. The 12 soil orders are based on the interaction of such features and processes as parent material, vegetation, and climate. It is beyond the scope of this chapter to go into detail about the 12 soil orders, but information about them is available from the NRCS, a branch of the U.S. Department of Agriculture.

Factors That Control Soil Formation

Complex interactions among several factors account for soil type, thickness, and fertility, but climate is the single most important factor. For example, soils that form in rather humid regions have most of their soluble minerals leached out of horizon A, which may be gray but is commonly dark because of abundant organic matter. Aluminum-rich clays and iron oxides tend to accumulate in horizon B. Soils that form in semi-arid to arid regions have much less organic matter and more unstable minerals in horizon A because so little water is available to leach them out. However, soluble minerals such as calcite ($CaCO_3$) are taken into solution and precipitate in horizon B as irregular masses of *caliche*. Precipitation of sodium salts in some arid-region soils where soil water evaporates yields *alkali soils* that are so alkaline that they support little or no vegetation (Figure 6.9a).

In the tropics, where chemical weathering is intense and leaching of most minerals is complete, a soil termed **laterite** forms (Figure 6.9b). These red soils extend to depths of many meters and are composed largely of aluminum hydroxides, iron oxides, and clay minerals. Laterite supports lush vegetation but is not very fertile because most plant nutrients have been leached out; the vegetation depends mostly on the surface layer of organic matter. In fact, when laterite is cleared of its vegetation and planted for crops, it can sustain farming for only a few years until the soil is depleted, in which case native farmers simply clear another area and repeat the process.

The same rock type can yield different soils in different climatic regimes, and in the same climatic regime the same soils can develop on different rock types. Thus, climate is more important than parent material in determining the type of soil. Nevertheless, rock type

Figure 6.9 Alkali Soil and Laterite

a. An alkali soil near Fallon, Nevada. The white material is sodium carbonate or potassium carbonate. Notice that only a few hardy plants grow in this soil.

b. Laterite, shown here in Madagascar, in a deep red soil that forms in response to intense chemical weathering in the tropics.

does exert some control. For example, the metamorphic rock quartzite will have a thin soil over it, because it is chemically stable, whereas an adjacent body of granite will have a much deeper soil.

Soils depend on organisms for their fertility, and in return they provide a suitable habitat for many organisms. Earthworms, ants, sowbugs, termites, centipedes, millipedes, and nematodes, along with fungi, algae, and single-celled organisms, make their homes in soil. All contribute to soil formation and provide humus when they die and decompose.

Much of the humus in soils comes from grasses or leaf litter that microorganisms decompose to obtain food. In so doing, they break down organic compounds in plants and release nutrients back into the soil. In addition, organic acids from decaying soil organisms are important in further weathering of parent materials and soil particles. Burrowing animals constantly churn and mix soils, and their burrows provide avenues for gases and water. Soil organisms, especially some types of bacteria, are extremely important in changing atmospheric nitrogen into a form of soil nitrogen suitable for use by plants.

The difference in elevation between high and low points in a region is called *relief*. Because climate is such an important factor in soil formation and climate changes with elevation, areas with considerable relief have different soils in mountains and adjacent lowlands. *Slope*, another important control, influences soil formation in two ways. One is *slope angle*; steep slopes have little or no soil, because weathered materials erode faster than soil-forming processes operate. The other factor is *slope direction*. In the Northern Hemisphere, north-facing slopes receive less sunlight than do south-facing slopes and have cooler internal temperatures, support different vegetation, and, if in a cold climate, remain snow covered or frozen longer.

How much time is needed to develop a centimeter of soil or a fully developed soil a meter or so deep? We can give no definite answer, because weathering proceeds at vastly different rates depending on climate and parent material, but an overall average might be about 2.5 cm per century. Nevertheless, a lava flow a few centuries old in Hawaii may have a well-developed soil on it, whereas a flow the same age in Iceland will have considerably less soil.

Soil Degradation

From the human perspective, soil forms so slowly that it is a nonrenewable resource. So any soil losses that exceed the rate of soil formation are viewed with alarm. Likewise, any reduction in soil fertility or production is cause for concern. Any process that removes soil or makes it less productive is defined as **soil degradation**, a serious problem that includes erosion, chemical deterioration, and physical changes.

> **soil degradation** Any process leading to a loss of soil productivity; may involve erosion, chemical pollution, or compaction.

Erosion is a natural process, but it is usually slow enough for soil formation to keep pace. Unfortunately, some human practices add to the problem. Removing natural vegetation by plowing, overgrazing, overexploitation for firewood, and deforestation all contribute to erosion by wind and running water. The Dust Bowl that developed in several U.S. Great Plains states during the 1930s is a poignant example of just how effective wind erosion is on soil that has been pulverized and exposed by plowing (Figure 6.10a).

Wind has caused considerable soil erosion in some areas, but running water is much more powerful. Some soil is removed by *sheet erosion*, which involves the removal of thin layers of soil more or less evenly over a broad, sloping surface. *Rill erosion*, in contrast, takes place when running water scours small, troughlike channels. Channels shallow enough to be eliminated by plowing are *rills*, but those too deep (about 30 cm) to be plowed over are *gullies* (Figure 6.10b, c). Where gullying is extensive, croplands can no longer be tilled and must be abandoned.

Soil undergoes chemical deterioration when its nutrients are depleted and its productivity decreases. Loss of soil nutrients is most notable in many of the populous developing nations, where soils are overused to maintain high levels of agricultural productivity. Chemical deterioration is also caused by insufficient use of fertilizers and by clearing soils of their natural vegetation.

Other types of chemical deterioration are pollution and *salinization*, which occurs when the concentration of salts increases in a soil, making it unfit for agriculture. Improper disposal of domestic and industrial wastes, oil and chemical spills, and the concentration of insecticides and pesticides in soils all cause pollution. Soil deteriorates physically when it is compacted by the weight of heavy machinery and livestock, especially cattle. Compacted soils are more costly to plow, and plants have a more difficult time emerging from them. Furthermore, water does not readily infiltrate, so

Figure 6.10 Soil Degradation Resulting from Erosion

a. The Dust Bowl of the 1930s was a time of drought and dust storms resulting in wind erosion of soil in Kansas, Oklahoma, Texas, New Mexico, and Colorado. This huge dust storm was photographed in 1934 at Lamar, Colorado.

James S. Monroe

b. Rill erosion in a field in Michigan during a rainstorm. The rill was later plowed over.

c. This gully in New Mexico is about 3 m deep.

James S. Monroe

more runoff occurs; this in turn accelerates the rate of water erosion.

Problems experienced in the past have stimulated the development of methods to minimize soil erosion on agricultural lands. Crop rotation, contour plowing, and terracing have all proved helpful. So has *no-till planting*, in which the residue from the harvested crop is left on the ground to protect the surface from the ravages of wind and water.

LO3 WEATHERING AND RESOURCES

Soils are certainly one of our most precious natural resources. Other aspects of soils are also important economically. We discussed the origin of laterite in response to intense chemical weathering in the tropics (Figure 6.9b). Laterite is not very productive for agriculture, but if the parent material is rich in aluminum, the ore of aluminum called *bauxite* accumulates in horizon B. Some bauxite is found in Arkansas, Alabama, and Georgia, but at present, it is cheaper to import rather than mine these deposits, so both the United States and Canada depend on foreign sources of aluminum ore.

Bauxite and other accumulations of valuable minerals form by the selective removal of soluble substances during chemical weathering and are known as *residual concentrations*. Bauxite is a good example, but other deposits that formed in a similar manner are those rich in iron, manganese, clays, nickel, phosphate, tin, diamonds, and gold. Some of the sedimentary iron deposits in the Lake Superior region of the United States and Canada were enriched by chemical weathering when soluble parts of the deposits were carried away. Some kaolinite deposits in the southern United States formed when chemical weathering altered feldspars in pegmatites or as residual concentrations of clay-rich limestones and dolostones.

LO4 SEDIMENT AND SEDIMENTARY ROCKS

Weathering is fundamental to the origin of **sediment** and **sedimentary rocks**, and so are *erosion* and *deposition*—that is, the movement of sediment from the weathering site by processes such as running water, wind, and glaciers, and its accumulation in some area (Figure 6.11). One important criterion for classifying detrital sediment is particle size. Particles described as *gravel* measure more than 2 mm, and *sand* measures between 1/16 and 2 mm. *Silt* applies to any particle from 1/256 to 1/16 mm. None of these designations imply anything about composition; most gravel is made up of rock fragments—that is, small pieces of granite, basalt, or any other type of rock—but most sand and silt grains are single minerals. Particles smaller than 1/256 mm are termed *clay*, but clay has two meanings. One is size designation, but the term also refers to certain sheet silicates called *clay minerals*. However, most clay minerals are also clay sized. Mixtures of silt and clay are called *mud*.

Remember that *detrital sediment* is solid particles derived from other rocks. *Chemical sediment*, in contrast, comes from chemicals that were derived from other rocks that were then extracted from solution by inorganic chemical processes, such as evaporating seawater, or by the activities of organisms. Clams, oysters, corals, and some plants make their skeletons of minerals, especially aragonite or calcite ($CaCO_3$) or silica (SiO_2). In any case, minerals form that may be converted into sedimentary rock.

All detrital sediment is transported some distance from its source, but chemical sediment forms in the area where it is deposited. During transport of detrital sediment, sand and gravel particles collide, and

> **sediment** Loose aggregate of solids derived by weathering from preexisting rocks, or solids precipitated from solution by inorganic chemical processes or extracted from solution by organisms.
>
> **sedimentary rock** Any rock composed of sediment, such as limestone and sandstone.

SOURCE OF SEDIMENTARY MATERIALS

© 2013 Cengage Learning

MECHANICAL WEATHERING
(gravel, sand, silt, clay–sized particles)

TRANSPORT

CHEMICAL WEATHERING
(clay minerals and ions, compounds in solution)

TO SITES OF DEPOSITION

Figure 6.11 Origin and Transport of Sediment
Whether derived from preexisting rocks by mechanical or chemical weathering, solid particles and ions and compounds in solution are transported and deposited elsewhere. If they are lithified, they become detrital and chemical sedimentary rocks.

depositional environment Any site such as a floodplain or beach where physical, biologic, and chemical processes yield a distinctive kind of sedimentary deposit.

lithification The process of converting sediment into sedimentary rock by compaction and cementation.

abrasion wears away the sharp corners and edges, a process called *rounding* (Figure 6.12). Transport also results in *sorting*, which refers to the size distribution of particles in sediment or sedimentary rocks, which are poorly sorted if a wide range of size is present and well sorted if all the particles are about the same size (Figure 6.12). Rounding

and sorting may seem unimportant, but both influence how readily groundwater, petroleum, and gas move through sediments and sedimentary rocks, which is essential to our efforts to recover these materials. They are also useful for determining how sediment deposition occurred, a topic covered more fully in a later section.

Regardless of how detrital sediment is transported or how chemical sediment forms, it is eventually deposited in some geographic area known as a **depositional environment** (Figure 6.13). Deposition may take place in a stream channel or on its floodplain, in a lake, on a beach, or on the deep seafloor where physical and biological processes impart distinctive characteristics to the accumulating sediment. The three broad depositional settings geologists recognize are *continental* (on the land), *transitional* (on or near seashores), and *marine* (in the seas), each with several specific depositional environments (Figure 6.13).

Figure 6.12 Rounding and Sorting in Sediments

a. Beginning students mistake rounding to mean ball-shaped or spherical. These three stones are all rounded, but only the one at the upper left is spherical, and none are ball-shaped.

Sue Monroe

b. Deposit of very poorly sorted but well rounded gravel.

How Does Sediment Become Sedimentary Rock?

Mud in lakes and sand and gravel in stream channels or on beaches are good examples of sediment. To convert these aggregates of particles into sedimentary rocks requires **lithification** by compaction, cementation, or both (Figure 6.14).

To illustrate the relative importance of compaction and cementation, consider detrital deposits of mud and sand. In both cases, the sediment consists of solid particles and *pore spaces*, the voids between

c. Angular, poorly sorted gravel. Note the quarter for scale.

James S. Monroe

Figure 6.13 Depositional Environments
Continental environments are shown in red type. The environments along the shoreline, shown in blue type, are transitional between continental and marine. The others, shown in black type, are marine environments.

particles. These deposits are subjected to **compaction** from their own weight and the weight of any additional sediment deposited on top of them, thereby reducing the amount of pore space and the volume of the deposit. Our hypothetical mud deposit may have 80% water-filled pore space, but after compaction, its volume is reduced by as much as 40% (Figure 6.14). The sand deposit, with as much as 50% pore space, is also compacted, so that the grains fit more tightly together (Figure 6.14).

Compaction alone is sufficient for lithification of mud, but for sand and gravel, **cementation** involving the precipitation of minerals in pore spaces is also necessary. The two most common chemical cements are calcium carbonate ($CaCO_3$) and silicon dioxide (SiO_2), but iron oxide and hydroxide cement, such as hematite (Fe_2O_3) and limonite [$FeO(OH) \cdot nH_2O$], are found in some sedimentary rocks. Recall that calcium carbonate readily dissolves in water that contains a small amount of carbonic acid, and chemical weathering of feldspars and other minerals yields silica in solution. Cementation takes place when minerals precipitate in the pore spaces of sediment from circulating water, thereby binding the loose particles together. Iron oxide and hydroxide cements account for the red, yellow, and brown sedimentary rocks found in many areas (see the chapter-opening photograph).

By far the most common chemical sediments are calcium carbonate mud and sand- and gravel-sized accumulations of calcium carbonate grains, such as shells and shell fragments. Compaction and cementation also take place in these sediments, converting them into limestone, but compaction is less effective, because cementation takes place soon after deposition. In any case, the cement is calcium carbonate derived by partial solution of some of the particles in the deposit.

compaction Reduction in the volume of a sedimentary deposit that results from its own weight and the weight of any additional sediment deposited on top of it.

cementation The process whereby minerals crystallize in the pore spaces of sediment and bind the loose particles together.

LO5 TYPES OF SEDIMENTARY ROCKS

Thus far, we have considered the origin of sediment, as well as its transport, deposition, and lithification. We now turn to the types of sedimentary rocks and how they are classified. The two broad classes or types of sedimentary rocks are *detrital* and *chemical*, although the latter has a subcategory known as *biochemical* (Table 6.1).

Detrital Sedimentary Rocks

Detrital sedimentary rocks are made up of solid particles such as gravel, sand, silt, and clay, and all have a *clastic texture*, meaning they are composed of particles or fragments known as *clasts*. The several varieties of detrital rocks are classified by the size of their constituent particles, although composition is used to modify some rock names.

Both *conglomerate* and *sedimentary breccia* are composed of gravel-sized particles (Figure 6.14 and Figure 6.15a, b), but conglomerate has rounded gravel, whereas sedimentary breccia has angular gravel. Conglomerate is common, but sedimentary breccia is rare, because gravel becomes rounded very quickly during transport. Considerable energy is needed to transport

gravel, so conglomerate is usually found in high-energy environments such as stream channels and beaches.

Sand is a size designation for particles between 1/16 mm and 2 mm, so any mineral or rock fragment can be in *sandstone*. Geologists recognize varieties of sandstone based on mineral content (Figure 6.14, 6.15a). *Quartz sandstone* is the most common and, as the name implies, is made up mostly of quartz sand. Another variety of sandstone, called *arkose*, contains at least 25% feldspar minerals. Sandstone is found in many depositional environments, including stream channels, sand dunes, beaches, barrier islands, deltas, and the continental shelf.

Mudrock is a general term that includes all detrital sedimentary rocks composed of silt- and clay-sized particles (Figure 6.14). Varieties include *siltstone* (mostly silt-sized particles), *mudstone* (a mixture of silt and clay), and *claystone* (primarily clay-sized particles). Some mudstones and claystones are designated *shale* if they are fissile, meaning that they break along closely spaced parallel planes (Figure 6.15c). Even weak currents transport silt- and clay-sized particles, and deposition takes place only where currents and fluid

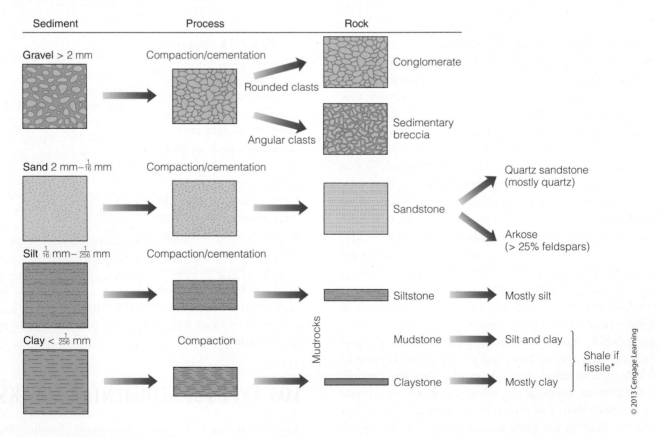

*Fissile refers to rocks that split along closely spaced planes.

Figure 6.14 Lithification and Classification of Detrital Sedimentary Rocks
Notice that little compaction takes place in gravel and sand.

© 2013 Cengage Learning

Table 6.1

Classification of Chemical and Biochemical Sedimentary Rocks

CHEMICAL SEDIMENTARY ROCKS		
TEXTURE	**COMPOSITION**	**ROCK NAME**
Varies	Calcite ($CaCO_3$)	Limestone } Carbonate rocks
Varies	Dolomite [$CaMg(CO_3)_2$]	Dolostone
Crystalline	Gypsum ($CaSO_4 \cdot 2H_2O$)	Rock gypsum } Evaporites
Crystalline	Halite ($NaCl$)	Rock salt
BIOCHEMICAL SEDIMENTARY ROCKS		
Clastic	Calcite ($CaCO_3$) shells	Limestone (various types, such as chalk and coquina)
Usually crystalline	Altered microscopic shells of SiO_2	Chert (various color varieties)
Amorphous	Carbon from altered land plants	Coal (lignite, bituminous, anthracite)

© 2013 Cengage Learning

turbulence are minimal, as in the quiet offshore waters of lakes or in lagoons.

Chemical and Biochemical Sedimentary Rocks

During chemical weathering several compounds and ions go into solution and provide the raw materials for chemical and biochemical sedimentary rocks. For example, seawater contains silica (SiO_2), calcium (Ca), carbonate (CO_3), sulfate (SO_4), potassium (K), sodium (Na), and chlorine (Cl), and many other substances that under certain conditions are extracted from the water to form minerals that make up **chemical sedimentary rocks**. Organisms play an important role in the origin of some of these rocks, which are designated **biochemical sedimentary rocks**. Some chemical sedimentary rocks have a crystalline texture, meaning that they are made up of a mosaic of interlocking mineral crystals as in rock salt. Others have a clastic texture as in some limestones composed of fragmented shells.

By far the most common chemical sedimentary rocks are the **carbonates**, so-called because they contain the carbonate radical (CO_3^{-2}). Several rocks meet this criterion but only two are common; limestone, composed of calcite ($CaCO_3$) and dolostone composed of dolomite [$CaMg(CO_3)_2$] (Table 6.1). The origin of limestone is fairly straightforward. Recall that calcite in the presence of acidic groundwater rapidly goes into solution, but the chemical reaction leading to solution is reversible, so under some conditions calcite can precipitate from solution. Thus, some limestone forms by inorganic chemical precipitation from seawater, more rarely lake water.

Most limestone is biochemical because organisms are so important in its origin. Indeed, the skeletons of sea-dwelling animals are common in many varieties of limestone (Figure 6.16). *Coquina* is a type of limestone composed almost entirely of fragmented seashells (Figure 6.16b), and *chalk* is a soft type of limestone that consists of microscopic shells. One distinctive variety of limestone contains small spherical grains called *ooids* that have a small nucleus around which concentric layers of calcite precipitate. Lithified deposits of ooids form *oolitic limestone* (Figure 6.16c).

Dolostone is similar to limestone, but it forms mostly by the alteration of limestone when magnesium replaces some of the calcium in calcite, thereby converting it to dolomite. This may take place in a lagoon where evaporation of seawater takes place, enriching the remaining seawater in magnesium that permeates limestone and brings about a chemical change.

Most of you are aware that seawater is salty—that is, it contains sodium and chlorine and several other compounds and ions in solution. If you were to take a glass of seawater and let it evaporate completely, you would find a layer of minerals on the bottom of the glass. Obviously evaporation is involved in the origin of these minerals and their corresponding rocks, which are collectively

chemical sedimentary rock Sedimentary rock made up of minerals that were dissolved during chemical weathering and later precipitated from seawater, more rarely lake water, or extracted from solution by organisms.

biochemical sedimentary rock Any sedimentary rock produced by the chemical activities of organisms.

carbonate rock Any rock, such as limestone and dolostone, made up mostly of carbonate minerals.

called **evaporites** (Table 6.1). The most familiar evaporites are *rock salt*, composed of the mineral halite (NaCl), and *rock gypsum*, made up of the mineral gypsum $(CaSO_4 \cdot 2H_2O)$ (Figure 6.17a, b). Compared with sandstone, mudrocks, and limestone, evaporites are not very common, and yet they are significant deposits in areas such as Michigan, Ohio, New York and the Gulf Coast region.

Chert is hard rock consisting of microscopic crystals of silica (SiO_2) (Table 6.1, Figure 6.17c). Perhaps you have heard of *flint*, which is simply chert colored black by inclusions of organic matter, or *jasper*, which is red or brown chert because of its iron oxide content. Because chert is hard and lacks cleavage, it can

be shaped to form sharp cutting edges for tools, spear points, and arrowheads. Some chert is found as irregular masses in other rocks, especially in limestone, whereas other chert is in distinct layers of *bedded chert* made up of tiny shells of silica-secreting organisms and thus is biochemical (Figure 6.17c).

We previously mentioned that *coal* consists of altered organic matter but is nevertheless a biochemical sedimentary rock (Table 6.1, Figure 6.17d). It forms from vegetation that accumulates in bogs and swamps where the water is deficient in oxygen. Bacteria that decompose vegetation can live without oxygen, but their wastes must be oxidized, and because little or no oxygen is present, wastes build up and kill the bacteria. Decomposition ceases and the vegetation forms organic muck, which if buried and compressed becomes *peat*, the first step in forming coal.

Where peat is abundant, as in Ireland and Scotland, it is used for fuel, but if it is altered further by deeper

Figure 6.15 Detrital Sedimentary Rocks

a. Sandstone overlain by conglomerate at Lillooet, British Columbia, Canada.

James S. Monroe

b. Sedimentary breccia in Death Valley, California. Notice the angular gravel-sized particles. The largest clast is about 12 cm across.

Sue Monroe

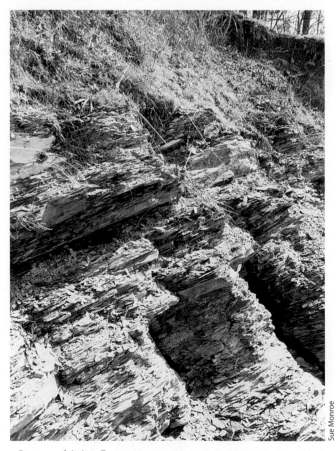

c. Exposure of shale in Tennessee.

Sue Monroe

Figure 6.16 Chemical Sedimentary Rocks—Limestone

a. Limestone with numerous fossil shells is called fossiliferous limestone.

b. Coquina is limestone made up entirely of broken shells.

c. This oolitic limestone is made up partly of ooids (see inset), which are rather spherical grains of calcium carbonate.

burial, and especially if it is heated, too, it becomes dull black coal called *lignite*. During the change from peat to lignite, the volatile elements are driven off, increasing the amount of carbon; peat has about 50% carbon whereas about 70% is present in lignite. Further changes yield *bituminous coal*, with about 80% carbon, which is dense, black, and so thoroughly altered that plant remains are rarely seen. The highest grade of coal is *anthracite*, a metamorphic type of coal with up to 98% carbon.

LO6 SEDIMENTARY FACIES

Long ago, geologists realized that when they traced layers of sediment or sedimentary rock laterally, they saw

that the layers changed in composition, texture, or both. They concluded that these changes resulted from the simultaneous operation of different processes in adjacent depositional environments. For example, sand may be deposited in a high-energy nearshore marine environment, whereas mud and carbonate sediments accumulate simultaneously in the laterally adjacent low-energy offshore environments (Figure 6.18). Deposition in each environment produces **sedimentary facies**, bodies of sediment that possess distinctive physical, chemical, and biological attributes.

sedimentary facies Any aspect of a sedimentary rock unit that makes it recognizably different from adjacent sedimentary rocks of the same or approximately the same age.

Figure 6.17 Evaporites, Chert, and Coal

a. This cylindrical core of rock salt was taken from an oil well in Michigan.

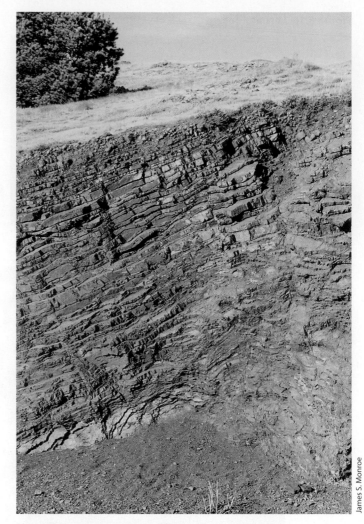

c. Bedded chert exposed in Marin County, California. Most of the layers are about 5 cm thick.

b. Rock gypsum. When deeply buried, gypsum ($CaSO_4 \cdot 2H_2O$) loses its water and is converted to anhydrite ($CaSO_4$).

d. Bituminous coal is the most common type of coal used for fuel.

Many sedimentary rocks in the interiors of continents show clear evidence of deposition in marine environments. The rock layers in Figure 6.18 (*left*), for example, consist of a sandstone facies that was deposited in a nearshore marine environment overlain by shale and limestone facies deposited in offshore environments. Geologists explain this vertical sequence of facies by deposition occurring during a time when sea level rose with respect to the continents. As sea level rises, the shoreline moves inland, giving rise to a **marine transgression** (Figure 6.18), and the depositional environments parallel to the shoreline migrate landward. As a result, offshore facies are superimposed over nearshore facies, thus accounting for the vertical succession of sedimentary facies. Even though the nearshore environment is long and narrow at any particular time, deposition takes place continuously as the environment migrates landward. The sand deposit may be tens to hundreds of meters thick but has horizontal dimensions of length and width measured in hundreds of kilometers.

The opposite of a marine transgression is a **marine regression** (Figure 6.18). If sea level falls with respect to a continent, the shoreline and environments that parallel the shoreline move seaward. The vertical sequence produced by a marine regression has facies of the nearshore environment superposed over facies of offshore environments.

marine transgression
The invasion of a coastal area or a continent by the sea, resulting from a rise in sea level or subsidence of the land.

marine regression
The withdrawal of the sea from a continent or coastal area, resulting in the emergence of the land as sea level falls or the land rises with respect to sea level.

Figure 6.18 Marine Transgressions and Regressions
Notice that the sandstone, shale, and limestone facies are deposited simultaneously in adjacent environments. Also note the vertical succession of facies that result from marine transgressions and regressions.

sedimentary structure Any feature in sedimentary rock that formed at or shortly after the time of deposition, such as cross-bedding, animal burrows, and mud cracks.

strata (singular, **stratum**) Refers to layering in sedimentary rocks.

bed An individual layer of rock, especially sediment or sedimentary rock.

cross-bedding A type of bedding in which layers are deposited at an angle to the surface on which they accumulate, as in sand dunes.

graded bedding Sedimentary layer in which a single bed shows a decrease in grain size from bottom to top.

ripple mark Wavelike (undulating) structure produced in granular sediment, especially sand, by unidirectional wind and water currents or by oscillating wave currents.

mud crack A crack in clay-rich sediment that forms in response to drying and shrinkage.

LO7 READING THE STORY PRESERVED IN SEDIMENTARY ROCKS

No one was present when ancient sediments were deposited, so geologists must evaluate those aspects of sedimentary rocks that allow them to make inferences about the original depositional environment. Sedimentary textures such as sorting and rounding can give clues to depositional processes. Windblown dune sands tend to be well sorted and well rounded, but poor sorting is typical of glacial deposits. The geometry, or three-dimensional shape, is another important aspect of sedimentary rock bodies.

Marine transgressions and regressions yield sediment bodies with a blanket or sheet-like geometry, but sand deposits in stream channels are long and narrow and are described as having a *shoestring* geometry. Sedimentary textures and geometry alone are usually insufficient to determine depositional environment, but when considered with other sedimentary rock properties, especially *sedimentary structures* and *fossils*, geologists can reliably determine the history of a deposit.

Sedimentary Structures

Physical and biological processes operating in depositional environments are responsible for features known as **sedimentary structures**. One of the most common consists of distinct layers known as **strata** or **beds** (Figure 6.19a), with individual layers of less than a millimeter up to many meters thick. These strata or beds are separated from one another by surfaces above and below in which the rocks differ in composition, texture, color, or a combination of features.

Many sedimentary rocks have **cross-bedding**, in which layers are arranged at an angle to the surface on which they were deposited (Figure 6.19b and c). Cross-beds are found in many depositional environments, such as sand dunes in deserts and along shorelines, as well as in stream-channel deposits. Invariably, cross-beds result from transport and deposition by wind or water currents, and the cross-beds are inclined downward in the same direction that the current flowed. Thus, ancient deposits with cross-beds inclined down toward the south, for example, indicate that the currents responsible for them flowed from north to south.

Some individual sedimentary rock layers show an upward decrease in grain size, termed **graded bedding**, mostly formed by turbidity current deposition. A *turbidity current* is an underwater flow of sediment and water with a greater density than sediment-free water. Because of its greater density, a turbidity current flows downslope until it reaches the relatively flat seafloor, or lake floor, where it slows and begins depositing large particles followed by progressively smaller ones (Figure 6.20).

The surfaces that separate layers in sand deposits commonly have **ripple marks**, small ridges with intervening troughs, giving them a corrugated appearance. Some ripple marks are asymmetrical in cross section, with a gentle slope on one side and a steep slope on the other. Currents that flow in one direction, as in stream channels, generate these so-called *current ripple marks* (Figure 6.21a, b). Because the steep slope of these ripples is on the downstream side, they are good indications of ancient current directions. In contrast, *wave-formed ripple marks* tend to be symmetrical in cross section and, as their name implies, are generated by the to-and-fro motion of waves (Figure 6.21c, d and the Virtual Field Trip in this chapter).

When clay-rich sediment dries, it shrinks and develops intersecting fractures called **mud cracks** (Figure 6.22). Mud cracks in ancient sedimentary rocks indicate that the sediment was deposited in an environment where periodic drying took place, such as on a river floodplain, near a lakeshore, or where muddy deposits are exposed along seacoasts at low tide.

Borisvian/Shutterstock.com

Figure 6.19 Bedding (Stratification) and Cross-Bedding

a. These rocks in Utah clearly show bedding, or stratification.

Wind or current direction

Sand moves up this slope

Deposition

© 2013 Cengage Learning

b. Origin of cross-bedding by deposition on the sloping surface of a desert dune. Cross-bedding is also common in dunelike structures in stream and river channels.

c. Cross-bedding in sandstone at Natural Bridges National Monument in Utah. The current moved from left to right.

Fossils: Remains and Traces of Ancient Life

Fossils, the remains or traces of once-living organisms, are interesting as evidence of prehistoric life (Figure 6.23) and are also important for determining depositional environments. The actual remains of organisms are known as *body fossils*, whereas any indication of organic activity, such as tracks and trails, are *trace fossils*. Most people are familiar with fossils of dinosaurs and some other land-dwelling animals but are unaware that fossils of invertebrates—animals lacking a segmented vertebral column, such as corals, clams, oysters, and a variety of microorganisms—are much more useful because they are so common.

It is true that the remains of land-dwelling creatures and plants can be washed into marine environments, but most are preserved in rocks deposited on land or perhaps transitional environments such as deltas. In contrast, fossils of corals tell us that the rocks in which they are preserved were deposited in the ocean.

Microfossils are particularly useful for environmental interpretations because hundreds or even thousands are recovered from small rock samples. When drilling for oil, geologists recover small rock chips called *well cuttings* that may contain numerous fossils of tiny organisms. These fossils are routinely used to determine depositional environments, and to match up or correlate rocks of the same relative age in different areas. In addition, fossils provide some of the evidence for organic evolution.

> **fossil** The remains or traces of once-living organisms.

Virtual Field Trip

Sedimentary Rocks

Ready to Go!

Now that you've read some of the key facts about sedimentary rocks, we are going to go "into the field" and see some up close. During this field trip, we will be visiting two of Utah's five national parks. Arches National Park and Capitol Reef National Park are geologically rich, making them ideal settings for learning about sedimentary rocks, including examining the effects of weathering, different types of sedimentary structures, dating strata by sight, and much more.

This field trip is designed to help you think about these fascinating places from a geologic perspective. Several types of sedimentary rocks are present in Arches National Park and Capitol Reef National Park, but most of them are sandstones. Here we will learn about the rocks' composition, textures, color, structures, and how they were deposited—all topics we talk about in this chapter, under the section titled "Sediment and Sedimentary Rocks." In addition, the parks have a fantastic variety of landforms such as windows, columns, pinnacles, and balanced rocks (which we learned about in the earlier pages of this chapter), all of which resulted from differential weathering and erosion. For an example of this phenomenon, revisit the chapter opening photo, which shows differential weathering and erosion in Red Canyon, Utah.

In addition to being great examples of sedimentary rocks, the rock exposures in these parks help you visualize some of the principles of deciphering geologic history, a topic that is discussed in greater depth in Chapter 16.

Onward!

DeerPoint, UT. topographic map, courtesy of USGS.
All photographs: Copyright and Photograph by Dr. Parvinder S. Sethi.

GOALS OF THE TRIP

Some of the goals of this field trip are to understand the following key concepts:

1. Strata (layered beds) are obvious at many rock exposures (outcrops) in both parks (see the photo to the left). Each change in the strata indicates a change in the deposition of successive layers of sediment.

2. In a succession of sedimentary rock layers the oldest layer is at the bottom with progressively younger layers upward. This is formalized as the "Principle of Superposition" which we further discuss in the section titled "Relative Dating Methods" in Chapter 16.

3. Most of the rocks are sandstones so we can reasonably infer that they are composed mostly of quartz because it is abundant, physically durable, and chemically stable. The sandstone strata are red because they contain iron oxide or hydroxide cement.

FOLLOW-UP QUESTIONS

1. How do windows form by differential weathering and erosion? Are there examples in the park of windows that are in their earliest stages of formation? If so, which one(s)?

2. The rocks in Arches and Capitol Reef have been only minimally deformed. Suppose that some of them were crumpled and fractured so that now some of the layers were upside down. What kinds of sedimentary structures would help you figure out if some layers had in fact been flipped over?

3. How can you use the Principles of Superposition and Original Horizontality to help you decipher the geologic history of Arches and Capitol Reef National Parks?

What to See When You Go

There is a lot to see when you visit these parks in Virtual Field Trips, but here are some of the highlights. . .

The photo above shows great examples of pillars, columns, and arches, all of which formed by differential weathering and erosion of the Entrada Sandstone. Also notice in the above photo that in addition to the Entrada Sandstone, the Navajo Sandstone is also present. Consider how it differs from the Entrada Sandstone, and in what environment It may have been deposited.

Sedimentary rocks possess several types of sedimentary structures that formed when the sediment was deposited or shortly thereafter and provide evidence about the environment of deposition. These are wave-formed ripple marks much like those shown in Figure 6.21c, d in this chapter.

Perhaps the most important observations you can make on a trip to these parks are those that illustrate that Earth is a complex system of interacting rocks, water, air, and life. It is also important to realize that rocks of any kind record events that happened in the far distant past. We discuss the methods geologists use to interpret geologic history in Chapter 16, but even now you can determine which of the rock layers in the above photos is oldest and youngest. You simply have to note where the rock layers are in a sequence of layers, because the oldest is at the bottom while the youngest is at the top. Also, note in the above photo that one layer is gray rather than red. The gray layer is made up of bentonite (a clay mineral), which indicates ancient volcanoes erupted large amounts of volcanic ash.

For the rocks in the photo above, the caption in the virtual field trip notes that the rocks from the bottom upward were deposited in tidal flat environments, slightly deeper water, and sand dunes. Geologists made these interpretations based on the features of the rocks, especially rock textures (sorting and rounding) and especially sedimentary structures. (See the section titled "Determining the Environment of Deposition" in this chapter.) Note that these rocks are records of events that took place in this area long before any humans were present to make records of their own.

Figure 6.20 Turbidity Currents and the Origin of Graded Bedding

a. A turbidity current flows downslope along the seafloor (or a lake bottom) because it is denser than sediment-free water.

b. The flow slows and deposits progressively smaller particles, thus forming a graded bed.

Figure 6.21 Current and Wave-Formed Ripple Marks

a. Current ripple marks form in response to currents that flow in one direction, as in a stream. The enlargement shows the cross-beds in an individual ripple.

c. The to-and-fro motion of waves in shallow water yield wave-formed ripple marks.

b. Current ripple marks that formed in a stream channel. Flow was from right to left.

d. Wave-formed ripple marks in sand in shallow seawater.

Determining the Environment of Deposition

What kinds of evidence would allow you to determine how a layer of sandstone was deposited? Certainly you would consider texture—that is, rounding and sorting—and also the kinds of sedimentary structures and fossils, if any. You might also compare features in the sandstone with those seen in sand deposits forming today. But are you justified in using present-day processes and deposits to infer what happened when no human observers were present (see the Virtual Field Trip in this chapter)?

Actually, you are familiar with the reasoning used to interpret events you did not witness. Skid marks on a

Figure 6.22 Mud Cracks

Mud cracks form in clay-rich sediments when they dry and contract.

a. Mud cracks in a present-day environment.

b. Ancient mud cracks in Glacier National Park in Montana. Note that the cracks have been filled in with sediment.

Figure 6.23 Fossils

a. Skull of dinosaur *Allosaurus* on display at the Natural History Museum in Vienna, Austria.

b. These fossil tribolites are on display in the Mesalands Community College's Dinosaur Museum in Tucumcari, New Mexico. Although distantly related to today's insects, spiders, crabs, and lobsters, tribolites went extinct at the end of the Paleozoic Era, 251 million years ago.

street, broken glass, and a damaged power pole almost certainly indicate that a vehicle hit the pole. If you see a badly burned, shattered tree in the forest you could conclude that a bomb damaged it, but in the absence of any bomb fragments or residue you would no doubt decide that the tree was hit by lightning. Geologists use exactly the same kind of reasoning—that is, their understanding of natural processes—when they evaluate evidence preserved in sedimentary rocks; when the rocks formed is irrelevant. Geologists simply rely on the principle of *uniformitarianism* in making these interpretations (see Chapter 1).

So what about that sandstone we mentioned at the beginning of this section? Suppose that it has symmetrical ripples and fossils of marine-dwelling organisms, in which case you would no doubt conclude that it was deposited in a shallow marine environment. If, on the other hand, in addition to symmetrical ripples it contained dinosaur and land-plant fossils you would probably conclude that it was deposited near a lakeshore. Many other features in sedimentary rocks are used in a similar fashion. Ooids (Figure 6.16c) form today in shallow marine environments with vigorous currents, and we have every reason to think that ancient ones formed in the same way. Glacial deposits are typically poorly sorted, show little stratification, and have other features that indicate glacial transport and deposition.

The Navajo Sandstone in the southwestern United States is made up of well sorted sand measuring 0.2 to 0.5 mm in diameter, it has large cross-beds and shows the footprints of land-dwelling animals (Figure 6.24a). Long ago geologists concluded that the Navajo Sandstone formed as ancient sand dunes. In fact, the inclination of the cross-beds indicates that the wind blew mostly from the northeast.

Evidence from the sequence of rocks exposed in the lower part of the Grand Canyon in Figure 6.24b indicates that they were deposited in shallow seas during a marine transgression as shown in Figure 6.18 (*left*).

LO8 IMPORTANT RESOURCES IN SEDIMENTARY ROCKS

Sand and gravel are essential to the construction industry; pure clay deposits are used for ceramics, and limestone is used in the manufacture of cement and in blast furnaces, where iron ore is refined to make steel. Evaporites are the source of table salt, as well as chemical

compounds, and rock gypsum is used to manufacture wallboard. Phosphate-bearing sedimentary rock is used in fertilizers and animal-feed supplements.

Placer deposits are surface accumulations resulting from the separation and concentration of materials of greater density from those of lesser density in streams or on beaches. Much of the gold recovered during the initial stages of the California gold rush (1849–1853) was mined from placer deposits.

Most coal mined in the United States has been bituminous coal from the Appalachian region that formed in coastal swamps during the Pennsylvanian Period. Huge lignite and subbituminous coal deposits in the western United States are becoming increasingly important. During 2009, about 975 million metric tons of coal were mined in this country, more than 70% of it from mines in Wyoming, West Virginia, and Kentucky.

Coke, a hard, gray substance consisting of the fused ash of bituminous coal, is used in blast furnaces where steel is produced. Synthetic oil and gas and several other products are also made from bituminous coal and lignite.

Petroleum and Natural Gas

Petroleum and natural gas are *hydrocarbons*, meaning that they are composed of hydrogen and carbon. The

Figure 6.24 Ancient Sedimentary Rocks and Their Interpretation

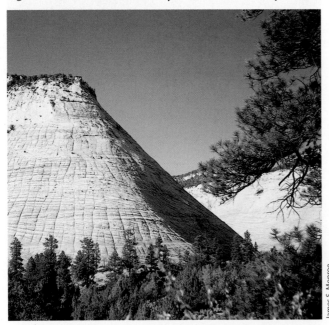

a. The Jurassic-age Navajo Sandstone in Zion National Park in Utah is a wind-blown dune deposit. Vertical fractures intersect cross-beds, hence the name Checkerboard Mesa for this rock exposure.

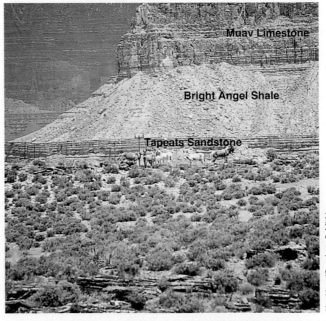

b. View of three formations in the Grand Canyon of Arizona. These rocks were deposited during a marine transgression. Compare with the vertical sequence of rocks in Figure 6.18.

remains of microscopic organisms settle to the seafloor, or lake floor in some cases, where little oxygen is present to decompose them. If buried beneath layers of sediment, they are heated and transformed into petroleum and natural gas. The rock in which hydrocarbons form is *source rock*, but for them to accumulate in economic quantities, they must migrate from the source rock into some kind of *reservoir rock*. Finally, the reservoir rock must have an overlying, nearly impervious *cap rock*; otherwise, the hydrocarbons would eventually reach the surface and escape (Figure 6.25a, b). Effective reservoir rocks must have appreciable pore space and good *permeability*, the capacity to transmit fluids; otherwise, hydrocarbons cannot be extracted from them in reasonable quantities.

Many hydrocarbon reservoirs consist of nearshore marine sandstones with nearby fine-grained, organic-rich source rocks. These are called *stratigraphic traps*, because they owe their existence to variations in the strata (Figure 6.25a). Indeed, some of the oil in the Persian Gulf region and Michigan is trapped in ancient reefs that are also good stratigraphic traps. *Structural traps* result when rocks are deformed by folding, fracturing, or both. In sedimentary rocks that have been deformed into a series of folds, hydrocarbons migrate to the high parts of these structures (Figure 6.25b).

Uranium

Most of the uranium used in nuclear reactors in North America comes from the complex potassium-, uranium-, vanadium-bearing mineral *carnotite*, which is found in some sedimentary rocks. Some uranium is also derived from *uraninite* (UO_2), a uranium oxide in granitic rocks and hydrothermal veins. Uraninite is easily oxidized and dissolved in groundwater, transported elsewhere, and chemically reduced and precipitated in the presence of organic matter.

The richest uranium ores in the United States are widespread in the Colorado Plateau area of Colorado and adjoining parts of Wyoming, Utah, Arizona, and New Mexico. These ores, consisting of fairly pure masses and encrustations of carnotite, are associated with plant remains in sandstones that formed in ancient stream channels. Although most of these ores are associated with fragmentary plant remains, some petrified trees also contain large quantities of uranium.

Large reserves of low-grade uranium ore are also found in the Chattanooga Shale. The uranium is finely disseminated in this black, organic-rich mudrock that underlies large parts of several states, including Illinois, Indiana, Ohio, Kentucky, and Tennessee. Canada is the world's largest producer and exporter of uranium.

Banded Iron Formation

The chemical sedimentary rock known as *banded iron formation* consists of alternating thin layers of chert and iron minerals, mostly the iron oxides hematite and magnetite. Banded iron formations are present on all continents and account for most of the iron ore mined in the world today. Vast banded iron formations are present in the Lake Superior region of the United States and Canada and in the Labrador Trough of eastern Canada.

Figure 6.25 Oil and Natural Gas.
The arrows in parts (a) and (b) indicate the direction of migration.

© 2013 Cengage Learning

a. Two examples of stratigraphic traps; one in sand within shale and the other in a buried reef.

b. Two examples of structural traps: one formed by folding and the other by faulting.

7

Metamorphism and Metamorphic Rocks

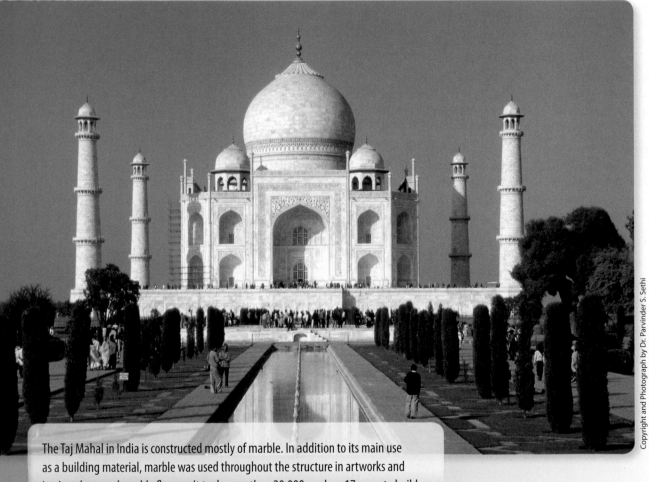

The Taj Mahal in India is constructed mostly of marble. In addition to its main use as a building material, marble was used throughout the structure in artworks and intricately carved marble flowers. It took more than 20,000 workers 17 years to build the Taj Mahal from 1631 to 1648.

Introduction

"A useful analogy for metamorphism is baking a cake: the outcome of each depends on the ingredients, their proportions, how they are mixed together, the temperature, and time."

Marble is a remarkable stone that has a variety of uses. Formed from limestone or dolostone by the metamorphic processes of heat and pressure, it comes in a variety of colors and textures. Marble has been used by sculptors and architects for many centuries in statuary and monuments. For example, the marble statue *Aphrodite of Melos*, also known as *Venus de Milo*, is one of the most recognizable works of art in the world. And the Peace Monument at Pennsylvania Avenue on the west side of the Capitol in Washington, DC, is constructed from white marble from Carrara, Italy, a locality famous for its marble. Marble is also used as a facing and main stone in many buildings and various other structures. The Taj Mahal in India is constructed mostly of Makrana marble quarried from hills just southwest of Jaipur in Rajasthan (see the chapter opening photo). Moreover, marble is used in floor tiling and other ornamental and structural uses. In addition, ground marble is used in toothpaste and it is a source of lime in agricultural fertilizers.

Metamorphic rocks (from the Greek *meta*, "change," and *morphe*, "shape") like marble are the third major group of rocks we will be examining. They result from the transformation of other rocks by processes that typically occur beneath Earth's surface (see Figure 1.14). During **metamorphism**, rocks are subjected to sufficient heat, pressure, and fluid activity to change their mineral composition, texture, or both, thus forming new rocks that usually do not look anything like the original rock before it was metamorphosed. These transformations take place below the melting temperature of the rock; otherwise, an igneous rock would result.

A useful analogy for metamorphism is baking a cake. Just like a metamorphic rock, the resulting cake depends on the ingredients, their proportions, how they are mixed together, how much water or milk is added, and the temperature and length of time used for baking the cake.

Except for marble and slate, most people are not familiar with metamorphic rocks. We are frequently asked by students why it is important to study metamorphic rocks and processes. Our answer is always, "Just look around you."

A large portion of Earth's continental crust is composed of metamorphic and igneous rocks. Together, they form the crystalline basement rocks underlying the sedimentary rocks of a continent's surface. Some of the oldest known rocks, dated at about 4.0 billion years (Figure 7.1), are metamorphic, which means that they formed from even older rocks! In addition, many metamorphic minerals and rocks, such as garnets, talc, asbestos, marble, and slate, are also economically important and useful.

metamorphic rock Any rock that has been changed from its original condition by heat, pressure, and the chemical activity of fluids, as in marble and slate.

metamorphism The phenomenon of rocks changing as a result of heat, pressure, and fluids, so that they are in equilibrium with a new set of environmental conditions.

Learning Outcomes

After reading this unit, you should be able to do the following:

LO1 Identify the agents of metamorphism

LO2 Identify the three types of metamorphism

LO3 Explain how metamorphic rocks are classified

LO4 Recognize the difference between metamorphic zones and facies

LO5 Understand how plate tectonics affects metamorphism

LO6 Understand the relationship between metamorphism and natural resources

Reed Wicander

The softness of marble, its uniform texture, and its varying colors have made it the favorite rock of builders and sculptors throughout history.

Figure 7.1 Acasta Gneiss
This metamorphic rock, found in Canada, is estimated to be about 4.0 bullion years old, making it one of the oldest known rocks on Earth. Gneiss is a foliated metamorphic rock.

LO1 THE AGENTS OF METAMORPHISM

heat An agent of metamorphism.

lithostatic pressure Pressure exerted on rocks by the weight of overlying rocks.

differential pressure Pressure that is not applied equally to all sides of a rock body.

The three principal agents of metamorphism are *heat*, *pressure*, and *fluid activity*. Time is also important to the metamorphic process, because chemical reactions proceed at different rates and thus require different amounts of time to complete. Reactions involving silicate compounds are particularly slow, and because most metamorphic rocks are composed of silicate minerals, it is thought that metamorphism is a very slow geologic process.

During metamorphism, the original rock, which was in equilibrium with its environment—meaning that it was chemically and physically stable under those conditions—undergoes changes to achieve equilibrium with its new environment. These changes may result in the formation of new minerals, a change in the texture of the rock, or both. In some instances, the change is minor, and features of the original rock can still be recognized. In other cases, the rock changes so much that the identity of the original rock can be determined only with great difficulty, if at all.

Heat

Heat is an important agent of metamorphism, because it increases the rate of chemical reactions that may produce minerals different from those in the original rock. Heat may come from lava, magma, or as a result of deep burial in the crust due to subduction along a convergent plate boundary.

When rocks are intruded by bodies of magma, they are subjected to intense heat that affects the surrounding rock. The most intense heating usually occurs adjacent to the magma body and gradually decreases with distance from the intrusion. The zone of metamorphosed rocks that forms in the country rock adjacent to an intrusive igneous body is usually distinct and easy to recognize.

It is known that temperature increases with depth. Some rocks that form at the surface may be transported to great depths by subduction along a convergent plate boundary and thus subjected to increasing temperature and pressure. During subduction, some minerals may be transformed into other minerals that are more stable under the higher temperature and pressure conditions.

Pressure

During burial, rocks are subjected to increasingly greater pressure, just as you feel greater pressure the deeper you dive into a body of water. Whereas the pressure you feel is known as *hydrostatic pressure*, because it comes from the water surrounding you, rocks undergo **lithostatic pressure**, which means that the *stress* (force per unit area) on a rock in Earth's crust is the same in all directions (Figure 7.2a). A similar situation occurs when an object is immersed in water. For example, the deeper a cup composed of Styrofoam™ is submerged in the ocean, the smaller it gets, because pressure increases with depth and is exerted on the cup equally in all directions, thereby compressing the cup (Figure 7.2b).

Along with lithostatic pressure resulting from burial, rocks may also experience **differential pressure** (Figure 7.3). In this case, the stresses are not equal in all directions, but are stronger from some directions than from others. Differential pressures typically occur when two plates collide, thus producing distinctive metamorphic textures and features.

Fluid Activity

In almost every region of metamorphism, water and carbon dioxide (CO_2) are present in varying amounts along mineral grain boundaries or in the pore spaces of

Figure 7.2 Lithostatic Pressure

Vertical pressure (kbar)

From C. Gillen, Metamorphic Geology, Figure 4.4, p. 73. Copyright 1982 Kluwer Academic Publishers. Reprinted by permission of the author.

1 kilobar (kbar) = 1,000 bars
Atmospheric pressure at sea level = 1 bar

a. Lithostatic pressure is applied equally in all directions in Earth's crust due to the weight of overlying rocks. Thus, pressure increases with depth, as indicated by the sloping black line. From C. Gillen, *Metamorphic Geology*, Figure 4.4, p. 73. Copyright 1982 Kluwer Academic Publishers. Reprinted by permission of the author.

David J. and Jane M. Matty

b. A similar situation occurs when 200-ml cups composed of Styrofoam™ are lowered to ocean depths of approximately 750 m and 1,500 m. Increased water pressure is exerted in all directions on the cups, and they consequently decrease in volume while maintaining their general shape.

rocks. These fluids, which may contain ions in solution, enhance metamorphism by increasing the rate of chemical reactions. Under dry conditions, most minerals react very slowly, but when even small amounts of fluid are introduced, reaction rates increase. The reaction rates increase mainly because ions can move readily through the fluid and thus enhance chemical reactions and the formation of new minerals.

Courtesy of Eric Johnson

Figure 7.3 Differential Pressure

Differential pressure results from stress that is unequally applied to an object. Rotated garnets are a good example of the effects of differential pressure applied to a rock during metamorphism. In this example from a schist in northeast Sardinia, stress was applied in opposite directions on the left and right side of the garnet (*center*), causing it to rotate.

The following reaction provides a good example of how new minerals can be formed by **fluid activity**. Seawater moving through hot basaltic rock in the oceanic crust transforms olivine into the metamorphic mineral serpentine:

> **fluid activity** An agent of metamorphism in which water and carbon dioxide promote metamorphism by increasing the rate of chemical reactions.

$$2Mg_2SiO_4 + 2H_2O \longrightarrow Mg_3Si_2O_5(OH)_4 + MgO$$

olivine   water   serpentine   carried away
                                            in solution

The chemically active fluids important in the metamorphic process come primarily from three sources: (1) water trapped in the pore spaces of sedimentary rocks as they form, (2) the volatile fluid within magma, and (3) the dehydration of water-bearing minerals such as gypsum ($CaSO_4 \cdot 2H_2O$) and some clays.

LO2 THE THREE TYPES OF METAMORPHISM

Geologists recognize three major types of metamorphism: (1) *contact (thermal) metamorphism*, in which magmatic heat and fluids act to produce change; (2) *dynamic metamorphism*, which is principally the result of high differential pressures associated with intense deformation; and (3) *regional metamorphism*, which occurs within a large area and is associated with major mountain-building episodes. Even though we will discuss each type of metamorphism separately, the boundary between them is not always distinct and depends largely on which of the three metamorphic agents was dominant.

Contact Metamorphism

Contact (thermal) metamorphism takes place when a body of magma alters the surrounding country rock. At shallow depths, intruding magma raises the temperature of the surrounding rock, causing thermal alteration. Furthermore, the release of hot fluids into the country rock by the cooling intrusion can aid in the formation of new minerals.

Important factors in contact metamorphism are the initial temperature, the size of the intrusion, and the fluid content of the magma and the country rock. Sometimes all three of these factors are involved. The initial temperature of an intrusion depends, in part, on its composition; mafic magmas are hotter than felsic magmas (see Chapter 4) and hence have a greater thermal effect on the rocks surrounding them. The size of the intrusion is also important. In the case of small intrusions, such as dikes and sills, usually only those rocks in immediate contact with the intrusion are affected. Because large intrusions, such as batholiths, take a long time to cool, the increased temperature in the surrounding rock may last long enough for a larger area to be affected.

The area of metamorphism surrounding an intrusion is an **aureole**, and the boundary between an intrusion and its aureole may be either sharp or transitional (Figure 7.4). Metamorphic aureoles vary in width depending on the size, temperature, and composition of the intruding magma, as well as the mineralogy of the surrounding country rock. Aureoles range from a few centimeters wide, bordering small dikes and sills, to several hundred meters or even several kilometers wide around large plutons.

Figure 7.4 **Metamorphic Aureole**

A metamorphic aureole, the area surrounding an intrusion, consists of zones that reflect the degree of metamorphism. The metamorphic aureole associated with this idealized granite pluton contains three zones of mineral assemblages reflecting the decrease in temperature with distance from the intrusion. An inner andalusite–cordierite hornfels zone forms adjacent to the pluton and is reflective of the high temperatures near the intrusion. This is followed by an intermediate zone of extensive recrystallization in which some biotite develops, and farthest from the intrusion is an outer zone characterized by spotted slates.

Unaltered country rock

Outer zone of spotted slates

Intermediate zone with some biotite

Inner andalusite - cordierite–hornfels zone

Granite pluton

AUREOLE

© 2013 Cengage Learning

The degree of metamorphic change within an aureole generally decreases with distance from the intrusion, reflecting the decrease in temperature from the original heat source. The region, or zone, closest to the intrusion, and hence subject to the highest temperatures, commonly contains high-temperature metamorphic minerals (that is, minerals in equilibrium with the higher-temperature environment) such as sillimanite. The outer zones, that is, those farthest from the intrusion, are typically characterized by lower-temperature metamorphic minerals such as chlorite, talc, and epidote.

Contact metamorphism can result not only from igneous intrusions, but also from lava flows, either along mid-oceanic ridges or from lava flowing over land and thermally altering the underlying rocks (Figure 7.5). Whereas recognizing a recent lava flow and the resulting contact metamorphism of the rocks below is easy, it is less obvious whether an igneous body is intrusive or extrusive in a rock outcrop where sedimentary rocks occur above and below the igneous body. Recognizing which sedimentary rock units have been metamorphosed enables geologists to determine whether the igneous body is intrusive (such as a sill or dike) or extrusive (lava flow). Such a determination is critical in reconstructing the geologic history of an area and may have important economic implications as well.

Fluids also play an important role in contact metamorphism. Magma is usually wet and contains hot, chemically active fluids that may emanate into the surrounding rock. These fluids can react with the rock and aid in the formation of new minerals. In addition, the country rock may contain pore fluids that, when heated by magma, increase reaction rates.

Because heat and fluids are the primary agents of contact metamorphism, two types of contact metamorphic rocks are generally recognized: those resulting from baking of country rock and those altered by hot solutions. Many of the rocks that result from contact metamorphism have the texture of porcelain; that is, they are hard and fine grained. This is particularly true for rocks with a high clay content, such as shale. Such texture results because the clay minerals in the rock are baked, just as a clay pot is baked when fired in a kiln.

During the final stages of cooling, when an intruding magma begins to crystallize, large amounts of hot, watery solutions are often released. These solutions may react with the country rock and produce new metamorphic minerals. This process, which usually occurs near Earth's surface, is called *hydrothermal alteration* (from the Greek *hydro*, "water," and *therme*, "heat") and may result in valuable mineral deposits.

Dynamic Metamorphism

Most **dynamic metamorphism** is associated with *faults* (fractures along which movement has occurred) or fault zones, where rocks are subjected to high levels of differential pressure. The metamorphic rocks that result from pure dynamic metamorphism are called *mylonites*, and typically they are restricted to narrow zones adjacent to faults. Mylonites are hard, dense, fine-grained rocks, many of which are characterized by thin laminations

> **dynamic metamorphism**
> Metamorphism in fault zones where rocks are subjected to high differential pressure.

Figure 7.5 Contact Metamorphism from a Lava Flow

A highly weathered basaltic lava flow near Susanville, California, has altered an underlying rhyolitic volcanic ash by contact metamorphism. The red zone below the lava flow has been baked by the heat of the lava when it flowed over the ash layer. The lava flow displays spheroidal weathering, a type of weathering common in fractured rocks (see Chapter 6).

Lava flow

Baked zone

Ash layer

James S. Monroe

(Figure 7.6). Tectonic settings where mylonites occur include the Moine Thrust Zone in northwest Scotland, the Adirondack Highlands in New York, and portions of the San Andreas fault in California (see Chapter 2).

Regional Metamorphism

Most metamorphic rocks result from **regional metamorphism**, which occurs over a large area and is usually caused by tremendous temperatures, pressures, and deformation, all occurring together within the deeper portions of the crust. Regional metamorphism is most obvious along convergent plate boundaries, where rocks are intensely deformed and recrystallized during convergence and subduction. Within these metamorphic rocks there is usually a gradation of metamorphic intensity from areas that were subjected to the most intense pressures or the highest temperatures, or both, to areas of lower pressures and temperatures. Such a gradation in metamorphism can be recognized by the metamorphic minerals that are present.

Regional metamorphism is not, however, confined only to convergent margins. It can also occur in areas where plates diverge, although usually at much shallower depths because of the high geothermal gradient associated with these areas.

Index Minerals and Metamorphic Grade

From field studies and laboratory experiments, certain minerals are known to form only within specific temperature and pressure ranges. Such minerals are known as **index minerals**, because their presence allows geologists to recognize low-, intermediate-, and high-grade metamorphism (Figure 7.7).

Metamorphic grade is a term that generally characterizes the degree to which a rock has undergone metamorphic change (Figure 7.7). Although the boundaries between the different metamorphic grades are not sharp, the distinction is nonetheless useful for communicating in a general way the degree to which rocks have been metamorphosed. The presence of index minerals thus helps determine metamorphic grade. For example, when a clay-rich rock such as shale undergoes regional metamorphism, the mineral chlorite first begins to crystallize under

Figure 7.6 Mylonite
An outcrop of mylonite from the Adirondack Highlands, New York. Mylonites result from dynamic metamorphism, where rocks are subjected to high levels of differential pressure. Note the thin laminations (closely spaced layers) that are characteristic of many mylonites.

Courtesy of Eric Johnson

relatively low temperatures of about 200°C. Its presence in these rocks thus indicates low-grade metamorphism. If temperatures and pressures continue to increase, new minerals form to replace chlorite, because they are more stable under those new conditions. Thus, there is a progression in the appearance of new minerals from chlorite—whose presence indicates low-grade metamorphism—to biotite and garnet, which are good index minerals for intermediate-grade metamorphism, and then to sillimanite, whose presence indicates high-grade metamorphism and temperatures exceeding 500°C (Figure 7.7).

Different rock compositions develop different sets of index minerals. For example, clay-rich rocks such as shale will develop the index minerals shown in Figure 7.7, whereas a sandy dolomite will produce a different set of index minerals as metamorphism progresses, because it has a different mineral composition than shale. Thus, a particular set of index minerals will form based on the original composition of the parent rock undergoing metamorphism.

LO3 HOW ARE METAMORPHIC ROCKS CLASSIFIED?

For purposes of classification, metamorphic rocks are commonly divided into two groups: those exhibiting a *foliated texture* (from the Latin *folium*, "leaf") and those with a *nonfoliated texture* (Table 7.1).

Figure 7.7 Metamorphic Grade
Change in mineral assemblage and rock type with increasing metamorphism of shale. When a clay-rich rock such as shale is subjected to increasing metamorphism, new minerals form, as shown by the colored bars. The progressive appearance of certain minerals, known as index minerals, allows geologists to recognize low-, intermediate-, and high-grade metamorphism.

TABLE 7.1
Classification of Common Metamorphic Rocks

TEXTURE	METAMORPHIC ROCK	TYPICAL MINERALS	METAMORPHIC GRADE	CHARACTERISTICS OF ROCKS	PARENT ROCK
Foliated	Slate	Clays, micas, chlorite	Low	Fine-grained, splits easily into flat pieces	Mudrocks, volcanic ash
	Phyllite	Fine-grained quartz, micas, chlorite	Low to medium	Fine-grained, glossy or lustrous sheen	Mudrocks
	Schist	Micas, chlorite, quartz, talc, hornblende, garnet, staurolite, graphite	Low to high	Distinct foliation, minerals visible	Mudrocks, carbonates, mafic igneous rocks
	Gneiss	Quartz, feldspars, hornblende, micas	High	Segregated light and dark bands visible	Mudrocks, sandstones, felsic igneous rocks
	Amphibolite	Hornblende, plagioclase	Medium to high	Dark, weakly foliated	Mafic igneous rocks
	Migmatite	Quartz, feldspars, hornblende, micas	High	Streaks or lenses of granite intermixed with gneiss	Felsic igneous rocks mixed with sedimentary rocks
Nonfoliated	Marble	Calcite, dolomite	Low to high	Interlocking grains of calcite or dolomite, reacts with HCl	Limestone or dolostone
	Quartzite	Quartz	Medium to high	Interlocking quartz grains, hard, dense	Quartz sandstone
	Greenstone	Chlorite, epidote, hornblende	Low to high	Fine-grained, green	Mafic igneous rocks
	Hornfels	Micas, garnets, andalusite, cordierite, quartz	Low to medium	Fine-grained, equidimensional grains, hard, dense	Mudrocks
	Anthracite	Carbon	High	Black, lustrous, subconcoidal fracture	Coal

© 2013 Cengage Learning

Foliated Metamorphic Rocks

Rocks subjected to heat and differential pressure during metamorphism typically have minerals arranged in a parallel fashion, giving them a **foliated texture** (Figure 7.8). The size and shape of the mineral grains determine whether the foliation is fine or coarse. Low-grade metamorphic rocks, such as slate, have a finely foliated texture in which the mineral grains are so small that they cannot be distinguished without magnification. High-grade foliated rocks, such as gneiss, are coarse-grained, such that the individual grains can easily be seen with the unaided eye. Foliated metamorphic rocks can be arranged in order of increasingly coarse grain size and perfection of foliation.

Slate is a very fine-grained foliated metamorphic rock that commonly exhibits *slaty cleavage* (Figure 7.9). It results from regional metamorphism of shale or, more rarely, volcanic ash. Because it can easily be split along cleavage planes into flat pieces, slate is an excellent rock for roofing and floor tiles, billiard and pool table tops, and blackboards. The different colors of slate are caused by minute amounts of graphite (black), iron oxide (red and purple), and chlorite (green).

Phyllite is similar in composition to slate but is coarser grained. The minerals, however, are still too small to be identified without magnification. Phyllite can be distinguished from slate by its glossy or lustrous sheen (Figure 7.10) and represents an intermediate grain size between slate and schist.

Schist is most commonly produced by regional metamorphism. The type of schist formed depends on the intensity of metamorphism and the character of the original rock (Figure 7.11). Metamorphism of many rock types can yield schist, but most schist appears to have formed from clay-rich sedimentary rocks (Table 7.1). All schists contain more than 50% platy and elongated minerals, all of which are large enough to be clearly visible. Their mineral composition imparts a schistosity, or schistose foliation, to the rock that usually produces a wavy type of parting when split. Schistosity is common in low- to high-grade metamorphic environments, and each type of schist is known by its most conspicuous mineral or minerals, such as mica schist, chlorite schist, or garnet-mica schist (Figure 7.11).

Gneiss is a high-grade metamorphic rock that is streaked, having segregated bands of light and dark minerals. Gneisses are composed of granular minerals such as quartz, feldspar, or both, with lesser percentages of platy or elongated minerals such as micas or amphiboles (Figure 7.12). Quartz and feldspar are the principal minerals composing the light-colored bands

Figure 7.8 Foliated Texture

Random arrangement of elongated minerals before pressure is applied to two sides

Elongated minerals arranged in a parallel manner as a result of pressure applied to two sides

a. When rocks are subjected to differential pressure, the mineral grains are typically arranged in a parallel manner, producing a foliated texture.

Elongated minerals arranged in a parallel manner as a result of shear

© 2013 Cengage Learning

Reed Wicander

b. Photomicrograph of a metamorphic rock with a foliated texture showing the parallel arrangement of mineral grains.

Figure 7.9 Slate

a. Hand specimen of red slate.

b. Slate of Chalet Enzian, Switzerland.

of minerals, whereas biotite and hornblende comprise the dark mineral bands. Gneiss typically breaks in an irregular manner, much like coarsely crystalline nonfoliated rocks.

Most gneiss probably results from recrystallization of clay-rich sedimentary rocks during regional metamorphism (Table 7.1). Gneiss also can form from igneous rocks such as granite or older metamorphic rocks.

Another fairly common foliated metamorphic rock is *amphibolite*. A dark rock, it is composed mainly of hornblende and plagioclase. The alignment of the hornblende crystals produces a slightly foliated texture. Many amphibolites result from intermediate- to high-grade metamorphism of basalt and ferromagnesian-rich mafic rocks.

In some areas of regional metamorphism, exposures of "mixed rocks" called *migmatites*, having both igneous and high-grade metamorphic characteristics, are present (Figure 7.13). Migmatites are thought to result from the extremely high temperatures produced during metamorphism. However, part of the problem in determining the origin of migmatites is explaining

how the granitic component formed. According to one model, the granitic magma formed in place by the partial melting of rock during intense metamorphism. Such an origin is possible provided that the host rocks

Figure 7.11 Schist

a. Almandine garnet crystals in a mica schist.

b. Hornblende—mica schist.

Figure 7.10 Phyllite
Hand specimen of phyllite. Note the lustrous sheen, as well as the bedding (upper left to lower right) at an angle to the cleavage of the specimen.

contained quartz and feldspars and that water was present.

Others argue that the characteristic layering or wavy appearance of migmatites arises by the redistribution of minerals during recrystallization in the solid state—that is, through purely metamorphic processes.

Figure 7.12 Gneiss
Gneiss is characterized by segregated bands of light and dark minerals. This folded gneiss is exposed at Wawa, Ontario, Canada.

Reed Wicander

Nonfoliated Metamorphic Rocks

In some metamorphic rocks, the mineral grains do not show a discernable preferred orientation. Instead, these rocks consist of a mosaic of roughly equidimensional minerals and are characterized as having a **nonfoliated texture** (Figure 7.14). Most nonfoliated metamorphic rocks result from contact or regional metamorphism of rocks with no platy or elongate minerals. Frequently, the only indication that a granular rock has been metamorphosed is the large grain size resulting from recrystallization.

Nonfoliated metamorphic rocks are generally of two types: those composed of mainly one mineral—for example, marble or quartzite—and those in which the different mineral grains are too small to be seen without magnification, such as greenstone and hornfels.

Marble is a well-known metamorphic rock composed predominantly of calcite or dolomite; its grain size ranges from fine to coarsely granular. Marble results from either contact or regional metamorphism of limestones or dolostones (Figure 7.15 and Table 7.1). Pure marble is snowy white or bluish; however, many color varieties exist because of the presence of mineral impurities in the original sedimentary rock.

Quartzite is a hard, compact rock typically formed from quartz sandstone under intermediate- to high-grade metamorphic conditions during contact or regional metamorphism (Figure 7.16). Because recrystallization

Copyright and Photograph by Dr. Parvinder S. Sethi

Figure 7.13 Migmatite
A migmatite boulder in Rocky Mountain National Park, near Estes Park, Colorado. Migmatites consist of high-grade metamorphic rock intermixed with streaks or lenses of granite.

Figure 7.14 Nonfoliated Texture
Nonfoliated textures are characterized by a mosaic of roughly equidimensional minerals, as in this photomicrograph of marble.

Reed Wicander

is so complete, metamorphic quartzite is of uniform strength and therefore usually breaks across the component quartz grains rather than around them when it is struck. Pure quartzite is white; however, iron and other impurities commonly impart a pinkish-red or other color to it. Quartzite is commonly used as foundation material for road and railway beds.

The name *greenstone* is applied to any compact, dark green, altered, mafic igneous rock that formed under low- to high-grade metamorphic conditions. The green color results from the presence of chlorite, epidote, and hornblende.

Hornfels is a common, fine-grained, nonfoliated metamorphic rock resulting from contact metamorphism and composed of various equidimensional mineral grains. The composition of hornfels depends directly on the composition of the original rock, and many compositional varieties are known. Most hornfels, however, are apparently derived from contact metamorphism of clay-rich sedimentary rocks or impure dolostones.

Anthracite is a black, lustrous, hard coal that contains a high percentage of fixed carbon and a low percentage of volatile matter and is highly valued by people who burn coal for heating and power. Anthracite usually forms from the metamorphism of lower-grade coals by heat and pressure, and many geologists consider it to be a metamorphic rock (Table 7.1).

LO4 METAMORPHIC ZONES AND FACIES

While mapping the 400- to 440-million-year-old Dalradian schists of Scotland in the late 1800s, George Barrow and other British geologists made the first systematic study of metamorphic zones. Here, clay-rich sedimentary rocks have been subjected to regional metamorphism, and the resulting metamorphic rocks can be divided into different zones based on the presence of distinctive silicate mineral assemblages. These mineral assemblages, each recognized by the presence of one or more index minerals, indicate different degrees of metamorphism. The index minerals that Barrow and his associates chose to represent increasing metamorphic intensity were chlorite, biotite, garnet, staurolite, kyanite, and sillimanite (Figure 7.7), which we now know all result from the recrystallization of clay-rich sedimentary rocks.

The successive appearance of metamorphic index minerals indicates gradually increasing or decreasing intensity of metamorphism. Going from lower- to higher-grade metamorphic zones, the first appearance of a particular index mineral indicates the location of the minimum temperature and pressure conditions needed for the formation of that mineral. When the locations of the first appearances of that index mineral are connected on a map, the result is a line of equal metamorphic intensity,

Limestone

Metamorphism →

Marble

Copyright and Photographs by Dr. Parvinder S. Sethi

Figure 7.15 Marble
Metamorphism of the sedimentary rock limestone or dolostone yields marble.

Metamorphism →

Quartz sandstone

Quartzite

Copyright and Photographs by Dr. Parvinder S. Sethi

Figure 7.16 Quartzite
Metamorphism of the sedimentary rock quartz sandstone yields quartzite.

---

metamorphic zone The region between lines of equal metamorphic intensity known as isograds.

metamorphic facies A group of metamorphic rocks characterized by particular minerals that formed under the same broad temperature and pressure conditions.

or an *isograd*. The region between two adjacent isograds makes up a single **metamorphic zone**—a belt of rocks displaying the same general degree of metamorphism. By mapping adjoining metamorphic zones, geologists can reconstruct metamorphic conditions throughout an entire area (Figure 7.17).

Not long after Barrow and his coworkers completed their work, geologists in Norway and Finland came up with a different method of mapping metamorphism that was more useful than the metamorphic zone approach. A **metamorphic facies** is defined as a group of metamorphic rocks characterized by particular mineral assemblages formed under broadly similar temperature and pressure conditions (Figure 7.18). Each facies is named after its most characteristic rock or mineral. For example, the green metamorphic mineral chlorite, which forms under relatively low temperatures and pressures, yields rocks belonging to the *greenschist facies*. Under increasingly higher temperatures and pressures, mineral assemblages indicative of the *amphibolite* and *granulite facies* develop.

Although usually applied to areas where the original rocks were clay-rich, the concept of metamorphic facies can be used with modification in other situations. It cannot, however, be used in areas where the original rocks were pure quartz sandstones or pure limestones or dolostones. Such rocks, regardless

of the imposed temperature and pressure conditions, will yield only quartzites and marbles, respectively. In such cases, all one can say is that "metamorphism happened."

LO5 PLATE TECTONICS AND METAMORPHISM

Although metamorphism is associated with all three types of plate boundaries, it is most common along convergent plate margins. Metamorphic rocks form at

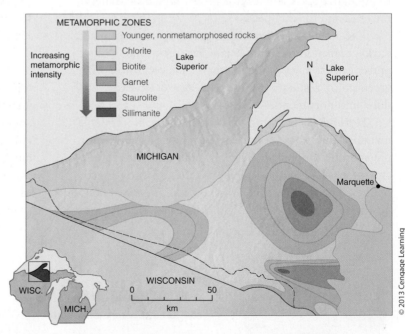

Figure 7.17 Metamorphic Zones in the Upper Peninsula of Michigan
The zones in this region are based on the presence of distinctive silicate mineral assemblages resulting from the metamorphism of sedimentary rocks during an interval of mountain building and minor granitic intrusion that occurred during the Proterozoic Eon, approximately 1.5 billion years ago. The lines separating the different metamorphic zones are isograds.

convergent plate boundaries because temperature and pressure increase there as a result of plate collisions.

Figure 7.19 illustrates the various metamorphic facies produced along a typical oceanic–continental

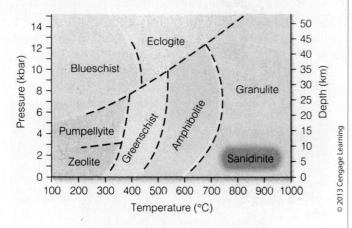

Figure 7.18 Metamorphic Facies and Their Associated Temperature–Pressure Conditions.

A temperature–pressure diagram showing under what conditions various metamorphic facies occur. A metamorphic facies is characterized by a particular mineral assemblage that formed under the same broad temperature–pressure conditions. Each facies is named after its most characteristic rock or mineral.

convergent plate boundary. When an oceanic plate collides with a continental plate, tremendous pressure is generated as the oceanic plate is subducted. Because rock is a poor heat conductor, the cold descending oceanic plate heats slowly, and metamorphism is caused mostly by increasing pressure with depth. Metamorphism in such an environment produces rocks typical of the blueschist facies (low temperature, high pressure). Geologists use the presence of blueschist facies rocks as evidence of ancient subduction zones.

As subduction along the oceanic–continental convergent plate boundary continues, both temperature and pressure increase with depth and yield high-grade metamorphic rocks.

Eventually, the descending plate begins to melt and generate magma that moves upward. This rising magma may alter the surrounding rock by contact metamorphism, producing migmatites in the deeper portions of the crust and hornfels at shallower depths. High temperatures and low to medium pressures characterize such an environment.

Although metamorphism is most common along convergent plate margins, many divergent plate boundaries are characterized by contact metamorphism. Rising magma at mid-oceanic ridges heats the adjacent

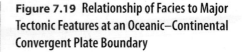

Figure 7.19 Relationship of Facies to Major Tectonic Features at an Oceanic–Continental Convergent Plate Boundary

A = amphibolite facies
BS = blueschist facies
CM = contact metamorphic zone shown in green
E = eclogite facies
EA = epidote-amphibolite facies
GR = granulite facies
GS = greenschist facies
P = prehnite-pumpellyite facies
Z = zeolite facies

Remember that metamorphic zones are identified by the appearance of a single index mineral within rocks of the same general composition that occur throughout an area. However, rocks of greatly different composition within an area can belong to the same metamorphic facies, because each facies has its own characteristic assemblage of minerals whose presence indicates metamorphism with the same broad temperature–pressure range unique to that facies.

rocks, producing contact metamorphic minerals and textures. In addition, fluids emanating from the rising magma—and its reaction with seawater—very commonly produce metal-bearing hydrothermal solutions that may precipitate minerals of economic value, such as the copper ores of Cyprus.

LO6 METAMORPHISM AND NATURAL RESOURCES

Many metamorphic minerals and rocks are valuable natural resources. Although these resources include various types of ore deposits, the two most familiar and widely used metamorphic rocks are marble (Figure 7.15) and slate (Figure 7.9), which have been used for centuries in a variety of ways.

Many ore deposits result from contact metamorphism during which hot, ion-rich fluids migrate from igneous intrusions into the surrounding rock, thereby producing rich ore deposits. The most common sulfide ore minerals associated with contact metamorphism are bornite (copper), chalcopyrite (copper), galena (lead), pyrite (iron), and sphalerite (zinc); two common iron oxide ore minerals are hematite and magnetite. Tin and tungsten are also important ores associated with contact metamorphism.

Other economically important metamorphic minerals include asbestos, used for insulation and fireproofing in buildings and building materials (Figure 7.20), talc for talcum powder, graphite for pencils and dry lubricants, and garnets and corundum, which are used as abrasives or gemstones, depending on their quality.

Copyright and Photograph by Dr. Parvinder S. Sethi

Figure 7.20 Specimen of Chrysotile Asbestos from Arizona
Chrysotile is the fibrous form of serpentine asbestos and the most commonly used in buildings and other structures.

4LTR Press solutions are designed for today's learners through the continuous feedback of students like you. Tell us what you think about **GEOL2** and help us improve the learning experience for future students.

YOUR FEEDBACK MATTERS.

Complete the Speak Up survey in CourseMate at www.cengagebrain.com

 Follow us at www.facebook.com/4ltrpress

8 | Earthquakes and Earth's Interior

Kyodo/Landov

The collapsed remains of the King's Education Language School following the 6.3-magnitude earthquake that struck Christchurch, New Zealand on February 23, 2011. In addition to the tremendous damage to buildings and infrastructure, 181 people were killed in this earthquake, making it the second deadliest earthquake to strike New Zealand.

Introduction

"Rocks subjected to intense forces bend until **they break and then return** to their original position, releasing **energy in the** process."

In the afternoon of January 12, 2010, a magnitude 7.0 earthquake struck the island nation of Haiti. According to official estimates, 222,570 people died, at least 300,000 were injured, and more than 285,000 residences and businesses were destroyed or severely damaged. Widespread devastation occurred in the capital of Port-au-Prince and elsewhere throughout the region, exacerbated by an almost total collapse of the vital infrastructure needed to respond to such a disaster, including medical, transportation, and communication systems.

Learning Outcomes

After reading this unit, you should be able to do the following:

LO1 Explain the elastic rebound theory

LO2 Describe seismology

LO3 Identify where and how often earthquakes occur

LO4 Identify the different types of seismic waves

LO5 Discuss how earthquakes are located

LO6 Explain how the strength of an earthquake is measured

LO7 Describe the destructive effects of earthquakes

LO8 Discuss earthquake prediction methods

LO9 Discuss earthquake control methods

LO10 Describe Earth's interior

LO11 Explain the features of Earth's core

LO12 Describe the structure, density, and composition of Earth's mantle

LO13 Explain the source of Earth's internal heat

LO14 Describe the features of Earth's crust

A little more than a year later, on March 11, 2011, a 9.0-magnitude earthquake and tsunami struck Japan, causing more than 13,000 deaths and also resulting in tremendous destruction, including severe damage to a nuclear power plant in the northeastern part of the island. Neither of these earthquakes is the first, nor will they be the last major devastating earthquakes in these regions of the world.

> **earthquake** Vibrations caused by the sudden release of energy, usually as a result of displacement of rocks along faults.

Earthquakes, along with volcanic eruptions, are manifestations of Earth's dynamic and active makeup. As one of nature's most frightening and destructive phenomena, earthquakes have always aroused feelings of fear and have been the subject of myths and legends. What makes an earthquake so frightening is that when it begins, there is no way to tell how long it will last or how violent it will be. Approximately 13 million people have died in earthquakes during the past 4,000 years, with about 3 million of these deaths occurring during the last century (Table 8.1).

Geologists define an **earthquake** as the shaking or trembling of the ground caused by the sudden release of energy, usually as a result of *faulting*, which involves the displacement of rocks along fractures (we discuss the different types of faults in Chapter 9). After an earthquake, continuing adjustments along a fault may generate a series of earthquakes known as *aftershocks*. Most aftershocks are smaller than the main shock, but they can still cause considerable damage to already weakened structures.

Why should you study earthquakes? The obvious reason is that they are destructive and cause many deaths and injuries to people living in earthquake-prone areas. Earthquakes also affect the economies of many countries in terms of cleanup costs, lost jobs, and lost business revenues. From a purely personal standpoint, you should be interested in earthquakes because you may be caught in one someday. Even if you don't live in an area that is subject to earthquakes, you may travel to places where the threat of earthquakes exists, and you should know what to do if you experience one. Such knowledge may help you avoid serious injury or even death.

TABLE 8.1

Some Significant Earthquakes

YEAR	LOCATION	MAGNITUDE (ESTIMATED BEFORE 1935)	DEATHS (ESTIMATED)
1556	China (Shanxi Province)	8.0	1,000,000
1755	Portugal (Lisbon)	8.6	70,000
1906	U.S.A. (San Francisco, California)	8.3	3,000
1923	Japan (Tokyo)	8.3	143,000
1960	Chile	9.5	5,700
1964	U.S.A (Anchorage, Alaska)	8.6	131
1976	China (Tangshan)	8.0	242,000
1985	Mexico (Mexico City)	8.1	9,500
1988	Armenia	6.9	25,000
1990	Iran	7.3	50,000
1993	India	6.4	30,000
1995	Japan (Kobe)	7.2	6,000+
1999	Turkey	7.4	17,000
2001	India	7.9	14,000+
2003	Iran	6.6	43,000
2004	Indonesia	9.0	>220,000
2005	Pakistan	7.6	>86,000
2006	Indonesia	6.3	>6,200
2008	China (Sichuan Province)	7.9	>69,000
2010	Haiti	7.0	>220,000
2011	New Zealand	6.3	181
2011	Japan	9.0	>13,000

© 2013 Cengage Learning

elastic rebound theory An explanation for the sudden release of energy that causes earthquakes when deformed rocks fracture and rebound to their original undeformed condition.

L01 ELASTIC REBOUND THEORY

Based on studies conducted after the 1906 San Francisco earthquake, H. F. Reid of The Johns Hopkins University proposed the **elastic rebound theory** to explain how energy is released during earthquakes. Reid studied three sets of measurements taken across a portion of the San Andreas fault that had broken during the 1906 earthquake. The measurements revealed that points on opposite sides of the fault had moved 3.2 m during the 50-year period

prior to breakage in 1906, with the west side moving northward (Figure 8.1).

According to Reid, rocks on opposite sides of the San Andreas fault had been storing energy and bending slightly for at least 50 years before the 1906 earthquake. Any straight line, such as a fence or road that crossed the San Andreas fault, was gradually bent, because rocks on one side of the fault moved relative to rocks on the other side (Figure 8.1). Eventually, the strength of the rocks was exceeded, and they then fractured. After fracturing, the rocks on both sides of the fault rebounded, or "snapped back," to their former undeformed shape, and the energy stored was released as earthquake waves radiating outward from the break (Figure 8.1). Additional field and laboratory studies conducted by Reid and others have confirmed that elastic rebound is the mechanism by which energy is released during earthquakes.

The energy stored in rocks undergoing deformation is analogous to the energy stored in a tightly wound watch spring. The tighter the spring is wound, the more energy is stored, thus making more energy available for release. If the spring is wound so tightly that it breaks, then the stored energy is released as the spring rapidly unwinds and partially regains its original shape.

© Stockbyte/Getty Images

Figure 8.1 The Elastic Rebound Theory

Fault
Fence

Original position

Deformation

a. According to the elastic rebound theory, rocks experiencing deformation store energy as they bend. When the initial strength of the rocks is exceeded, they rupture, releasing their accumulated energy, and "snap back" or rebound to their former undeformed shape. This sudden release of energy is what causes an earthquake.

Rupture and release of energy

Rocks rebound to original undeformed shape

© 2013 Cengage Learning

Copyright and Photograph by Dr. Parvinder S. Sethi

b. During the 1906 San Francisco earthquake, this fence in Marin County was displaced by almost 5 m. Whereas many people would see just a broken fence, a geologist sees that the fence has moved, or been displaced, and would look for evidence of a fault. A geologist would also notice that the ground has been displaced toward the right side, relative to his or her view. Regardless of what side of the fence you stand on, you must look to the right to see the other part of the fence. Try it!

A useful analogy is that of bending a long, straight stick over your knee. As the stick bends, it deforms and eventually reaches the point at which it breaks. When this happens, the two pieces of the original stick snap back into their original straight position. Likewise, rocks subjected to intense forces bend until they break, and they then return to their original position, releasing energy in the process.

LO2 SEISMOLOGY

Seismology, the study of earthquakes, emerged as a true science during the 1880s with the development of **seismographs**, instruments that detect, record, and measure the vibrations produced by an earthquake

(Figure 8.2). The record made by a seismograph is called a *seismogram*. Modern seismographs have electronic sensors and record movements precisely using computers, rather than simply relying on the drum strip charts commonly used on older seismographs.

When an earthquake occurs, energy in the form of *seismic waves* radiates out from the point of release (Figure 8.3a). These waves are somewhat analogous to the ripples that move out concentrically from the point where a stone is thrown into a pond. Unlike waves on

seismology The study of earthquakes.

seismograph An instrument that detects, records, and measures the various waves produced by earthquakes.

Figure 8.2 Seismographs

a. Seismographs record ground motion during an earthquake. The record produced is a seismogram. The seismograph shown here records earthquakes on a strip of paper attached to a rotating drum.

b. A horizontal-motion seismograph. Because of its inertia, the suspended mass that contains the marker remains stationary while the rest of the structure moves along with the ground during an earthquake. As long as the length of the arm is not parallel to the direction of ground movement, the marker will record the earthquake waves on the rotating drum. This seismograph would record waves from west or east, but to record waves from the north or south, another seismograph at right angles to this one is needed.

c. A vertical-motion seismograph. This seismograph operates on the same principle as the horizontal-motion instrument and records vertical ground movement.

a pond, however, seismic waves move outward in all directions from their source.

Earthquakes take place because rocks are capable of storing energy, but their strength is limited, so if enough force is present, they rupture and thus release their stored energy. In other words, most earthquakes result when movement occurs along faults, most of which are related, at least indirectly, to plate movements. Once rupturing begins, it moves along the fault at several kilometers per second for as long as conditions for failure exist. The longer the fault along which

movement occurs, the more time it takes for the stored energy to be released, and therefore the longer the ground will shake. During some very large earthquakes, the ground might shake for 3 minutes, a seemingly brief time, but interminable if you are experiencing the earthquake firsthand!

The Focus and Epicenter of an Earthquake

The location within Earth's lithosphere where rupturing begins—that is, the point at which energy is first

Figure 8.3 The Focus and Epicenter of an Earthquake

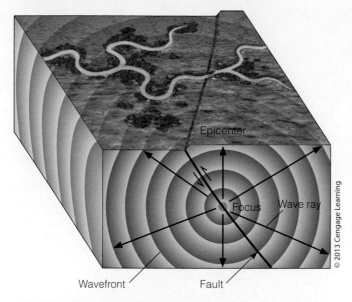

Epicenter

Focus

Wave ray

Wavefront

Fault

© 2013 Cengage Learning

a. The focus of an earthquake is the location where the rupture begins and energy is released. The place on the surface vertically above the focus is the epicenter. Seismic wavefronts move out in all directions from their source, the focus of an earthquake.

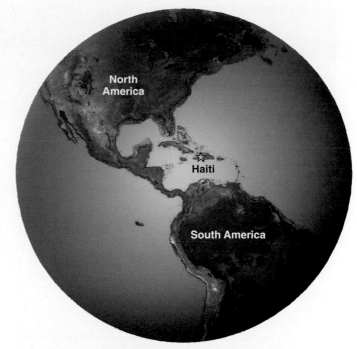

North America

Haiti

South America

b. The epicenter of the magnitude 7.0 January 12, 2010, earthquake that devastated Haiti was located approximately 25 km west-southwest of Port-au-Prince, and had a focal depth of 13 km.

released—is an earthquake's **focus**, or *hypocenter*. What we usually hear in news reports, however, is the location of the **epicenter**, the point on Earth's surface directly above the focus (Figure 8.3a).

Seismologists recognize three categories of earthquakes based on focal depth. *Shallow-focus* earthquakes have focal depths of less than 70 km from the surface, whereas those with foci between 70 and 300 km are *intermediate-focus*, and the foci of those characterized as *deep-focus* are more than 300 km deep. However, earthquakes are not evenly distributed among these three categories. Approximately 90% of all earthquake foci are at depths of less than 100 km, whereas only about 3% of all earthquakes are deep-focus. Shallow-focus earthquakes are, with few exceptions, the most destructive, because the energy they release has little time to dissipate before reaching the surface.

A definite relationship exists between earthquake foci and plate boundaries. Earthquakes generated along divergent or transform plate boundaries are invariably shallow-focus, whereas many shallow-focus earthquakes and nearly all intermediate- and deep-focus earthquakes occur along convergent margins (Figure 8.4). Furthermore, a pattern emerges when the focal depths of earthquakes near island arcs and their adjacent ocean trenches are plotted. Notice in Figure 8.5 that the focal depth increases beneath the Tonga Trench in a narrow, well-defined zone that dips approximately 45 degrees. Dipping seismic zones, called *Benioff* or *Benioff–Wadati zones*, are common to convergent plate boundaries where one plate is subducted beneath another. Such dipping seismic zones indicate the angle of plate descent along a convergent plate boundary.

LO3 WHERE DO EARTHQUAKES OCCUR, AND HOW OFTEN?

No place on Earth is immune to earthquakes, but almost 95% take place in seismic belts corresponding to plate boundaries where plates converge, diverge, and slide past each other. The relationship between plate margins and the distribution of earthquakes is readily

focus The site within Earth where an earthquake originates and energy is released.

epicenter The point on Earth's surface directly above an earthquake's focus.

Deep-focus earthquake • Intermediate-focus earthquake • Shallow-focus earthquake

Figure 8.4 Earthquake Epicenters and Plate Boundaries

This map of earthquake epicenters shows that most earthquakes occur within seismic zones that correspond closely to plate boundaries. Approximately 80% of earthquakes occur within the circum-Pacific belt, 15% within the Mediterranean—Asiatic belt, and the remaining 5% within plate interiors and along oceanic spreading ridges. The dots represent earthquake epicenters and are divided into shallow-, intermediate-, and deep-focus earthquakes. Along with shallow-focus earthquakes, nearly all intermediate- and deep-focus earthquakes occur along convergent plate boundaries.

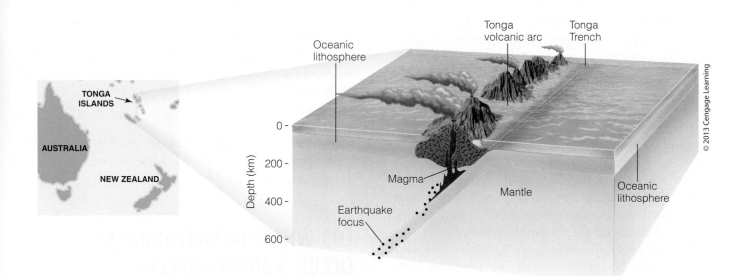

Figure 8.5 Benioff Zones

Focal depth increases in a well-defined zone that dips approximately 45 degrees beneath the Tonga volcanic arc in the South Pacific. Dipping seismic zones are called *Benioff* or *Benioff–Wadati zones*.

apparent when the locations of earthquake epicenters are superimposed on a map showing the boundaries of Earth's plates (Figure 8.4).

The majority of all earthquakes (approximately 80%) occur in the *circum-Pacific belt*, a zone of seismic activity nearly encircling the Pacific Ocean basin. Most of these earthquakes result from convergence along plate margins, as in the case of the 2011 Japanese earthquake. The earthquakes along the North American Pacific Coast, especially in California, are also in this belt, but here plates slide past one another rather than converge. The January 17, 1994, Northridge earthquake (Figure 8.6) occurred along this plate boundary.

The second major seismic belt, accounting for 15% of all earthquakes, is the *Mediterranean–Asiatic belt*. This belt extends westward from Indonesia through the Himalayas, across Iran and Turkey, and westward through the Mediterranean region of Europe. The 2005 earthquake in Pakistan that killed more than 86,000

people is a recent example of the destructive earthquakes that strike this region (Table 8.1).

The remaining 5% of earthquakes occur mostly in the interiors of plates and along oceanic spreading-ridge systems. Most of these earthquakes are not strong, although several major intraplate earthquakes of note have occurred. For example, the 1811 and 1812 earthquakes near New Madrid, Missouri, killed approximately 20 people and nearly destroyed the town. So strong were these earthquakes that they were felt from the Rocky Mountains to the Atlantic Ocean and from the Canadian border to the Gulf of Mexico. Within the immediate area, numerous buildings were destroyed and forests were flattened. The land sank several meters in some areas, causing flooding, and reportedly the Mississippi River reversed its flow during the shaking and changed its course slightly.

The cause of intraplate earthquakes is not well understood, but geologists think that they arise from localized stresses caused by the compression that most plates experience along their margins. A useful analogy is moving a house. Regardless of how careful the movers are, moving something so large without its internal parts shifting slightly is impossible. Similarly, plates are not likely to move without some internal stresses that occasionally cause earthquakes.

More than 900,000 earthquakes are recorded annually by the worldwide network of seismograph stations. Many of these are too small to be felt, but are nonetheless recorded. These small earthquakes result from the energy released as continual adjustments take place between the various plates. However, on average, more than 31,000 earthquakes per year are strong enough to be felt and can cause various amounts of damage, depending on how strong they are and where they occur.

LO4 SEISMIC WAVES

Many people have experienced an earthquake, but most are probably unaware that the shaking they feel and the damage to structures are caused by the arrival of *seismic waves*, a general term encompassing all waves generated by an earthquake. When movement on a fault takes place, energy is released in the form of two kinds of seismic waves that radiate outward in all directions from an earthquake's focus. *Body waves*, so called because they travel through the solid body of Earth, are somewhat like sound waves, and *surface waves*, which travel along the ground surface, are analogous to undulations or waves on water surfaces.

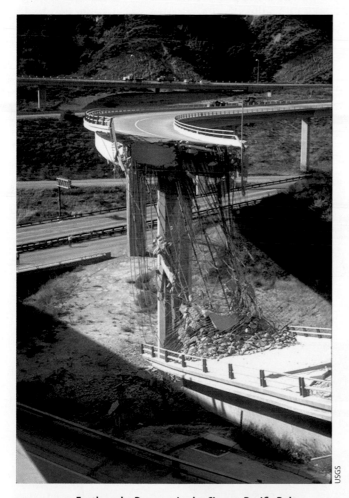

Figure 8.6 Earthquake Damage in the Circum-Pacific Belt
One of the several elevated freeway spans that collapsed during the January 1994 Northridge, California, earthquake in which 61 people were killed.

P-wave A compressional, or push-pull, wave; the fastest seismic wave and one that can travel through solids, liquids, and gases; also called a primary wave.

S-wave A shear wave that moves material perpendicular to the direction of travel, thereby producing shear stresses in the material it moves through; also known as a secondary wave; S-waves travel only through solids.

Body Waves

An earthquake generates two types of body waves: P-waves and S-waves (Figure 8.7). **P-waves**, or *primary waves*, are the fastest seismic waves and can travel through solids, liquids, and gases. P-waves are *compressional*, or *push-pull*, *waves* and are similar to sound waves in that they move material forward and backward along a line in the same direction that the waves themselves are moving (Figure 8.7b). Thus, the material through which a P-wave travels is expanded and compressed as the waves move through it and returns to its original size and shape after the wave passes by.

S-waves, or *secondary waves*, are somewhat slower than P-waves and can travel only through solids. S-waves are *shear waves* because they move the material perpendicular to the direction of travel, thereby producing shear stresses in the material they move through (Figure 8.7c). Because liquids (as well as gases) are not rigid, they have no shear strength, and S-waves cannot be transmitted through them.

The velocities of P- and S-waves are determined by the density and elasticity of the materials through which they travel. For example, seismic waves travel more slowly through rocks of greater density but more rapidly through rocks with greater elasticity. *Elasticity* is a property of solids, such as rocks, and means that once they have been deformed by an applied force, they return to their original shape when the force is no longer present. Because P-wave velocity is greater than S-wave velocity in all materials, P-waves always arrive at seismic stations first.

Surface Waves

Surface waves travel along the surface of the ground or just below it and are slower than body waves (Figure 8.8). Unlike the sharp jolting and shaking that body waves cause, surface waves generally produce a rolling or swaying motion, much like the experience of being on a boat.

Seismologists recognize several types of surface waves. The two most important are Rayleigh waves

Figure 8.7 Primary and Secondary Seismic Body Waves
Body waves travel through Earth.

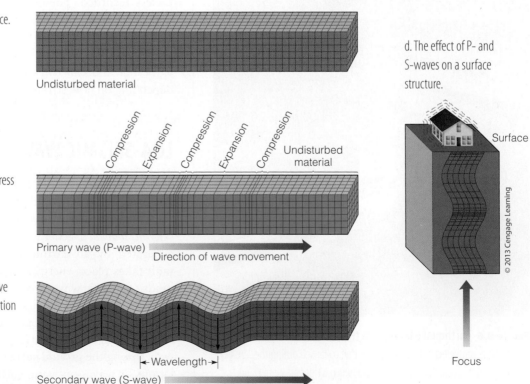

a. Undisturbed material for reference.

Undisturbed material

b. Primary waves (P-waves) compress and expand material in the same direction that they travel.

Compression Expansion Compression Expansion Compression

Undisturbed material

Primary wave (P-wave) → Direction of wave movement

c. Secondary waves (S-waves) move material perpendicular to the direction of wave movement.

|← Wavelength →|

Secondary wave (S-wave) →

d. The effect of P- and S-waves on a surface structure.

Surface

© 2013 Cengage Learning

Focus

Figure 8.8 Rayleigh and Love Seismic Surface Waves

Surface waves travel along Earth's surface or just below it.

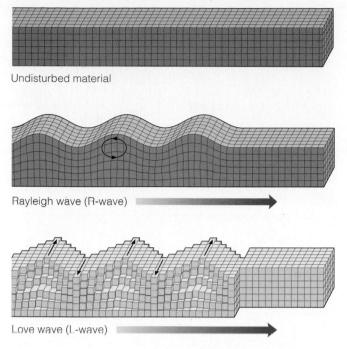

Undisturbed material

Rayleigh wave (R-wave)

Love wave (L-wave)

a. Undisturbed material for reference.

b. Rayleigh waves (R-waves) move material in an elliptical path in a plane oriented parallel to the direction of wave movement.

Rayleigh wave   Love wave

© 2013 Cengage Learning

c. Love waves (L-waves) move material back and forth in a horizontal plane perpendicular to the direction of wave movement.

d. The arrival of R- and L-waves causes the surface to undulate and shake from side to side.

and Love waves, named after the British scientists who discovered them, Lord Rayleigh and A. E. H. Love. **Rayleigh waves (R-waves)** are generally the slower of the two and behave like water waves in that they move forward while the individual particles of material move in an elliptical path within a vertical plane oriented in the direction of wave movement (Figure 8.8b). The motion of a **Love wave (L-wave)** is similar to that of an S-wave, but the individual particles of the material move only back and forth in a horizontal plane perpendicular to the direction of wave travel (Figure 8.8c).

LO5 LOCATING AN EARTHQUAKE

We mentioned that news articles commonly report an earthquake's epicenter, but just how is the location of an epicenter determined? Once again, geologists rely on the study of seismic waves. We know that P-waves travel faster than S-waves, nearly twice as fast in all substances, so P-waves arrive at a seismograph station first, followed some time later by S-waves. Both P- and S-waves travel directly from the focus to the seismograph station through Earth's interior, but L- and R-waves arrive last, because they are the slowest,

and they also travel the longest route along the surface (Figure 8.9a). However, only the P- and S-waves need concern us here, because they are the ones important in finding an epicenter.

Seismologists, geologists who study seismology, have accumulated a tremendous amount of data over the years and now know the average speeds of P- and S-waves for any specific distance from their source. These P- and S-wave travel times are published in *time–distance graphs* that illustrate the difference between the arrival times of the two waves as a function of the distance between a seismograph and an earthquake's focus (Figure 8.9b). That is, the farther the waves travel, the greater the *P–S time interval*, which is simply the time difference between the arrivals of P- and S-waves (Figure 8.9a, b).

> **Rayleigh wave (R-wave)** A surface wave in which individual particles of material move in an elliptical path within a vertical plane oriented in the direction of wave movement.
>
> **Love wave (L-wave)** A surface wave in which the individual particles of material move only back and forth in a horizontal plane perpendicular to the direction of wave travel.

Figure 8.9 Determining the Distance from an Earthquake

a. A schematic seismogram showing the arrival order and pattern produced by P-, S-, and L-waves. When an earthquake occurs, body and surface waves radiate out from the focus at the same time. Because P-waves are the fastest, they arrive at a seismograph first, followed by S-waves, and then by surface waves, which are the slowest waves. The difference between the arrival times of the P- and S-waves is the *P–S time interval*; it is a function of the distance of the seismograph station from the focus.

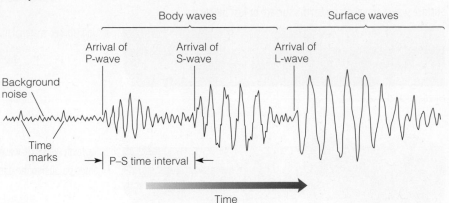

b. A time–distance graph showing the average travel times for P- and S-waves. The farther away a seismograph station is from the focus of an earthquake, the longer the interval between the arrival of the P- and S-waves, and hence the greater the distance between the P- and S-wave curves on the time–distance graph as indicated by the P–S time interval. For example, let's assume the difference in arrival times between the P- and S-waves is 10 minutes (P–S time interval). Using the Travel time (minutes) scale, measure how long 10 minutes is (P–S time interval), and move that distance between the S-wave curve and the P-wave curve until the line touches both curves as shown. Then draw a line straight down to the Distance from focus (km) scale. That number is the distance the seismograph is from the earthquake's focus. In this example, the distance is almost 9,000 km.

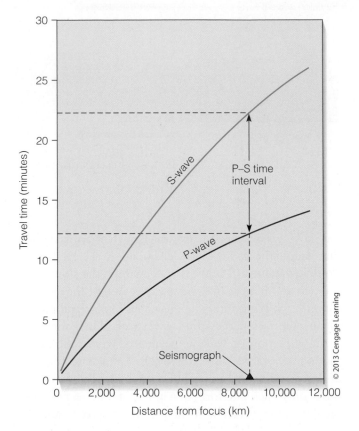

If the P–S time intervals are known from at least three seismograph stations, then the epicenter of any earthquake can be determined (Figure 8.10). Here is how it works: Subtracting the arrival time of the first P-wave from the arrival time of the first S-wave gives the P–S time interval for each seismic station. Each of these time intervals is plotted on a time–distance graph, and a line is drawn straight down to the distance axis of the graph, thus giving the distance from the focus to each seismic station (Figure 8.9b). Next, a circle whose radius equals the distance shown on the time–distance graph from each of the seismic stations is drawn on a map (Figure 8.10). The intersection of the three circles is the location of the earthquake's epicenter. It should be obvious from Figure 8.10 that P–S time intervals from at least three seismic stations are needed. If only one were used, the epicenter could be at any location on the circle drawn around that station, and using two stations would give two possible locations for the epicenter.

Determining the focal depth of an earthquake is much more difficult and considerably less precise than finding its epicenter. The focal depth is usually found by making computations based on several assumptions, comparing the results with those obtained at other seismic stations, and then recalculating and approximating the depth as closely as possible.

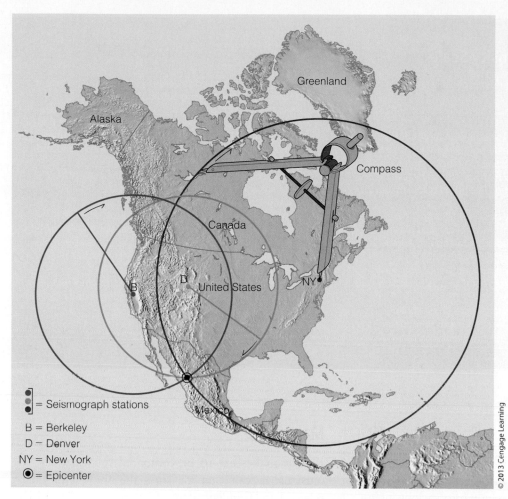

Figure 8.10 Determining the Epicenter of an Earthquake
Three seismograph stations are needed to locate the epicenter of an earthquake. The P–S time interval is plotted on a time–distance graph for each seismograph station to determine the distance that station is from the epicenter. A circle with that radius is drawn from each station, and the intersection of the three circles is the epicenter of the earthquake.

= Seismograph stations

B = Berkeley
D = Denver
NY = New York
⊙ = Epicenter

LO6 MEASURING THE STRENGTH OF AN EARTHQUAKE

Following any earthquake that causes extensive damage, fatalities, and injuries, graphic reports of the quake's violence and human suffering are common. Although such descriptions of fatalities and damage give some indication of the size of an earthquake, geologists are interested in more reliable methods for determining an earthquake's size.

Two measures of an earthquake's strength are commonly used. One is *intensity*, a qualitative assessment of the kinds of damage done by an earthquake. The other, *magnitude*, is a quantitative measure of the amount of energy released by an earthquake. Each method provides important information that can be used to prepare for future earthquakes.

Intensity

Intensity is a subjective or qualitative measure of the kind of damage done by an earthquake as well as people's reaction to it. Since the mid-19th century, geologists have used intensity as a rough approximation of the size and strength of an earthquake. The most common intensity scale used in the United States is the **Modified Mercalli Intensity Scale**, which has values ranging from I to XII (Table 8.2).

intensity The subjective measure of the kind of damage done by an earthquake, as well as people's reaction to it.

Modified Mercalli Intensity Scale A scale with values from I to XII used to characterize earthquakes based on damage.

TABLE 8.2
Modified Mercalli Intensity Scale

I   Not felt except by a very few under especially favorable circumstances.

II Felt by only a few people at rest, especially on upper floors of buildings.

III Felt quite noticeably indoors, especially on upper floors of buildings, but many people do not recognize it as an earthquake. Standing automobiles may rock slightly.

IV During the day felt indoors by many, outdoors by few. At night some awakened. Sensation like heavy truck striking building, standing automobiles rocked noticeably.

V   Felt by nearly everyone, many awakened. Some dishes, windows, etc. broken, a few instances of cracked plaster. Disturbance of trees, poles, and other tall objects sometimes noticed.

VI Felt by all, many frightened and run outdoors. Some heavy furniture moved. A few instances of fallen plaster or damaged chimneys. Damage slight.

VII Everybody runs outdoors. Damage negligible in buildings of good design and construction; slight to moderate in well-built ordinary structures; considerable in poorly built or badly designed structures; some chimneys broken. Noticed by people driving automobiles.

VIII Damage slight in specially designed structures; considerable in normally constructed buildings with possible partial collapse; great in poorly built structures. Fall of chimneys, monuments, walls. Heavy furniture overturned. Sand and mud ejected in small amounts.

IX Damage considerable in specially designed structures. Buildings shifted off foundations. Ground noticeably cracked. Underground pipes broken.

X   Some well-built wooden structures destroyed; most masonry and frame structures with foundations destroyed; ground badly cracked. Rails bent. Landsides considerable from river banks and steep slopes. Water splashed over river banks.

XI Few, if any (masonry) structures remain standing. Bridges destroyed. Broad fissures in ground. Underground pipelines completely out of service.

XII Damage total. Waves seen on ground surface. Objects thrown upward into the air.

Source: U.S. Geological Survey.

Richter Magnitude Scale An open-ended scale that measures the amount of energy released during an earthquake.

magnitude The total amount of energy released by an earthquake at its source.

Intensity maps can be constructed for regions hit by earthquakes by dividing the affected region into various intensity zones. The intensity value given for each zone is the maximum intensity that the earthquake produced for that zone. Even though intensity maps are not precise, because of the subjective nature of the measurements, they do provide geologists with a rough approximation of the location of the earthquake, the kind and extent of the damage done, and the effects of local geology on different types of building construction. Because intensity is a measure of the kind of damage done by an earthquake, insurance companies still classify earthquakes on the basis of intensity.

Generally, a large earthquake will produce higher intensity values than a small earthquake, but many other factors besides the amount of energy released by an earthquake also affect its intensity. These include distance from the epicenter, focal depth of the earthquake, population density, geology of the area, type of building construction employed, and duration of shaking.

Magnitude

If earthquakes are to be compared quantitatively, we must use a scale that measures the amount of energy released and is independent of intensity. Charles F. Richter, a seismologist at the California Institute of Technology, developed such a scale in 1935. The **Richter Magnitude Scale** measures earthquake **magnitude**, which is the total amount of energy released by an earthquake at its source. It is an open-ended scale with values beginning at zero. The largest magnitude recorded was a 9.5-magnitude earthquake in Chile on May 22, 1960 (Table 8.1).

Scientists determine the magnitude of an earthquake by measuring the amplitude of the largest seismic wave as recorded on a seismogram (Figure 8.11). To avoid large numbers, Richter used a conventional base-10 logarithmic scale to convert the amplitude of the largest recorded seismic wave to a numerical magnitude value. Therefore, each whole-number increase in magnitude represents a 10-fold increase in wave amplitude. For example, the amplitude of the largest seismic wave for an earthquake of magnitude 6 is 10 times that produced by an earthquake of magnitude 5, 100 times as large as a magnitude 4 earthquake, and 1,000 times that of an earthquake of magnitude 3 ($10 \times 10 \times 10 = 1,000$).

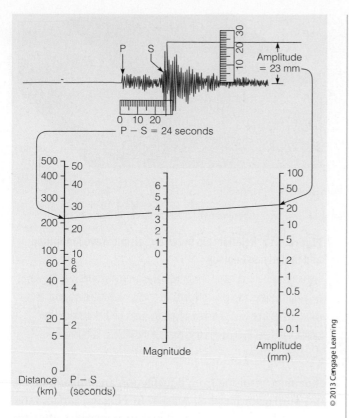

Figure 8.11 Richter Magnitude Scale

The Richter Magnitude Scale measures the total amount of energy released by an earthquake at its source. The magnitude is determined by measuring the maximum amplitude of the largest seismic wave and marking it on the right-hand scale. The difference between the arrival times of the P- and S waves (recorded in seconds) is marked on the left-hand scale. When a line is drawn between the two points, the magnitude of the earthquake is the point at which the line crosses the center scale.

A common misconception about the size of earthquakes is that an increase of one unit on the Richter Magnitude Scale—a 7 versus a 6, for instance—means a 10-fold increase in size. It is true that each whole-number increase in magnitude represents a 10-fold increase in the wave amplitude, but each magnitude increase of one unit corresponds to a roughly 30-fold increase in the amount of energy released (actually, it is 31.5, but 30 is close enough for our purposes). This means, for example, that the 2011 Japanese earthquake, with a magnitude of 9.0, released nearly 900 times more energy than the 2010 Haiti earthquake, with a magnitude 7.0 (30 × 30 = 900)!

The Richter Magnitude Scale was devised to measure earthquake waves on a particular seismograph and at a specific distance from an earthquake. One of its limitations is that it underestimates the energy of very large earthquakes because it measures the highest peak on a seismogram, which represents only an instant during an earthquake. For large earthquakes, though, the energy might be released over several minutes and along hundreds of kilometers of a fault. For example, during the 1857 Fort Tejon, California earthquake, the ground shook for longer than 2 minutes, and energy was released for 360 km along the fault.

Seismologists now commonly use a somewhat different scale to measure magnitude. Known as the *seismic-moment magnitude scale*, this scale takes into account the strength of the rocks, the area of a fault along which rupture occurs, and the amount of movement of rocks adjacent to the fault. Because larger earthquakes rupture more rocks than smaller earthquakes, and rupture usually occurs along a longer segment of a fault and therefore for a longer duration, these very large earthquakes release more energy. For example, the December 26, 2004, Sumatra, Indonesia, earthquake that generated the devastating tsunami created the longest fault rupture and had the longest duration ever recorded.

Thus, magnitude is now frequently given in terms of both Richter magnitude and seismic-moment magnitude. For example, the 1964 Alaska earthquake is given a Richter magnitude of 8.6 and a seismic-moment magnitude of 9.2. Because the Richter Magnitude Scale is most commonly used in the news, we will use that scale here.

LO7 WHAT ARE THE DESTRUCTIVE EFFECTS OF EARTHQUAKES?

The number of deaths and injuries, as well as the amount of property damage, depends on several factors. Generally speaking, earthquakes that occur

Great earthquakes (those with a magnitude greater than 8.0) occur, on average, once every year.

during working and school hours in densely populated urban areas are the most destructive and cause the most fatalities and injuries. However, magnitude, duration of shaking, distance from the epicenter, geology of the affected region, and the type of structures are also important considerations. Given these variables, it should not be surprising that a comparatively small earthquake can have disastrous effects whereas a much larger one might go largely unnoticed, except perhaps by seismologists.

The destructive effects of earthquakes include ground shaking, fire, seismic sea waves, and landslides, as well as panic, disruption of vital services, and psychological shock.

Ground Shaking

Ground shaking, the most obvious and immediate effect of an earthquake, varies depending on the earthquake's magnitude, distance from the epicenter, and type of underlying materials in the area—unconsolidated sediment or fill versus bedrock, for instance. Certainly, ground shaking is terrifying, and it may be violent enough for fissures to open in the ground. Nevertheless, contrary to popular myth, fissures do not swallow up people and buildings and then close on them. And although California will no doubt have big earthquakes in the future, rocks cannot store enough energy to displace a landmass as large as California into the Pacific Ocean as some alarmists claim.

The effects of ground shaking, such as collapsing buildings, falling building facades and window glass, and toppling monuments and statues, cause more damage and result in more loss of life and injuries than any other earthquake hazard. Structures built on solid bedrock generally suffer less damage than those built on poorly consolidated material such as water-saturated sediments or artificial fill (Figure 8.12).

Structures built on poorly consolidated or water-saturated material are subjected to ground shaking of longer duration and greater S-wave amplitude than structures built on bedrock (Figure 8.12). In addition, fill and water-saturated sediments tend to liquefy, or behave as a fluid, a process known as *liquefaction*. When shaken, the individual grains lose cohesion, and the ground flows. Two dramatic and classic examples

© 2013 Cengage Learning

Water

Seismic wave

Bedrock | Well-consolidated sediments | Poorly consolidated sediments | Bay mud (water-saturated)

Figure 8.12 Relationship between Seismic Wave Amplitude and Underlying Geology

The amplitude and duration of seismic waves generally increase as the waves pass from bedrock to poorly consolidated or water-saturated material. Thus, structures built on weaker material typically suffer greater damage than similar structures built on bedrock, because the shaking lasts longer.

of damage resulting from liquefaction are Niigata, Japan, and Turnagain Heights, Alaska. In Niigata, Japan, large apartment buildings were tipped to their sides after the water-saturated soil of the hillside collapsed (Figure 8.13). In Turnagain Heights, Alaska, many homes were destroyed when the Bootlegger Cover Clay lost all of its

National Geophysical Data Center

Figure 8.13 Liquefaction

The effects of ground shaking on water-saturated soil are dramatically illustrated by the collapse of these buildings in Niigata, Japan, during a 1964 earthquake. The buildings, which were designed to be earthquake resistant, fell over on their sides intact when the ground below them underwent liquefaction.

168   CHAPTER 8: EARTHQUAKES AND EARTH'S INTERIOR

strength when it was shaken by the 1964 earthquake (see Figure 10.17).

Besides the magnitude of an earthquake and the underlying geology, the material used and the type of construction also affect the amount of damage done. Adobe and mud-walled structures are the weakest and almost always collapse during an earthquake. Unreinforced brick structures and poorly built concrete structures are also particularly susceptible to collapse, such as was the case in the 1999 Turkey earthquake in which an estimated 17,000 people died (Figure 8.14). The 1976 earthquake in Tangshan, China, completely leveled the city, because hardly any structures were built to resist seismic forces. In fact, most had unreinforced brick walls, which have no flexibility, and consequently they collapsed during the shaking.

Fire

In many earthquakes, particularly in urban areas, fire is a major hazard. Nearly 90% of the damage done in the 1906 San Francisco earthquake was caused by fire. The shaking severed many of the electrical and gas lines, which touched off flames and started fires

throughout the city. Because the earthquake ruptured water mains, there was no effective way to fight the fires that raged out of control for three days, destroying much of the city.

Eighty-three years later, during the 1989 Loma Prieta earthquake, a fire broke out in the Marina district of San Francisco. This time, however, the fire was contained within a small area because San Francisco had a system of valves throughout its water and gas pipeline

Figure 8.14 Ground Shaking
Most of the buildings collapsed or were severely damaged as a result of ground shaking during the August 17, 1999, Turkey earthquake, which killed more than 17,000 people.

system so that lines could be isolated from breaks.

During the September 1, 1923, earthquake in Japan, fires destroyed 71% of the houses in Tokyo and practically all of the houses in Yokohama. In all, 576,262 houses were destroyed by fire, and 143,000 people died, many as a result of fire.

Tsunami: Killer Waves

On December 26, 2004, a magnitude 9.0 earthquake struck 160 km off the west coast of northern Sumatra, Indonesia, generating the deadliest tsunami in history. Within hours, walls of water as high as 10.5 m pounded the coasts of Indonesia, Sri Lanka, India, Thailand, Somalia, Myanmar, Malaysia, and the Maldives, killing more than 220,000 people and causing billions of dollars in damage.

Following the 9.0-magnitude earthquake that struck Japan on March 11, 2011, a massive tsunami was generated that resulted in tremendous property damage and loss of life along the coastline of Japan. Within minutes after the earthquake, walls of water, some as high as 37 m, inundated low-lying areas along the Japanese coast and extended as far as 10 km inland. Boats, vehicles, and structures were destroyed or swept aside as if they were toys, leaving a wake of destruction as the waters finally receded (Figure 8.15).

Both of these earthquakes generated what is popularly called a "tidal wave," but is more correctly termed a *seismic sea wave* or **tsunami**, a Japanese term meaning "harbor wave." The term *tidal wave* nevertheless persists in popular literature and some news accounts, but these waves are not caused by, or related to, tides. Indeed, tsunami are destructive sea waves generated when the sea floor undergoes sudden, vertical movements. Many result from submarine earthquakes, but volcanoes at sea or submarine landslides can also cause them. For example, the 1883 eruption of Krakatau between Java and Sumatra generated a large sea wave that killed 36,000 on nearby islands.

Jetkat/Shutterstock.com

Tsunami can barrel through the ocean at 965 km per hour. That's as fast as the cruising speed of a jet plane.

Once a tsunami is generated, it can travel across an entire ocean and cause devastation far from its source. In the open sea, tsunami travel at several hundred kilometers per hour and commonly go unnoticed as they pass beneath ships because they are usually less

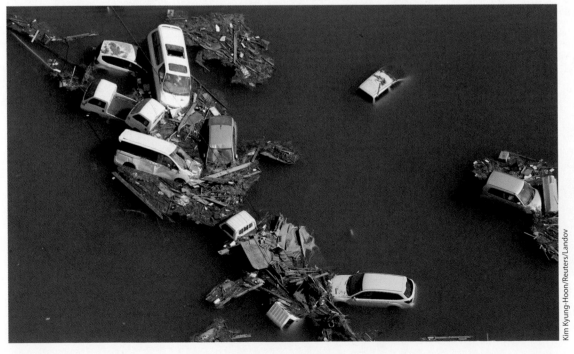

Figure 8.15 Japanese Tsunami
Vehicles, tossed like toys, mix with the rubble in floodwaters near the coastal town of Sendai following the tsunami that ravaged the northeastern portion of Japan following the 9.0-magnitude earthquake on March 11, 2011.

Kim Kyung-Hoon/Reuters/Landov

than 1 m high and the distance between wave crests is typically hundreds of kilometers. When they enter shallow water, however, the waves slow down and water piles up to heights anywhere from a meter or two to many meters high. The 1946 tsunami that struck Hilo, Hawaii, was 16.5 m high, and the one that struck Japan in 2011 was reported to have reached more than twice that height in some areas. In any case, the tremendous energy possessed by a tsunami is concentrated on a shoreline when it hits, either as a large breaking wave or, in some cases, what appears to be a very rapidly rising tide.

One of nature's warning signs of an approaching tsunami is a sudden withdrawal of the sea from a coastal region. In fact, the sea might withdraw so far that it cannot be seen and the seafloor is laid bare over a huge area. On more than one occasion, people have rushed out to inspect exposed reefs or to collect fish and shells only to be swept away when the tsunami arrived.

Following the tragic 1946 tsunami that hit Hilo, Hawaii, the U.S. Coast and Geodetic Survey established a Pacific Tsunami Early Warning System in Ewa Beach, Hawaii. This system combines seismographs and instruments that detect earthquake-generated sea waves. Whenever a strong earthquake takes place anywhere within the Pacific basin, its location is determined, and instruments are checked to see whether a tsunami has been generated. If it has, a warning is sent out to evacuate people from low-lying areas that may be affected.

Unfortunately, no such warning system exists for the Indian Ocean. If one had been in place, it is possible that the death toll from the December 26, 2004, tsunami would have been significantly lower.

Ground Failure

Earthquake-triggered landslides are particularly dangerous in mountainous regions and have been responsible for tremendous amounts of damage and many deaths. The 1959 Hebgen Lake earthquake in Madison Canyon, Montana, for example, caused a huge rock slide (Figure 8.16), and the 1970 Peru earthquake caused an avalanche that destroyed the town of Yungay and killed an estimated 66,000 people.

Figure 8.16 Ground Failure
On August 17, 1959, an earthquake with a Richter magnitude of 7.3 shook southwestern Montana and a large area in adjacent states.

a. The fault scarp in this image was produced when the block in the background moved up several meters compared to the one in the foreground.

b. The earthquake also triggered a landslide (visible in the distance) that blocked the Madison River in Montana and created Earthquake Lake (foreground). The slide entombed approximately 26 people in a campground at the very bottom.

Most of the 100,000 deaths from the 1920 earthquake in Gansu, China, resulted when cliffs composed of loess (wind-deposited silt) collapsed.

LO8 EARTHQUAKE PREDICTION

A successful prediction must include a time frame for the occurrence of an earthquake, its location, and its strength. Despite the tremendous amount of information geologists have gathered about the cause of earthquakes, successful predictions are still rare. Nevertheless, if reliable predictions can be made, they can greatly reduce the number of deaths and injuries.

From an analysis of historic records and the distribution of known faults, geologists construct *seismic risk maps* that indicate the likelihood and potential severity of future earthquakes based on the intensity of past earthquakes. An international effort by scientists from several countries resulted in the publication of the first Global Seismic Hazard Assessment Map in December 1999 (Figure 8.17). Although such maps cannot be used to predict when an earthquake will take place in any particular area, they are useful in anticipating future earthquakes and helping people plan and prepare for them.

Earthquake Precursors

Studies conducted during the past several decades indicate that most earthquakes are preceded by both short-term and long-term changes within Earth. Such changes are called *precursors* and may be useful in earthquake prediction.

One long-range prediction technique used in seismically active areas involves plotting the location of major earthquakes and their aftershocks to detect areas that have had major earthquakes in the past but are currently inactive. Such regions are said to be *locked* and not releasing energy. Nevertheless, pressure is continuing to accumulate in these regions because of plate motions, making these *seismic gaps* prime locations for future earthquakes. Several seismic gaps along the San Andreas fault have the potential for future major earthquakes (Figure 8.18). A major earthquake that damaged Mexico City in 1985 occurred along a seismic gap in the convergence zone along the west coast of Mexico.

Earthquake precursors that may be useful in making short-term predictions include slight changes in elevation and tilting of the land surface, fluctuations in the water level in wells, changes in Earth's magnetic field, and the electrical resistance of the ground.

Figure 8.17 Global Seismic Hazard Assessment Map

The Global Seismic Hazard Assessment Program published this seismic hazard map showing peak ground accelerations. The values are based on a 90% probability that the indicated horizontal ground acceleration during an earthquake is not likely to be exceeded in 50 years. The higher the number, the greater the hazard is. As expected, the greatest seismic risks are in the circum-Pacific belt and the Mediterranean—Asiatic belt.

Figure 8.18 Earthquake Precursors

Seismic gaps are one type of earthquake precursor that can indicate a potential earthquake in the future. These seismic gaps are regions along a fault that are locked, meaning they are not moving and releasing energy. Three gaps are evident in this cross section along the San Andreas fault from north of San Francisco to south of Parkfield. The first is between San Francisco and Portola Valley, the second near Loma Prieta Mountain, and the third is southeast of Parkfield. The top section shows the epicenters of earthquakes between January 1969 and July 1989. The bottom section shows the southern Santa Cruz Mountains gap after it was filled by the October 17, 1989, Loma Prieta earthquake (*open circle*) and its aftershocks.

Earthquake Prediction Programs

Currently, only a handful of nations—such as the United States, Japan, Russia, and China—have government-sponsored earthquake prediction programs. These programs include laboratory and field studies of rock behavior before, during, and after large earthquakes, as well as monitoring activity along major active faults. Most earthquake prediction work in the United States is done by the U.S. Geological Survey (USGS) and involves research into all aspects of earthquake-related phenomena.

The Chinese have perhaps the most ambitious earthquake prediction program in the world, which is understandable considering their long history of destructive earthquakes. Their earthquake prediction program was initiated soon after two large earthquakes occurred at Xingtai (300 km southwest of Beijing) in 1966, and includes extensive study and monitoring of all possible earthquake precursors. Chinese seismologists successfully predicted the 1975 Haicheng earthquake, but unfortunately, they failed to predict the devastating 1976 Tangshan earthquake that killed at least 242,000 people, and the 2008 Sichuan earthquake that killed more than 69,000 people.

Progress is being made toward dependable, accurate earthquake predictions, and studies are underway to assess public reactions to long-, medium-, and short-term earthquake warnings. However, unless short-term warnings are actually followed by an earthquake, most people will probably ignore the warnings, as they frequently do now for hurricanes, tornadoes, and tsunami.

LO9 EARTHQUAKE CONTROL

Reliable earthquake prediction is still in the future, but can anything be done to control, or at least partly control, earthquakes? Because of the tremendous energy involved, it seems unlikely that humans will ever be able to prevent earthquakes. However, it may be possible to gradually release the energy stored in rocks, thus decreasing the probability of a large earthquake and extensive damage.

During the early- to mid-1960s, Denver, Colorado, experienced numerous small earthquakes, which was surprising because Denver had not been prone to earthquakes in the past. Geologist David M. Evans suggested that the earthquakes were directly related to the injection of contaminated wastewater into a 3,674-m-deep disposal well at the Rocky Mountain Arsenal, northeast of Denver. A USGS study concluded that the pumping of waste fluids into the fractured rocks beneath the disposal well decreased the friction on opposites sides of the fractures—in effect lubricating them so that movement occurred, thus causing the earthquakes that Denver experienced.

Interestingly, a high degree of correlation was found when comparing the number of earthquakes in Denver against the average amount of contaminated fluids injected into the disposal well per month (Figure 8.19). Furthermore, when no waste fluids were injected, earthquake activity decreased dramatically.

In a similar situation to Denver, the number of earthquakes in the Dallas–Ft. Worth area of Texas has dramatically increased, resulting in more earthquakes during 2008 and 2009 than in the previous 30 years combined. Beginning in 2004, thousands of gas wells have been drilled in the area, and almost all of them have undergone hydraulic fracturing, in which large amounts of a high-pressure water mix are injected into the wells to open up the preexisting fractures to allow gas to flow into the wells. Although no one is claiming there is a link between the hydraulic fracturing associated with the drilling and the onset of small earthquakes, geologists are looking closely at the situation to see if there is a cause-and-effect relationship between the two events.

Based on these results, some geologists have proposed that fluids be pumped into the locked segments, or seismic gaps, of active faults to cause small- to moderate-sized earthquakes. They think that this would relieve the pressure on the fault and prevent a major earthquake from occurring.

Although this plan is intriguing, it has many potential problems. For instance, there is no guarantee that only a small earthquake might result. Instead, a major earthquake might occur, causing tremendous property damage and loss of life, especially in a densely populated area (Figure 8.20). Who would be responsible? Certainly, a great deal more research is needed before such an experiment is performed, even in an area of low population density.

LO10 WHAT IS EARTH'S INTERIOR LIKE?

During most of historic time, Earth's interior was perceived as an underground world of vast caverns, heat, and sulfur gases populated by demons. By the 1860s, scientists knew the average density of Earth and knew that pressure and temperature increased with depth. And even though Earth's interior is hidden from direct observation, scientists now have a reasonably good idea of its internal structure and composition.

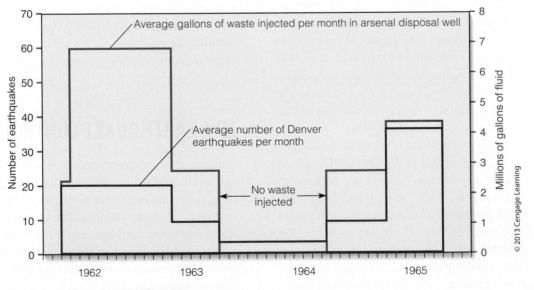

Figure 8.19 Controlling Earthquakes

This graph shows the relationship between the amount of wastewater injected into the Rocky Mountain Arsenal well per month and the average number of Denver earthquakes per month. There have been no significant earthquakes in Denver since injection of wastewater into the disposal well ceased in 1966.

Earth is generally depicted as consisting of concentric layers that differ in composition and density, separated from adjacent layers by rather distinct boundaries (Figure 8.21). Recall that the outermost layer, or *crust*, is Earth's thin skin. Below the crust and extending about halfway to Earth's center is the *mantle*, which comprises more than 80% of Earth's volume. The central part of Earth consists of a *core*, which is divided into a solid inner portion and a liquid outer part (Figure 8.21).

The behavior and travel times of P- and S-waves provide geologists with information about Earth's internal structure. Seismic waves travel outward as wavefronts from their source areas, although it is most convenient to depict them as *wave rays*, which are lines showing the direction of movement of small parts of wavefronts (Figure 8.3).

As we noted earlier, P- and S-wave velocity is determined by the density and elasticity of the materials they travel through, both of which increase with depth. Wave velocity is slowed by increasing density but increases in materials with greater elasticity. Because elasticity increases with depth faster than density, a general increase in seismic wave velocity takes place as the waves penetrate to greater depths. P-waves travel faster than S-waves under all circumstances, but unlike P-waves, S-waves are not transmitted through liquids, because liquids have no shear strength (rigidity); liquids simply flow in response to shear stress.

Because Earth is not a homogeneous body, seismic waves travel from one material into another of different density and elasticity, and thus their velocity and direction of travel change. That is, the waves are bent, a phenomenon known as *refraction*, in much the same way as light waves are refracted as they pass from air into more-dense water. Because seismic waves pass through materials of differing density and elasticity, they are continually refracted, so that their paths are curved. Wave rays travel in a straight line only when their direction of travel is perpendicular to a boundary (Figure 8.22).

Figure 8.20 Population Density of San Francisco
Downtown San Francisco sits atop the active plate boundary between the Pacific plate and the North American plate. The high density of population poses a risk to residents in the event of an earthquake. Although pumping seismic fluids may relieve the pressure of a fault and prevent major earthquakes from occurring, people in populated areas are reluctant to risk earthquake control, lest a major earthquake be initiated by the process.

In addition to refraction, seismic rays are *reflected*, much as light is reflected from a mirror. When seismic rays encounter a boundary separating materials of different density or elasticity, some of a wave's energy is reflected back to the surface (Figure 8.22). If we know the wave velocity and the time required for the wave to travel from its source to the boundary and back to the surface, we can calculate the depth of the reflecting boundary. Such information is useful in determining not only Earth's internal structure, but also the depths of sedimentary rocks that may contain petroleum. Seismic reflection is a common tool used in petroleum exploration.

Although changes in seismic wave velocity occur continuously with depth, P-wave velocity increases suddenly at the base of the crust and decreases abruptly at a depth of approximately 2,900 km (Figure 8.23). These marked changes in seismic wave velocity indicate a boundary called a **discontinuity**, across which a significant change in Earth materials or their properties occurs. Discontinuities are the basis for subdividing Earth's interior into concentric layers.

discontinuity A boundary across which seismic wave velocity or direction of travel changes abruptly, such as the mantle–core boundary.

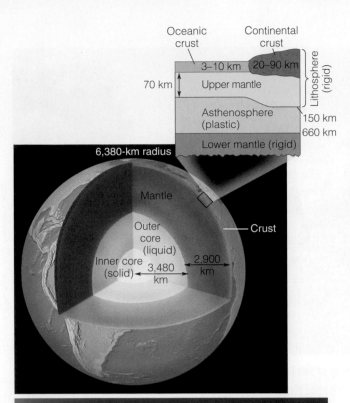

Figure 8.22 **Refraction and Reflection of Seismic Waves**

Refraction and reflection of P-waves as they encounter boundaries separating materials of different density or elasticity. Notice that the only wave ray not refracted is the one perpendicular to boundaries.

Earth's Composition and Density

	COMPOSITION	DENSITY (g/cm3)
Continental crust	Average composition of granodiorite	≈2.7
Oceanic crust	Upper part basalt, lower part gabbro	≈3.0
Mantle	Peridotite (made up of ferromagnesian silicates)	3.3–5.7
Outer core	Iron with perhaps 12% sulfur, silicon, oxygen, nickel, and potassium	9.9–12.2
Inner core	Iron with 10–20% nickel	12.6–13.0
Earth		5.5

© 2013 Cengage Learning

Figure 8.21 **Earth's Internal Structure**

The inset shows Earth's outer part in more detail. The asthenosphere is solid but behaves plastically and flows.

LO11 THE CORE

In 1906, R. D. Oldham of the Geological Survey of India realized that seismic waves arrived later than expected at seismic stations more than 130 degrees from an earthquake focus. He postulated that Earth has a core that transmits seismic waves more slowly than shallower Earth materials. We now know that P-wave velocity decreases markedly at a depth of 2,900 km, which indicates an important discontinuity now recognized as the core–mantle boundary (Figure 8.23).

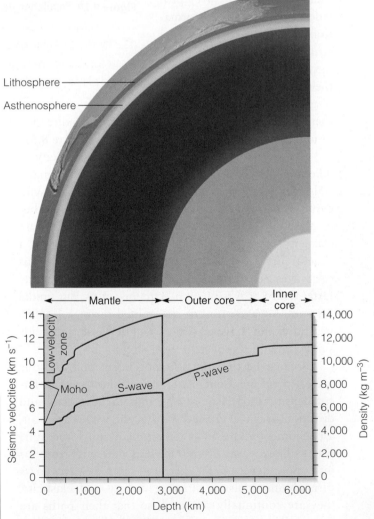

Figure 8.23 **Seismic Wave Velocities**

Profiles showing seismic wave velocities versus depth. Several discontinuities are shown, across which seismic wave velocities change rapidly.

HTTP://SCIENCE.JRANK.ORG/PAGES/47534/EARTH-STRUCTURE.HTML.

Because of the sudden decrease in P-wave velocity at the core–mantle boundary, P-waves are refracted in the core. Consequently, little P-wave energy reaches the surface in the area between 103 degrees and 143 degrees from an earthquake focus (Figure 8.24a). This **P-wave shadow zone**, as it is called, is an area in which little P-wave energy is recorded by seismographs.

The P-wave shadow zone is not a perfect shadow zone because some weak P-wave energy is recorded within it. Scientists proposed several hypotheses to account for this observation, but all were rejected by the

Figure 8.24 P-Wave and S-Wave Shadow Zones

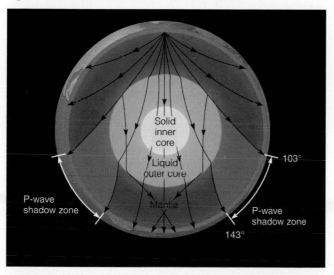

a. P-waves are refracted such that no direct P-wave energy reaches the surface in the P-wave shadow zone.

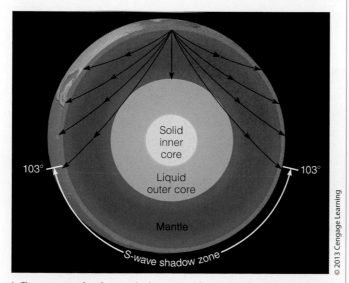

b. The presence of an S-wave shadow zone indicates that S-waves are being blocked within Earth.

Danish seismologist Inge Lehmann, who in 1936 postulated that the core is not entirely liquid as previously thought. She proposed that seismic wave reflection from a solid inner core accounts for the arrival of weak P-wave energy in the P-wave shadow zone, a proposal that was quickly accepted by seismologists.

In 1926, the British physicist Harold Jeffreys realized that S-waves were not simply slowed by the core, but were completely blocked by it. So, besides a P-wave shadow zone, a much larger and more complete **S-wave shadow zone** also exists (Figure 8.24b). At locations greater than 103 degrees from an earthquake focus, no S-waves are recorded, which indicates that S-waves cannot be transmitted through the core. S-waves will not pass through a liquid, so it seems that the outer core must be liquid or behave as a liquid. The inner core, however, is thought to be solid, because P-wave velocity increases at the base of the outer core.

Density and Composition of the Core

The core constitutes 16.4% of Earth's volume and nearly one-third of its mass. Geologists can estimate the core's density and composition by using seismic evidence and laboratory experiments. Furthermore, meteorites, which are thought to represent remnants of the material from which the solar system formed, are used to make estimates of density and composition. From these studies, the density of the outer core has been determined to vary from 9.9 to 12.2 g/cm³. At Earth's center, the pressure is equivalent to approximately 3.5 million times normal atmospheric pressure.

The core cannot be composed of minerals common at the surface, because, even under the tremendous pressures at great depth, they would still not be dense enough to yield an average density of 5.5 g/cm³ for Earth. Both the outer and inner cores are thought to be composed largely of iron, but pure iron is too dense to be the sole constituent of the outer core. It must be "diluted" with elements of lesser density. Laboratory experiments and comparisons with iron meteorites indicate that perhaps 12% of the outer core consists of sulfur and possibly some silicon, oxygen, nickel, and potassium (Figure 8.21).

P-wave shadow zone An area between 103 and 143 degrees from an earthquake focus where little P-wave energy is recorded by seismographs.

S-wave shadow zone Those areas more than 103 degrees from an earthquake focus where no S-waves are recorded.

In contrast, pure iron is not dense enough to account for the estimated density of the inner core, so perhaps 10%–20% of the inner core consists of nickel. These metals form an iron–nickel alloy thought to be sufficiently dense under the pressure at that depth to account for the density of the inner core.

LO12 EARTH'S MANTLE

Another significant discovery about Earth's interior was made in 1909 when the Yugoslavian seismologist Andrija Mohorovičić detected a seismic discontinuity at a depth of about 30 km. While studying the arrival times of seismic waves from Balkan (part of southeastern Europe) earthquakes, Mohorovičić noticed that seismic stations a few hundred kilometers from an earthquake's epicenter were recording two distinct sets of P- and S-waves.

From his observations, Mohorovičić concluded that a sharp boundary separates rocks with different properties at a depth of about 30 km. He postulated that P-waves below this boundary travel at 8 km per second, whereas those above the boundary travel at 6.75 km per second. When an earthquake occurs, some waves travel directly from the focus to a seismic station, whereas others travel through the deeper layer, and some of their energy is refracted back to the surface (Figure 8.25). The waves traveling through the deeper layer (the mantle) travel farther to a seismic station, but they do so more rapidly and arrive before those that travel more slowly in the shallower layer.

The boundary identified by Mohorovičić separates the crust from the mantle and is now called the **Mohorovičić discontinuity**, or simply the **Moho**. It is present everywhere except beneath spreading ridges. However, its depth varies: beneath continents, it ranges from 20 to 90 km, with an average of 35 km; beneath the sea floor, it is 5 to 10 km deep.

The Mantle's Structure, Density, and Composition

Although seismic wave velocity in the mantle increases with depth, several discontinuities exist. Between depths of 100 and 250 km, both P- and S-wave velocities decrease markedly (Figure 8.23). This 100- to 250-km-deep layer is the *low-velocity zone*, which corresponds closely to the *asthenosphere* (Figure 8.21), a layer in which the rocks are close to their melting point and are less elastic, accounting for the observed decrease in seismic wave velocity. The asthenosphere is an important zone because it is where most magma is generated, especially under the ocean basins. Furthermore, it lacks strength, flows plastically, and is thought to be the layer over which the plates of the outer, rigid *lithosphere* move.

Other discontinuities are also present at deeper levels within the mantle. But unlike those between the crust and mantle or between the mantle and core, these probably represent structural changes in minerals rather than compositional changes. In other words, geologists

Figure 8.25 Seismic Discontinuity
Andrija Mohorovičić studied seismic waves and detected a seismic discontinuity at a depth of about 30 km. The deeper, faster seismic waves arrive at seismic stations first, even though they travel farther. This discontinuity, now known as the Moho, is between the crust and mantle.

think that the mantle is composed of the same material throughout, but the structural states of minerals such as olivine change with depth.

Although the mantle's density, which varies from 3.3 to 5.7 g/cm³, can be inferred rather accurately from seismic waves, its composition is less certain. The igneous rock *peridotite*, containing mostly ferromagnesian silicates, is considered the most likely component of the upper mantle. Laboratory experiments indicate that it possesses physical properties that account for the mantle's density and observed rates of seismic wave transmissions. Peridotite also forms the lower parts of igneous rock sequences thought to be fragments of the oceanic crust and upper mantle emplaced on land.

LO13 EARTH'S INTERNAL HEAT

During the 19th century, scientists realized that the temperature in deep mines increases with depth, and this same trend has been observed in deep drill holes. This temperature increase with depth, or **geothermal gradient**, is approximately 25°C/km near the surface. In areas of active or recently active volcanism, the geothermal gradient is greater than in adjacent nonvolcanic areas, and temperature rises faster beneath spreading ridges than elsewhere beneath the seafloor.

Much of Earth's internal heat is generated by radioactive decay, especially the decay of isotopes of uranium and thorium and, to a lesser degree, potassium-40. When these isotopes decay, they emit energetic particles and gamma rays that heat surrounding rocks. Because rock is such a poor conductor of heat, it takes little radioactive decay to build up considerable heat, given enough time.

Unfortunately, the geothermal gradient is not useful for estimating temperatures at great depth. If we were simply to extrapolate from the surface downward, the temperature at 100 km would be so high that, despite the great pressure, all known rocks would melt. Current estimates of the temperature at the base of the crust are 800°C to 1,200°C. The latter figure seems to be an upper limit; if it were any higher, melting would be expected. Furthermore, fragments of mantle rock, thought to have come from depths of 100–300 km, appear to have reached equilibrium at these depths at a temperature of approximately 1,200°C. At the core–mantle boundary,

the temperature is probably between 2,500°C and 5,000°C; the wide range of values indicates the uncertainties of such estimates. If these figures are reasonably accurate, the geothermal gradient in the mantle is only about 1°C/km.

geothermal gradient Earth's temperature increase with depth; it averages 25°C/km near the surface but varies from area to area.

LO14 EARTH'S CRUST

Our main concern in the latter part of this chapter is Earth's interior; however, to be complete, we must briefly discuss the crust, which, along with the upper mantle, constitutes the lithosphere.

Continental crust is complex, consisting of all rock types, but it is usually described as "granitic," meaning that its overall composition is similar to that of granitic rocks. With the exception of metal-rich rocks such as iron ore deposits, most rocks of the continental crust have densities between 2.5 g/cm³ and 3.0 g/cm³, with the average density of the crust being about 2.7 g/cm³. P-wave velocity in continental crust is about 6.75 km/sec, but at the base of the crust, P-wave velocity abruptly increases to about 8 km/sec. Continental crust averages 35 km thick, but its thickness varies from 20 km to 90 km. Beneath mountain ranges such as the Rocky Mountains, the Alps in Europe, and the Himalayas in Asia, continental crust is much thicker than it is in adjacent areas. In contrast, continental crust is much thinner than average beneath the Rift Valleys of East Africa and in a large area called the Basin and Range Province in the western United States and northern Mexico. The crust in these areas has been stretched and thinned in what appear to be the initial stages of rifting (see Chapter 2).

In contrast to continental crust, oceanic crust is simpler, consisting of gabbro in its lower part and overlain by basalt. It is thinnest, about 5 km, at spreading ridges, and nowhere is it thicker than 10 km. Its average density of 3.0 g/cm³ accounts for the fact that it transmits P-waves at about 7 km/sec. In fact, this P-wave velocity is what one would expect if oceanic crust is composed of basalt and gabbro.

9

Deformation, Mountain Building, and the Continents

View from Pakistan of the Karakoram Range, which is part of the Himalayan mountain system. This range lies on the borders of Pakistan, India, and China. The Karakoram Range has more peaks over 8,000 m high than any other mountain range, including K2, which at 8,611 m is the second highest peak on Earth. In addition, the range has a large concentration of glaciers.

Persis Sturges

Introduction

"**Earth** is an active planet with processes **driven by internal heat, particularly** plate movements."

"**S**olid as a rock" implies permanence and durability, but you already know that when rocks weather they disaggregate and decompose (see Chapter 6), and they behave very differently during metamorphism (see Chapter 7). Under the tremendous pressure and high temperature at several kilometers below the surface, rock layers yield to forces and actually crumple or fold, yet remain solid. And at shallower depths they may fracture, or fold and fracture (Figure 9.1). In any case, dynamic forces within Earth cause **deformation**, a general term for all changes in the shape or volume of rocks.

The fact that dynamic forces continue to operate within Earth is obvious from seismic activity, volcanism, plate movements, and the evolution of mountains in South America, Asia, and elsewhere. In short, Earth is an active planet with several processes driven by internal heat, especially plate movements; much of Earth's seismic activity, volcanism, deformation, and mountain building take place at convergent plate boundaries. The other terrestrial planets and Earth's moon, with the possible exception of Venus, show little evidence of ongoing deformation, volcanism, and so on.

The origin of Earth's truly large mountain ranges on the continents involves tremendous deformation—usually accompanied by emplacement of plutons,

volcanism, and metamorphism—at convergent plate boundaries. And, in some cases, this activity continues even now (see the chapter-opening photo). Thus, deformation and mountain building are closely related topics, and accordingly we consider both in this chapter.

> **deformation** A general term for any change in shape or volume, or both, of rocks in response to stress; involves folding and fracturing.

The past and continuing evolution of continents involves not only deformation at continental margins, but also additions of new material to existing continents, a phenomenon known as *continental accretion* (see Chapter 17). North America, for instance, has not always had its present shape and area. Indeed, it began evolving during the Archean Eon (4.6 to 2.5 billion years ago) as new material was added to the continent at deformation belts along its margins.

Much of this chapter is devoted to a review of *geologic structures*, such as folded and fractured rock layers resulting from deformation; their descriptive terminology; and the forces responsible for them. There are several practical reasons to study deformation and mountain building. For one thing, deformed rock layers provide a record of the kinds and intensities of forces that operated during the past. Thus, interpretations of these structures allow us to satisfy our curiosity about Earth history, and, in addition, such studies are essential in engineering endeavors such as choosing sites for dams, bridges, and nuclear power plants, especially if they are in areas of ongoing deformation. Also, many aspects of mining and exploration for petroleum and natural gas rely on correctly identifying geologic structures.

Learning Outcomes

After reading this unit, you should be able to do the following:

LO1 Explain how rock deformation occurs

LO2 Understand strike and dip—the orientation of deformed rock layers

LO3 Identify the types of deformation and geologic structures

LO4 Understand deformation and the origin of mountains

LO5 Describe Earth's continental crust

LO1 ROCK DEFORMATION: HOW DOES IT OCCUR?

We defined the term *deformation* as any change in the volume or shape of rocks, but our reference to rocks means rock layers as depicted in many images in this text, not isolated stones you might find in your driveway or on a streambed. The type of rock is irrelevant,

Figure 9.1 Many Rocks Show the Effects of Deformation

a. These rock layers in the Balkan Mountains of Serbia have been folded. The vertical dimension of this image is about 10 m.

James S. Monroe

b. Deformation of rocks in Utah by fracturing. Notice that the continuity of the rock layers has been disrupted along a fracture, or what geologists call a fault.

Sue Monroe

stress The force per unit area applied to a material such as rock.

strain Deformation caused by stress.

compression Stress resulting when materials are squeezed by external forces directed toward one another.

tension A type of stress in which forces act in opposite directions but along the same line, thus tending to stretch an object.

shear stress The result of forces acting parallel to one another but in opposite directions; results in deformation by displacement of adjacent layers along closely spaced planes.

although layered rocks show the effects of deformation most clearly. In any case, rock layers may be crumpled into folds or fractured as a result of **stress**, which results from force applied to a given area of rock. The rock's internal strength resists stress, but if the stress is great enough, the rock undergoes **strain**, which is simply deformation caused by stress. The terminology is a little confusing at first, but the following discussion and reference to Figure 9.2 will help you understand these terms.

Stress and Strain

Remember that stress is the force applied to a given area of rock, usually expressed in kilograms per square centimeter (kg/cm²). For example, the stress, or force, exerted by a person walking on an ice-covered pond is a function of the person's weight and the area beneath her or his feet. The ice's internal strength resists the stress unless the stress is too great, in which case the ice may bend or crack as it is strained (deformed) (Figure 9.2). To avoid breaking through the ice, the person may lie down; this does not reduce the person's weight on the ice, but it does distribute their weight over a larger area, thus reducing the stress per unit area.

Although stress is force per unit area, it comes in three varieties: *compression, tension,* and *shear.* In **compression**, materials are squeezed, or compressed, by forces directed toward one another along the same line, as when you squeeze a rubber ball in your hand. Rock layers in compression tend to be shortened in the direction of stress by either folding or fracturing (Figure 9.1). **Tension** results from forces acting along the same line, but in opposite directions. Tension tends to lengthen rocks or pull them apart. Incidentally, rocks are much stronger in compression than they are in tension. In **shear stress**, forces act parallel to one another, but in opposite directions, resulting in deformation by displacement along closely spaced planes.

Figure 9.2 Stress and Strain Exerted on an Ice-Covered Pond

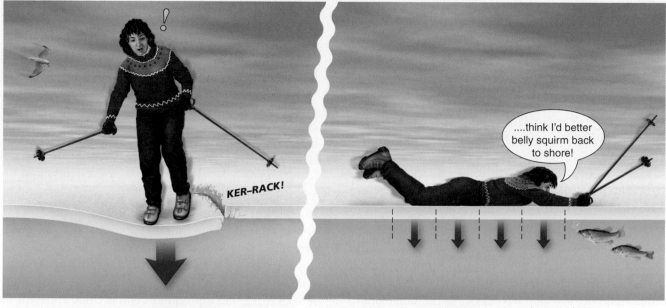

a. The woman weighs 65 kg (6,500 g). Her weight is imparted to the ice through her feet, which have a contact area of 120 cm². The stress she exerts on the ice (6,500/120 cm²) is 54 g/cm². This is sufficient stress to cause the ice to crack.

b. To avoid plunging into the freezing water, the woman lies flat, thereby decreasing the stress she exerts on the ice. Her weight remains the same, but her contact area with the ice is 3,150 cm², so the stress is only about 2 g/cm² (6,500 g/3,150 cm²), which is well below the threshold needed to crack the ice.

Like people, rocks under stress and strain either bounce back, bend, or break.

Types of Strain

Geologists characterize strain as **elastic strain** if deformed rocks return to their original shape when the deforming forces are relaxed. In Figure 9.2, the ice on the pond may bend under a person's weight but return to its original shape once the person leaves. As you might expect, rocks are not very elastic, but Earth's crust behaves elastically when loaded by glacial ice and depressed into the mantle.

As stress is applied, rocks respond first by elastic strain, but when strained beyond their elastic limit, they undergo **plastic strain** as when they yield by folding, or they behave like brittle solids and **fracture**. In either folding or fracturing, the strain is permanent; that is, the rocks do not recover their original shape or volume even if the stress is removed.

> **elastic strain** A type of deformation in which the material returns to its original shape when stress is relaxed.
>
> **plastic strain** Permanent deformation of a solid with no failure by fracturing.
>
> **fracture** A break in rock resulting from intense applied pressure.

strike The direction of a line formed by the intersection of an inclined plane and a horizontal plane.

dip A measure of the maximum angular deviation of an inclined plane from horizontal.

geologic structure Any feature in rocks that results from deformation, such as folds, joints, and faults.

Whether strain is elastic, plastic, or produces fracture depends on the kind of stress applied, pressure and temperature, rock type, and the length of time rocks are subjected to stress. A small stress applied over a long period, as on a mantelpiece supported only at its ends, will cause the rock to sag; that is, the rock deforms plastically. By contrast, a large stress applied rapidly to the same object, as when struck by a hammer, results in fracture. Rock type is important, because not all rocks have the same internal strength and thus respond to stress differently. Some rocks are *ductile*, whereas others are *brittle*. Brittle rocks show little or no plastic strain before they fracture, but ductile rocks exhibit a great deal.

Many rocks show the effects of plastic strain that must have taken place deep within the crust. At or near the surface, rocks commonly behave like brittle solids and fracture, but at depth, they become more ductile with increasing pressure and temperature.

LO2 STRIKE AND DIP: THE ORIENTATION OF DEFORMED ROCK LAYERS

One concept in geology is the *principle of original horizontality*, meaning that sediments accumulate in horizontal or nearly horizontal layers (see Figure 16.2). Thus, if we observe steeply inclined sedimentary rocks, we are justified in inferring that they were deposited nearly horizontally, lithified, and then tilted into their present position. Rock layers deformed by folding, faulting, or both are no longer in their original position, so geologists use *strike* and *dip* to describe their orientation with respect to a horizontal plane.

By definition, **strike** is the direction of a line formed by the intersection of a horizontal plane and an inclined plane. The surfaces of the rock layers in Figure 9.3 are good examples of inclined planes, whereas the land surface is a horizontal plane. The direction of the line formed at the intersection of these planes is the strike of the rock layers. The strike line's orientation is determined by using a compass to measure its angle with respect to north. **Dip** is a measure of an inclined plane's deviation from horizontal, so it must be measured at right angles to strike direction (Figure 9.3).

Geologic maps showing the age, aerial distribution, and geologic structures of rocks use a special symbol to indicate strike and dip. A long line oriented in the appropriate direction indicates strike, and a short line perpendicular to the strike line shows the direction of dip (Figure 9.3). Adjacent to the strike and dip symbol is a number corresponding to the dip angle. The usefulness of strike and dip symbols will become apparent in the sections on folds and faults.

LO3 DEFORMATION AND GEOLOGIC STRUCTURES

Remember that *deformation* and its synonym *strain* refer to changes in the shape or volume of rocks. During deformation, rocks might be crumpled into folds, or they might be fractured, or perhaps folded and fractured. Any of these features resulting from deformation is referred to as a **geologic structure**. Geologic

Figure 9.3 Strike and Dip of Deformed Rock Layers
We can infer that the sedimentary rocks in this illustration were deposited horizontally or nearly so, lithified, and then deformed. To describe their orientation, geologists use the terms *strike* and *dip*. Strike is the line formed by the intersection of a horizontal plane with an inclined plane. Dip is the maximum angular deviation of the inclined plane from horizontal. Notice the symbol that shows strike and dip.

structures are found almost everywhere that rock exposures are present, and many are detected far below the surface by drilling and several geophysical techniques.

Folded Rock Layers

If you place your hands on a tablecloth and move them toward one another, the tablecloth crumples into a series of up- and down-arched folds. Rock layers behave similarly when in compression as they deform into **folds**, but in this case the folding is permanent. That is, plastic strain has taken place, so even if the stress is removed, the rock layers remain folded. Most folding probably occurs deep in the crust where pressure and temperature are high and rocks are more ductile than they are at or near the surface. The configuration of folds and the intensity of folding vary, but there are only three basic types of folds: *monoclines*, *anticlines*, and *synclines*.

Monoclines
A simple bend or flexure in otherwise horizontal or uniformly dipping rock layers is a monocline. The large monocline in Figure 9.4 formed when the Bighorn Mountains in Wyoming rose vertically along a fracture. The fracture did not penetrate to the surface, so as uplift of the mountains proceeded, the near-surface rocks were bent so that they now appear to be draped over the margin of the uplifted block.

Anticlines and Synclines
An anticline is an up-arched or convex upward fold with the oldest rock layers in its core, whereas a syncline is a down-arched

Monocline

Sue Monroe

Figure 9.4 Monocline
A monocline in the Bighorn Mountains in Wyoming.

Figure 9.5 **Anticlines and Synclines**

Folded rocks in the Calico Mountains of southeastern California. Compression was responsible for these folds, which are, from left to right, a syncline, an anticline, and a syncline.

or concave downward fold in which the youngest rock layers are in its core (Figure 9.5). Anticlines and synclines have an axial plane connecting the points of maximum curvature of each folded layer (Figure 9.6); the axial plane divides folds into halves, each half being a *limb*. Because folds are most often found in a series of anticlines alternating with synclines, an anticline and adjacent syncline share a limb.

Folds are commonly exposed to view in areas of deep erosion, but even where eroded, strike and dip and the relative ages of the folded rock layers easily distinguish anticlines from synclines. Notice in Figure 9.7 that in the surface view of the anticline, each limb dips outward, or away from the center of the fold, and the oldest

> **It is important to remember that anticlines and synclines are simply folded rock layers and do not necessarily correspond to high and low areas at the surface.**

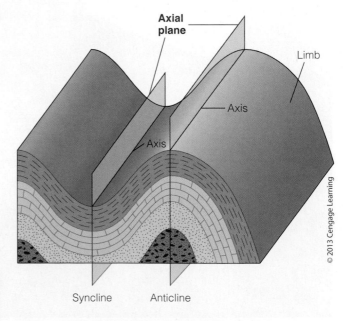

Figure 9.6 **Syncline and Anticline Axial Planes**

Syncline and anticline showing the axial plane, axis, and fold limbs.

Oldest exposed rock unit

Youngest exposed rock unit

50 50 50 50 50 50

Figure 9.7 Eroded Anticlines and Synclines
Geologists identify eroded anticlines and synclines by strike and dip and the relative ages of the folded rock layers.

© 2013 Cengage Learning

exposed rocks are in the fold's core. In an eroded syncline, though, each limb dips inward toward the fold's center, where the youngest exposed rocks are found.

The folds described so far are *upright*, meaning that their axial planes are vertical, and both fold limbs dip at the same angle (Figure 9.7). In many folds, the axial plane is not vertical, the limbs dip at different angles, and the folds are characterized as *inclined* (Figure 9.8a). If both limbs dip in the same direction, the fold is *overturned*. That is, one limb has been rotated more than 90 degrees from its original position so that it is now upside down (Figure 9.8b). In some areas, deformation has been so intense that axial planes of folds are now horizontal, giving rise to what geologists call *recumbent folds* (Figure 9.8c). Overturned and recumbent folds are particularly common in mountains resulting from compression at convergent plate boundaries.

Plunging Folds
In some folds, the fold axis—a line formed by the intersection of the axial plane with the folded layers—is horizontal and the folds are *nonplunging* (Figure 9.7). Much more commonly, though, fold axes are inclined so that they appear to plunge beneath adjacent rocks, and the folds are said to be *plunging* (Figure 9.9).

It might seem that with this additional complication, differentiating plunging anticlines from plunging synclines would be much more difficult. However, you can use exactly the same criteria that you used for nonplunging folds. Therefore, all rock layers dip away from the fold axis in plunging anticlines and toward the axis in plunging synclines. The oldest exposed rocks are in the core of an eroded plunging anticline, whereas the youngest exposed rock layers are found in the core of an eroded plunging syncline (Figure 9.9b).

In Chapter 6, we noted that anticlines form one type of structural trap in which petroleum and natural gas might accumulate. As a matter of fact, most of the world's petroleum production comes from anticlines, although other geologic structures and stratigraphic traps are also important.

dome A rather circular geologic structure in which all rock layers dip away from a central point and the oldest exposed rocks are in the dome's center.

basin A circular fold in which all strata dip inward toward a central point and the youngest exposed strata are in the center.

Domes and Basins
Anticlines and synclines are elongate structures, meaning that their length greatly exceeds their width. In contrast, folds that are nearly equidimensional (i.e., circular) are *domes* and *basins*. In a **dome**, all of the folded strata dip outward from a central point (as opposed to outward from a line as in an anticline), and the oldest exposed rocks are at the center of the fold (Figure 9.10a). In contrast, a **basin** has all strata dipping inward toward a central point, and the youngest exposed rocks are at the fold's center (Figure 9.10b).

Unfortunately, the terms *dome* and *basin* are also used to distinguish high and low areas of Earth's surface, but domes and basins as defined here do not necessarily

Figure 9.8 Inclined, Overturned, and Recumbent Folds

Inclined

Overturned

a. An inclined fold. The axial plane is not vertical, and the fold limbs dip at different angles.

b. Overturned folds. Both fold limbs dip in the same direction, but one limb is inverted. Notice the special strike and dip symbol to indicate overturned beds.

Recumbent

c. Recumbent folds are folds in which the axial planes are horizontal.

© 2013 Cengage Learning

joint A fracture along which no movement has occurred or where movement is perpendicular to the fracture surface.

correspond with mountains or valleys. In some of the following discussions, we will use these terms in other contexts, but we will try to be clear when we refer to surface elevations as opposed to geologic structures.

Joints

Joints are fractures along which no movement has taken place parallel with the fracture surface (Figure

9.11), although they may open up; that is, joints may show movement perpendicular to the fracture. Remember that rocks near the surface are brittle and therefore commonly fail by fracturing when subjected to stress. In fact, almost all near-surface rocks have joints that form in response to compression, tension, and shearing. They vary from minute fractures to those extending for many kilometers and are often arranged in two or perhaps three prominent sets.

We already discussed columnar joints that form when lava or magma in some shallow plutons cools and contracts (see Figure 5.5). A different type of jointing previously discussed is sheet joints that form in response to pressure release (see Figure 6.3).

Figure 9.9 Plunging Folds

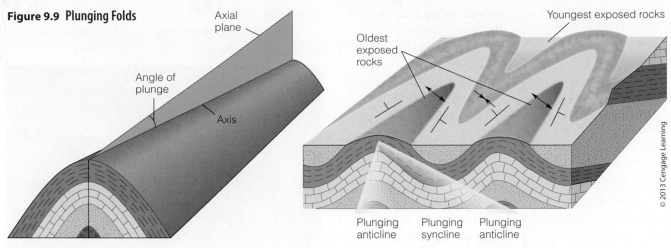

a. A plunging fold.

b. Surface and cross-sectional views of plunging folds. The long arrow is the geologic symbol for a plunging fold; it shows the direction of plunge.

Figure 9.10 Domes and Basins

Dome

a. Notice that in a dome, the oldest exposed rocks are in the center and all rocks dip outward from a central point.

Basin

b. In a basin, the youngest exposed rocks are in the center and all rocks dip inward toward a central point.

Faults

Joints and faults are both fractures, but on joints no movement occurs parallel with the fracture surface, whereas on **faults**, blocks on opposite sides of the fracture move parallel with the fracture surface, which is a **fault plane** (Figure 9.12a). Faults that penetrate to the surface might show a *fault scarp*, a bluff or cliff formed by vertical movement (Figure 9.12b). In some cases, the fault plane is scratched and polished, but in others the movement of the blocks on opposite sides of the fault grind and pulverize the rock into a zone of *fault breccia*.

Notice in Figure 9.12a that the rocks overlying the fault make up the **hanging wall block**, whereas those beneath the fault constitute the **footwall block**. You can recognize these two blocks on any fault with the exception of one that is vertical—that is, one that dips at 90 degrees. In order to identify some faults, you must identify these two blocks and determine the direction of relative movement of the blocks. By *relative movement*, we mean which one appears to have moved up or down the fault plane. For example, in Figure 9.12a, the footwall block may have moved up, the hanging wall may have moved down, or both blocks may have moved. Nevertheless, the hanging wall block appears to have

fault A fracture along which rocks on opposite sides of the fracture have moved parallel with the fracture surface.

fault plane A fault surface that is more or less planar.

hanging wall block The block of rock that overlies a fault plane.

footwall block The block of rock that lies beneath a fault plane.

dip-slip fault A fault on which all movement is parallel with the dip of the fault plane.

normal fault A dip-slip fault on which the hanging wall block has moved downward relative to the footwall block.

reverse fault A dip-slip fault on which the hanging wall block has moved upward relative to the footwall block.

thrust fault A type of reverse fault in which a fault plane dips less than 45 degrees.

strike-slip fault A fault involving horizontal movement of blocks of rock on opposite sides of a fault plane.

moved down relative to the footwall block.

Remember that geologists use *strike* and *dip* to define the orientation of dipping rock layers. Fault planes are also dipping planes, so this same concept applies to them. As a matter of fact, two types of faults are recognized based on whether the blocks on opposite sides of the fault moved parallel to the direction of dip (dip-slip faults) or along the direction of strike (strike-slip faults).

Dip-Slip Faults

All movement on **dip-slip faults** takes place parallel with the fault's dip; that is, movement is vertical, either up or down the fault plane. In Figure 9.13a, for example, the hanging wall block moved down relative to the footwall block, giving rise to a **normal fault** (Figure 9.14a). In contrast, in a **reverse fault**, the hanging wall block moves up relative to the footwall block (Figure 9.13b, 9.14b). In Figure 9.13c, the hanging wall block also moved up relative to the footwall block, but the fault has a dip of less than 45 degrees and is a special variety of reverse fault known as a **thrust fault**.

Numerous normal faults are present along one or both sides of mountain ranges in the Basin and Range Province of the western United States where the crust is being stretched and thinned. The Sierra Nevada at the western margin of the Basin and Range is bounded by normal faults, and the range has risen along these faults so that it now stands more than 3,000 m above the lowlands to the east.

Large-scale examples of reverse and thrust faults are both found in mountain ranges that formed at convergent plate margins, where one would expect compression (Figure 9.13b, c).

Strike-Slip Faults

Strike-slip faults, resulting from shear stresses, show horizontal movement with blocks on opposite sides

Figure 9.11 Joints

Fractures along which no movement has taken place parallel with the fracture surface are called joints.

C. G. Tillman

a. Joints intersecting at right angles yield this rectangular pattern in Wales.

James S. Monroe

b. The Jurassic-age Navajo Sandstone in Zion National Park, Utah. Vertical joints intersect cross-beds, hence the name Checkerboard Mesa.

of the fault sliding past one another (Figure 9.13d). In other words, all movement is in the direction of the fault plane's strike. Several large strike-slip faults are known, but the best studied is the San Andreas fault, which cuts through coastal California. Recall from Chapter 2 that the San Andreas fault is called a *transform fault* in plate tectonics terminology.

Strike-slip faults are characterized as right-lateral or left-lateral, depending on the apparent direction of offset. In Figure 9.13d, for example, observers looking

Figure 9.12 Faults

Fractures along which movement has occurred parallel to the fracture surface are called faults.

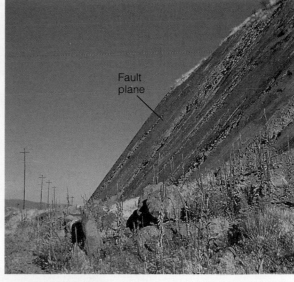

a. Terms used to describe the orientation of a fault plane. Striae are scratch marks that form when one block slides past another. You can measure offset or displacement on a fault wherever the truncated end of one feature (point A) can be related to its equivalent along the fault (B).

b. A polished, scratched fault plane and fault scarp near Klamath Falls, Oregon.

at the block on the opposite side of the fault from their location notice that it appears to have moved to the left. Accordingly, this is a *left-lateral strike-slip fault*. If it had been a *right-lateral strike-slip fault*, the block across the fault from the observers would appear to have moved to the right.

Oblique-Slip Faults

The movement on most faults is primarily dip slip or strike-slip, but on **oblique-slip faults**, both types of movement take place. Strike-slip movement might be accompanied by a component of dip-slip, giving rise to a combined movement that includes left-lateral and reverse, or right-lateral and normal (Figure 9.13e).

LO4 DEFORMATION AND THE ORIGIN OF MOUNTAINS

Mountain is a designation for any area of land that stands significantly higher, at least 300 m, than the surrounding country and has a restricted summit area. Some mountains are single, isolated peaks, but most are parts of linear associations of peaks and ridges known as *mountain ranges* that are related in age and origin. A *mountain system*, a complex linear zone of deformation and crustal

thickening, on the other hand, consists of several or many mountain ranges. The Teton Range in Wyoming is one of many ranges in the Rocky Mountains. Mountains form in several ways, but the truly large mountains on continents result mostly from compression-induced deformation at convergent plate boundaries (see Chapter 2).

oblique-slip fault A fault showing both dip-slip and strike-slip movement.

Mountain Building

Block-faulting is one way that mountains form, which is caused by movement on normal faults so that one or more blocks are elevated relative to adjacent blocks (Figure 9.15). A classic example is the Basin and Range Province, which is centered on Nevada but extends into adjacent areas. Differential movement on faults has produced uplifted blocks called *horsts* and down-dropped blocks called *grabens*. Erosion of the horsts has produced mountainous topography.

Volcanic outpourings form chains of volcanic mountains such as the Hawaiian Islands, where a plate moves over a hot spot (see Figure 2.22). Some mountains such as the Cascade Range of the Pacific Northwest are made up almost entirely of volcanic rocks, and the mid-ocean ridges are also mountains. However, most mountains

Figure 9.13 Types of Faults

(overhang collapses into rubble)

.a. Normal fault—hanging wall block (HW) moves down relative to the footwall block (FW).

b. Reverse fault—hanging wall block moves up relative to the footwall block.

(wide zone of rubble and broken ground)

c. A thrust is a type of reverse fault with a fault plane dipping at less than 45 degrees.

Both dip and strike slip components of movement.

© 2013 Cengage Learning

d. Left-lateral strike-slip fault.

e. An oblique-slip fault involves a combination of dip-slip and strike-slip movements.

---

orogeny An episode of mountain building involving deformation, usually accompanied by igneous activity, and crustal thickening.

on land are composed of all rocks types and show clear evidence of deformation by compression.

Plate Tectonics and Mountain Building

Any theory that accounts for mountain building, or what geologists call **orogeny**, must adequately explain the characteristics of mountain ranges, such as their geometry and location; they tend to be long and narrow and at or near plate margins. Mountains also show intense deformation, especially compression-induced overturned and recumbent folds, as well as reverse and thrust faults. Furthermore, granitic plutons and regional metamorphism characterize the interiors or cores of mountain ranges. Another feature is sedimentary rocks that are now far above sea level but that were deposited in shallow and deep marine environments.

Figure 9.14 Dip-Slip Faults

James S. Monroe

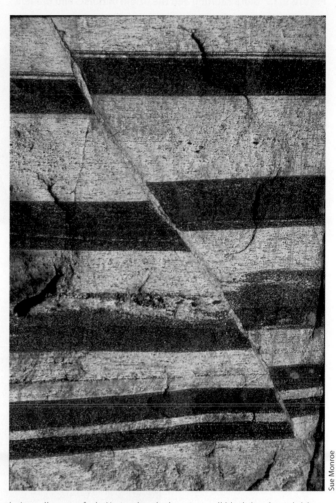

Sue Monroe

a. A normal fault at Mt. Carmel Junction, Utah. The hanging wall block (on the right) has moved down about 2 m relative to the footwall block.

b. A small reverse fault. Notice that the hanging wall block (on the right) has moved up relative to the footwall block.

Deformation and associated activities at convergent plate boundaries are certainly important processes in mountain building. They account for a mountain system's location and geometry, as well as complex geologic structures, plutons, and metamophism. Yet, the present-day topographic expression of mountains is also related to several surface processes, such as mass wasting (gravity-driven processes including landslides), glaciers, and running water. In other words, erosion also plays an important role in the evolution of mountains.

Orogenies at Oceanic–Oceanic Plate Boundaries
Deformation, igneous activity, and the origin of a volcanic island arc characterize orogenies that take place where oceanic lithosphere is subducted beneath oceanic lithosphere (see Figure 2.18b). Sediments derived from the island arc are deposited in an adjacent oceanic trench and then deformed and scraped off against the landward

side of the trench (Figure 9.16). These deformed sediments are part of a subduction complex, or an *accretionary wedge*, of intricately folded rocks cut by numerous thrust faults. In addition, orogenies in this setting are characterized by low-temperature, high-pressure metamorphism of the blueschist facies (see Figure 7.18).

Deformation caused largely by the emplacement of plutons also takes place in the island arc system, where many rocks show evidence of high-temperature, low-pressure metamorphism. The overall effect of an island arc orogeny is the origin of two more-or-less parallel orogenic belts consisting of a landward volcanic island arc underlain by batholiths and a seaward belt of deformed trench rocks (Figure 9.16). The Japanese Islands are a good example.

In the area between an island arc and its nearby continent, the back-arc basin, volcanic rocks and sediments derived from the island arc and the adjacent

Figure 9.15 Block Faulting and the Origin of Horsts and Grabens

a. Block faulting and the origin of horsts and grabens. Many of the mountain ranges in the Basin and Range Province of the western United States and northern Mexico formed in this manner.

© 2013 Cengage Learning

James S. Monroe

b. The Stillwater Range in Nevada is a horst bounded by normal faults.

continent are also deformed as the plates continue to converge. The sediments are intensely folded and displaced toward the continent along low-angle thrust faults. Eventually, the entire island arc complex is fused to the edge of the continent, and the back-arc basin sediments are thrust onto the continent, forming a thick stack of thrust sheets (Figure 9.16).

Orogenies at Oceanic–Continental Plate Boundaries

The Andes of South America are the best example of continuing orogeny at an oceanic–continental plate boundary (see Figure 2.18b). Among the ranges of the Andes are the highest mountain peaks in the Americas and many active volcanoes. Furthermore, the west coast of South America is an extremely active segment of the circum-Pacific earthquake belt, and one of Earth's great oceanic trench systems, the Peru–Chile Trench, lies just off the coast.

Before 200 million years ago, the western margin of South America was a broad continental margin where sediments accumulated much as they do now along the East Coast of North America. However, when Pangaea split apart along what is now the Mid-Atlantic Ridge, the South American plate moved westward. As a consequence, the oceanic lithosphere

Figure 9.16 Orogeny and the Origin of a Volcanic Island Arc at an Oceanic–Oceanic Plate Boundary

Volcanic island arc

Back-arc basin

a. Subduction of an oceanic plate and the origin of a volcanic island arc and a back-arc basin.

Back-arc spreading

b. Continued subduction and back-arc spreading.

Deformation of back-arc basin and continental margin

c. Back-arc basin begins to close, resulting in deformation of back-arc basin and continental margin deposits.

Island arc "plastered" onto continental margin

Accretionary wedge

Fore-arc basin

Back-arc fold and thrust belt

Metamorphism

© 2013 Cengage Learning

d. Thrusting of back-arc sediments onto the adjacent continent and suturing of the island arc to the continent.

west of South America began subducting beneath the continent (Figure 9.17). Subduction resulted in partial melting of the descending plate, which produced the andesitic volcanic arc of composite volcanoes, and the West Coast became an active continental margin. Felsic magmas, mostly of granitic composition, were emplaced as large plutons beneath the arc (Figure 9.17).

As a result of the events just described, the Andes Mountains consist of a central core of granitic rocks capped by andesitic volcanoes. To the west of this central core along the coast are the deformed rocks of the

Figure 9.17 The Andes Mountains in South America

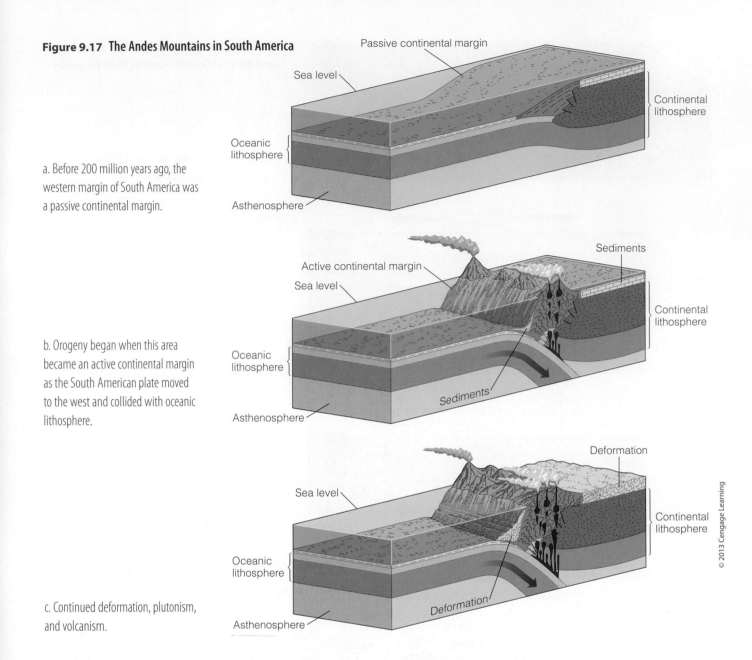

a. Before 200 million years ago, the western margin of South America was a passive continental margin.

b. Orogeny began when this area became an active continental margin as the South American plate moved to the west and collided with oceanic lithosphere.

c. Continued deformation, plutonism, and volcanism.

© 2013 Cengage Learning

accretionary wedge. And to the east of the central core are intensely folded sedimentary rocks that were thrust eastward onto the continent (Figure 9.17). Present-day subduction, volcanism, and seismicity along South America's west coast indicate that the Andes Mountains are still forming.

Orogenies at Continental–Continental Plate Boundaries

The best example of an orogeny along a continental–continental plate boundary is the Himalayas of Asia (see Figure 2.18c). The Himalayas began forming when India collided with Asia about 40 to 50 million years ago. Before that time, India was far south of Asia and separated from it by an ocean basin (Figure 9.18a).

As the Indian plate moved northward, a subduction zone formed along the southern margin of Asia where oceanic lithosphere was consumed. Magma rose to form a volcanic arc, and large granite plutons were emplaced into what is now Tibet. At this stage, the activity along Asia's southern margin was similar to what is now occurring along the west coast of South America.

The ocean separating India from Asia continued to close, and India eventually collided with Asia (Figure 9.18a). As a result, two continental plates became welded, or sutured, together. Thus, the Himalayas are now within a continent rather than along a continental margin. The leading margin of India was thrust beneath Asia, causing crustal thickening, thrusting, and uplift. Sedimentary rocks that had been deposited in the sea south of Asia were thrust northward, and two major thrust faults carried rocks of Asian origin onto the Indian plate. Rocks deposited in the shallow seas along India's northern margin now form the higher parts of the Himalayas (Figure 9.18b). Since its collision with Asia, India has been thrust horizontally about 2,000 km beneath Asia and now moves north at several centimeters per year.

Other mountain systems also formed as a result of collisions between two continental plates. The Urals in Russia and the Appalachians of North America formed by such collisions. In addition, the Arabian plate is now colliding with Asia along the Zagros Mountains of Iran.

Terranes and the Origin of Mountains

In the preceding section, we discussed orogenies along convergent plate boundaries that result in adding material to a continent, a process termed **continental accretion**. Much of the material added to continental margins is eroded older continental crust, but some plutonic and volcanic rocks are new additions. During the 1970s and 1980s, however, geologists discovered that parts of many mountain systems are also made up of small, accreted lithospheric blocks that clearly originated elsewhere. These **terranes*** are fragments of seamounts, island arcs, and small pieces of

continental accretion Orogenics along convergent plate boundaries that result in adding material to a continent.

terrane Fragments of seamounts, island arcs, and small pieces of continents that were carried on oceanic plates that collided with continental plates.

*Some geologists prefer the terms *suspect terrane*, *exotic terrane*, or *displaced terrane*. Notice also the spelling of *terrane* as opposed to the more familiar *terrain*, the latter a geographic term indicating a particular area of land.

Figure 9.18 Orogeny at a Continental–Continental Plate Boundary and the Origin of the Himalayas of Asia

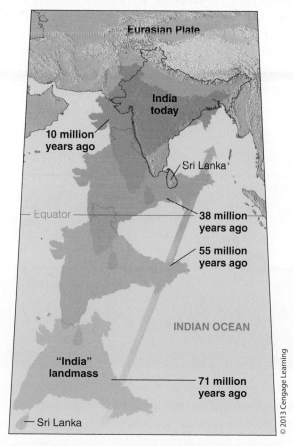

© 2013 Cengage Learning

a. During its long journey north, India moved 15 to 20 cm per year; however, beginning 40 to 50 million years ago, its rate of movement decreased markedly as it collided with the Eurasian plate.

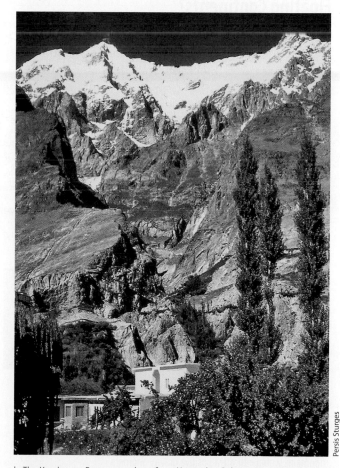

Persis Sturges

b. The Karakoram Range seen here from Karimaba, Pakistan, is within the Himalayan origin. The range lies on the border of Pakistan, China, and India.

continents that were carried on oceanic plates that collided with continental plates, thus adding them to the continental margins.

LO5 EARTH'S CRUST

Continental crust stands higher than oceanic crust, but why should this be so? Also, why do mountains stand higher than surrounding areas? To answer these questions, we must examine Earth's crust in more detail. You already know that continental crust is granitic, with an overall density of 2.7 g/cm³, whereas oceanic crust is made up of basalt and gabbro, and its density is 3.0 g/cm³ (see Figure 8.21). In most places, continental crust is about 35 km thick, except beneath mountain systems, where it is much thicker. The oceanic crust, in contrast, varies from only 5 to 10 km thick. So, differences in composition, as well as variations in crustal thickness, account for why mountains stand high and why continents stand higher than ocean basins.

Floating Continents?

How is it possible for a solid (continental crust) to float in another solid (the mantle)? Floating brings to mind a ship at sea or a block of wood in water; however, continents do not behave in this manner. Or do they? Actually, they do float, in a manner of speaking, but a complete answer requires more discussion on the concept of gravity and on the *principle of isostasy*.

Isaac Newton formulated the law of universal gravitation in which the force of gravity (*F*) between two masses (m_1 and m_2) is directly proportional to the products of their masses and inversely proportional to the square of the distance between their centers of mass. This means that an attractive force exists between any two objects, and the magnitude of that force varies depending on the masses of the objects and the distance between their centers.

Gravitational attraction would be the same everywhere on the surface if Earth were perfectly spherical, homogeneous throughout, and not rotating. But because Earth varies in all of these aspects, the force of gravity varies from area to area.

Principle of Isostasy

Geologists realized long ago that mountains are not simply piles of materials on Earth's surface, and in 1865, George Airy proposed that, in addition to projecting high above sea level, mountains also project far below the surface and thus have a low-density root (Figure 9.19). In effect, he was saying that the thicker

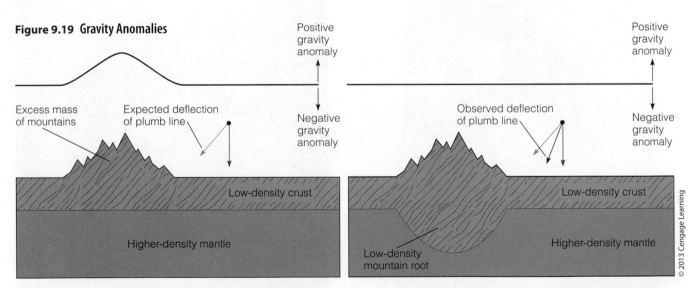

Figure 9.19 Gravity Anomalies

a. A plumb line (a cord with a suspended weight) is normally vertical, pointing to Earth's center of gravity. Near a mountain range, the plumb line should be deflected as shown if the mountains are simply thicker, low-density material resting on denser material, and a gravity survey across the mountains would indicate a positive gravity anomaly.

b. The actual deflection of the plumb line as measured by British surveyors during a survey in India was less than expected. It was explained by postulating that the Himalayas have a low-density root. A gravity survey, in this case, would show no anomaly because the mass of the mountains above the surface is compensated for at depth by low-density material displacing denser material.

crust of mountains float on denser rock at depth, with their excess mass above sea level compensated for by low-density material at depth. Another explanation was proposed by J. H. Pratt, who thought that mountains were high because they were composed of rocks of lower density than those in adjacent regions.

Actually, both Airy and Pratt were correct, because there are places where density or thickness accounts for differences in the level of the crust. For example, Pratt's hyphothesis was confirmed because (1) continental crust is thicker and less dense than oceanic crust and thus stands high, and (2) the mid-oceanic ridges stand high because the crust there is hot and less dense than cooler crust elsewhere. Airy, on the other hand, was correct in his claim that the crust, continental or oceanic, "floats" on the mantle, which has a density of 3.3g/cm^3 in its upper part. However, we have not yet explained what we mean by one solid floating in another solid.

This phenomenon of Earth's crust floating in the denser mantle is now known as the **principle of isostasy**, which is easy to understand by an analogy to an iceberg (Figure 9.20). Ice is slightly less dense than water, so it floats. According to Archimedes' principle of buoyancy, an iceberg sinks in water until it displaces a volume of water whose weight is equal to that of the ice. When the iceberg has sunk to an equilibrium position, only about 10% of its volume is above water level. If some of

the ice above water level should melt, the iceberg rises to maintain equilibrium with the same proportion of ice above and below the water.

Earth's crust is similar to the iceberg in that it sinks into the mantle to its equilibrium level. Where the crust is thickest, as beneath mountains, it sinks farther down into the mantle, and it also rises higher above the surface. And because continental crust is thicker and less dense than oceanic crust, it stands higher than the ocean basins. Remember, the mantle is hot, yet solid, and under tremendous pressure, so it behaves in a fluid-like manner.

Some of you might realize that crust floating on the mantle raises an apparent contradiction. In Chapter 8, we said that the mantle is a solid because it transmits S-waves, which do not move through a fluid. But according to the principle of isostasy, the mantle behaves as a fluid. When considered in terms of the brief time required for S-waves to pass through it, the mantle is indeed solid. But when subjected to stress over long periods, it yields by flowage; thus, at this time scale, it is regarded as a viscous fluid.

Isostatic Rebound

What happens when a ship is loaded with cargo and then later unloaded? Of course, it first sinks lower in the water and then rises, but it always finds its equilibrium position. Earth's crust responds similarly to loading and unloading, but much more slowly. For example, if the crust is loaded, as when widespread glaciers accumulate, the crust sinks farther into the mantle to maintain equilibrium. The crust behaves similarly in areas where huge quantities of sediment accumulate.

If loading by glacial ice or sediment depresses Earth's crust farther into the mantle, it follows that when vast glaciers melt or where deep erosion takes place, the crust should rise back up to its equilibrium level—and in fact it does. This phenomenon, known as **isostatic rebound**, is taking place in Scandinavia, which was covered by a thick ice sheet until about 10,000 years ago; it is now rebounding at about 1 m per century. In fact, coastal cities in Scandinavia have rebounded rapidly enough that docks constructed several centuries ago are now far from shore. Isostatic rebound has also

© 2013 Cengage Learning

Figure 9.20 The Principle of Isostasy
An iceberg sinks to an equilibrium level with 10% of its mass above water level. The larger iceberg sinks farther below and rises higher above the water surface than the smaller one. If some of the ice above water level melts, the icebergs will rise to maintain the same proportions of ice above and below water level.

In isostatic rebound, Earth's crust returns to equilibrium.

occurred in eastern Canada, where the crust has risen as much as 100 m in the last 6,000 years.

Figure 9.21 shows the response of Earth's continental crust to loading and unloading as mountains form and evolve. Recall that during an orogeny, emplacement of plutons, metamorphism, and general thickening of the crust accompany deformation. However, as the mountains erode, isostatic rebound takes place and the mountains rise, whereas adjacent areas of sedimentation subside (Figure 9.21). If continued long enough, the mountains will disappear and then can be detected only by the plutons and metamorphic rocks that show their former existence.

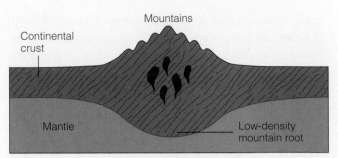

a. The crust and mantle before erosion and deposition.

b. Erosion of the mountains and deposition in adjacent areas. Isostatic rebound begins.

c. Continuing erosion, deposition, and isostatic rebound.

Figure 9.21 Isostatic Rebound

A diagrammatic representation showing the isostatic response of the crust to erosion (unloading) and widespread deposition (loading).

USE THE TOOLS.

- Rip out the Review Cards in the back of your book to study.

Or Visit CourseMate to:

- Study famous geologic locations throughout the Unites States with Virtual Field Trips
- Read, search, highlight, and take notes in the Interactive eBook
- Review Flashcards (Print or Online) to master key terms
- Test yourself with Auto-Graded Quizzes
- Bring concepts to life with Games, Videos, and Animations!

Go to CourseMate for **GEOL2** to begin using these tools. Access at **www.cengagebrain.com**

Complete the Speak Up survey in CourseMate at www.cengagebrain.com

f Follow us at www.facebook.com/4ltrpress

©iStockphoto.com/A-Digit

10 | Mass Wasting

Floods and landslides in January 2011 devastated numerous towns in the mountainous areas around Rio de Janeiro, Brazil, killing more than 800 people and leaving at least 14,000 people homeless. As a result of several days of heavy rains, portions of this hillside in Nova Friburgo gave way, causing a landslide that sent mud, vegetation, and buildings sliding down its slope.

AP Photo/Felipe Dana

Introduction

"**Although** water can play an important role, **the relentless pull** of **gravity is** the major force behind mass wasting."

In January 2011, southeast Brazil was struck by devastating floods and landslides triggered by unusually heavy torrential rains that dumped a month's worth of rain in just a few days (see the chapter opening photo). The hardest hit area was the mountainous Serrana region north of Rio de Janeiro, where relentless flooding and landslides destroyed hundreds of homes, killing more than 800 people and leaving at least 14,000 people homeless. The death toll easily surpassed that of the 1967 mudslides in Caraguatatuba, making this Brazil's worst natural disaster in more than four decades. Collapsed roads and bridges further exacerbated the situation, hampering rescue operations and emergency aid, and lack of power and telephone service contributed further to the misery of the survivors. Although the floods and landslides swept away homes of both rich and poor alike, the poorer rural areas suffered the most because many structures were built in unstable areas.

This terrible tragedy illustrates how geology affects all of our lives. The underlying causes of the mudslides in Brazil can be found anywhere in the world. In fact, *landslides* (a general term for mass movements of Earth materials) cause, on average, between 25 and 50 deaths and more than $2 billion in damage annually in the United States. By being able to recognize and understand how landslides occur and what the results may be, we can find ways to reduce the hazards and minimize damage in terms of both human life and property damage.

Mass wasting (also called *mass movement*) is defined as the downslope movement of material under the direct influence of gravity. Most types of mass wasting are aided by weathering and usually involve surficial material. The material moves at rates ranging from almost imperceptible, as in the case of creep, to extremely fast, as in a rockfall or slide. Although water can play an important role, the relentless pull of gravity is the major force behind mass wasting.

> **mass wasting** The downslope movement of Earth materials under the influence of gravity.
>
> **shear strength** The resisting forces that help maintain a slope's stability.

Learning Outcomes

After reading this unit, you should be able to do the following:

LO1 List the factors that influence mass wasting

LO2 Describe the types of mass wasting

LO3 Understand how to recognize and minimize the effects of mass wasting

LO1 FACTORS THAT INFLUENCE MASS WASTING

Mass wasting is an important geologic process that can occur at any time and almost any place. Although all major landslides have natural causes, many smaller ones are the result of human activity and could have been prevented or their damage minimized.

When the gravitational force acting on a slope exceeds its resisting force, slope failure (mass wasting) occurs. The resisting forces that help maintain slope stability include the slope material's strength and cohesion, the amount of internal friction between grains, and any external support of the slope (Figure 10.1a). These factors collectively define a slope's **shear strength**.

Opposing a slope's shear strength is the force of gravity. Gravity operates vertically but also has a component of force acting parallel to the slope, thereby causing instability (Figure 10.1a). The steeper a slope's angle is, the greater the component of force acting parallel to the slope, and the greater the chance for mass wasting. The steepest angle that a slope can maintain without collapsing is its *angle of repose* (Figure 10.1b). At this angle, the shear strength of the slope's material exactly counterbalances the force of gravity. For unconsolidated material, the angle of repose normally ranges

Figure 10.1 Slope Shear Strength

Gravitational Force

Shear Strength

Cohesiveness of grains

Compression of grains

Internal friction between grains

Component of gravitational force acting parallel to slope

© 2013 Cengage Learning

a. A slope's shear strength depends on the slope material's strength and cohesion, the amount of internal friction between grains, and any external support of the slope. These factors promote slope stability. The force of gravity operates vertically but also has a component of force acting parallel to the slope. When this force, which promotes instability, exceeds a slope's shear strength, slope failure occurs.

b. The angle of repose is a function of sheer strength. Dry sand usually has an angle of repose of about 30 degrees. With damp sand, shear strength is increased, and much steeper angles of repose are possible.

Copyright and Photograph by Dr. Parvinder S. Sethi

from 25 to 40 degrees. Slopes steeper than 40 degrees usually consist of unweathered solid rock.

All slopes are in a state of *dynamic equilibrium*, which means that they are constantly adjusting to new conditions. Although we tend to view mass wasting as a disruptive and usually destructive event, it is one of the ways that a slope adjusts to new conditions. Whenever a building or road is constructed on a hillside, the equilibrium of that slope is affected. The slope must then adjust, sometimes by mass wasting, to this new set of conditions.

Many factors can cause mass wasting: a change in slope angle, weakening of material by weathering, increased water content, changes in the vegetation cover, and overloading. Although most of these processes are interrelated, we will examine them separately for ease of discussion, but we will also show how they individually and collectively affect a slope's equilibrium.

Slope Angle

Slope angle is probably the major cause of mass wasting. Generally speaking, the steeper the slope, the less stable it is. Therefore, steep slopes are more likely to experience mass wasting than gentle ones.

Several processes can oversteepen a slope. One of the most common is undercutting by stream or wave action (Figure 10.2). This process removes the slope's base, increases the slope angle, and thereby increases the gravitational force acting parallel to the slope. Wave action, especially during storms, often results in mass movements along the shores of oceans or large lakes (Figure 10.3).

Excavations for road cuts and hillside building sites are another major cause of slope failure (Figure 10.4). Grading the slope too steeply or cutting into its side increases the stress in the rock or soil until it is no longer strong enough to remain at the steeper angle, and mass movement ensues.

Figure 10.2 Undercutting a Slope's Base by Stream Erosion

a. Undercutting by stream erosion removes a slope's base,

Original slope profile

Scarp

Surface of failure

b. which increases the slope angle, and can lead to slope failure.

© 2013 Cengage Learning

Scarp

Slump block

Reed Wicander

c. Undercutting by stream erosion caused slumping along this stream near Weidman, Michigan. Notice the scarp, which is the exposed surface of the underlying material following slumping.

Weathering and Climate

Mass wasting is more likely to occur in loose or poorly consolidated slope material than in bedrock. As soon as rock is exposed at Earth's surface, weathering begins

Mass wasting is more likely to occur in loose or poorly consolidated slope material than in bedrock.

to disintegrate and decompose it, reducing its shear strength and increasing its susceptibility to mass wasting. The deeper the weathering zone extends, the greater the likelihood of some type of mass movement.

Recall that some rocks are more susceptible to weathering than others and that climate plays an important role in the rate and type of weathering (see Chapter 6). In the tropics, where temperatures are high and considerable rain falls, the effects of weathering extend to depths of several tens of meters, and mass movements most commonly occur in the deep weathering zone. In arid and semiarid regions, the weathering zone is usually considerably shallower. Nevertheless, intense, localized cloudbursts can drop large quantities of water on an area in a short time. With little vegetation to absorb this water, runoff is rapid and frequently results in mudflows.

Water Content

The amount of water in rock or soil influences slope stability. Large quantities of water from melting snow or heavy rainfall greatly increase the likelihood of slope failure. The additional weight that water adds to a slope can be enough to cause mass movement. Furthermore, water percolating through a slope's material helps to decrease friction between grains, contributing to a loss of cohesion. For example, slopes composed of dry clay are usually quite stable, but when wetted, they quickly lose cohesiveness and internal friction and become an unstable slurry. This occurs because clay, which can hold large quantities of water, consists of platy particles that easily slide over each other when wet. For this reason, clay beds are frequently the slippery layer along which overlying rock units slide downslope.

Figure 10.3 Undercutting a Slope's Base by Wave Action

This sea cliff north of Bodega Bay, California, was undercut by waves during the winter of 1997–1998. As a result, part of the land slid into the ocean, damaging several houses.

Figure 10.4 Highway Excavation and Mass Wasting

a. Highway excavations disturb the equilibrium of a slope by

b. removing a portion of its support, as well as oversteepening it at the point of excavation, which can result in

c. landslides along the highway.

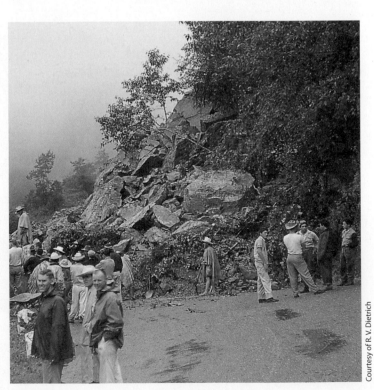

d. Cutting into the hillside to construct this portion of the Pan-American Highway in Mexico resulted in a rockfall that completely blocked the road.

Triggering Mechanisms

Volcanic eruptions, explosions, and even loud claps of thunder may be enough to trigger a landslide if the slope is sufficiently unstable. Many *avalanches*, which are rapid movements of snow and ice down steep mountain slopes, are triggered by a loud gunshot or, in rare cases, even a person's shout.

Vegetation

Vegetation affects slope stability in several ways. By absorbing the water from a rainstorm, vegetation decreases water saturation of a slope's material that would otherwise lead to a loss of shear strength. Vegetation's root system also helps stabilize a slope by binding soil particles together and holding the soil to bedrock.

The removal of vegetation by either natural or human activity is a major cause of many mass movements. Summer brush and forest fires in southern California frequently leave the hillsides bare of vegetation. Fall rainstorms saturate the ground, causing mudslides that do tremendous damage and cost millions of dollars to clean up.

Overloading

Overloading is almost always the result of human activity and typically results from the dumping, filling, or piling up of material. Under natural conditions, a material's load is carried by its grain-to-grain contacts, with the friction between the grains maintaining a slope. The additional weight created by overloading increases the water pressure within the material, which in turn decreases its shear strength, thereby weakening the slope material. If enough material is added, the slope will eventually fail, sometimes with tragic consequences.

Geology and Slope Stability

The relationship between the topography of an area and its geology is important in determining slope stability

(Figure 10.5). If the rocks underlying a slope dip in the same direction as the slope, mass wasting is more likely to occur than if the rocks are horizontal or dip in the opposite direction. When the rocks dip in the same direction as the slope, water can percolate along the various bedding planes and decrease the cohesiveness and friction between adjacent rock units. This is particularly true when clay layers are present, because clay becomes slippery when wet.

Even if the rocks are horizontal or dip in a direction opposite to that of the slope, joints may dip in the same direction as the slope. Water migrating through them weathers the rock and expands these openings until the weight of the overlying rock causes it to fall.

Triggering Mechanisms

The factors discussed thus far all contribute to slope instability. Most, though not all, rapid mass movements are triggered by a force that temporarily disturbs slope equilibrium. The most common triggering mechanisms are strong vibrations from earthquakes and excessive amounts of water from a winter snow melt or a heavy rainstorm (Figure 10.6).

LO2 TYPES OF MASS WASTING

Mass movements are generally classified on the basis of three major criteria (Table 10.1): (1) rate of movement (rapid or slow); (2) type of movement (primarily falling, sliding, or flowing); and (3) type of material involved

Figure 10.5 Geology, Slope Stability, and Mass Wasting

Rocks dipping in the same direction as a hill's slope are particularly susceptible to mass wasting.

1. Water percolates through soil into clay-rich layers [ ] that become slippery, and may swell, weakening the overlying rock [ ].

2. The clay-rich layer dips in the same direction as the even more steeply dipping slope. Gravity can therefore turn it into a skid surface, or potential landslide plane.

3. Undercutting by the stream at the foot of the slope exposes another watery, weak clay layer underlying a heavy, strong limestone bed [ ]. The heavy limestone is now prone to slide across the clay, carrying the rest of the overlying slope with it.

4. Layers on this side of the valley dip in an opposite direction from the slope. Thus, gravity cannot easily act to destabilize them, even if water percolation is deep and undercutting occurs.

© 2013 Cengage Learning

rapid mass movement Any kind of mass wasting that involves a visible downslope displacement of material.

slow mass movement Mass movement that advances at an imperceptible rate and is usually detectable only by the effects of its movement.

(rock, soil, or debris). Even though many slope failures are combinations of different materials and movements, the resulting mass movements are typically classified according to their dominant behavior.

Rapid mass movements involve a visible movement of material. Such movements usually occur quite suddenly, and the material moves quickly downslope. Rapid mass movements are potentially dangerous and frequently result in loss of life and property damage. Most rapid mass movements occur on relatively steep slopes and can involve rock, soil, or debris.

Slow mass movements advance at an imperceptible rate and are usually detectable only by the effects of their movement, such as tilted trees and power poles or cracked foundations. Although rapid mass movements are more dramatic, slow mass movements are responsible for the downslope transport of a much greater volume of weathered material.

Figure 10.6 Landslide Triggered by Heavy Rains, La Conchita, California

La Conchita, California, is located at the base of a steep-sloped terrace. Heavy rains and irrigation of an avocado orchard (visible at the top of the terrace) contributed to the landslide that destroyed nine homes in 1995. Ten years later (2005), similar factors caused another massive landslide in the same area.

Figure 10.7 Rockfalls

Talus

a. Rockfalls result from failure along cracks, fractures, or bedding planes in the bedrock and are common features in areas of steep cliffs.

b. A recent rockfall of granite in Yosemite National Park, California.

Falls

Rockfalls are a common type of extremely rapid mass movement in which rocks of any size may fall through the air (Figure 10.7). Rockfalls occur along steep canyons, cliffs, and road cuts, and they build up accumulations of loose rocks and rock fragments at their base called *talus*.

Rockfalls result from failure along joints or bedding planes in the bedrock and are commonly triggered by natural or human undercutting of slopes or by earthquakes. Many rockfalls in cold climates are the result of frost wedging (see Figure 6.2). Chemical weathering caused by water percolating through the fissures in carbonate rocks (limestone, dolostone, and marble) is also responsible for many rockfalls.

Rockfalls range in size from small rocks falling from a cliff to massive falls involving millions of cubic meters of debris that destroy buildings, bury towns, and block highways. Rockfalls are a particularly common hazard in mountainous areas where roads have been built by blasting and grading through steep hillsides of bedrock (Figure 10.7b). Slopes that are particularly susceptible to rockfalls are sometimes covered with wire mesh in an effort to prevent dislodged rocks from falling onto the road below (Figure 10.8).

rockfall A type of extremely fast mass wasting in which rocks fall through the air.

Figure 10.8 Minimizing Damage from Rock Falls
Wire mesh has been used to cover this steep slope near Narvik in northern Norway. This is a common practice in mountainous areas to prevent rocks from falling on the road.

Copyright and Photograph by Dr. Parvinder S. Sethi

Slides

slide Mass wasting involving movement of material along one or more surfaces of failure.

slump Mass wasting that takes place along a curved surface of failure and results in the backward rotation of the slump mass.

A **slide** involves movement of material along one or more surfaces of failure. The type of material may be soil, rock, or a combination of the two, and it may break apart during movement or remain intact. A slide's rate of movement can vary from extremely slow to very rapid (Table 10.1).

Two types of slides are generally recognized: (1) slumps, or rotational slides, in which movement occurs along a curved surface; and (2) rock or block slides, which move along a more or less planar surface.

A **slump** involves the downward movement of material along a curved surface of rupture and is characterized by the backward rotation of the slump block (Figure 10.9). Slumps usually occur in unconsolidated or

TABLE 10.1

Classification of Mass Movements and Their Characteristics

TYPE OF MOVEMENT	SUBDIVISION	CHARACTERISTICS	RATE OF MOVEMENT
Falls	Rockfall	Rocks of any size fall through the air from steep cliffs, canyons, and road cuts	Extremely rapid
Slides	Slump	Movement occurs along a curved surface of rupture; most commonly involves unconsolidated or weakly consolidated material	Extremely slow to moderate
	Rock slide	Movement occurs along a generally planar surface	Rapid to very rapid
Flows	Mudflow	Consists of at least 50% silt- and clay-sized particles and up to 30% water	Very rapid
	Debris flow	Contains larger-sized particles and less water than mudflows	Rapid to very rapid
	Earthflow	Thick, viscous, tongue-shaped mass of wet regolith	Slow to moderate
	Quick clays	Composed of fine silt and clay particles saturated with water; when disturbed by a sudden shock, lose their cohesiveness and flow like a liquid	Rapid to very rapid
	Solifluction	Water-saturated surface sediment	Slow
	Creep	Downslope movement of soil and rock	Extremely slow
Complex movements		Combination of different movement types	Slow to extremely rapid

Fractures
Scarp
Slump block
Surface of rupture

© 2013 Cengage Learning

Figure 10.9 Slumping

In a slump, material moves downward along the curved surface of a rupture, causing the slump block to rotate backward. Most slumps involve unconsolidated or weakly consolidated material and are typically caused by erosion along the slope's base.

weakly consolidated material and range in size from small individual sets, such as occur along stream banks (Figure 10.2c), to massive, multiple sets that affect large areas and cause considerable damage.

Slumps can be caused by a variety of factors, but the most common is erosion along the base of a slope, which removes support for the overlying material. This local steepening may be caused naturally by stream erosion along its banks (Figure 10.2c) or by wave action at the base of a coastal cliff (Figure 10.10).

Slope oversteepening can also be caused by human activity, such as the construction of highways and housing developments. Slumps are particularly prevalent along highway cuts, where they are generally the most frequent type of slope failure observed.

Although many slumps are merely a nuisance, large-scale slumps in populated areas and along highways can cause extensive damage. Such is the case in coastal southern California, where slumping and sliding have been a constant problem, resulting in the destruction of many homes and the closing and relocation of numerous roads and highways (Figure 10.10).

A **rock slide**, or *block slide*, occurs when rocks move downslope along a more or less planar surface. Most rock slides take place because the local slopes and rock layers dip in the same direction (Figures 10.5 and 10.11), although they can also occur along fractures par-

Copyright and Photograph by Dr. Parvinder S. Sethi

Figure 10.10 Slumping in Point Fermin, California

Homes dangerously close to an oversteepened, eroding coastal bluff in Point Fermin, California. It is just a matter of time before these homes are destroyed by the effects of erosion or slumping of the unstable bluff.

allel to a slope. Rock slides are common occurrences along the southern California coast, such as Point Fermin, where seaward-dipping rocks with interbedded slippery clay layers are undercut by waves, causing numerous slides.

rock slide Rapid mass wasting in which rocks move downslope along a more or less planar surface.

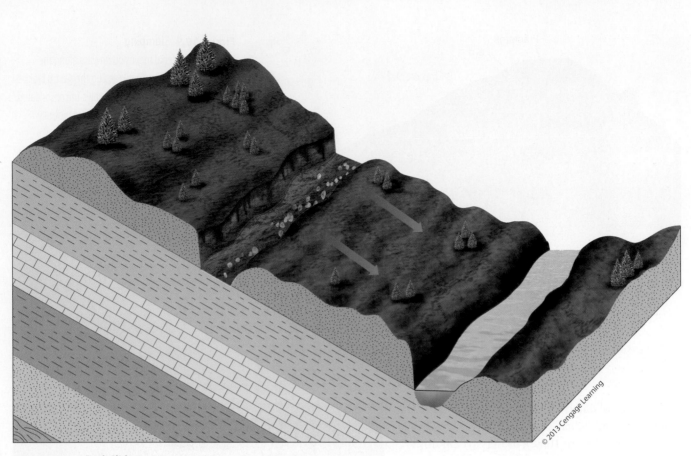

Figure 10.11 Rock Slide

Rock slides occur when material moves downslope along a generally planar surface. Most rock slides result when the underlying rocks dip in the same general angle as the slope of the land. Undercutting along the base of the slope and clay layers beneath porous rock or soil layers increase the chance of rock slides.

Farther south from Point Fermin is the town of Laguna Beach, where residents were hit by rock slides and mudslides in 1978, 1998, and as recently as 2005 (Figure 10.12). Just as at Point Fermin, the rocks at Laguna Beach dip about 25 degrees in the same direction as the slope of the canyon walls and contain clay beds that "lubricate" the overlying rock layers, causing the rocks and the houses built on them to slide. Percolating water from heavy rains wets subsurface clayey siltstone, thus reducing its shear strength and helping to activate the slide. In addition, these slides are part of a larger ancient slide complex.

Not all rock slides are the result of rocks dipping in the same direction as a hill's slope. The rock slide at Frank, Alberta, Canada, on April 29, 1903, illustrates how nature and human activity can combine to create a situation with tragic results (Figure 10.13).

It would appear at first glance that the coal-mining town of Frank, lying at the base of Turtle Mountain, was in no danger from a landslide (Figure 10.13). After

all, many of the rocks dipped away from the mining valley. The joints in the massive limestone composing Turtle Mountain, however, dip steeply toward the valley and are essentially parallel with the slope of the mountain itself. Furthermore, Turtle Mountain is supported by weak siltstones, shales, and coal layers that underwent slow plastic deformation from the weight of the overlying massive limestone. Coal mining along the base of the valley also contributed to the stress on the rocks by removing some of the underlying support. All of these factors, as well as the frost action and chemical weathering that widened the joints, finally resulted in a massive rock slide. Approximately 40 million m^3 of rock slid down Turtle Mountain along joint planes, killing 70 people and partially burying the town of Frank.

Flows

Mass movements in which material flows as a viscous fluid or displays plastic movement are termed *flows*.

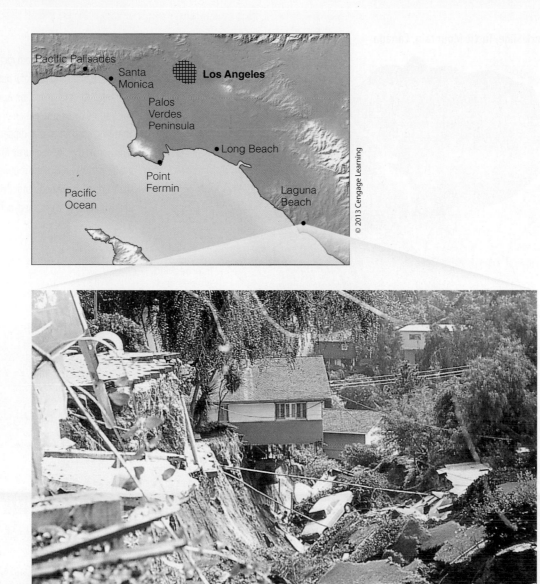

Figure 10.12 Rock Slide, Laguna Beach, California

A combination of interbedded clay layers that become slippery when wet, rocks dipping in the same direction as the slope of the sea cliffs, and undercutting of the sea cliffs by wave action activated a rock slide at Laguna Beach, California, that destroyed numerous homes and cars on October 2, 1978. This same area was hit by another rock slide in 2005.

Their rate of movement ranges from extremely slow to extremely rapid (Table 10.1). In many cases, mass movements begin as falls, slumps, or slides, and change into flows farther downslope.

Of the major mass movement types, **mudflows** are the most fluid and move most rapidly (at speeds up to 80 km per hour). They consist of at least 50% silt- and clay-sized material combined with a significant amount of water (up to 30%). Mudflows are common in arid and semi-arid environments, where they are triggered by heavy rainstorms that quickly saturate the regolith, turning it into a raging flow of mud that engulfs everything in its path. Mudflows can also occur in mountain regions (Figure 10.14) and in areas covered by volcanic ash, where they can be particularly destructive

mudflow A flow consisting mostly of clay- and silt-sized particles and up to 30% water that moves downslope under the influence of gravity.

Figure 10.13 Rock Slide, Turtle Mountain, Canada

a. The tragic Turtle Mountain rock slide that killed 70 people and partially buried the town of Frank, Alberta, Canada, on April 29, 1903, was caused by a combination of factors. These included joints that dipped in the same direction as the slope of Turtle Mountain, a fault partway down the mountain, weak shale and siltstone beds underlying the base of the mountain, and mined-out coal seams.

b. Results of the 1903 rock slide at Frank.

(see Chapter 5). Because mudflows are so fluid, they generally follow preexisting channels until the slope decreases or the channel widens, at which point they fan out.

Debris flows are composed of larger particles than mudflows and do not contain as much water. Consequently, they are usually more viscous than mudflows, typically do not move as rapidly, and rarely

Figure 10.14 Mudflow, Rocky Mountain National Park
Mudflows move swiftly downslope, engulfing everything in their path. Note how this mudflow in Rocky Mountain National Park has fanned out at the base of the hill. Also note the small lake adjacent to the mudflow that was formed after this mudflow created a dam across the stream.

are confined to preexisting channels. Debris flows can be just as damaging, though, because they can transport large objects (Figure 10.15).

Earthflows move more slowly than either mudflows or debris flows. An earthflow slumps from the upper part of a hillside, leaving a scarp, and flows slowly downslope as a thick, viscous, tongue-shaped mass of wet regolith (Figure 10.16). Like mudflows and debris flows, earthflows can be of any size and are frequently destructive. They occur most commonly in humid climates on grassy, soil-covered slopes following heavy rains.

Some clays spontaneously liquefy and flow like water when they are disturbed. Such **quick clays** have caused serious damage and loss of lives in Sweden, Norway, eastern Canada, and Alaska. Quick clays are composed of fine silt and clay particles made by the grinding action of glaciers. Geologists think that these fine sediments were originally deposited in a marine environment, where their pore space was filled with saltwater. The ions in saltwater helped establish strong bonds between the clay particles, thus stabilizing and strengthening the clay. When the clays were subsequently uplifted above sea level, the saltwater was flushed out by fresh groundwater, reducing the effectiveness of the ionic bonds between the clay particles and thereby reducing the overall strength and cohesiveness of the

clay. Consequently, when the clay is disturbed by a sudden shock or shaking, it essentially turns to a liquid and flows.

An excellent example of the damage that can be done by quick clays occurred in the Turnagain Heights area of Anchorage, Alaska, in 1964 (Figure 10.17). Underlying most of the Anchorage area is the Bootlegger Cove Clay, a massive clay unit of poor permeability. Because the Bootlegger Cove Clay forms a barrier that prevents groundwater from flowing through the adjacent glacial deposits to the sea, considerable hydraulic pressure builds up on the landward side of the clay. Some of this water has flushed out the saltwater in the clay and has saturated the lenses of sand and silt associated with the clay beds. When the 8.6-magnitude Good Friday earthquake

debris flow A type of mass wasting that involves a viscous mass of soil, rock fragments, and water that moves downslope; debris flows have larger particles than mudflows and contain less water.

earthflow A mass wasting process involving the downslope movement of water-saturated soil.

quick clay A clay deposit that spontaneously liquefies and flows like water when disturbed.

Figure 10.15 Debris Flow, Ophir Creek, Nevada
A debris flow and damaged house in lower Ophir Creek, western Nevada. Note the many large boulders that are part of the debris flow. Debris flows do not contain as much water as mudflows and typically are composed of larger particles.

Figure 10.16 Earthflow

a. Earthflows form tongue-shaped masses of wet regolith that move slowly downslope. They occur most commonly in humid climates on grassy, soil-covered slopes.

b. An earthflow near Baraga, Michigan.

Figure 10.17 Quick Clay Slide, Anchorage, Alaska

© 2013 Cengage Learning

a. Ground shaking by the 1964 Alaska earthquake turned parts of the Bootlegger Cove Clay into a quick clay, causing numerous slides.

Figure 18-A, U.S. Geological Survey Circular 491

b. Low-altitude photograph of the Turnagain Heights subdivision of Anchorage shows some of the numerous landslide fissures that developed, as well as the extensive damage to buildings in the area. The remains of the Four Seasons apartment building can be seen in the background.

struck on March 27, 1964, the shaking turned parts of the Bootlegger Cove Clay into a quick clay and precipitated a series of massive slides in the coastal bluffs that destroyed most of the homes in the Turnagain Heights subdivision (Figure 10.17b).

Solifluction is the slow downslope movement of water-saturated surface sediment. Solifluction can occur in any climate where the ground becomes saturated with water, but is most common in areas of permafrost.

Permafrost, ground that remains permanently frozen, covers nearly 20% of the world's land surface (Figure 10.18). During the warmer season, when the upper portion of the permafrost thaws, water and surface sediment form a soggy mass that flows by solifluction and produces a characteristic lobate topography (Figure 10.18b).

As might be expected, many problems are associated with construction in a permafrost environment.

For example, when an uninsulated building is constructed directly on permafrost, heat escapes through the floor, thaws the ground below, and turns it into a soggy, unstable mush. Because the ground is no longer solid, the building settles unevenly into the ground, and numerous structural problems result (Figure 10.19).

Creep, the slowest type of flow, is the most widespread and significant mass wasting process in terms of the total amount of material moved downslope and the monetary damage it does annually. Creep involves extremely slow downhill movement of soil or rock. Although it can occur anywhere and in any climate, it has the most effect and is most significant as a geologic agent in humid regions.

Because the rate of movement is essentially imperceptible, we are frequently unaware of creep's existence until we notice its effects: tilted trees and power poles, broken streets and sidewalks, or cracked retaining walls or foundations (Figure 10.20). Creep usually involves the whole hillside and probably occurs, to some extent, on any weathered or soil-covered sloping surface.

Creep is not only difficult to recognize, but also to control. Although engineers can sometimes slow or stabilize creep, many times the only course of action is to simply avoid the area if at all possible or, if the zone of creep is relatively thin, design structures that can be anchored into the bedrock.

Complex Movements

Recall that many mass movements are combinations of different movement types. When one type is dominant, the movement can be classified as one of those described thus far. If several types are more or less equally involved, however, it is called a **complex movement**. The most common type of complex movement is the *slide-flow*, in which there is sliding at the

solifluction Mass wasting involving the slow downslope movement of water-saturated surface materials; occurs especially at high elevations or high latitudes where the flow is underlain by frozen soil.

permafrost Ground that remains permanently frozen.

creep A widespread type of mass wasting in which soil or rock moves slowly downslope.

complex movement A combination of different types of mass movements in which no single type is dominant; usually involves sliding and flowing.

Figure 10.18 Permafrost and Solifluction

a. Distribution of permafrost areas in the Northern Hemisphere.

b. Solifluction flows in Kluane National Park, Yukon Territory, Canada, show the lobate topography that is characteristic of solifluction conditions.

head and then some type of flowage farther along its course. Most slide-flow landslides involve well-defined slumping at the head, followed by a debris flow or earthflow (Figure 10.21). Any combination of different mass movement types is a complex movement.

LO3 RECOGNIZING AND MINIMIZING THE EFFECTS OF MASS WASTING

The most important factor in eliminating or minimizing the damaging effects of mass wasting is a thorough geologic investigation of the region in question. In this way, former landslides and areas susceptible to mass movements can be identified and perhaps avoided. By assessing the risks of possible mass wasting before construction begins, engineers can take steps to eliminate or minimize the effects of such events.

Figure 10.19 Permafrost Damage

This house, south of Fairbanks, Alaska, has settled unevenly because the underlying permafrost in fine-grained silts and sands has thawed.

Figure 10.20 Creep

a. Some evidence of creep: (A) curved tree trunks, (B) displaced monuments, (C) tilted power poles, (D) displaced and tilted fences, (E) roadways moved out of alignment, (F) hummocky surface.

b. Trees bent by creep, Wyoming.

c. Creep has bent these sandstone and shale beds of the Haymond Formation near Marathon, Texas.

d. Stone wall tilted due to creep in Champion, Michigan.

Figure 10.21 Complex Movement

A complex movement is one in which several types of mass wasting are involved. In this example, slumping occurs at the head, followed by an earthflow.

Scarp

Slumping

Fractures

Earthflow

Slump block

Surface of rupture

Toe

Soil

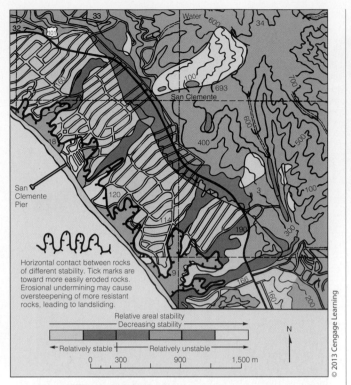

Figure 10.22 Slope-Stability Map

This slope-stability map of part of San Clemente, California, shows areas delineated according to relative stability. Such maps help planners and developers make decisions about where to site roads, utility lines, buildings, and other structures.

Identifying areas with a high potential for slope failure is important in any hazard assessment study; these studies include identifying former landslides as well as sites of potential mass movement. Scarps, open fissures, displaced or tilted objects, a hummocky surface, and sudden changes in vegetation are some of the features that indicate former landslides or an area susceptible to slope failure. The effects of weathering, erosion, and vegetation may, however, obscure the evidence of previous mass wasting.

The information derived from a hazard assessment study can be used to produce *slope-stability maps* of the area (Figure 10.22). These maps allow planners and developers to make decisions about where to site roads, utility lines, and housing or industrial developments based on the relative stability or instability of a particular location. The maps also indicate the extent of an area's landslide problem and the type of mass movement that may occur.

Although most large mass movements usually cannot be prevented, geologists and engineers can use various methods to minimize the danger and damage resulting from them. Because water plays such an important role in many landslides, one of the most effective and inexpensive ways to reduce the potential for slope failure or to increase existing slope stability is surface and subsurface drainage of a hillside. Drainage serves two purposes. It (1) reduces the weight of the material likely to slide and (2) increases the shear strength of the slope material by lowering pore pressure.

Surface waters can be drained and diverted by ditches, gutters, or culverts designed to direct water away from slopes. Drainpipes perforated along one surface and driven into a hillside can help remove subsurface water (Figure 10.23). Finally, planting vegetation on hillsides helps stabilize slopes by holding the soil together and reducing the amount of water in the soil.

Another way to help stabilize a hillside is to reduce its slope. Recall that overloading and oversteepening by grading are common causes of slope failure. Reducing the angle of a hillside decreases the potential for slope failure. Two methods are usually employed to reduce a slope's angle. In the *cut-and-fill* method, material is removed from the upper part of the slope and used as fill at the base, thus providing a flat surface for construction and reducing the slope (Figure 10.24).

The second method, which is called *benching*, involves cutting a series of benches, or steps, into a hillside (Figure 10.25). This process reduces the overall average slope, and the benches serve as collecting sites for small landslides or rockfalls that might occur. Benching is most commonly used on steep hillsides in conjunction with a system of surface drains to divert runoff.

In some situations, retaining walls are constructed to provide support for the base of the slope (Figure 10.26). The walls are usually anchored well into bedrock, backfilled with crushed rock, and provided with drain holes to prevent the buildup of water pressure in the hillside.

Recognition, prevention, and control of landslide-prone areas are expensive, but not nearly as expensive as the damage can be when such warning signs are ignored or not recognized. Unfortunately, there are numerous examples of landfill and dam collapses that serve as tragic reminders of the price paid in loss of lives and property damage when the warning signs of impending disaster are ignored.

Figure 10.23 Using Drainpipes to Remove Subsurface Water

a. Driving drainpipes that are perforated on one side into a hillside, with the perforated side up, can remove some subsurface water and help stabilize a hillside.

b. A drainpipe driven into the hillside at Point Fermin, California, helps to reduce the amount of subsurface water in these porous beds.

Reed Wicander

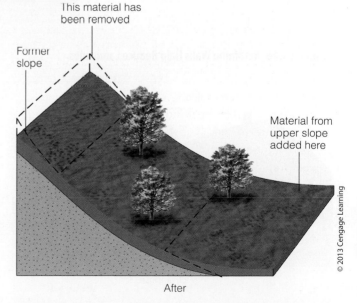

Before

After

Figure 10.24 Stabilizing a Hillside by the Cut-and-Fill Method

One common method used to help stabilize a hillside and reduce its slope is the cut-and-fill method. Material from the steeper upper part of the hillside is removed, thereby decreasing the slope angle, and is used to fill in the base. This provides some additional support at the base of the slope.

Figure 10.25 Stabilizing a Hillside by Benching

Before

Former slope

After

a. Another common method used to stabilize a hillside and reduce its slope is benching. This process involves making several cuts along a hillside to reduce the overall slope. As another result, individual slope failures are limited in size, and the material collects on the benches.

b. Benching is used in many road cuts and can be clearly seen in this photograph.

© John D. Cunningham/Visuals Unlimited

Figure 10.26 Retaining Walls Help Reduce Landslides

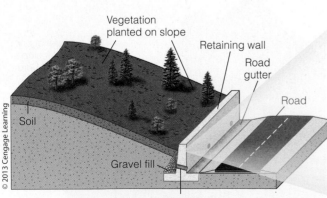

Vegetation planted on slope

Retaining wall

Road gutter

Road

Soil

Gravel fill

Drainpipe

© 2013 Cengage Learning

Copyright and Photograph by Dr. Parvinder S. Sethi

a. Retaining walls anchored into bedrock, backfilled with gravel, and provided with drainpipes can support a slope's base and reduce landslides.

b. A steel retaining wall built to stabilize the slope and keep falling and sliding rocks off the highway.

THE IN-CROWD

Share your 4LTR Press story on Facebook at www.facebook.com/4ltrpress for a chance to win.

To learn more about the In-Crowd opportunity, 'like' us on Facebook.

11 Running Water

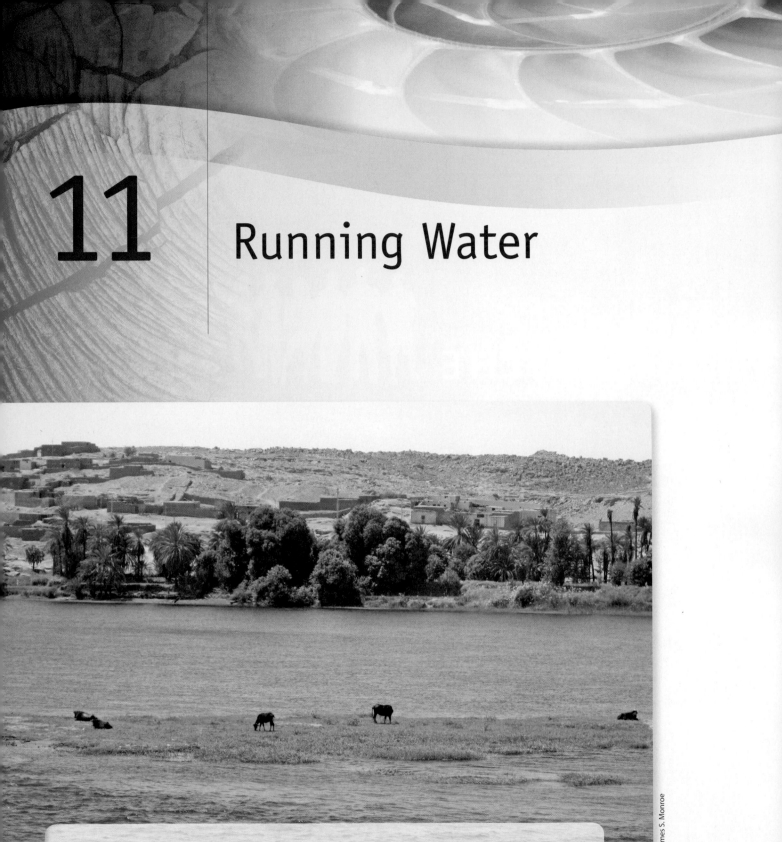

View of the Nile River in Egypt. North is to the right, which is the direction the river flows. The Nile River floodplain and its delta are nearly the only areas in Egypt that support agriculture. At this point, between the cities of Luxor and Aswan, the floodplain is narrow, and nowhere is it more than 20 km across. Notice the abrupt change from the irrigated floodplain to desert. The Nile River, at 6,650 km long, is the world's longest. It is an unusual river in that after it enters Egypt it has no tributaries.

James S. Monroe

Introduction

"The water in channels is, with few exceptions, the most important geologic agent in modifying the land surface."

All the terrestrial planets have a similar early history of accretion, differentiation, and volcanism, but their subsequent histories differ considerably. Indeed, now Earth is the only terrestrial planet with an oxygen-rich atmosphere and abundant surface water. Mercury's small size and its distance from the Sun, and Venus's runaway greenhouse effect, preclude the possibility of water on those planets. Mars is too small and too cold for liquid water, although it has some frozen water and small amounts of water vapor in its atmosphere. Furthermore, satellite imagery reveals winding valleys and canyons that indicate running water during the planet's early history. In contrast, 71% of Earth's surface is covered by oceans and seas, its atmosphere contains up to 4.0% water vapor, and the continents receive precipitation that feeds streams, rivers, lakes, and the groundwater system.

The hydrosphere consists of all water on Earth, including water frozen in glaciers, although most of it (97.2%) is in the oceans. Our main concern here is with the 0.0001% of Earth's water in stream and river channels. Keep in mind, though, that water is continually cycled through channels, and, in fact, running water has a tremendous impact on much of the land surface.

National Park Service

Figure 11.1 Aftermath of the Johnstown, Pennsylvania, Flood
On May 31, 1889, an 18-m-high wall of water destroyed Johnstown and killed at least 2,200 people.

Only the most arid deserts and areas covered by glaciers show little or no effects of running water.

You have probably experienced the power of running water if you have ever swam or canoed in a fast-flowing stream or river, but the vivid accounts of floods truly impress us with the energy of moving water. At 4:07 p.m. on May 31, 1889, residents of Johnstown, Pennsylvania, heard "a roar like thunder" and within 10 minutes the town was devastated by an 18-m-high wall of water that tore through the town at 60 km/hr, sweeping up debris, houses, and entire families (Figure 11.1). Heavy rainfall and the failure of a dam upstream from the town caused the flood in which at least 2,200 people were killed, making it the most deadly river flood in U.S. history.

In July and August 2010, because of exceptionally heavy monsoon rains, the Indus River in Pakistan flooded about 20% of the entire country, accounting for at least 2,000 deaths. And even more recently, extensive flooding has occurred in Brisbane, Australia, on the Lower Mississippi River, and as of this writing (June 2011) the Missouri River is threatening to burst its levees and flood.

Every year floods cause extensive property damage and fatalities, and yet we derive many benefits from running water, even from some floods. Before the completion of the Aswan High Dam in 1970, Egyptian farmers depended on silt deposited on the Nile River floodplain to replenish their croplands (see the chapter opening photo). Indeed, in ancient Egypt, taxes were levied depending on the level of the Nile River. In addition, running water is one source of freshwater for agriculture, industry, domestic use, and recreation,

Learning Outcomes

After reading this unit, you should be able to do the following:

LO1 Identify sources of water on Earth

LO2 Describe the role of running water

LO3 Explain how running water erodes and transports sediment

LO4 Describe deposition by running water

LO5 Question whether floods can be predicted and controlled

LO6 Understand drainage systems

LO7 Recognize the significance of base level

LO8 Understand the evolution of valleys

© iStockphoto.com/Olga Lyubkina

All water derived from the oceans eventually makes it back to the oceans and can thus begin the hydrologic cycle again.

and about 8% of all electricity used in North America is generated at hydroelectric plants. When Europeans first explored North America, they followed the St. Lawrence, Mississippi, Missouri, and Ohio Rivers. These and similar waterways throughout the world are also major avenues of commerce.

LO1 WATER ON EARTH

Most of Earth's 1.33 billion km^3 of water is in the oceans, and nearly all of the rest is frozen in glaciers on land. Only 0.65% of all Earth's water is in the atmosphere, groundwater, lakes, swamps, bogs, and a tiny but important amount in stream and river channels. Nevertheless, the water in channels is, with few exceptions, the most important geologic agent in modifying the land surface.

Much of our discussion of running water is descriptive, but always be aware that streams and rivers are dynamic systems that must continuously respond to change. For example, paving in urban areas increases surface runoff to waterways, and other human activities such as building dams and impounding reservoirs also alter the dynamics of stream and river systems. Natural changes, too, affect the complex interacting parts of these systems.

The Hydrologic Cycle

The connection between precipitation and clouds is obvious, but where does the moisture for rain and snow come from in the first place? You might immediately suspect that the oceans are the ultimate source of precipitation. In fact, water is continuously recycled from the oceans, through the atmosphere, to the continents, and back to the oceans. This **hydrologic cycle**, as it is called (Figure 11.2), is powered by solar radiation and is possible because water changes easily from liquid to gas (water vapor). About 85% of all water entering the atmosphere evaporates from the oceans. The remaining 15% comes from water on land, but this water originally came from the oceans as well.

Regardless of its source, water vapor rises into the atmosphere where the complex processes of cloud formation and condensation take place. About 80% of all precipitation falls directly back into the oceans, so the hydrologic cycle is a three-step process of evaporation, condensation, and precipitation. For the 20% of precipitation that falls on land, the hydrologic cycle is more complex, involving evaporation, condensation, movement of water vapor from the oceans to land, precipitation, and runoff. Some precipitation evaporates as it falls and reenters the cycle, but about 36,000 km^3 of the precipitation falls on land and returns to the oceans by **runoff**, the surface flow in streams and rivers.

Some precipitation is temporarily stored in lakes and swamps, snowfields and glaciers, or seeps below the surface where it enters the groundwater system. Water might remain in these reservoirs for thousands of years, but eventually glaciers melt, lakes and groundwater feed streams and rivers, and this water returns to the oceans. Even the water used by plants evaporates, a process known as *evapotranspiration*, and returns to the atmosphere.

Fluid Flow

Solids are rigid substances that retain their shapes unless deformed by a force, but fluids—that is, liquids and gases—have no strength, so they flow in response to any force, no matter how slight. Liquid water flows downslope in response to gravity.

Runoff during a rainstorm depends on **infiltration capacity**, the maximum rate at which surface materials

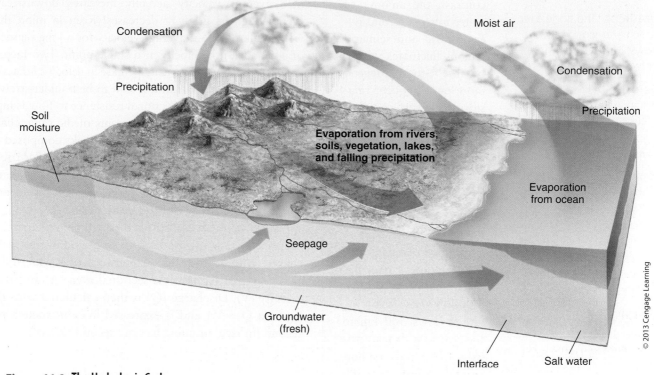

Figure 11.2 The Hydrologic Cycle

During the hydrologic cycle, water evaporates from the oceans and rises as water vapor to form clouds that release their precipitation over the oceans or over land. Some of the precipitation falling on land enters the groundwater system, but much of it returns to the oceans by surface runoff, thus completing the cycle.

absorb water. Several factors control infiltration capacity, including intensity and duration of rainfall. If rain is absorbed as fast as it falls, no surface runoff takes place. Loosely packed dry soil absorbs water faster than tightly packed wet soil, and thus more rain must fall on loose dry soil before runoff begins. Regardless of the initial condition of surface materials, once they are saturated, excess water collects on the surface and, if on a slope, it moves downhill.

LO2 RUNNING WATER

The term *running water* applies to any surface water that moves from higher to lower areas in response to gravity. We have already noted that running water is very effective in modifying Earth's land surface by erosion and that it is the primary geologic process responsible for sediment transport and deposition in many areas. Indeed, it is responsible for everything from the tiniest rills in farmer's fields to scenic wonders such as the Grand Canyon in Arizona, as well as vast deposits such as the Mississippi River delta.

Sheet Flow and Channel Flow

Even on steep slopes, flow is initially slow and hence causes little or no erosion. As water moves downslope, though, it accelerates and may move by *sheet flow*, a more or less continuous film of water flowing over the surface. Sheet flow is not confined to depressions, and it accounts for *sheet erosion*, a particular problem on some agricultural lands.

In *channel flow*, surface runoff is confined to trough-like depressions that vary in size from tiny rills with a trickling stream of water to huge river channels. We describe flow in channels with terms such as *rill*, *brook*, *creek*, *stream*, and *river*, most of which are distinguished by size and volume. Here we use the terms *stream* and *river* more or less interchangeably, although the latter usually refers to a larger body of running water.

Streams and rivers receive water from sheet flow and rain that falls directly into their channels, but far more important is the water supplied by soil moisture and groundwater. In areas where groundwater is plentiful, streams and rivers maintain a fairly stable flow year-round, because their water supply is continuous. In

gradient The slope over which a stream or river flows, expressed in m/km or ft/mi.

velocity A measure of distance traveled per unit of time, as in the flow velocity in a stream or river.

discharge The volume of water in a stream or river moving past a specific point in a given interval of time; expressed in cubic meters per second (m³/sec) or cubic feet per second (ft³/sec).

dissolved load The part of a stream's load consisting of ions in solution.

hydraulic action The removal of loose particles by the power of moving water.

contrast, the amount of water in streams and rivers of arid and semiarid regions fluctuates widely, because they depend more on infrequent rainstorms and surface runoff for their water.

Gradient, Velocity, and Discharge

Water in any channel flows downhill over a slope known as its **gradient**. Suppose a river has its headwaters (source) 1,000 m above sea level, and it flows 500 km to the sea, so it drops vertically 1,000 m over a horizontal distance of 500 km. Its gradient is found by dividing the vertical drop by the horizontal distance, which in this example is 1,000 m/500 km = 2 m/km on average (Figure 11.3a).

The **velocity** of running water is a measure of the downstream distance water travels in a given time. It is usually expressed in meters per second (m/sec) or feet per second (ft/sec), and it varies across a channel's width as well as along its length. Water moves more slowly and with greater turbulence near a channel's bed and banks, because friction is greater there than it is some distance from these boundaries (Figure 11.3b). Channel shape and roughness also influence flow velocity. Broad, shallow channels and narrow, deep channels have proportionately more water in contact with their perimeters than do channels with semicircular cross sections (Figure 11.3c). As one would expect, rough channels, such as those strewn with boulders, offer more frictional resistance to flow than do channels with a bed and banks composed of sand or mud.

Intuitively, you might think that the gradient is the most important control on velocity—the steeper the gradient, the greater the velocity. In fact, a chan-

nel's average velocity actually increases downstream even though its gradient decreases! Keep in mind that we are talking about average velocity for a long segment of a channel, not velocity at a single point. Two factors account for this downstream increase in velocity. First, the upstream reaches of channels tend to be boulder-strewn, broad, and shallow, so frictional resistance to flow is high there, whereas downstream segments of the same channels are more semicircular and have banks composed of finer materials. And second, the number of smaller tributaries joining a larger channel increases downstream. Thus, the total volume of water (discharge) increases, and increasing discharge results in greater velocity.

Specifically, **discharge** is the volume of water that passes a particular point in a given period of time. Discharge is found from the dimensions of a water-filled channel—that is, its cross-sectional area (A) and flow velocity (V). Discharge (Q) is then calculated with the formula $Q = VA$ and is expressed in cubic meters per second (m³/sec) or cubic feet per second (ft³/sec).

LO3 RUNNING WATER, EROSION, AND SEDIMENT TRANSPORT

Streams and rivers possess two kinds of energy: potential and kinetic. *Potential energy* is the energy of position, such as the energy of water at high elevation. During stream flow, potential energy is converted to *kinetic energy*, the energy of motion. Much of this kinetic energy is used up in fluid turbulence, but some is available for erosion and transport. The materials transported by a stream include a *dissolved load*, and a load of solid particles (mud, sand, and gravel).

Because the **dissolved load** of a stream is invisible, it is commonly overlooked, but it is an important part of the total sediment load. Some of it is acquired from a stream's bed and banks, where soluble rocks such as limestone are present, but much of it is carried into waterways by sheet flow and by groundwater. A stream's solid load is made up of particles ranging from clay sized (> 1/256 mm) to huge boulders, much of it supplied by mass wasting, but some is eroded directly from a stream's bed and banks. The direct impact of running water, **hydraulic action**, is sufficient to set particles in motion.

© Bananastock/Jupiterimages

Figure 11.3 Gradient and Flow Velocity

a. The average gradient of this stream is 2 m/km; however, gradient can be calculated for any segment of a stream, as shown in this example. Notice that the gradient is steepest in the headwaters area and decreases downstream.

b. The maximum flow velocity is near the center and top of a straight channel, where the least friction takes place. The arrows are proportional to velocity.

	Broad, shallow channel	Narrow, deep channel	Semicircular channel
Cross-sectional area	10 m²	10 m²	10 m²
Perimeter in contact with water	12 m	12 m	7.9 m

© 2013 Cengage Learning

c. These three differently shaped channels have the same cross-sectional area; however, the semicircular one has less water in contact with its perimeter and thus less frictional resistance to flow.

Running water carrying sand and gravel erodes by **abrasion**, as exposed rock is worn and scraped by the impact of these particles (Figure 11.4a). Circular to oval depressions called *potholes* in streambeds are one manifestation of abrasion (Figure 11.4a, b). They form where swirling currents with sand and gravel eroded the rock.

Once materials are eroded, they are transported from their source and eventually deposited. The dissolved load is transported in the water itself, but the load of solid particles moves as *suspended load* or *bed load*. The **suspended load** consists of the smallest particles

of silt and clay, which are kept suspended above the channel's bed by fluid turbulence (Figure 11.5).

The **bed load** of larger particles, mostly sand and gravel, cannot be kept suspended by fluid turbulence, so that it is transported along the bed. However, some of the sand may

abrasion The process whereby rock is worn smooth by the impact of sediment transported by running water, glaciers, waves, or wind.

suspended load The smallest particles (silt and clay) carried by running water, which are kept suspended by fluid turbulence.

bed load The part of a stream's sediment load, mostly sand and gravel, transported along its bed.

Figure 11.4 Abrasion By Running Water Carrying Sand and Gravel

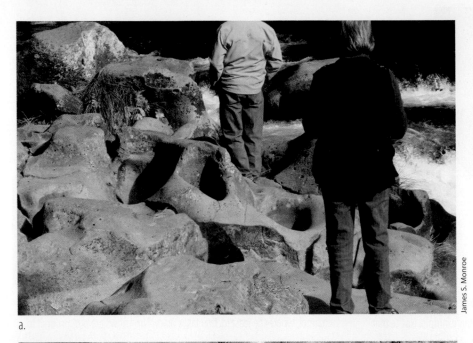

a.

a. Potholes in the bed of the McCloud River in California. Also notice the rock surface that has been smoothed by abrasion.

b.

b. View into one of the potholes showing the sand and gravel that swirled around to erode the pothole.

Figure 11.5 Sediment Transport

Sediment transport as bed load, suspended load, and dissolved load. Flow velocity is highest near the surface, but gravel- and sand-sized particles are too large to be lifted far from the streambed, so they make up the bed load, whereas silt and clay are in the suspended load.

be temporarily suspended by currents that swirl across the streambed and lift grains into the water. The grains move forward with the water, but also settle and finally come to rest and then again move by the same process of intermittent bouncing and skipping, a phenomenon known as *saltation* (Figure 11.5). Particles too large to be even temporarily suspended are transported by traction; that is, they simply roll or slide along a channel's bed.

LO4 DEPOSITION BY RUNNING WATER

Rivers and streams constantly erode, transport, and deposit sediment, but they do most of their geologic work when they flood. Consequently, their deposits, collectively called **alluvium**, do not represent the day-to-day activities of running water, but rather the periodic sedimentation that takes place during floods. Recall from Chapter 6 that sediments accumulate in continental, transitional, and marine *depositional environments* (see Figure 6.13). Deposits of rivers and streams are found mostly in the first two of these settings; however, much of the detrital sediment found on continental margins is derived from the land and transported to the oceans by running water.

The Deposits of Braided and Meandering Channels

A **braided stream** has an intricate network of dividing and rejoining channels separated from one another by sand and gravel bars (Figure 11.6). Braided channels develop when the sediment supply exceeds the transport capacity of running water, resulting in the deposition of sand and gravel bars. Braided streams have broad, shallow channels and are characterized as bed-load transport streams, because they transport and deposit mostly sand and gravel.

Meandering streams have a single sinuous channel with broadly looping curves known as *meanders* (Figure 11.7). Channels of meandering streams are semicircular in cross section along straight reaches, but markedly asymmetric at meanders, where they vary from shallow to deep across the meander. The deeper side of the channel is known as the *cut bank*, because greater velocity and fluid turbulence erode it (see the Virtual Field Trip in this chapter). In contrast, flow velocity is at a minimum

alluvium A collective term for all detrital sediment transported and deposited by running water.

braided stream A stream with multiple dividing and rejoining channels.

meandering stream A stream that has a single, sinuous channel with broadly looping curves.

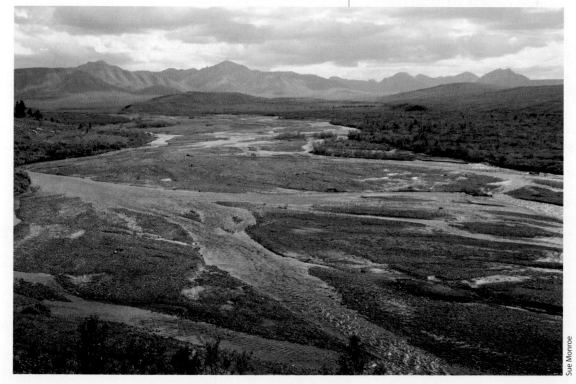

Figure 11.6 A Braided Stream and Its Deposits
This braided stream is in Denali National Park in Alaska. Its deposits are mostly gravel and sand.

Sue Monroe

Figure 11.7 Diagrammatic View of a Meandering River

© 2013 Cengage Learning

---

point bar The sediment body deposited on the gently sloping side of a meander loop.

oxbow lake A cutoff meander filled with water.

floodplain A low-lying, flat area adjacent to a channel that is partly or completely water-covered when a stream or river overflows its banks.

natural levee A ridge of sandy alluvium deposited along the margins of a channel during floods.

on the opposite bank, which slopes gently into the channel. As a result of this unequal distribution of flow velocity across meanders, the cut bank erodes, and a **point bar** is deposited on the gently sloping inner bank (Figure 11.8) (see the Virtual Field Trip in this chapter).

Meanders commonly become so sinuous that the thin neck of land between adjacent ones is cut off during a flood. Many of the floors of valleys with meandering channels are marked by

crescent-shaped **oxbow lakes**, which are simply cutoff meanders (Figures 11.7 and 11.9). Oxbow lakes may persist for a long time, but they eventually fill with organic matter and fine-grained sediments carried by floods.

Floodplain Deposits

Streams and rivers periodically receive more water than their channels can handle, so they overflow their banks and spread across adjacent low-lying, fairly flat **floodplains** (Figure 11.10). Sand and gravel might be deposited on floodplains where a stream or river bursts from it banks, but much more commonly the deposits are silt and clay, or simply mud. During a flood, a stream or river overtops its banks and water pours onto the floodplain, but as it does so, its depth and velocity rapidly decrease. As a result, ridges of sandy alluvium known as **natural levees** are deposited along the channel margins,

Figure 11.8 Point Bars

Two point bars of sand in Otter Creek in Yellowstone National Park in Wyoming. Point bars form on the gently sloping side of meanders, where flow velocity is lowest. Note that the point bars are inclined into the deeper part of the channel.

Point bars

Cut bank

James S. Monroe

Figure 11.9 Oxbow Lakes

© 2013 Cengage Learning

Erosion   Oxbow lake   Cutoff meander

Oxbow lake

a. A meandering stream showing the stages in the evolution of oxbow lakes. A future oxbow will form when the meander on the left side of the illustration is cut off.

b. Oxbow lakes along the Mississippi River in Minnesota.

Jake Rajs/Stone/Getty Images

and mud is carried beyond the natural levees into the floodplain, where it settles from suspension.

Another feature found on floodplains is oxbow lakes; recall that oxbow lakes are cut-off meanders (Figure 11.9). Once isolated from the main channel, oxbow lakes receive water mostly from periodic floods, although groundwater may also contribute. When they fill in with sediments brought in by floods and accumulating organic matter they are called *meander scars*, some of which are visible in the upper part of the image in Figure 11.9b.

Deltas

Where a river or stream flows into a lake or the ocean, its flow velocity rapidly diminishes, and any sediment in transport is deposited. Under some circumstances, this deposition creates a **delta**, an alluvial deposit that causes the shoreline to build outward into the lake or sea; the process is called *progradation*. The simplest prograding

delta A deposit of sediment where a stream or river enters a lake or the ocean.

Figure 11.10 Floodplain Deposits

a. The origin of vertical accretion deposits. During floods, streams deposit natural levees, and silt and mud settle from suspension on the floodplain.

b. After flooding.

© 2013 Cengage Learning

alluvial fan A cone-shaped accumulation of mostly sand and gravel deposited where a stream flows from a mountain valley onto an adjacent lowland.

deltas have a characteristic vertical sequence of *bottomset beds* overlain successively by foreset beds and *topset beds* (Figure 11.11). This vertical sequence develops when a river or stream enters another body of water where the finest sediment (silt and clay) is carried some distance out into the lake or sea, where it settles to form bottomset beds. Nearer shore, foreset beds are deposited as gently inclined layers, and topset beds, consisting of the coarsest sediments, are deposited in a network of *distributary channels* traversing the top of the delta (Figure 11.13).

Small deltas in lakes may have the three-part sequence just described, but deltas deposited along seacoasts are much larger, far more complex, and considerably more important as potential areas of natural resources. Depending on the relative importance of running water, waves, and tides, geologists identify three main types of marine deltas (Figure 11.12). *Stream-dominated deltas* have long fingerlike sand bodies, each deposited in a distributary channel that progrades far seaward. *Wave-dominated deltas* also have distributary channels, but the seaward margin of the delta consists of islands reworked by waves, and the entire margin of the delta progrades. *Tide-dominated deltas* are continuously modified into tidal sand bodies that parallel the direction of tidal flow.

Alluvial Fans

Fan-shaped deposits of alluvium on land known as **alluvial fans** form best on lowlands with adjacent highlands in arid and semiarid regions where little vegetation exists to stabilize surface materials (Figure 11.13). During periodic rainstorms, surface materials are quickly saturated, and surface runoff is funneled into a mountain canyon leading to adjacent lowlands. In the mountain canyon, the runoff is confined so that it cannot spread laterally, but when it discharges

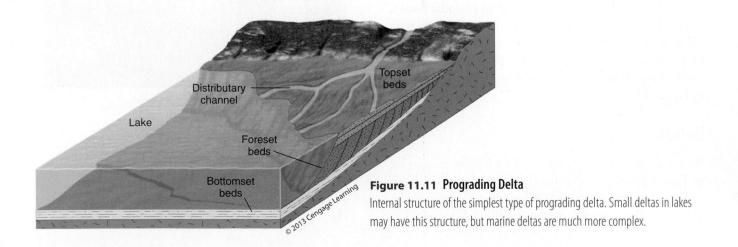

Figure 11.11 Prograding Delta

Internal structure of the simplest type of prograding delta. Small deltas in lakes may have this structure, but marine deltas are much more complex.

© 2013 Cengage Learning

Figure 11.12 Stream-, Wave-, and Tide-Dominated Deltas

a. The Mississippi River delta on the U.S. Gulf Coast is stream-dominated.

Image courtesy of the Image Science & Analysis Laboratory, NASA Johnson Space Center (http://eol.jsc.nasa.gov)

c. The Ganges-Brahmaputra delta of Bangladesh is tide-dominated.

Image courtesy of USGS National Center for EROS and NASA Landsat Project Science Office

b. The Nile River delta of Egypt is wave-dominated.

Jacques Descloitres, MODIS Rapid Response Team, NASA/GSFC

onto the lowlands, it quickly spreads out, its velocity diminishes, and deposition ensues. Repeated episodes of sedimentation result in the accumulation of a fan-shaped body of alluvium.

Deposition by running water in the manner just described is responsible for many alluvial fans. In some cases, though, the water flowing through a canyon picks up so much sediment that it becomes a viscous debris flow. Consequently, some alluvial fans consist mostly of debris-flow deposits that show little or no layering.

Figure 11.13 Alluvial Fans and Their Deposits
These alluvial fans at the base of the Panamint Range in Death Valley, California, were deposited where streams discharged from mountain canyons onto adjacent lowlands.

Copyright and photograph by Dr. Parvinder S. Sethi

LO5 CAN FLOODS BE CONTROLLED AND PREDICTED?

When any waterway receives more water than its channel can handle, it floods, occupying part or all of its floodplain. Indeed, floods are so common that unless they cause extensive property damage or fatalities, they rate little more than a passing notice in the news. Dozens of floods occur in the United States every year, but the most extensive river flooding in recent history was The Great Flood of 1993. Large parts of nine Midwestern states were inundated, but particularly hard hit were Missouri and Iowa (Figure 11.14).

Flooding is a fact of life, but there are several practices that can protect people and their property and minimize the impact of floods. Unfortunately, none of these measures is foolproof and most are expensive. One is to simply move all or parts of small communities to higher ground. Accordingly, following The Great Flood of 1993, parts of some small communities in Missouri and Illinois were moved to higher ground. Martin, Kentucky, has been flooded 37 times since 1862, and parts of the downtown area are being moved.

Another common flood control practice is to construct *dams* that impound excess water during floods (Figure 11.15a). Of course dams are expensive, they eventually fill in with sediment unless dredged, and some fail. In 1976 the Teton Dam in Idaho collapsed, killing 11 people and 13,000 cattle.

Another practice is to construct *levees* that raise the banks of streams and rivers, thereby increasing a channel's capacity (Figure 11.15b). Unfortunately, deposition within channels raises streambeds, making the levees less effective unless they are raised too. Levees along the Huang He in China caused the streambed to rise more than 20 m above its surrounding floodplain in 4,000 years. In 1887, when the river breached its levees, more than 1 million people were killed.

In some areas, state or federal agencies build *floodways*, which are channels used to divert excess water around a community or an area of economic importance (Figure 11.15c). And yet another practice is to build *floodwalls* to protect vulnerable areas. These are vertical structures placed along stream or river banks where levees are not practical, such as in cities (Figure 11.15d). Reforestation of cleared land also reduces the potential for flooding because vegetated soil absorbs water and decreases runoff.

Flooding streams and rivers account for more than $5 billion in property damage each year in the United States, not including coastal flooding from hurricanes. And even though more and more flood control projects are completed, the damages from flooding are not decreasing. The combination of fertile soils, level surfaces, and proximity to water for agriculture, industry, and domestic use, makes floodplains popular places for development. Unfortunately, urbanization increases surface runoff because soils are compacted or covered by asphalt or concrete, reducing infiltration capacity.

As for predicting floods, the best that can be done is to monitor streams, evaluate their past behavior, and anticipate floods of a given size in a specified period. Most people have heard of 10-year floods, 20-year floods, and so on, but how are such determinations made? The U.S. Geologic Survey, as well as state agencies, record and analyze stream behavior through time in order to anticipate floods of a specified size. So a 20-year flood, for example, is the period during which a flood of a given magnitude can be expected. It does not mean that the river in question will have a flood of that size every 20 years, only that over a long period of time, it will average 20 years. Or we can say that the chances of a 10-year flood taking place in any one year are 1 in 10 (1/10). In fact, it is possible that two 10-year floods could take place in successive years, but then not occur again for several decades.

LO6 DRAINAGE SYSTEMS

Thousands of waterways that are parts of larger drainage systems flow directly or indirectly into the oceans. The only exceptions are some rivers and streams that flow into desert basins surrounded by higher areas. But even these are parts of larger systems consisting of a main channel with all its tributaries—that is, streams

In August 2007, a tragic flood took place in Southeast Asia, particularly in India and Bangladesh. More than 2,100 people perished, and millions were displaced by swollen rivers, overflowing from monsoon rains. As displaced people fled to high ground, so did the wildlife, including poisonous snakes that claimed dozens of lives.

Figure 11.14 The Flood of '93

b. Flood waters in Portage des Sioux covered the 5.5-m high pedestal of this statue on the bank of the Mississippi River.

Missouri Highway and Transportation Department/USGS

Sue Monroe

a. This view shows flooding on July 30, 1993, in the vicinity of the Jefferson City Memorial Airport just north of Jefferson City, Missouri. Iowa and Missouri were particularly hard hit by this flood, but large areas in nine states were flooded.

that contribute water to another stream. The Mississippi River and its tributaries, such as the Ohio, Missouri, Arkansas, and Red rivers and thousands of smaller ones, carry runoff from an area known as a **drainage basin**. A topographically high area called a **divide** separates a drainage basin from adjoining ones (Figure 11.16). The continental divide along the crest of the Rocky Mountains in North America, for instance, separates drainage in opposite directions; drainage to the west goes to the Pacific, whereas drainage to the east eventually reaches the Gulf of Mexico.

The arrangements of channels within an area are types of **drainage patterns**. The most common is *dendritic drainage*, which consists of a network of channels resembling tree branching (Figure 11.17a). It develops on gently sloping surfaces composed of materials that respond more or less homogeneously to erosion, such as areas underlain by nearly horizontal sedimentary rocks.

In dendritic drainage, tributaries join larger channels at various angles, but *rectangular drainage* is characterized by right-angle bends and tributaries joining larger channels at right angles (Figure 11.17b). Such regularity in channels is strongly controlled by geologic structures, particularly regional joint systems that intersect at right angles.

Trellis drainage, consisting of a network of nearly parallel main streams with tributaries joining them at right angles, is common in some parts of the eastern United States. In Virginia and Pennsylvania, erosion of folded sedimentary rocks developed a landscape of alternating ridges on resistant rocks and valleys underlain by easily eroded rocks. Main waterways follow the valleys, and short tributaries flowing from the nearby ridges join the main channels at nearly right angles (Figure 11.17c).

In *radial drainage*, streams flow outward in all directions from a central high point, such as a large volcano (Figure 11.17d). Many of the volcanoes in the Cascade Range of western North America have radial drainage patterns.

In the types of drainage mentioned so far, some kind of pattern is easily recognized. *Deranged drainage*, in contrast, is characterized by irregularity, with streams flowing into and out of swamps and lakes, streams with only a few short tributaries, and vast swampy areas between channels (Figure 11.17e). This kind of drainage developed recently and has not yet formed a fully organized drainage system. In parts of Minnesota, Wisconsin, and Michigan, where glaciers obliterated the previous drainage, only 10,000 years have elapsed since the glaciers melted. As a result, drainage systems have not yet fully developed, and large areas remain undrained.

drainage basin The surface area drained by a stream or river and its tributaries.

divide A topographically high area that separates adjacent drainage basins.

drainage pattern The regional arrangement of channels in a drainage system.

Figure 11.15 Flood Control

Dams and reservoirs, levees, floodways, and floodwalls are some of the structures used to control floods.

a. Oroville Dam in California, at 235 m high, is the highest dam in the United States. It helps control floods, provides water for irrigation, and produces electricity at its power plant.

b. This levee, an artificial embankment along a waterway, helps protect nearby areas from floods. A university campus lies out of view just to the right of the levee.

c. This floodway carries excess water from a river (not visible) around a small community.

d. This floodwall on the bank of the Danube River at Mohács, Hungary, helps protect the city from floods.

LO7 THE SIGNIFICANCE OF BASE LEVEL

base level The level below which a stream or river cannot erode; sea level is ultimate base level.

Base level is the lowest limit to which a stream or river can erode. With the exception of streams that flow into closed depressions in deserts, all others are restricted ultimately to sea level. That is, they can erode no lower than sea level, because they must have some gradient to maintain flow. So *ultimate base level* is sea level, which is simply the lowest level of erosion for any waterway that flows into the sea (Figure 11.18). Ultimate base level applies to an entire stream or river system, but channels may also have a *local* or *temporary base level*. For example, a local base level may be a lake or another stream, or where a stream or river flows across particularly resistant rocks and a waterfall develops (Figure 11.18a).

Ultimate base level is sea level, but suppose that sea level dropped or rose with respect to the land, or suppose that the land rose or subsided? In these cases, base level would change and bring about changes in stream and river systems. During the maximum extent of Pleistocene glaciers, sea level was about 130 m lower than it is now, and streams adjusted by eroding deeper valleys and extending well out onto the continental shelves. Rising sea level at the end of the Ice Age accounted for a rising base level, decreased stream gradients, as well as deposition within channels.

Geologists and engineers are well aware that building a dam to impound a reservoir creates a local base level. A stream entering a reservoir deposits sediment, so unless they are dredged, reservoirs eventually fill with sediment.

Draining a lake may seem like a small change and well worth the time and expense to expose dry land for agriculture or commercial development. But draining a lake eliminates a local base level, and a stream that origi-

Figure 11.16 Drainage Basins

A detailed view of the Wabash River's drainage basin, a tributary of the Ohio River. All tributary streams within the drainage basin, such as the Vermilion River, have their own smaller drainage basins. Divides are shown by red lines.

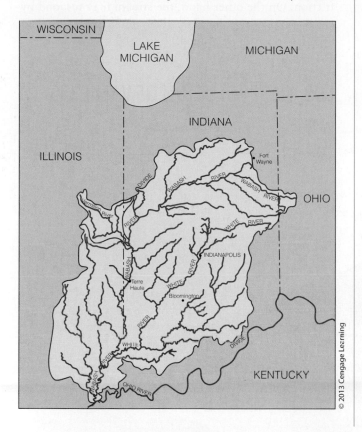

nally flowed into the lake responds by rapidly eroding a deeper valley as it adjusts to a new base level.

What Is a Graded Stream?

The *longitudinal profile* of any waterway shows the elevations of a channel along its length as viewed in cross section (Figure 11.18). For some rivers and streams, the longitudinal profile is smooth, but others show irregularities such as lakes and waterfalls, all of which are local base levels. Over time, these irregularities tend to be eliminated, because deposition takes place where the gradient is insufficient to maintain sediment transport, and erosion decreases the gradient where it is steep. So, given enough time, rivers and streams develop a smooth, concave longitudinal profile of equilibrium, meaning that all parts of the system dynamically adjust to one another.

A **graded stream** is one with an equilibrium profile in which a delicate balance exists among gradient, discharge, flow velocity, channel shape, and sediment load so that neither significant erosion nor deposition takes place within its channel (Figure 11.18b). Such a delicate balance is rarely attained, so the concept

> **graded stream** A stream that has an equilibrium profile in which a delicate balance exists among gradient, discharge, flow velocity, channel characteristics, and sediment load so that neither significant deposition nor erosion takes place within its channel.

Figure 11.17 Drainage Patterns

a. Dendritic drainage.

b. Rectangular drainage.

c. Trellis drainage.

d. Radial drainage.

e. Deranged drainage.

of a graded stream is an ideal. Nevertheless, the graded condition is closely approached in many streams, although only temporarily and not necessarily along their entire lengths.

Even though the concept of a graded stream is an ideal, we can anticipate the response of a graded stream to changes that alter its equilibrium. For instance, a change in base level would cause a stream to adjust as previously discussed. Increased rainfall in a stream's drainage basin would result in greater discharge and flow velocity. In short, the stream would now possess greater energy—energy that must be dissipated within the stream system by, for example, a change from a semicircular to a broad, shallow channel that would dissipate more energy by friction. On the other hand, the stream may respond by eroding a deeper valley, effectively reducing its gradient until it is once again graded.

LO8 THE EVOLUTION OF VALLEYS

Valleys are low areas on land bounded by higher land, and most of them have a river or stream running their

Figure 11.18 Base Level and Graded Streams

Erosion

Deposition

Erosion

Deposition

a. Sea level is ultimate base level, but a resistant rock layer over which a waterfall plunges is a local base level. Also, this stream has several irregularities in its profile, so it is ungraded.

Graded stream (concave-upward profile)

b. Erosion and deposition along its course eliminate irregularities and the stream becomes graded as it develops a smooth, concave profile of equilibrium.

© 2013 Cengage Learning

length, with tributaries draining the nearby high areas (see the Virtual Field Trip in this chapter). With few exceptions, valleys form and evolve in response to erosion by running water, although other processes, especially mass wasting, contribute. The shapes and sizes of valleys vary from small, steep-sided *gullies* to those that are broad with gently sloping valley walls (Figure 11.19). Steep-walled, deep valleys of vast size are *canyons*, and particularly narrow and deep ones are *gorges*.

A valley might start to erode where runoff has sufficient energy to dislodge surface materials and excavate a small rill. Once formed, a rill collects more runoff and becomes deeper and wider and continues to do so until a full-fledged valley develops. *Downcutting* takes place when a river or stream has more energy than it needs to transport sediment, so some of its excess energy is used to deepen its valley. In most cases, the valley walls are simultaneously undercut, a process called *lateral erosion*, creating unstable slopes that may fail by *mass wasting*. Furthermore, erosion by sheetwash and erosion by tributary streams carry materials from the valley walls into the main stream in the valley.

Valleys not only become deeper and wider, but they also become longer by *headward erosion*, a phenomenon involving erosion by entering runoff at the upstream end of a valley (Figure 11.20a). Continued headward erosion may result in *stream piracy*, the breaching of a drainage divide and diversion of part of the drainage of another stream (Figure 11.20b).

Stream Terraces

Rather flat surfaces paralleling a stream or river but at a higher level than the present-day floodplain are **stream terraces**. These surfaces represent an older floodplain that was adjacent to the waterway when it flowed at a higher level but subsequently eroded down to a lower level (Figure 11.21). Some streams have several steplike surfaces above their present floodplains, indicating the terraces formed more than once.

> **stream terrace** An erosional remnant of a floodplain that formed when a stream was flowing at a higher level.

Figure 11.19 Gullies and Valleys

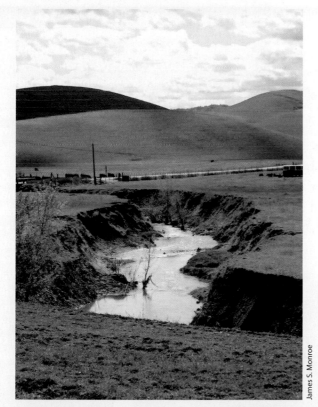

a. Gullies are small valleys, but they are steep-sided. This gully measures about 15 m across.

James S. Monroe

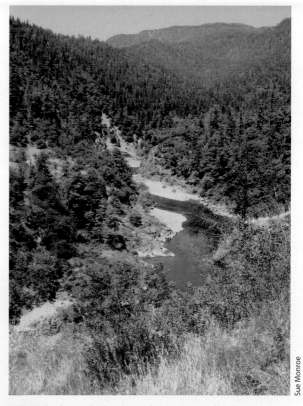

b. This valley has walls that descend to a narrow valley bottom.

Sue Monroe

Virtual Field Trip

Running Water

DeerPoint, UT. topographic map, courtesy of USGS.
All photographs: Copyright and Photograph by Dr. Parvinder S. Sethi

Ready to Go!

Zion National Park in southwestern Utah was made a national park in November, 1919. Majestic cliffs and towers made up of colorful rocks and deep canyons dominate the landscape. Indeed, the rocks seem to change color throughout the day. Several geologic processes, such as mass wasting and mechanical and chemical weathering, have contributed to the magnificent scenery in the park, but erosion and deposition by running water is our focus in this virtual field trip. The Virgin River is the main waterway in the park, and it, along with its tributaries, is largely responsible for erosion of the mostly sandstone terrain.

The rocks exposed in the park's cliffs constitute the geologic record for this area and preserve evidence of ancient shallow seas, deposition in rivers, lakes, and ponds, as well as coastal sand dunes. The Navajo Sandstone consists of a lower reddish-brown, iron-rich sandstone and an overlying light-colored, iron-poor sandstone. All of the park's rock formations are Mesozoic in age and represent about 150 million years of geologic time. About 13 million years ago, uplift of the entire region steepened the gradient of the Virgin River which began downcutting, a process that would eventually yield the cliffs and deep canyons now present in Zion National Park.

Let's start our trip trough Zion National Park where we can make many observations relevant to our discussions of running water in this chapter.

GOALS OF THE TRIP

There are many geologic features to see in Zion National Park, but our main goals on this trip are to study:

1. Erosion.
2. Sediment transport.
3. Deposition by running water.
4. The origin and evolution of the park's canyons.

FOLLOW-UP QUESTIONS

1. Why do meandering streams form an alternating series of cut banks and point bars along their course?
2. Give a brief history of the origin and development of the canyons in Zion National Park.
3. What processes, other than downcutting by the Virgin River, contribute to the ongoing evolution of the canyons in Zion National Park?

What to See When You Go

This image shows both erosion and deposition by a meandering stream, that is, a stream with a single, sinuous channel (see Figure 11.7 in this chapter). Notice the steep bank on the right side of the image, which is called the cut bank because, as discussed in this chapter's section "The Deposits of Braided and Meandering Channels," the water velocity is highest here and erosion takes place. On the opposite bank, however, velocity is least and deposition occurs, forming a point bar deposit mostly of sand that slopes gently toward the cut bank. Point bars are the most distinctive features of ancient meandering stream deposits. It is important to note that further downstream the maximum velocity switches from right to left, thus forming a succession alternating point bars and cut banks (see Figure 11.8 in this chapter).

In the photo showing a point bar (on the previous page) notice that the water in the river is turbid (murky) because it is carrying mud (silt and clay) and probably sand along its bed. The photo to the right shows a deposit of gravel and sand. Some of the gravel is of boulder size (greater than 25.6 cm). From observations here we can conclude that the stream was much higher when this deposit formed, and it must have been flowing very rapidly to transport such large particles. Also the gravel particles are well rounded, meaning that their sharp corners and edges were worn smooth by abrasion during transport. And finally, the deposit is poorly sorted because it contains a wide range of particle sizes.

Although this canyon and the slot canyon discussed previously were both eroded by running water, this one has a very different profile. As opposed to being narrow and deep, this canyon is broad and deep with alternating steep cliffs and gentle slopes. The lower, gentle slopes formed on softer rocks, mostly shale and siltstone, whereas the steep cliffs developed in the rust-colored Navajo Sandstone.

Great view of a slot canyon, that is, a canyon that is much deeper than it is wide and usually with vertical or near vertical sides. Most slot canyons form where running water erodes into sandstone (as here) or limestone, although they may form in other rocks, too. Past floods are indicated by the scour marks and polished rock surface several meters above water level. Be very careful if you hike in a slot canyon, because a summer thunderstorm can cause the water level to rise several meters in just a few minutes.

As we learned in this chapter in the section "The Evolutions of Valleys," downcutting by running water accounts for canyon deepening, but canyons also widen by a combination of processes including erosion by tributaries, weathering, and mass wasting. This blind arch, as it is called, formed when the softer rock below was eroded and part of the overlying sandstone collapsed. In any event, the debris from this event eventually makes it to the river only to be carried away and deposited elsewhere.

The formation of stream terraces is preceded by an episode of deposition followed by erosion as the stream or river begins eroding downward. For example, suppose that a stream is graded so that it is adjusted to its present gradient, volume of water, and sediment load. If the land over which the stream flows is uplifted or sea level becomes lower, the stream's gradient is steeper and it has more energy to deepen its valley. When this stream once again reaches a level at which it is graded, downcutting ceases, and it begins to erode laterally, thus establishing a new floodplain at a lower level. Several such episodes account for the multiple stream terraces seen in the valleys of some streams.

Although changes in base level probably account for many stream terraces, a change in climate can have the same result. If the amount of precipitation in a stream's drainage basin increases, the stream has a greater volume of water and the capacity to erode a deeper valley, perhaps leaving remnants of an older floodplain as stream terraces.

Figure 11.21 Origin of Stream Terraces

a. A stream has a broad floodplain.

b. The stream erodes downward and establishes a new floodplain at a lower level. Remnants of its old, higher floodplain are stream terraces.

c. Another level of stream terraces forms as the stream erodes downward again.

© 2013 Cengage Learning

Figure 11.20 Two Stages in the Evolution of a Valley

a. The stream widens its valley by lateral erosion and mass wasting while simultaneously extending its valley by headward erosion.

b. As the larger stream continues to erode headward, stream piracy takes place when it captures some of the drainage of the smaller stream. Notice also that the valley is wider in (b) than it was in (a).

© 2013 Cengage Learning

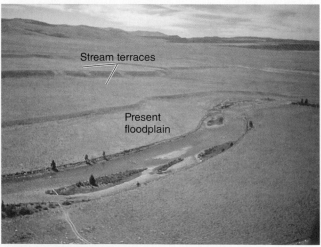

d. Stream terraces along the Madison River in Montana.

J.R. Stacy/USGS

Incised Meanders

Some streams are restricted to deep, meandering canyons cut into bedrock, where they form features called **incised meanders** (Figure 11.22). Streams restricted by rock walls usually cannot erode laterally; thus, they lack a floodplain and occupy the entire width of the canyon floor.

It is not difficult to understand how a stream can cut downward into rock, but how a stream forms a meandering pattern in bedrock is another matter. Because lateral erosion is inhibited once downcutting begins, one must infer that the meandering course was established when the stream flowed across an area covered by alluvium. For example, suppose that a stream near base level has established a meandering pattern. If the land that the stream flows over is uplifted, then erosion begins and the meanders become incised into the underlying bedrock.

Superposed Streams

Water flows downhill in response to gravity, so the direction of flow in streams and rivers is determined by topography. Yet a number of waterways seem, at first glance, to have defied this fundamental control. For instance, several rivers in the eastern United States flow in valleys that cut directly through ridges that lie in their paths. These are examples of **superposed streams**, all of which once flowed on a surface at a higher level, but as they eroded downward, they eroded into resistant rocks and cut narrow canyons, or what geologists call *water gaps* (Figure 11.23).

A water gap has a stream flowing through it, but if the stream is diverted elsewhere, perhaps by stream piracy, the abandoned gap is then called a *wind gap*. The Cumberland Gap in Kentucky is a good example; it was the avenue through which settlers migrated from Virginia to Kentucky from 1790 until well into the 1800s. Furthermore, several water gaps and wind gaps played important strategic roles during the Civil War (1861–1865).

incised meander A deep, meandering canyon cut into bedrock by a stream or river.

superposed stream A stream that once flowed on a higher surface and eroded downward into resistant rocks while maintaining its course.

Figure 11.23 Origin of a Superposed Stream

a. As a stream erodes down and removes the surface layers of rock, it is lowered onto ridges that form when resistant rocks in the underlying structure are exposed.

b. The narrow valleys through the ridges are water gaps.

© 2013 Cengage Learning

c. View of a water gap cut by the Jefferson River in Montana.

Marli Bryant Miller

Figure 11.22 Incised Meanders
The Colorado River at Dead Horse State Park in Utah is incised to a depth of 600 m.

James S. Monroe

12 Groundwater

In April 2000, giant mineral crystals were found in a silver and lead mine in Mexico. One cavity in the mine is lined with hundreds of giant gypsum crystals, more than 1 m long. There are also gypsum crystals dubbed "crystal moonbeams" (shown here) that are at least 15 m long and 1 m in diameter.

© Richard D. Fisher

Introduction

"As the world's population and industrial **development expand**, the demand for water, particularly groundwater, **will increase**."

Within the limestone region of western Kentucky lies the largest cave system in the world. In 1941, approximately 51,000 acres were set aside and designated as Mammoth Cave National Park. In 1981, it became a World Heritage Site. From ground level, the topography of the area is unimposing, with gently rolling hills. Beneath the surface, however, are more than 540 km of interconnected passageways whose spectacular geologic features have been enjoyed by millions of cave explorers and tourists.

During the War of 1812, approximately 180 metric tons of saltpeter (potassium nitrate—KNO_3) used in the manufacture of gunpowder, were mined from Mammoth Cave. At the end of the war, the saltpeter market collapsed, and Mammoth Cave was developed as a tourist attraction, easily overshadowing the other caves in the area. During the next 150 years, the discovery of new passageways and links to other caverns helped establish Mammoth Cave as the world's premier cave and the standard against which all others were measured.

The colorful cave deposits are the primary reason that millions of tourists have visited Mammoth Cave

Learning Outcomes

After reading this unit, you should be able to do the following:

LO1 Describe groundwater and its role in the hydrologic cycle

LO2 Define porosity and permeability

LO3 Describe the water table

LO4 Understand patterns of groundwater movement

LO5 Identify the differences between springs, water wells, and artesian systems

LO6 Discuss erosion and deposition caused by groundwater

LO7 Describe the effects of modifications to the groundwater system

LO8 Describe hydrothermal activity

---

groundwater
Underground water stored in the pore spaces of soil, sediment, and rock.

over the years. Hanging down from the ceiling and growing up from the floor are spectacular icicle-like structures, as well as columns and curtains in a variety of colors. Moreover, intricate passageways connect rooms of various sizes. The cave is also home to more than 200 species of insects and other animals, including about 45 blind species.

In addition to the beautiful caves, caverns, and cave deposits produced by groundwater movement, groundwater is also an important natural resource. Although groundwater constitutes only 0.625% of the world's water, it is nonetheless a significant source of freshwater for agricultural, industrial, and domestic use. More than 65% of the groundwater used in the United States each year goes for irrigation, with industrial use second, followed by domestic needs. These demands have severely depleted the groundwater supply in many areas and have led to such problems as ground subsidence and saltwater contamination. In other areas, pollution from landfills, toxic waste, and agriculture has rendered the groundwater supply unsafe.

It is becoming vitally important that people become aware of what a valuable resource groundwater is, so they can ensure that future generations have a clean and adequate supply of water from this source.

As the world's population and industrial development expand, the demand for water, particularly groundwater, will increase. Not only must new groundwater sources be located, but once found, these sources must be protected from pollution and managed properly to ensure that users do not withdraw more water than can be replenished.

LO1 GROUNDWATER AND THE HYDROLOGIC CYCLE

Groundwater—water that fills open spaces in rocks, sediment, and soil beneath Earth's surface—is one reservoir in the hydrologic cycle (see Figure 11.2),

porosity The percentage of a material's total volume that is pore space.

permeability A material's capacity to transmit fluids.

accounting for approximately 22% (8.4 million km³) of the world's supply of freshwater. Like all other water in the hydrologic cycle, the ultimate source of groundwater is the oceans; however, its more immediate source is the precipitation that infiltrates the ground and seeps down through the voids in soil, sediment, and rocks. Groundwater may also come from water infiltrating from streams, lakes, swamps, artificial recharge ponds, and water-treatment systems.

Regardless of its source, groundwater moving through the tiny openings between soil and sediment particles and the spaces in rocks filters out many impurities, such as disease-causing microorganisms and many pollutants. However, not all soils and rocks are good filters, and sometimes so much undesirable material may be present that it contaminates the groundwater. Groundwater movement and its recovery from wells depend on two critical aspects of the materials that it moves through: *porosity* and *permeability*.

LO2 POROSITY AND PERMEABILITY

Porosity and permeability are important physical properties of Earth materials and are largely responsible for the amount, availability, and movement of groundwater. Water soaks into the ground because soil, sediment, and rock have open spaces, or pores. **Porosity** is the percentage of a material's total volume that is pore space. Porosity most often consists of the spaces between particles in soil, sediment, and sedimentary rocks, but other types of porosity include cracks, fractures, faults, and vesicles in volcanic rocks (Figure 12.1).

Porosity varies among different rock types and is dependent on the size, shape, and arrangement of the material composing the rock (Table 12.1). Most igneous and metamorphic rocks, as well as many limestones and dolostones, have very low porosity, because they consist of tightly interlocking crystals. Their porosity can be increased, however, if they have been fractured or weathered by groundwater. This is particularly true for massive limestone and dolostone, whose fractures can be enlarged by acidic groundwater.

TABLE 12.1

Porosity Values for Different Materials

MATERIAL	PERCENTAGE POROSITY
UNCONSOLIDATED SEDIMENT	
Soil	55
Gravel	20–40
Sand	25–50
Silt	35–50
Clay	50–70
ROCKS	
Sandstone	5–30
Shale	0–10
Solution activity in limestone, dolostone	10–30
Fractured basalt	5–40
Fractured granite	10

Source: U.S. Geological Survey, Water Supply Paper 2220 (1983) and others.

Groundwater provides 80% of the water used for rural livestock and domestic use, as well as providing 40% of public water supplies.

© Don Mason/Brand X Pictures/JupiterImages

By contrast, detrital sedimentary rocks composed of well-sorted and well-rounded grains can have high porosity, because any two grains touch at only a single point, leaving relatively large open spaces between the grains (Figure 12.1a). Poorly sorted sedimentary rocks, on the other hand, typically have low porosity, because smaller grains fill in the spaces between the larger grains, further reducing porosity (Figure 12.1b). In addition, the amount of cement between grains can decrease porosity.

Porosity determines the amount of groundwater that Earth materials can hold, but it does not guarantee that the water can be easily extracted. So, in addition to being porous, Earth materials must have the capacity to transmit fluids, a property known as **permeability**. Thus, both porosity and permeability play important roles in groundwater movement and recovery.

Figure 12.1 Porosity

A rock's porosity depends on the size, shape, and arrangement of the material composing the rock.

Pore space

a. A well-sorted sedimentary rock has high porosity, whereas

b. a poorly sorted one has lower porosity.

Openings resulting from solution

Fractures

© 2013 Cengage Learning

c. In soluble rocks such as limestone, porosity can be increased by solution, whereas

d. crystalline metamorphic and igneous rocks can be rendered porous by fracturing.

Permeability is dependent not only on porosity, but also on the size of the pores or fractures and their interconnections. For example, deposits of silt or clay are typically more porous than sand or gravel, but they have low permeability, because the pores between the particles are very small and molecular attraction between the particles and water is great, thereby preventing movement of the water. In contrast, the pore spaces between grains in sandstone and conglomerate are much larger, and molecular attraction on the water is therefore low. Chemical and biochemical sedimentary rocks, such as limestone and dolostone, and many igneous and metamorphic rocks that are highly fractured, can also be very permeable provided that the fractures are interconnected.

A permeable layer transporting groundwater is an *aquifer*, from the Latin *aqua*, "water." The most effective aquifers are deposits of well-sorted and well-rounded sand and gravel. Limestones in which fractures and bedding planes have been enlarged by solution are also good aquifers. Shales and many igneous and metamorphic rocks make poor aquifers because they are typically impermeable unless fractured. Rocks such as these and any other materials that prevent the movement of groundwater are *aquicludes*.

LO3 THE WATER TABLE

Some of the precipitation on land evaporates, and some enters streams and returns to the oceans by surface runoff; the remainder seeps into the ground. As this water moves down from the surface, a small amount adheres to the material it moves through and halts its downward progress. With the exception of this *suspended water*, however, the rest seeps further downward and collects until it fills all of the available pore spaces. Thus, two zones are defined by whether their pore spaces contain mostly air, the **zone of aeration**, or mostly water, the underlying **zone of saturation**. The surface that separates these two zones is the **water table** (Figure 12.2).

The base of the zone of saturation varies from place to place, but it usually extends to a depth where an impermeable layer is encountered or to a depth where confining pressure closes all open space. Extending irregularly upward a few centimeters to several meters from the zone of saturation is the *capillary fringe*. Water moves upward in this region because of surface tension, much as water moves upward through a paper towel.

In general, the configuration of the water table is a subdued replica of the overlying land surface; that is, it rises beneath hills and has its lowest elevations beneath valleys. Several factors contribute to the overall configuration of a region's water table, including regional differences in the amount of rainfall, permeability, and rate of groundwater movement. However, in arid and semiarid regions, the water table tends to be quite flat regardless of the overlying land surface.

LO4 GROUNDWATER MOVEMENT

Gravity provides the energy for the downward movement of groundwater. Water entering the ground moves through the zone of aeration to the zone of saturation (Figure 12.3). When water reaches the water table, it continues to move through the zone of saturation from

zone of aeration The zone above the water table that contains both air and water within the pore spaces of soil, sediment, or rock.

zone of saturation The area below the water table in which all pore spaces are filled with water.

water table The surface that separates the zone of aeration from the underlying zone of saturation.

Figure 12.2 Water Table

The zone of aeration contains both air and water within its pore spaces, whereas all pore spaces in the zone of saturation are filled with groundwater. The water table is the surface separating the zones of aeration and saturation. Within the capillary fringe, water rises by surface tension from the zone of saturation into the zone of aeration.

Suspended water

Zone of aeration

Water table

Zone of saturation

Capillary fringe

Groundwater

Zone of aeration

Capillary fringe

Water table

Zone of saturation

© 2013 Cengage Learning

---

spring A place where groundwater flows or seeps out of the ground.

areas where the water table is high toward areas where it is lower, such as streams, lakes, or swamps. Only some of the water follows the direct route along the slope of the water table. Most of it takes longer curving paths down and then enters a stream, lake, or swamp from below, because it moves from areas of high pressure toward areas of lower pressure within the saturated zone.

Groundwater velocity varies greatly and depends on many factors. Velocities range from 250 m per day in some extremely permeable material to less than a few centimeters per year in nearly impermeable material. In most ordinary aquifers, the average velocity of groundwater is a few centimeters per day.

LO5 SPRINGS, WATER WELLS, AND ARTESIAN SYSTEMS

You can think of the water in the zone of saturation much like a reservoir whose surface rises or falls depending on additions as opposed to natural and artificial withdrawals. *Recharge*—that is, additions to the zone of saturation—comes from rainfall or melting snow, or water might be added artificially at wastewater-treatment plants or recharge ponds constructed for just this purpose. But if groundwater is discharged naturally or withdrawn at wells without sufficient recharge, the water table drops—just as a savings account diminishes if withdrawals exceed deposits. Withdrawals from the groundwater system take place where groundwater flows laterally into streams, lakes, or swamps, where it discharges at the surface as *springs*, and where it is withdrawn from the system at water wells.

Springs

Places where groundwater flows or seeps out of the ground as **springs** have always fascinated people. The water flows out of the ground for no apparent reason and from no readily identifiable source. So it is not surprising that springs have long been regarded with superstition and revered for their supposed medicinal value and healing powers. Nevertheless, there is nothing mystical or mysterious about springs.

Although springs can occur under a wide variety of geologic conditions, they all form in basically the same way (Figure 12.4a). When percolating water reaches the water table or an impermeable layer, it flows laterally, and if this flow intersects the surface, the water discharges as a spring (Figure 12.4b).

Springs can also develop wherever a *perched water table*—a local aquiclude present within a larger aquifer, such as a lens of shale within sandstone—intersects the surface (Figure 12.3). As water migrates through the zone of aeration, it is stopped by the local aquiclude, and a localized zone of saturation "perched" above the main water table forms. Water moving laterally along the perched water table may intersect the surface to produce a spring.

Figure 12.3 Groundwater Movement
Groundwater moves down through the zone of aeration to the zone of saturation. Then some of it moves along the slope of the water table, and the rest moves through the zone of saturation from areas of high pressure toward areas of low pressure. Some water might collect over a local aquiclude, such as a shale layer, thus forming a perched water table.

Water Wells

Water wells are openings made by digging or drilling down into the zone of saturation. Once the zone of saturation has been penetrated, water percolates into the well, filling it to the level of the water table. A few wells are free-flowing; however, for most, the water must be brought to the surface by pumping.

When groundwater is pumped from a well, the water table in the area around the well is lowered, forming a **cone of depression** (Figure 12.5). This happens when the rate of water withdrawal from the well exceeds the rate of water inflow to the well, thus lowering the water table around the well. A cone of depression's gradient—that is, whether it is steep or gentle—depends to a great extent on the permeability of the aquifer being pumped. A highly permeable aquifer produces a gentle gradient in the cone of depression, whereas a low-permeability aquifer results in a steep cone of depression, because water cannot easily flow to the well to replace the water being withdrawn.

The formation of a cone of depression does not normally pose a problem for the average domestic well provided that the well is drilled deep enough into the zone of saturation. However, the tremendous amounts of water used by industry and for irrigation may create a large cone of depression that lowers the water table sufficiently to cause shallow wells in the immediate area to go dry (Figure 12.5). This situation is not uncommon and frequently results in lawsuits by the owners of the shallow dry wells.

water well A well made by digging or drilling into the zone of saturation.

cone of depression A cone-shaped depression around a well where water is pumped from an aquifer faster than it can be replaced.

In many parts of the United States and Canada, one can see windmills from times past that used wind power to pump water. Most of these are no longer in use, having been replaced by more efficient electric pumps.

Figure 12.4 Springs

Springs form wherever laterally moving groundwater intersects Earth's surface.

© John Karachewski/Geoscapes Photography

a. Most commonly, springs form when percolating water reaches an impermeable layer and migrates laterally until it seeps out at the surface.

b. Thunder River Spring in the Grand Canyon, Arizona, issues from rocks along a wall of the Grand Canyon. Water percolating downward through permeable rocks is forced to move laterally when it encounters an impermeable zone, and thus gushes out along this cliff. Notice the vegetation parallel to and below the springs, indicating that enough water flows from springs along the cliff wall to support the vegetation.

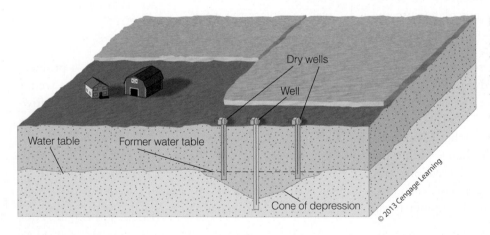

Figure 12.5 Cone of Depression

A cone of depression forms whenever water is withdrawn from a well. If water is withdrawn faster than it can be replenished, the cone of depression will grow in depth and circumference, lowering the water table in the area and causing nearby shallow wells to go dry.

Lowering of the regional water table because of more groundwater being withdrawn than is being replenished is becoming a serious problem in many areas, particularly in the southwestern United States, where rapid growth has placed tremendous demands on the groundwater system. Unrestricted withdrawal of groundwater cannot continue indefinitely, and the rising costs and decreasing supply of groundwater should soon limit the growth of some regions in the United States, such as the aforementioned Southwest.

Artesian Systems

The word *artesian* comes from the French town and province of Artois (called Artesium during Roman times) near Calais, where the first European artesian

Area of recharge for aquifer

Artesian-pressure surface

Flowing artesian well

Nonflowing artesian well

Shale aquicludes

Sandstone aquifer

© 2013 Cengage Learning

Figure 12.6 Artesian System

An artesian system must have an aquifer confined above and below by aquicludes, the aquifer must be exposed at the surface, and there must be sufficient precipitation in the recharge area to keep the aquifer filled. The elevation of the water table in the recharge area, which is indicated by a sloping dashed line (the artesian pressure surface), defines the highest level to which well water can rise.

well was drilled in 1126 and is still flowing today. The term **artesian system** can be applied to any system in which groundwater is confined and builds up high hydrostatic (fluid) pressure (Figure 12.6). Water in such a system is able to rise above the level of the aquifer if a well is drilled through the confining layer, thereby reducing the pressure and forcing the water upward. An artesian system can develop when (1) an aquifer is confined above and below by aquicludes; (2) the rock sequence is (usually) tilted to build up hydrostatic pressure; and (3) the aquifer is exposed at the surface, thus enabling it to be recharged.

The elevation of the water table in the recharge area and the distance of the well from the recharge area determine the height to which artesian water rises in a well. The surface defined by the water table in the recharge area, called the *artesian-pressure surface*, is indicated by the sloping dashed line in Figure 12.6. Because friction slightly reduces the pressure of the aquifer water, and consequently the level to which artesian water rises, the artesian-pressure surface therefore slopes.

An artesian well will flow freely at the ground surface only if the wellhead is at an elevation below the artesian-pressure surface. In this situation, the water flows out of the well, because it rises toward the artesian-

pressure surface, which is at a higher elevation than the wellhead. In a nonflowing artesian well, the wellhead is above the artesian-pressure surface, and the water will rise in the well only as high as the artesian-pressure surface.

> **artesian system** A confined groundwater system with high hydrostatic pressure that causes water to rise above the level of the aquifer.

One of the best-known artesian systems in the United States underlies South Dakota and extends southward to central Texas. The majority of the artesian water from this system is used for irrigation. The aquifer of this artesian system, the Dakota Sandstone, is recharged where it is exposed along the margins of the Black Hills of South Dakota. Originally, the hydrostatic pressure in this system was great enough to produce free-flowing wells and to operate waterwheels. However, because of the extensive use of this groundwater for irrigation over the years, the hydrostatic pressure in many of the wells is so low that they are no longer free-flowing, and the water must be pumped to the surface.

As a final comment on artesian systems, we should mention that it is not unusual for advertisers to tout the quality of artesian water as somehow being superior to other groundwater. Some artesian water might

sinkhole A depression in the ground that forms by the solution of the underlying carbonate rocks or by the collapse of a cave roof.

indeed be of excellent quality, but its quality is not dependent on the fact that water rises above the surface of an aquifer. Rather, its quality is a function of dissolved minerals and any introduced substances, so artesian water really is no different from any other groundwater. The myth of its superiority probably arises from the fact that people have always been fascinated by water that flows freely from the ground.

LO6 GROUNDWATER EROSION AND DEPOSITION

When rainwater begins to seep into the ground, it immediately starts to react with the minerals it contacts, weathering them chemically. In an area underlain by soluble rock, groundwater is the principal agent of erosion and is responsible for the formation of many major features of the landscape.

Limestone, a common sedimentary rock composed primarily of the mineral calcite ($CaCO_3$), underlies large areas of Earth's surface (Figure 12.7). Although limestone is practically insoluble in pure water, it readily dissolves if a small amount of acid is present. Carbonic acid (H_2CO_3) is a weak acid that forms when carbon dioxide combines with water ($H_2O + CO_2 \rightarrow H_2CO_3$). Because the atmosphere contains a small amount of carbon dioxide (0.03%) and carbon dioxide is also produced in soil by the decay of organic matter, most groundwater is slightly acidic. When groundwater percolates through the various openings in limestone, the slightly acidic water readily reacts with the calcite to dissolve the rock by forming soluble calcium bicarbonate, which is carried away in solution (see Chapter 6).

Sinkholes and Karst Topography

In regions underlain by soluble rock, the ground surface may be pitted with numerous depressions that vary in size and shape. These depressions, called **sinkholes** or merely *sinks*, mark areas with underlying soluble rock (Figure 12.8). Most sinkholes form in one of two ways. The first is when soluble rock below the soil is dissolved by seeping water, and openings in the rock are enlarged and filled in by the overlying soil. As the groundwater continues to dissolve the rock, the soil is eventually removed, leaving shallow depressions with gently sloping sides. When adjacent sinkholes merge, they form

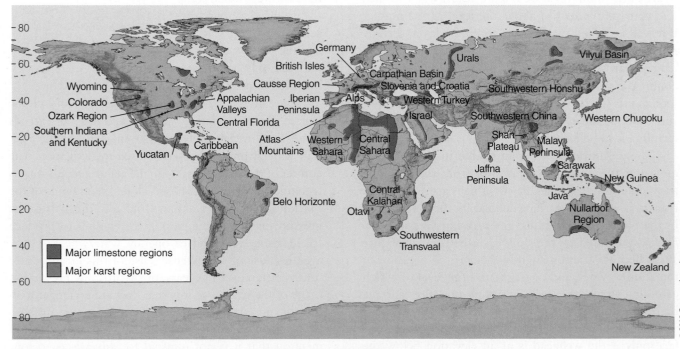

© 2013 Cengage Learning

Figure 12.7 Distribution of the Major Limestone and Karst Areas of the World
Karst topography develops largely by groundwater erosion in areas underlain by soluble rocks.

a network of larger, irregular, closed depressions called *solution valleys*.

Sinkholes also form when a cave's roof collapses, usually producing a steep-sided crater. Sinkholes formed in this way are a serious hazard, particularly in populated areas. In regions prone to sinkhole formation, extensive geologic and hydrogeologic investigations must be done to determine the depth and extent of underlying cave systems prior to any site development to ensure that the underlying rocks are thick enough to support planned structures.

Karst topography, or simply *karst*, develops largely by groundwater erosion in many areas underlain by soluble rocks (Figure 12.9). The name *karst* is derived from the plateau region of the border area of Slovenia, Croatia, and northeastern Italy, where this type of topography is well developed. In the United States, regions of karst topography include large areas of southwestern Illinois, southern Indiana, Kentucky, Tennessee, southern Missouri, Alabama, and

central and northern Florida (Figure 12.7).

Karst topography is characterized by numerous caves, springs, sinkholes,

Figure 12.8 Sinkholes
This sinkhole formed on May 8 and 9, 1981, in Winter Park, Florida. It formed in previously dissolved limestone following a drop in the water table. The 100-m-wide, 35-m-deep sinkhole destroyed a house, numerous cars, and a municipal swimming pool.

AP Photo

karst topography
Landscape consisting of numerous caves, sinkholes, and solution valleys formed by groundwater solution of rocks such as limestone and dolostone.

Solution valleys

Springs

Sinkholes

Disappearing streams

Deeply entrenched permanent stream

Cave

© 2013 Cengage Learning

Figure 12.9 Features of Karst Topography
Erosion of soluble rock by groundwater produces karst topography. Features commonly found include solution valleys, springs, sinkholes, and disappearing streams.

cave A natural subsurface opening generally connected to the surface and large enough for a person to enter.

solution valleys, and disappearing streams (Figure 12.9). *Disappearing streams* are so named because they typically flow only a short distance at the surface and then disappear into a sinkhole. The water continues flowing underground through fractures or caves until it surfaces again at a spring or other stream.

Karst topography varies from the spectacular high-relief landscapes of China (Figure 12.10) to the subdued and pockmarked landforms of Kentucky. Common to all karst topography, though, is the presence of thick-bedded, readily soluble rock at the surface or just below the soil, and enough water for solution activity to occur. Karst topography is therefore typically restricted to humid and temperate climates.

Caves and Cave Deposits

Caves are perhaps the most spectacular examples of the combined effects of weathering and erosion by groundwater. As groundwater percolates through carbonate rocks, it dissolves and enlarges fractures and openings to form a complex interconnecting system of crevices, caves, caverns, and underground streams. A **cave** is usually defined as a naturally formed subsurface open-

Jesse James (left) and members of his outlaw band (probably two of the Younger brothers) pose in this ca. 1870s photo. Jesse James and his gang repeatedly used Meramec Caverns, Missouri, as a hideout in the 1870s.

© Bettmann/Corbis

ing that is generally connected to the surface and is large enough for a person to enter. A *cavern* is a very large cave or a system of interconnected caves.

More than 17,000 caves are known in the United States. Some of the more famous ones are Mammoth Cave, Kentucky; Carlsbad Caverns, New Mexico; Lewis and Clark Caverns, Montana; Lehman Cave, Nevada; and Meramec Caverns, Missouri, which Jesse James and his outlaw band often used as a hideout. Canada also has many famous caves, including the 536-m-deep Arctomys Cave in Mount Robson Provincial Park, British Columbia, the deepest known cave in North America. And Mexico is known for its Cueva de los Cristales, or Cave of Crystals, where some of the world's largest crystals of gypsum and selenite have been found (see the chapter opening photo).

Caves and caverns form as a result of the dissolution of carbonate rocks by weakly acidic groundwater (Figure 12.11). Groundwater percolating through the zone of aeration slowly dissolves the carbonate rock and enlarges its fractures and bedding planes. On reaching the water table, the groundwater

Figure 12.10 Karst Landscape southeast of Kunming, China
The Stone Forest, 125 km southeast of Kunming, China, is a high-relief karst landscape formed by the dissolution of carbonate rocks.

Reed Wicander

256   CHAPTER 12: GROUNDWATER

Figure 12.11 Cave Formation

a. As groundwater percolates through the zone of aeration and flows through the zone of saturation, it dissolves the carbonate rocks and gradually forms a system of passageways.

b. As the stream erodes more deeply, groundwater moves along the surface of the lower water table, forming a system of horizontal passageways through which dissolved rock is carried to the surface streams, thus enlarging the passageways.

c. As the surface streams erode deeper valleys, the water table drops, and the abandoned channelways form an interconnecting system of caves and caverns.

migrates toward the region's surface streams. As the groundwater moves through the zone of saturation, it continues to dissolve the rock and gradually forms a system of horizontal passageways through which the dissolved rock is carried to the streams (Figure 12.11a). As the surface streams erode deeper valleys, the water table drops in response to the lower elevation of the streams (Figure 12.11b). The water that flowed through the system of horizontal passageways now percolates to the lower water table, where a new system of passageways begins to form. The abandoned channelways form an interconnecting system of caves and caverns. Caves eventually become unstable and collapse, littering the floor with fallen debris.

When most people think of caves, they think of the seemingly endless variety of colorful and bizarre-shaped deposits found in them. Although a great many different types of cave deposits exist, most form in essentially the same manner and are collectively known as *dripstone*. As water seeps into a cave, some of the dissolved carbon dioxide in the water escapes, and a small amount of calcite is precipitated. In this manner, the various dripstone deposits are formed (Figure 12.11c).

Stalactites are icicle-shaped structures hanging from cave ceilings that form as a result of precipitation from dripping water (Figure 12.12). With each drop of water, a thin layer of calcite is deposited over the previous layer, forming a cone-shaped projection that grows down from the ceiling. The water that drips from a cave's ceiling also precipitates a small amount of calcite

Figure 12.12 Cave Deposits
Stalactites are the icicle-shaped structures hanging from the cave's ceiling, whereas the upward-pointing structures on the floor are stalagmites. Columns result when stalactites and stalagmites meet. A drip curtain is a vertical sheet of rock formed by groundwater seeping from a crack in the cave's ceiling. All four structures are present in Mammoth Cave, Kentucky.

when it hits the floor. As additional calcite is deposited, an upward-growing projection called a *stalagmite* forms (Figure 12.12). If a stalactite and stalagmite meet, they form a *column*. Groundwater seeping from a crack in a cave's ceiling may form a vertical sheet of rock called a *drip curtain*, and water flowing across a cave's floor may produce *travertine terraces* (Figure 12.11c).

LO7 MODIFICATIONS OF THE GROUNDWATER SYSTEM AND ITS EFFECTS

Groundwater is a valuable natural resource that is rapidly being exploited with seemingly little regard to the effects of overuse and misuse. Currently, approximately 20% of all water used in the United States is groundwater. This percentage is rapidly increasing, and unless this resource is used more wisely, sufficient amounts of clean groundwater will not be available in the future. Modifications of the groundwater system may have many consequences, including (1) lowering of the water table, causing wells to dry up; (2) saltwater incursion; (3) subsidence; and (4) contamination.

Lowering the Water Table

Withdrawing groundwater at a significantly greater rate than it is replaced by either natural or artificial recharge can have serious effects. For example, the High Plains aquifer is one of the most important aquifers in the United States. It underlies more than 450,000 km², including most of Nebraska, large parts of Colorado and Kansas, portions of South Dakota, Wyoming, and

New Mexico, as well as the panhandle regions of Oklahoma and Texas, and accounts for approximately 30% of the groundwater used for irrigation in the United States (Figure 12.13).

Although the High Plains aquifer has contributed to the high agricultural productivity of the region, it cannot continue to provide the quantities of water that it has in the past. In some parts of the High Plains, from 2 to 100 times more water is being pumped annually than is being recharged, causing a substantial drop in the water table in many areas (Figure 12.13).

It must be noted, however, that much of the High Plains aquifer water infiltrated during the wetter glacial climates more than 10,000 years ago. Consequently, most of the water being pumped is fossil water that is not being replenished at anywhere near the same rate as when it formed during the Pleistocene Epoch.

Figure 12.13 High Plains Aquifer
The geographic extent of the High Plains aquifer and changes in water level from predevelopment through 1993. Irrigation from the High Plains aquifer is largely responsible for the region's agricultural productivity.

Most users of the aquifer realize that they cannot continue to withdraw the quantities of groundwater that they have in the past, and thus are turning to greater conservation, monitoring of the aquifer, and using new technologies to try to better balance withdrawal with recharge rates.

Water supply problems certainly exist in many areas, but on the positive side, water use in the United States actually declined during the five years following 1980 and has remained nearly constant since then, even though the population has increased. This downturn in demand resulted largely from improved techniques in irrigation, more efficient industrial water use, and a general public awareness of water problems coupled with conservation practices. Nevertheless, the rates of withdrawal of groundwater from some aquifers still exceed their rates of recharge, and population growth in the arid and semiarid Southwest is continuing to put significant demands on an already limited water supply.

Saltwater Incursion

The excessive pumping of groundwater in coastal areas has resulted in *saltwater incursion*, which has become a major problem in many rapidly growing coastal communities where greater demand for groundwater creates an even greater imbalance between withdrawal and recharge. Along coastlines where permeable rocks or sediments are in contact with the ocean, the fresh groundwater, being less dense than seawater, forms a lens-shaped body above the underlying saltwater (Figure 12.14a). The weight of the freshwater exerts pressure on the underlying saltwater. As long as rates of recharge equal rates of withdrawal, the contact between the fresh groundwater and the seawater remains the same.

If excessive pumping occurs, however, a deep cone of depression forms in the fresh groundwater (Figure 12.14b). Because some of the pressure from the overlying freshwater has been removed, saltwater forms a *cone of ascension* as it rises to fill the pore space that formerly contained freshwater. When this occurs, wells become contaminated with saltwater and remain contaminated until recharge by freshwater restores the former level of the fresh-groundwater water table.

To counteract the effects of saltwater incursion, recharge wells are often drilled to pump water back into the groundwater system (Figure 12.14c). Recharge ponds that allow large quantities of fresh surface water to infiltrate the groundwater supply may also be constructed.

Figure 12.14 Saltwater Incursion

a. Because freshwater is not as dense as saltwater, it forms a lens-shaped body above the underlying saltwater.

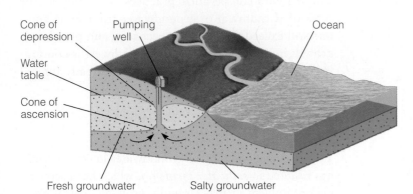

b. If excessive pumping occurs, a cone of depression develops in the fresh groundwater, and a cone of ascension forms in the underlying salty groundwater, which may result in saltwater contamination of the well.

c. Pumping water back into the groundwater system through recharge wells can help lower the interface between the fresh groundwater and the salty groundwater and reduce saltwater incursion.

© 2013 Cengage Learning

Subsidence

As excessive amounts of groundwater are withdrawn from poorly consolidated sediments and sedimentary rocks, the water pressure between grains is reduced, and the weight of the overlying materials causes the grains to pack more closely together, resulting in *subsidence* of the ground. As greater amounts of groundwater are pumped to meet the burgeoning needs of agriculture, industry, and population growth, subsidence is becoming more prevalent, and an issue that is receiving increasingly more attention.

The San Joaquin Valley of California is a major agricultural region that relies largely on groundwater for irrigation. Between 1925 and 1977, groundwater withdrawals in parts of the valley caused subsidence of nearly 9 m. Other areas in the United States that have experienced subsidence due to groundwater withdrawal are New Orleans, Louisiana, and Houston, Texas, both of which have subsided more than 2 m, and Las Vegas, Nevada, where 8.5 m of subsidence has taken place.

Looking elsewhere in the world, the tilt of the Leaning Tower of Pisa in Italy is partly a result of groundwater withdrawal (Figure 12.15). The tower started tilting soon after construction began in 1173 because of differential compaction of the foundation. During the 1960s, the city of Pisa withdrew evergreater amounts of groundwater, causing the ground to subside further; as a result, the tilt of the tower increased until it was in danger of falling over. Strict control of groundwater withdrawal, stabilization of the foundation, and recent renovations have reduced the amount of tilting to about 1 mm per year, thus ensuring that the tower should stand for several more centuries.

A spectacular example of continuing subsidence is taking place in Mexico City, which is built on a former lake bed. As groundwater is removed for the increasing needs of the city's 21.2 million people, the water table has been lowered by up to 10 m. As a result, the fine-grained lake deposits are compacting, and Mexico City is slowly and unevenly subsiding, with some areas having subsided as much as 7.5 m! The fact that 72% of the city's water comes from the aquifer beneath the metropolitan area ensures that problems of subsidence will continue.

The extraction of oil can also cause subsidence. Long Beach, California, has subsided 9 m as a result of many decades of oil production. More than $100 million in damages was done to the pumping, trans-

Fedor Selivanov/Shutterstock.com

Toxic Waste: Dump now,
pay the piper later!

iStockphoto.com/Valentin Casarsa

Figure 12.15 The Leaning Tower of Pisa, Italy

The tilting is partly the result of subsidence due to the removal of groundwater. Strict control of groundwater withdrawal, recent stabilization of the foundation, and renovation of the structure itself have ensured that the Leaning Tower will continue leaning for many more centuries.

portation, and harbor facilities in this area because of subsidence and encroachment of the sea. Once water was pumped back into the oil reservoir, thus stabilizing it, subsidence virtually stopped.

Groundwater Contamination

A major problem facing our society is the safe disposal of the numerous pollutant by-products of an industrialized economy. We are becoming increasingly aware that streams, lakes, and oceans are not unlimited reservoirs for waste and that we must find new, safe ways to dispose of pollutants.

The most common sources of groundwater contamination are sewage, landfills, toxic waste disposal sites, and agriculture. Once pollutants get into the groundwater system, they spread wherever groundwater travels, which can make their containment difficult, and because groundwater moves so slowly, it takes a long time to cleanse a groundwater reservoir once it has become contaminated.

In many areas, septic tanks are the most common way of disposing of sewage. A septic tank slowly releases sewage into the ground, where it is decomposed by oxidation and microorganisms and filtered by the sediment as it percolates through the zone of aeration. In most situations, by the time the water from the sewage reaches the zone of saturation, it has been cleansed of any impurities and is safe to use (Figure 12.16a). If the water table is close to the surface, or if the rocks are very permeable, water entering the zone of saturation may still be contaminated and unfit to use.

Landfills are also potential sources of groundwater contamination (Figure 12.16b). Not only does liquid waste seep into the ground, but rainwater also carries dissolved chemicals and other pollutants down into the groundwater reservoir. Unless the landfill is carefully designed and lined with an impermeable layer such as clay, many toxic compounds such as paints, solvents, cleansers, pesticides, and battery acid will find their way into the groundwater system.

Toxic waste sites where dangerous chemicals are either buried or pumped underground are an increasing source of groundwater contamination. The United States alone must dispose of several thousand metric tons of hazardous chemical waste per year. Unfortunately, much of this waste has been, and still is, being improperly dumped and is contaminating the surface water, soil, and groundwater.

LO8 HYDROTHERMAL ACTIVITY

Hydrothermal is a term referring to hot water. Some geologists restrict the meaning to include only water heated by magma, but here we use it to refer to any hot subsurface water and the surface activity that results from its discharge. One manifestation of hydrothermal activity in areas of active or recently active volcanism is the discharge of gases, such as steam, at vents known as *fumeroles* (Figure 12.17). Of more immediate concern here is the groundwater that rises to the surface as *hot springs* or *geysers* (see the Virtual Field Trip in this chapter). It may be heated by its proximity

Figure 12.16 **Groundwater Contamination**

a. A septic system slowly releases sewage into the zone of aeration. Oxidation, bacterial degradation, and filtering usually remove impurities before they reach the water table. However, if the rocks are very permeable or the water table is too close to the septic system, contamination of the groundwater can result.

b. Unless there is an impermeable barrier between a landfill and the water table, pollutants can be carried into the zone of saturation and contaminate the groundwater supply: (1) Infiltrating water leaches contaminants from the landfill; (2) the polluted water enters the water table and moves away from the landfill; (3) wells may tap the polluted water and thus contaminate drinking water supplies; and (4) the polluted water may emerge into streams and other water bodies downslope from the landfill.

Figure 12.17 Fumeroles

Gases emitted from vents (fumeroles) at Yellowstone National Park, Wyoming.

to magma or by Earth's geothermal gradient because it circulates deeply.

Hot Springs

A **hot spring** (also called a *thermal spring* or *warm spring*) is any spring in which the water temperature is higher than 37°C, the temperature of the human body (Figure 12.18). Some hot springs are much hotter, with temperatures up to the boiling point in many instances. Of the approximately 1,100 known hot springs in the United States, more than 1,000 are in the far West, with the others in the Black Hills of South Dakota, Georgia, the Ouachita region of Arkansas, and the Appalachian region.

Figure 12.18 Sunset Lake Hot Spring, Yellowstone National Park, Wyoming

The bands of color surrounding Sunset Lake hot spring are mats of heat-loving cyanobacteria and algae, known as thermophiles. Each color represents a certain temperature range that allows for specific bacterial species to thrive in this extreme environment.

Hot springs are also common in other parts of the world. One of the most famous is in Bath, England, where shortly after the Roman conquest of Britain in A.D. 43, numerous bathhouses and a temple were built around the hot springs (Figure 12.19).

> **hot spring** A spring in which the water temperature is warmer than the temperature of the human body (37°C).

The heat for most hot springs comes from magma or cooling igneous rocks. The geologically recent igneous activity in the western United States accounts for the large number of hot springs in that region. The water in some hot springs, however, circulates deep into Earth, where it is warmed by the normal increase in temperature, the geothermal gradient. For example, the spring water of Warm Springs, Georgia, is heated in this manner. This hot spring was a health and bathing resort long before the Civil War (1861–1865); later, with the establishment of the Georgia Warm Springs Foundation, it was used to help treat polio victims.

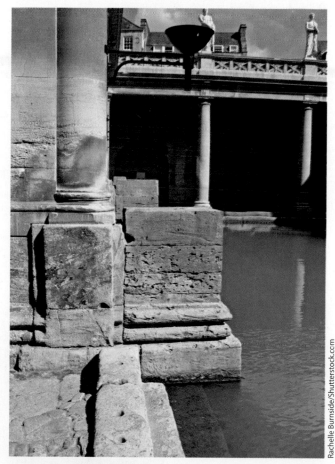

Figure 12.19 Bath, England

One of the many bathhouses in Bath, England, that were built around hot springs shortly after the Roman conquest in A.D. 43.

Virtual Field Trip

Hydrothermal Activity

Ready to Go!

Hydrothermal is a term referring to hot water, and when geologists talk about hydrothermal activity, they are referring to the interaction between hot water and Earth. Most people associate hydrothermal activity with geysers, such as Old Faithful in Yellowstone National Park, Wyoming, or hot springs where one can go to bathe or relax. Hydrothermal activity also can result in the production of geothermal energy.

There are three areas in the world known for their hydrothermal activity. These are Iceland, New Zealand, and Yellowstone National Park, Wyoming. In this virtual field trip, we'll be visiting what is probably the best known area to view hydrothermal activity and its results. So, get ready to visit the crown-jewel of the U.S. National Park System, Yellowstone National Park.

DeerPoint, UT. topographic map, courtesy of USGS.
All photographs: Copyright and Photograph by Dr. Parvinder S. Sethi

GOALS OF THE TRIP

In this field trip, you will learn about a number of hydrothermal features and what processes are involved in their formation. Here are the features we will see:

1. Hot springs, their formation, and the denizens that live in them.
2. Geysers and how they form.
3. Deposits associated with hot springs and geysers.

FOLLOW-UP QUESTIONS

1. Why do you think Old Faithful erupts on such a regular basis?
2. Can the color of hot springs deposits serve as an indicator of water chemistry and temperature?
3. Do you think the hydrothermal activity in Yellowstone National Park could be tapped as a potential source of geothermal energy? What are the pros and cons of doing so?

What to See When You Go

A *hot spring* (**also called a** *thermal spring* **or** *warm spring*) **is any spring in which the water temperature is higher than the temperature of the human body. The heat for most hot springs comes from magma or cooling igneous rock. The large number of hot springs in the western United States is due to geologically recent igneous activity. The underground system of fractures and opening that are associated with hot springs are not as constrictive, or usually as deep, as in a geyser, hence the water can bubble up and spill out onto the surface.**

Nevertheless, steam and volcanic gases are still associated with hot springs, as seen in this photo below.

The photograph on the right shows the vent of the Pump Geyser. Hot, acidic water containing silica and dissolved gases flows down the slope from the vent, providing an environment for *extremophiles* (organisms that live in extreme environments), in this case *thermophiles* (species that thrive in extremely hot water, and are typically bacteria or cyanobacteria).

Both hot springs and geysers typically contain large quantities of dissolved minerals. When the highly mineralized waters of hot springs and geysers reach the surface, some of the dissolved mineral matter is precipitated, forming various types of deposits. The amount and type of precipitated minerals depends on the solubility and composition of the groundwater flowing through the underground fractures.

The beautiful red colors are rust-colored iron-oxide deposits resulting from the metabolism of iron by some of these bacteria. The yellow-colored deposits are sulfur, which is formed by sulfur-loving species of bacteria that reduce the hydrogen sulfide gas being emitted at the vent. The brownish bacterial mats contain bacteria that live in the cooler waters (below 60°C).

When hot water containing dissolved silica is erupted from a geyser, it cools and deposits silica (commonly called *sinter*) around the vent. Sinter comes in a variety of shapes and sizes. One of the largest deposits in the world is found at The Castle Geyser, so named for its resemblance to a castle ruin. The Castle Geyser is thought to be thousands of years old and the sinter it is currently depositing is being laid down over even older and thicker deposits of sinter.

Hot springs that intermittently eject hot water and steam with tremendous force are known as *geysers*. Recall from your reading of this chapter that geysers are the surface expression of an extensive underground system of deep, interconnected fractures within hot rocks. Groundwater in these fractures is heated above the boiling point of water. Because of the pressure at depth, this water does not boil, but if there is a slight drop in pressure, such as from escaping gas, the hot water will instantly change to steam, which will then push the water upwards to the ground and into the air, producing a geyser.

Old Faithful Geyser, shown above in eruption, is probably the most famous geyser in the world. Eruptions occur approximately every 90 minutes.

The terraces you see above are composed of travertine ($CaCO_3$) as opposed to sinter (SiO_2). These deposits are the result of groundwater flowing through limestones and dissolving the calcium carbonate. As the water reaches the surface and begins to flow over it, calcium carbonate is precipitated as small crystals and thin sheets, which, over time, form the massive terraces seen here.

Geysers

Hot springs that intermittently eject hot water and steam with tremendous force are known as **geysers**. The word comes from the Icelandic *geysir*, "to gush" or "to rush forth." One of the most famous geysers in the world is Old Faithful in Yellowstone National Park, Wyoming (Figure 12.20 and the Virtual Field Trip in this chapter). With a thunderous roar, it erupts a column of hot water and steam every 30 to 90 minutes. Other well-known geyser areas are found in Iceland and New Zealand.

Geysers are the surface expression of an extensive underground system of interconnected fractures within hot igneous rocks (Figure 12.21). Groundwater percolating down into the network of fractures is heated as it comes into contact with the hot rocks. Because the water near the bottom of the fracture system is under higher pressure than the water near the top, it must be heated to a higher temperature before it will boil. Thus, when the deeper water is heated to near the boiling point, a slight rise in temperature or a drop in pressure, such as from escaping gas, will instantly change it to

steam. The expanding steam quickly pushes the water above it out of the ground and into the air, producing a geyser eruption. After the eruption, relatively cool groundwater starts to seep back into the fracture system, where it heats to near its boiling temperature and the eruption cycle begins again. Such a process explains how geysers erupt with some regularity.

Figure 12.21 Anatomy of a Geyser

a. The eruption of a geyser starts when groundwater percolates down into a network of interconnected openings and is heated by the hot igneous rocks. The water near the bottom of the fracture system is under higher pressure than the water near the top and consequently must be heated to a higher temperature before it will boil.

b. Any rise in the temperature of the water above its boiling point or a drop in pressure will cause the water to change to steam, which quickly pushes the water above it up and out of the ground, producing a geyser eruption.

© 2013 Cengage Learning

James S. Monroe

Figure 12.20 Old Faithful Geyser

Old Faithful Geyser in Yellowstone National Park, Wyoming, is one of the world's most famous geysers, erupting faithfully every 30 to 90 minutes and spewing water 32 to 56 m high.

Figure 12.22 Hot Spring Deposits in Yellowstone National Park, Wyoming

Minerva Terrace, formed when calcium carbonate-rich hot spring water cooled, precipitating travertine.

Hot spring and geyser water typically contains large quantities of dissolved minerals, because most minerals dissolve more rapidly in warm water than in cold water. Because of this high mineral content, some believe that the waters of many hot springs have medicinal properties. Numerous spas and bathhouses have been built at hot springs throughout the world to take advantage of these supposed healing properties.

When the highly mineralized water of hot springs or geysers cools at the surface, some of the material in solution is precipitated, forming various types of deposits. The amount and type of precipitated minerals depend on the solubility and composition of the material that the groundwater flows through. If the groundwater contains dissolved calcium carbonate ($CaCO_3$), then *travertine* or *calcareous tufa* (both of which are varieties of limestone) are precipitated. Spectacular examples of hot spring travertine deposits are found at Pamukhale in Turkey and at Mammoth Hot Springs in Yellowstone National Park (Figure 12.22). Groundwater containing dissolved silica will, upon reaching the surface, precipitate a soft, white, hydrated mineral called *siliceous sinter* or *geyserite*, which can accumulate around a geyser's opening (see the Virtual Field Trip in this chapter).

Geothermal Energy   **Geothermal energy** is any energy produced from Earth's internal heat. In fact, the term *geothermal* comes from *geo*, "Earth," and *thermal*, "heat." Several forms of internal heat are known, such as hot dry rocks and magma, but so far, only hot water and steam are used.

As oil reserves decline, geothermal energy is becoming an attractive alternative. It is estimated that approximately 1% to 2% of the world's current energy needs could be met by geothermal energy. In those areas where it is plentiful, geothermal energy can supply most, if not all, of the energy needs, sometimes at a fraction of the cost of other types of energy. Some of the countries currently using geothermal energy in one form or another are Iceland, the United States, Mexico, Italy, New Zealand, Japan, the Philippines, and Indonesia.

In the United States, the first commercial geothermal electricity-generating plant was built in 1960 at The Geysers, about 120 km north of San Francisco, California. Here, wells were drilled into the numerous near-vertical fractures underlying the region. As pressure on the rising groundwater decreases, the water changes to steam, which is piped directly to electricity-generating turbines and generators (Figure 12.23).

geothermal energy
Energy that comes from steam and hot water trapped within Earth's crust.

Figure 12.23 The Geysers, Sonoma County, California

Steam rising from one of the geothermal power plants at The Geysers in Sonoma County, California. Steam from wells drilled into the geothermal region, about 120 km north of San Francisco, is piped directly to electricity-generating turbines to produce electricity that is distributed throughout the area.

13 | Glaciers and Glaciation

View of the Jungfrau Firn in Switzerland, which merges with two other valley glaciers to form the Aletsch Glacier, the largest glacier in the Alps of Europe. From this viewpoint, the glacier's terminus is 23 km distant. The Aletsch Glacier covers 120 km^2, but here this tributary is only about 1.5 km wide. Valley glaciers like this one are present on all the continents except Australia, but during the Ice Age (2.6 million to 10,000 years ago), they were more numerous and much larger than they are now.

Sue Monroe

Introduction

"What causes an Ice Age?"

© iStockphoto.com/ Slavoljub Pantelic

At present, glaciers cover about 10% of Earth's land surface, but during the Ice Age, or what geologists call the Pleistocene Epoch, they were much more extensive. During that time, glaciers covered vast areas, especially on the Northern Hemisphere continents, and small glaciers in mountain valleys were much more numerous and larger than they are now (see the chapter opening photo). During this comparatively brief interval of geologic time (2.6 million to 10,000 years ago) glaciers waxed and waned several times, and as they did so they deeply eroded some areas and deposited huge quantities of sediments elsewhere.

Since the Ice Age, Earth has gone through several climatic changes. About 6,000 years ago, during the Holocene Maximum, average temperatures were slightly higher than they are now, and some of today's arid regions were much more humid. Today the only arable land in Egypt is along the Nile River (see chapter 11's opening photo), but during the Holocene Maximum much of North Africa was covered by grasslands, swamps, and lakes.

Following the Holocene Maximum was a time of cooler temperatures, but from about A.D. 1000 to 1300 Europe went through the Medieval Warm Period, when wine grapes grew 480 km farther north than they do now. Then a cooling trend began in about A.D. 1300 that led to the **Little Ice Age**, which lasted from 1500 to the middle or late 1800s. During the Little Ice Age, glaciers expanded to their greatest historic extent; summers were cooler and wetter; winters were colder; and sea ice persisted longer around Greenland, Iceland, and the Canadian Arctic islands. The cooler, wetter summers were a problem because the growing seasons were much shorter, resulting in several famines.

> **Little Ice Age** An interval from about 1500 to the mid- to late-1800s during which glaciers expanded to their greatest historic extent.

Many of you have probably heard of the Ice Age and have some idea of what a glacier is, but it is doubtful that you know about the dynamics of glaciers, how they form, and what causes ice ages. All *glaciers* are moving bodies of ice on land, and as moving solids they have a tremendous capacity to erode, transport, and deposit. Indeed, glaciers are responsible for many easily recognizable landforms and account for some of the most spectacular scenery in our National Parks. And even though they are more restricted than during the Ice Age, glaciers remain an important geologic agent for modifying Earth's surface, especially in high mountains and at high latitudes.

Why study glaciers? Of course, glaciers are part of the hydrologic cycle, illustrating again the complex interactions among Earth's systems. In addition, a large part of the water in some countries, such as Nepal, Tibet, and Pakistan, comes from melting glaciers. Even in the United States and Canada, some areas in the West depend partly on water stored in glaciers. And, finally, glaciers are very sensitive to climatic changes, so scientists are interested in fluctuations in glaciers as possible indications of global warming.

Learning Outcomes

After reading this unit, you should be able to do the following:

LO1 Identify the different kinds of glaciers

LO2 Recognize that glaciers are moving bodies of ice on land

LO3 Understand the glacial budget of accumulation and wastage

LO4 Identify features resulting from erosion and transport by glaciers

LO5 Identify the types of glacial deposits

LO6 Explain what scientists think caused the Ice Age

Figure 13.1 A Valley Glacier in Canada
This glacier is in a mountain valley in the Coast Mountains, Yukon Territory, Canada. Notice the tributaries that join to form a larger glacier.

LO1 THE KINDS OF GLACIERS

A **glacier** is a moving body of ice on land that flows downslope or outward from areas of accumulation. Our definition of a glacier excludes frozen seawater, as in the North Polar region, and sea ice that forms yearly adjacent to Greenland and Iceland. Drifting icebergs are not glaciers either, although they may have come from glaciers that flowed into lakes or the sea. The critical points in the definition are *moving* and *on land*. Accordingly, permanent snowfields in high mountains, though on land, are not glaciers because they do not move. All glaciers share several characteristics, but they differ enough in size and location for scientists to define two specific types, valley glaciers and continental glaciers, and several subvarieties.

glacier A mass of ice on land that moves by plastic flow and basal slip.

valley glacier A glacier confined to a mountain valley or an interconnected system of mountain valleys.

continental glacier A glacier that covers a vast area (at least 50,000 km²) and is not confined by topography; also called an *ice sheet*.

Valley Glaciers

Valley glaciers are confined to mountain valleys, where they flow from higher to lower elevations (Figure 13.1), whereas continental glaciers cover vast areas, are not confined by the underlying topography, and flow outward in all directions from areas of snow and ice accumulation. We use the term *valley glacier*, but some geologists prefer the synonyms *alpine glacier* and *mountain glacier*. Valley glaciers commonly have tributaries, just as streams do, thereby forming a network of glaciers in an interconnected system of mountain valleys.

Valley glaciers are common in the mountains of all continents except Australia. A valley glacier's shape is controlled by the shape of the valley it occupies, so it tends to be a long, narrow tongue of moving ice. Some valley glaciers flow into the ocean and are called *tidewater glaciers*; they differ from other valley glaciers only in that their terminus is in the sea rather than on land.

Valley glaciers are small compared with the much more extensive continental glaciers, but even so, they may be several kilometers across, 200 km long, and hundreds of meters thick. Erosion and deposition by valley glaciers account for much of the spectacular scenery in several U.S. and Canadian national parks.

Continental Glaciers

Continental glaciers, also known as *ice sheets*, are vast, covering at least 50,000 km², and they are unconfined by

Figure 13.2 Continental Glaciers and Ice Caps

a. The West and East Antarctic ice sheets merge to form a nearly continuous ice cover that averages 2,160 m thick. The blue lines are lines of equal thickness.

b. View of the Penny Ice Cap on Baffin Island, Canada. It covers about 6,000 km².

James S. Monroe

© 2013 Cengage Learning

topography; that is, their shape and movement are not controlled by the underlying landscape. As described earlier, valley glaciers are long, narrow tongues of ice that conform to the shape of the valley they occupy, and the existing slope determines their direction of flow. In contrast, continental glaciers flow outward in all directions from a central area or areas of accumulation in response to variations in ice thickness.

In Earth's two areas of continental glaciation, Greenland and Antarctica, the ice is more than 3,000 m thick and covers all but the highest mountains (Figure 13.2a). The continental glacier in Greenland covers about 1,800,000 km², and in Antarctica the East and West Antarctic Glaciers merge to form a continuous ice sheet covering more than 12,650,000 km². The glaciers in Antarctica flow into the sea, where the buoyant effect of water causes the ice to float in vast *ice shelves*; the Ross Ice Shelf alone covers more than 547,000 km² (Figure 13.2a).

ice cap A dome-shaped mass of glacial ice that covers less than 50,000 km².

glaciation Refers to all aspects of glaciers, including their origin, expansion, and retreat, and their impact on Earth's surface.

If all glacial ice were in the United States and Canada, it would form a continuous ice cover about 1.5 km thick!

An **ice cap**, a dome-shaped mass of glacial ice, is similar to but smaller than a continental glacier, covering less than 50,000 km² (Figure 13.2b). Some ice caps form when valley glaciers grow and overtop the divides and passes between adjacent valleys, then coalesce to form a continuous ice cover. They also form on fairly flat terrain in Iceland and some of the islands in the Canadian Arctic.

LO2 GLACIERS: MOVING BODIES OF ICE ON LAND

We use the term **glaciation** to indicate all glacial activity, including the origin, expansion, and retreat of glaciers, as well as their impact on Earth's surface. Presently, glaciers cover nearly 15 million km², or about 10% of Earth's land surface.

firn Granular snow formed by partial melting and refreezing of snow; transitional material between snow and glacial ice.

glacial ice Water in the solid state within a glacier; forms as snow partially melts and refreezes and compacts so that it is transformed first to firn and then to glacial ice.

At first glance, glaciers appear static. Even briefly visiting a glacier may not dispel this impression because, although glaciers move, they usually do so slowly. Nevertheless, they do move, and just like other geologic agents such as running water, glaciers are dynamic systems that continuously adjust to changes. For example, a glacier may flow slower or more rapidly depending on decreased or increased amounts of snow or the absence or presence of water at its base.

Glaciers: Part of the Hydrologic Cycle

Glaciers make up one reservoir in the hydrologic cycle, where water is stored for long periods, but even this water eventually returns to its original source, the oceans (see Figure 11.2). Many glaciers at high latitudes, as in Alaska, northern Canada, and Scandinavia, flow directly into the oceans (tidewater glaciers), where they melt, or icebergs break off (a process known as *calving*) and drift out to sea, where they eventually melt. At low latitudes or areas remote from the oceans, glaciers flow to lower elevations, where they melt and the liquid water enters the groundwater system (another reservoir in the hydrologic cycle), or it returns to the seas by surface runoff.

In addition to melting, glaciers lose water by *sublimation*, when ice changes to water vapor without an intermediate liquid phase. The water vapor so derived enters the atmosphere, where it may condense and fall as rain or snow, but in the long run, this water also returns to the oceans.

How Do Glaciers Originate and Move?

Ice is a crystalline solid with characteristic physical properties and a specific chemical composition, and thus is a mineral. Accordingly, *glacial ice* is a type of metamorphic rock, but one that is easily deformed. Glaciers form in any area where more snow falls than melts during the warmer seasons and a net accumulation takes place. Freshly fallen snow has about 80% air-filled pore space and 20% solids, but it compacts as it accumulates, partially thaws, and refreezes, convert-

Figure 13.3 Glacial Ice

© 2013 Cengage Learning

a. The conversion of freshly fallen snow to firn and then to glacial ice.

Copyright and Photograph by Parvinder Sethi

b. This iceberg in Portage Lake in Alaska shows the blue color of glacial ice. The longer wavelengths of white light are absorbed by the ice, but blue (short wavelength) is transmitted into the ice and scattered, accounting for the blue color.

ing to a granular type of snow known as **firn**. As more snow accumulates, the firn is buried and further compacted and recrystallized until it is transformed into **glacial ice**, consisting of about 90% solids and 10% air (Figure 13.3).

Now you know how glacial ice forms, but we still have not addressed how glaciers move. At this time, it is useful to recall some terms from Chapter 9. Remember that *stress* is force per unit area, and *strain* is a change in the shape or volume of solids. When accumulating snow and ice reach a critical thickness of about 40 m, the stress on the ice at depth is great enough to induce **plastic flow**, a type of permanent deformation involving no fracturing. Glaciers move mostly by plastic flow, but they may also slide over their underlying surface by **basal slip** (Figure 13.4). Liquid water facilitates basal slip because it reduces friction between a glacier and the surface over which it moves.

The total movement of a glacier in a given time is a result of plastic flow and basal slip, although the former occurs continuously, whereas the latter varies, depending on the season, latitude, and elevation. Indeed, if a glacier is solidly frozen to the surface below, as in the case of many polar environments, it moves only by plastic flow. Furthermore, basal slip is far more important in valley glaciers, as they flow from higher to lower elevations whereas continental glaciers need no slope for flow.

Although glaciers move by plastic flow, the upper 40 m or so of ice behaves like a brittle solid and fractures if subjected to stress. Large crevasses commonly develop in glaciers where they flow over an increase in slope of the underlying surface or where they flow around a corner (Figure 13.5). In either case, the ice is stretched (subjected to tension) and crevasses open, which extend down to the zone of plastic flow. In some cases, a glacier descends over such a steep precipice that crevasses break up the ice into a jumble of blocks and spires, and an icefall develops.

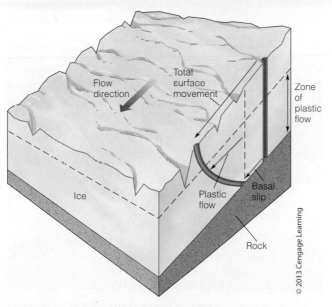

Figure 13.4 Part of a Glacier Showing Movement by a Combination of Plastic Flow and Basal Slip
Plastic flow involves internal deformation within the ice, whereas basal slip is sliding over the underlying surface. If a glacier is solidly frozen to its bed, it moves only by plastic flow. Notice that the top of the glacier moves farther in a given time than the bottom does.

Distribution of Glaciers

As you might suspect, the amount of snowfall and temperature are important factors in determining where glaciers form. Parts of northern Canada are cold enough to support glaciers but receive too little snowfall, whereas some mountain areas in California receive huge amounts of snow but are too warm for glaciers. Of course, temperature varies with elevation and latitude, so we would expect to find glaciers in high mountains and at high latitudes if these areas receive enough snow.

Many small glaciers are present in the Sierra Nevada of California, but only at elevations exceeding 3,900 m. In fact, the high mountains in California, Oregon, and Washington all have glaciers because they are high and receive so much snow. Mount Baker in Washington had almost 29 m of snow during the winter of 1998–1999, and average accumulations of 10 m or more are common in many parts of these mountains.

Glaciers are also found in the mountains along the Pacific Coast of Canada, which also receive considerable snowfall, and, of course, they are farther north. Some of the higher peaks in the Rocky Mountains in both the United States and Canada also support glaciers. At even higher latitudes, as in Alaska, northern Canada, and Scandinavia, glaciers exist at sea level.

plastic flow The flow that takes place in response to pressure and causes deformation with no fracturing.

basal slip Movement involving a glacier sliding over its underlying surface.

glacial budget The balance between expansion and contraction of a glacier in response to accumulation versus wastage.

zone of accumulation The part of a glacier where additions exceed losses and the glacier's surface is perennially covered with snow. Also refers to horizon B in soil where soluble material leached from horizon A accumulates as irregular masses.

LO3 THE GLACIAL BUDGET

We describe a glacier's behavior in terms of a **glacial budget**, which is essentially a balance sheet of accumulation and wastage. For instance, the upper part of a valley glacier is a **zone of accumulation**, where

Figure 13.5 Crevasses

Crevasses are common in the upper parts of glaciers when the ice is subjected to tension.

a. Crevasses open where the brittle part of a glacier is stretched as it moves over a steeper slope in its valley.

Zone of plastic flow

© 2013 Cengage Learning

additions exceed losses and the surface is perennially snow covered. In contrast, the lower part of the same glacier is a **zone of wastage**, where losses from melting, sublimation, and calving of icebergs exceed the rate of accumulation (Figure 13.6).

At the end of winter, a glacier's surface is covered with the accumulated seasonal snowfall. During the spring and summer, the snow begins to melt, first at lower elevations and then progressively higher up the glacier. The elevation to which snow recedes during a wastage season is the *firn limit* (Figure 13.6). You can easily identify the zones of accumulation and wastage by noting the location of the firn limit.

The firn limit on a glacier may change yearly, but if it does not change or shows only minor fluctuations, the glacier has a balanced budget. That is, additions in the zone of accumulation are exactly balanced by losses in the zone of wastage, and the distal end, or terminus, of the glacier remains stationary (Figure 13.6). If the firn limit moves up the glacier, indicating a negative budget, the glacier's terminus retreats. If the firn limit moves down the glacier, however, the glacier has a positive budget, additions exceed losses, and its terminus advances.

Even though a glacier may have a negative budget and a retreating terminus, the glacial ice continues to move toward the terminus by plastic flow and basal slip. If a negative budget persists long enough, though, the glacier continues to recede, and it thins until it is no longer thick enough to maintain flow. It then ceases moving and becomes a *stagnant glacier*; if wastage continues, the glacier eventually disappears.

We used a valley glacier as an example, but the same budget considerations control the flow of ice caps

b. These crevasses are on the Byron Glacier in Kenai Fjords National Park and Preserve, near Seward, Alaska. Note the two hikers for scale.

Copyright and Photograph by Parvinder Sethi

and continental glaciers as well. The entire Antarctic ice sheet is in the zone of accumulation, but it flows into the ocean, where wastage occurs.

Figure 13.6 The Glacial Budget

All glaciers have a zone of accumulation, where additions exceed losses and the glacier's surface is perennially covered by snow. They also have a zone of wastage, where losses from melting, calving of icebergs, evaporation, and sublimation exceed gains. If a glacier's budget is balanced, its terminus remains in the same position. However, should the budget be positive, its terminus advances, and if the budget is negative the terminus retreats.

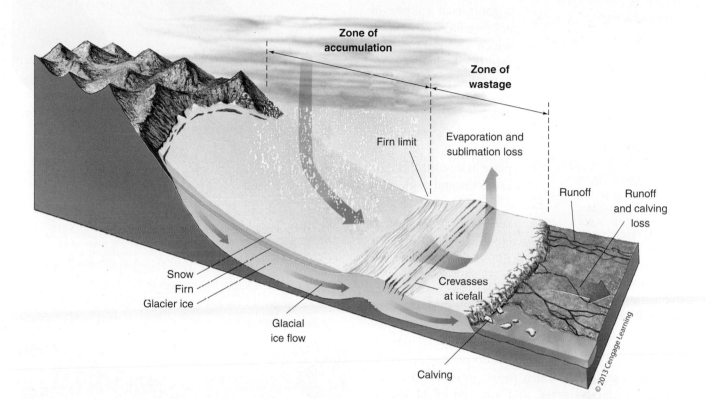

How Fast Do Glaciers Move?

Valley glaciers usually move more rapidly than continental glaciers, but the rates for both vary from centimeters to tens of meters per day. Valley glaciers moving down steep slopes flow more rapidly than glaciers of comparable size on gentle slopes, assuming that other variables are the same. The main glacier in a valley glacier system contains a greater volume of ice and thus has a greater discharge and flow velocity than its tributaries. Temperature exerts a seasonal control on valley glaciers because, although plastic flow remains rather constant year-round, basal slip is more important during warmer months when meltwater is abundant.

Flow rates also vary within the ice itself. For example, flow velocity increases downslope in the zone of accumulation until the firn limit is reached; from that point, the velocity becomes progressively lower toward the glacier's terminus. Valley glaciers are similar to streams in that the valley walls and floor cause frictional resistance to flow, so the ice in contact with these boundaries moves more slowly than the ice some distance away (Figure 13.7a).

Notice in Figure 13.7a that flow velocity in the interior of a glacier increases upward until the top few tens of meters of ice are reached, but little or no additional increase occurs after that point. This upper ice layer constitutes the rigid part of the glacier that is moving as a result of basal slip and plastic flow below.

Continental glaciers ordinarily flow at a rate of centimeters to meters per day. One reason continental glaciers move comparatively slowly is that they exist at higher latitudes and are frozen to the underlying surface much of the time, which limits the amount of basal slip. Nevertheless, some parts of continental glaciers manage to achieve extremely high flow rates. Near the margins of the Greenland ice sheet, the ice is forced between mountains in what are called *outlet glaciers*. In some of these outlets, flow velocities exceed 100 m per day.

In parts of the continental glacier covering West Antarctica, scientists have identified ice streams in which flow rates are considerably higher than in adjacent glacial ice. Drilling has revealed a 5-m-thick layer of water-saturated sediment beneath these ice streams, which acts to facilitate movement of the ice above. Some geologists think that geothermal heat from subglacial volcanism

glacial surge A time of greatly accelerated flow in a glacier. Commonly results in displacement of the glacier's terminus by several kilometers.

abrasion The process whereby rock is worn smooth by the impact of sediment transported by running water, glaciers, waves, or wind.

glacial polish A smooth, glistening rock surface formed by the movement of sediment-laden ice over bedrock.

glacial striation A straight scratch rarely more than a few millimeters deep on a rock caused by the movement of sediment-laden glacial ice.

melts the underside of the ice, thus accounting for the layer of water-saturated sediment.

A **glacial surge** is a short-lived episode of accelerated flow during which the glacier's surface breaks into a maze of crevasses and its terminus advances noticeably. These brief episodes are best known in valley glaciers, but they also take place in ice caps and even in continental glaciers. In 1995, a huge ice shelf in Antarctica broke apart and several ice streams from the Antarctic ice sheet surged toward the ocean.

During a surge, a glacier's terminus may advance several tens of meters per day for weeks or months and then return to its previous flow rate. Not many glaciers surge, and none do in the United States outside Alaska. Even in Canada, surges occur only in the Yukon Territory and the Queen Elizabeth Islands. Lowell Glacier in Kluane National Park in the Yukon Territory began surging in late 2009, and by May 2010 its terminus had advanced 1.5 km (Figure 13.7b). The fastest surge ever recorded was in 1953 in the Kutiah Glacier in Pakistan; the glacier's terminus advanced 12 km in three months, or 130 m per day on average.

LO4 EROSION AND SEDIMENT TRANSPORT BY GLACIERS

As moving solids, glaciers erode, transport, and eventually deposit huge quantities of sediment and soil. Indeed, they have the capacity to transport boulders the size of a house, as well as clay-sized particles. Important processes of erosion include bulldozing, plucking, and abrasion.

Although *bulldozing* is not a formal geologic term, it is fairly self-explanatory; glaciers shove or push unconsolidated materials in their paths. *Plucking*, also called *quarrying*, results when glacial ice freezes in the

Figure 13.7 Flow Velocity in Valley Glaciers

© 2013 Cengage Learning

a. Flow velocity in a valley glacier varies horizontally and vertically. Velocity is greatest at the top center because friction with the walls and floor of the trough slows the flow adjacent to these boundaries. The lengths of the arrows are proportional to velocity.

James S. Monroe

b. The terminus of the Lowell Glacier in Kluane National Park in the Yukon Territory of Canada. The glacier was surging when this image was taken on July 2, 2010. Its surge began probably in October 2009 when its terminus was more than 1.5 km farther up its valley.

cracks and crevices of a bedrock projection and eventually pulls it loose.

Bedrock over which sediment-laden glacial ice moves is effectively eroded by **abrasion** and develops a **glacial polish**, a smooth surface that glistens in reflected light (Figure 13.8a). Abrasion also yields **glacial striations**, rather straight scratches rarely more

Figure 13.8 Glacial Striations, Polish, and Rock Flour

a. Abrasion produced glacial polish and striations—the straight scratches—on this basalt at Devils Postpile National Monument in California.

b. The water in this stream in Switzerland is milky because it is discolored by rock flour, small particles yielded by glacial abrasion.

than a few millimeters deep on rock surfaces. Abrasion thoroughly pulverizes rocks, yielding an aggregate of clay- and silt-sized particles that have the consistency of flour—hence, the name *rock flour* (Figure 13.8b).

Continental glaciers derive sediment from mountains projecting through them, and windblown dust settles on their surfaces, but most of their sediment comes from the surface they move over. As a result, most sediment is transported in the lower part of the ice sheet. In contrast, valley glaciers carry sediment in all parts of the ice, but it is concentrated at the base and along the margins (Figure 13.9). Some of the marginal sediment is derived by abrasion and plucking, but much of it is supplied by mass wasting, as when soil, sediment, or rock falls or slides onto the glacier's surface.

Erosion by Valley Glaciers

When mountains are eroded by valley glaciers, they take on a unique appearance of angular ridges and peaks in the midst of broad, smooth valleys with near-vertical walls. The erosional landforms produced by valley glaciers are easily recognized and enable us to appreciate the tremendous erosive power of moving ice.

> **U-shaped glacial trough** A valley with steep or vertical walls and a broad, rather flat floor formed by the movement of a glacier through a stream valley.

U-Shaped Glacial Troughs

A **U-shaped glacial trough** is one of the most distinctive features of valley glaciation. Mountain valleys eroded by running water are typically V-shaped in cross section; that is, they have valley walls that descend to a narrow valley bottom (Figure 13.10a). In contrast, valleys scoured by glaciers are deepened, widened, and straightened so that they have very steep or vertical walls, but broad, rather flat valley floors; thus they exhibit a U-shaped profile (Figure 13.10).

During the Pleistocene, when glaciers were more extensive, sea level was as much as 130 m lower than at present, so glaciers flowing into the sea eroded their valleys below present sea level. When the glaciers melted at the end of the Pleistocene, sea level rose and the ocean filled the lower ends of the glacial troughs, so

fiord An arm of the sea extending into a glacial trough eroded below sea level.

hanging valley A tributary glacial valley whose floor is at a higher level than that of the main glacial valley.

cirque A steep-walled, bowl-shaped depression on a mountainside at the upper end of a glacial valley.

arête A narrow, serrated ridge between two glacial valleys or adjacent cirques.

that now they are long, steep-walled embayments called **fiords**.

Fiords are restricted to high latitudes where glaciers exist at low elevations, such as Alaska, western Canada, Scandinavia, Greenland, southern New Zealand, and southern Chile. Lower sea level during the Pleistocene was not entirely responsible for the formation of all fiords. Unlike running water, glaciers can erode a considerable distance below sea level. In fact, a glacier 500 m thick can stay in contact with the seafloor and effectively erode it to a depth of about 450 m before the buoyant effects of water cause the glacial ice to float!

Hanging Valleys
Some of the world's highest and most spectacular waterfalls are found in recently glaciated areas. Bridalveil Falls in Yosemite National Park, California, plunge 188 m from a **hanging valley**, which is a tributary valley whose floor is at a higher level than that of the main valley (Figure 13.11). Where the two valleys meet, the mouth of the hanging valley is perched far above the main valley's floor (Figure 13.10c). Accordingly, streams flowing through hanging valleys plunge over vertical or steep precipices.

Although not all hanging valleys form by glacial erosion, many do. As Figure 13.10 shows, the large glacier in the main valley vigorously erodes, whereas the smaller glaciers in tributary valleys are less capable of erosion. When the glaciers disappear, the smaller tributary valleys remain as hanging valleys.

Cirques, Arêtes, and Horns
Perhaps the most spectacular erosional landforms in areas of valley glaciation are at the upper ends of glacial troughs and along the divides that separate adjacent glacial troughs. Valley glaciers form and move out from steep-walled, bowl-shaped depressions called **cirques** at the upper end of their troughs (Figure 13.10c, 13.12a). Cirques

are typically steep-walled on three sides, but one side is open and leads into the glacial trough. Some cirques slope continuously into the glacial trough, but many have a lip, or threshold, at their lower end.

The details of cirque origin are not fully understood, but they probably form by erosion of a preexisting depression on a mountainside. As snow and ice accumulate in the depression, frost wedging and plucking, combined with glacial erosion, enlarge and transform the head of a steep mountain valley into a typical amphitheater-shaped cirque. Small lakes of meltwater, called *tarns*, often form on the floors of cirques.

Arêtes—narrow, serrated ridges—form in two ways. In many cases, cirques form on opposite sides of a ridge, and headward erosion reduces the ridge until only a thin partition of rock remains (Figure 13.10c, 13.12b). The same effect occurs when erosion in two parallel glacial troughs reduces the intervening ridge to a thin spine of rock.

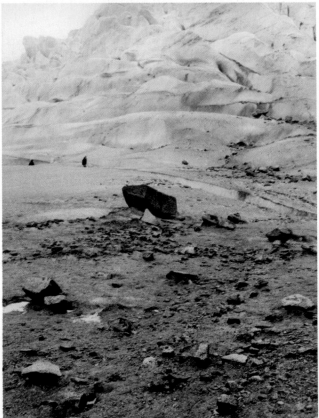

Figure 13.9 Sediment Transport by Valley Glaciers
Debris on the surface of the Mendenhall Glacier in Alaska. The largest boulder is about 2 m across. Notice the icefall in the background. The person left of center provides scale.

Sue Monroe

Figure 13.10 Erosional Landforms Produced by Valley Glaciers

a. A mountain area before glaciation.

b. The same area during the maximum extent of valley glaciers.

c. After glaciation.

© 2013 Cengage Learning

The most majestic of all mountain peaks are **horns**, steep-walled, pyramidal peaks formed by headward erosion of cirques. For a horn to form, a mountain peak must have at least three cirques on its flanks, all of which erode headward (Figure 13.10, 13.12c).

Continental Glaciers and Erosional Landforms

Areas eroded by continental glaciers tend to be smooth and rounded, because these glaciers bevel and abrade high areas that project into the ice. Rather than yielding the sharp, angular landforms typical of valley glaciation,

they produce a landscape of subdued topography interrupted by rounded hills, because they bury landscapes entirely during their development.

In a large part of Canada, particularly the vast Canadian shield region, continental glaciers have stripped off the soil and unconsolidated surface sediment, revealing extensive exposures of striated and polished bedrock. These areas have deranged drainage (see Figure 11.17e), numerous lakes and swamps, low relief, extensive bedrock exposures, and little or no soil. They are referred to as *ice-scoured plains* (Figure 13.13). Similar, but smaller, bedrock exposures are also widespread in the northern United States from Maine through Minnesota.

horn A steep-walled, pyramid-shaped peak formed by the headward erosion of at least three cirques.

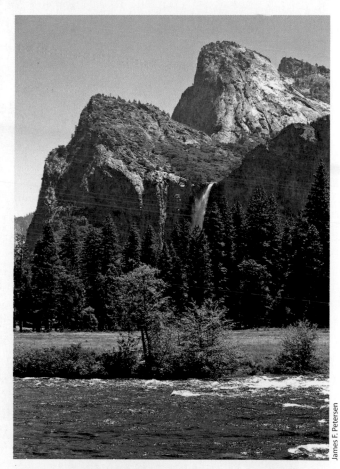

James F. Petersen

Figure 13.11 Hanging Valley

Bridalveil Falls plunge 188 m from a hanging valley in Yosemite National Park in California. The hanging valley was occupied by a glacier that was a tributary to a much larger glacier in Yosemite Valley in the foreground, which is a U-shaped glacial trough.

Figure 13.12 Cirques, Arêtes, and Horns

a. A cirque in the St. Elias Mountains of the Yukon Territory of Canada.

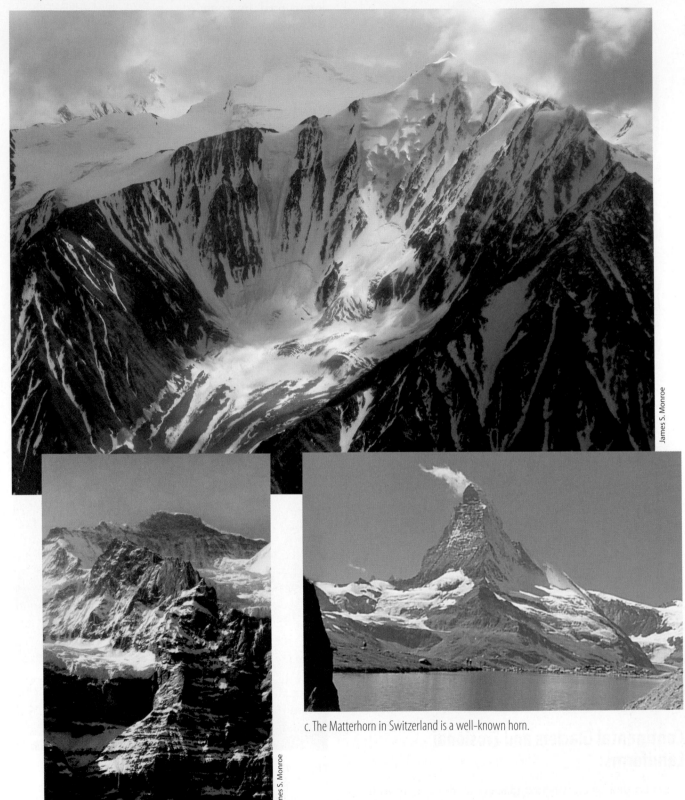

James S. Monroe

Swiss National Tourist Office

c. The Matterhorn in Switzerland is a well-known horn.

James S. Monroe

b. This view in Switzerland shows two small valley glaciers, the headwall of a cirque, and an arête.

Figure 13.13 An Ice-Scoured Plain in Northwest Territories of Canada

This low-relief surface is an ice-scoured plain in the Northwest Territories of Canada. Numerous lakes, little or no soil, and extensive bedrock exposures are typical of these areas eroded by continental glaciers.

LO5 DEPOSITS OF GLACIERS

Glacial Drift

You know that glaciers erode and transport gravel, sand, and mud; it follows that they must also deposit sediment. All glacial deposits go by the general term **glacial drift**, but geologists recognize two types of drift—*till* and *stratified drift*. **Till** is made up of any sediment deposited directly by glacial ice, as, for example, at the terminus of a glacier. Till is not sorted by particle size and it shows no stratification (Figure 13.14a). The till of both valley and continental glaciers is similar, but that of continental glaciers is much more extensive and usually has been transported much further. As opposed to till, **stratified drift**, as the name implies, is layered or stratified, and invariably it shows some degree of sorting by particle size. As a matter of fact, most stratified drift is layers of gravel and sand or mixtures thereof deposited in braided streams that discharge from melting glaciers.

The appearance of till and stratified drift may not be as inspiring as some landforms resulting from glacial erosion, but they are important groundwater reservoirs, and they are exploited for their sand and gravel. In fact, sand and gravel, mostly for construction, are the most valuable mineral commodities in many areas. It is true that all glaciers deposit stratified drift, but, as you would expect, those of continental glaciers are far more extensive. In contrast, the deposits of a valley glacier tend to be restricted to the lower parts of the valley occupied by the glacier.

One conspicuous aspect of glacial drift is rock fragments of various sizes that were obviously not derived from the underlying bedrock. These fragments, called **glacial erratics**, came from some distant source and were then transported and deposited in their present location (Figure 13.14b). Some glacial erratics are gigantic—the Madison Boulder in New Hampshire and Daggett Rock in Maine weigh about 4,550 and 7,270 metric tons, respectively.

Landforms Composed of Till

Landforms composed of till include several types of *moraines* and elongated hills known as *drumlins*.

glacial drift A collective term for all sediment deposited directly by glacial ice (till) and by meltwater streams (outwash).

till All sediment deposited directly by glacial ice.

stratified drift Glacial deposits that show both stratification and sorting.

glacial erratic A rock fragment carried some distance from its source by a glacier and usually deposited on bedrock of a different composition.

end moraine A pile or ridge of rubble deposited at the terminus of a glacier.

ground moraine The layer of sediment released from melting ice as a glacier's terminus retreats.

End Moraines If a glacier has a balanced budget, its terminus may become stabilized in one position for some period of time, perhaps a few years or even decades. When an ice front is stationary, flow within the glacier continues, and any sediment transported within or upon the ice is dumped as a pile of rubble at the glacier's terminus (Figure 13.15). These deposits are **end moraines**, which continue to grow as long as the ice front remains stationary. End moraines of valley glaciers are crescent-shaped ridges of till spanning the valley occupied by the glacier. Those of continental glaciers similarly parallel the ice front but are much more extensive.

Following a period of stabilization, a glacier may advance or retreat depending on changes in its budget. If it advances, the ice front overrides and modifies its former moraine. If it has a negative budget, though, the ice front retreats toward the zone of accumulation. As the ice front recedes, till is deposited as it is liberated from the melting ice and forms a layer of **ground moraine**.

recessional moraine An end moraine that forms when a glacier's terminus retreats, then stabilizes, and a ridge or mound of till is deposited.

terminal moraine An end moraine consisting of a ridge or mound of rubble marking the farthest extent of a glacier.

lateral moraine Ridge of sediment deposited along the margin of a valley glacier.

medial moraine A moraine carried on the central surface of a glacier; formed where two lateral moraines merge.

drumlin An elongate hill of till formed by the movement of a continental glacier or by floods.

Ground moraine has an irregular, rolling topography, whereas end moraine consists of long, ridgelike accumulations of sediment.

After a glacier has retreated for some time, its terminus may once again stabilize, and it deposits another end moraine. Because the ice front has receded, such moraines are called **recessional moraines**. The outermost end moraines, marking the greatest extent of the glaciers, go by the special name **terminal moraine**.

Lateral and Medial Moraines

Valley glaciers transport considerable sediment along their margins, much of it abraded and plucked from the valley walls, but a significant amount falls or slides onto the glacier's surface by mass wasting processes. In any case, this sediment is transported and deposited as long ridges of till called **lateral moraines** along the margin of the glacier (Figure 13.16).

Where two lateral moraines merge, as when a tributary glacier flows into a larger glacier, a **medial moraine** forms (Figure 13.16). A large glacier will often have several dark stripes of sediment on its surface, each of which is a medial moraine. One can determine how many tributaries a valley glacier has by the number of its medial moraines.

Drumlins

In many areas where continental glaciers deposited till, the till has been reshaped into elongated hills known as **drumlins** (Figure 13.17). Some drumlins are 50 m high and 1 km long, but most are much smaller. From the side, a drumlin looks like an inverted spoon, with the steep end on the side from which the glacial ice advanced and the gently sloping end pointing in the direction of ice movement. Drumlins are rarely found as single, isolated hills; instead,

Figure 13.14 Glacial Drift and Glacial Erratics

Copyright and Photograph by Dr. Parvinder S. Sethi

a. This glacial drift from the Matanuska Glacier, near Palmer, Alaska, is till, because it is unsorted and shows no stratification.

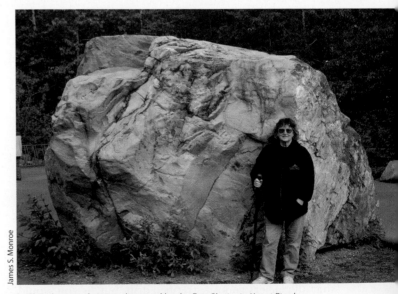

James S. Monroe

b. Glacial erratic that was deposited by the Exit Glacier in Kenai Fjords National Park near Seward, Alaska.

Figure 13.15 End Moraine

End moraines are accumulations of unsorted, nonstratified till deposited at the end of a glacier.

a. End moraine of the Salmon Glacier in the Coast Mountains of British Columbia, Canada.

b. This small end moraine was deposited in 1995 by the Exit Glacier in Kenai Fjords National Park, near Seward, Alaska.

they occur in *drumlin fields* that contain hundreds or thousands of drumlins.

According to one hypothesis, drumlins form when till beneath a glacier is reshaped into streamlined hills as the ice moves over it by plastic flow. Another hypothesis holds that huge floods of glacial meltwater modify till into drumlins.

Landforms Composed of Stratified Drift

Stratified drift exhibits sorting and layering, both indications that it was deposited by running water. In fact, it is deposited by streams discharging from valley and continental glaciers.

Outwash Plains and Valley Trains

Sediment-laden meltwater discharges from glaciers most of the time, except perhaps during the coldest months. This meltwater forms a series of braided streams that radiate out from the front of continental glaciers over a wide region. So much sediment is supplied to these streams that much of it is deposited within their channels as sand and gravel bars, thereby forming an **outwash plain**. Valley glaciers also discharge large amounts of meltwater and have braided streams extending from them. However, these streams are confined to the lower parts of glacial troughs, and their long, narrow deposits of stratified drift are known as **valley trains** (Figure 13.18).

Outwash plains, valley trains, and some moraines commonly contain numerous circular to oval depressions, many of which contain small lakes. These depressions, or *kettles*, form when a retreating glacier leaves a block of ice that is subsequently partly or wholly buried (Figures 13.17 and 13.18). When the ice block eventually melts, it leaves a depression, and if the depression extends below the water table, it becomes the site of a small lake. Some outwash plains have so many kettles that they are called *pitted outwash plains*.

Kames and Eskers

Kames are conical hills of stratified drift up to 50 m high (Figure 13.17). Many form when a stream deposits sediment in a depression on a glacier's surface; as the ice melts, the deposit is

outwash plain The sediment deposited by meltwater discharging from a continental glacier's terminus.

valley train A long, narrow deposit of stratified drift confined within a glacial valley.

kame Conical hill of stratified drift originally deposited in a depression on a glacier's surface.

esker A long, sinuous ridge of stratified drift deposited by running water in a tunnel beneath stagnant ice.

lowered to the land surface. Kames also form in cavities within or beneath stagnant ice.

Long sinuous ridges of stratified drift, many of which meander and have tributaries, are **eskers** (Figure 13.17). Most eskers have sharp crests and sides that slope at about 30 degrees. Some are as high as 100 m and can be traced for more than 500 km. The sorting and stratification of the sediments in eskers clearly indicate deposition by running water. The features of ancient eskers and observations of present-day glaciers show that they form in tunnels beneath stagnant ice.

Deposits in Glacial Lakes

Some lakes in areas of glaciation formed as a result of glaciers scouring out depressions; others occurred where a stream's drainage was blocked; and others are the result of water accumulating behind moraines or in kettles. Regardless of how they formed, glacial lakes, like all lakes, are areas of deposition. Sediment may be carried into them and deposited as small deltas, but of special interest are the fine-grained deposits. Mud deposits in glacial lakes are commonly finely laminated (having layers less than 1 cm thick) and consist of alternating light and dark layers known as *varves* (Figure 13.19), which represents an annual episode of deposition. The light layer formed during the spring and summer and consists of silt and clay; the dark layer formed during the winter when the smallest particles of clay and organic matter settled from suspension as the lake froze over.

Another distinctive feature of glacial lakes with varves is *dropstones* (Figure 13.19). These are pieces of gravel, some of boulder size, in otherwise very fine-grained deposits. The presence of varves indicates that currents and turbulence in these lakes were minimal; otherwise, clay and organic matter would not have settled from suspension. How then can we account for dropstones in a low-energy environment? Most of them were probably carried into the lakes by icebergs that eventually melted and released sediment contained in the ice.

LO6 WHAT CAUSES ICE AGES?

We discussed the conditions necessary for a glacier to form earlier in this chapter: More snow falls than melts during the warm season, thus accounting for a net accu-

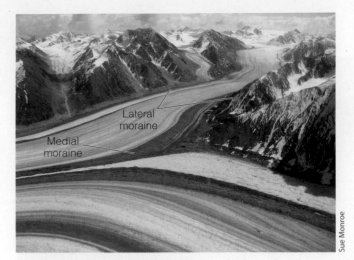

Figure 13.16 Types of Moraines
A moraine is a mound or ridge of unstratified till. Moraines are called end, lateral, and medial depending on their position. End moraines are deposited at a glacier's terminus (see Figure 13.5a), but a lateral moraine is found along the margin of a glacier, and a medial moraine, which forms where two lateral moraines merge, is on the more central part of a glacier.

mulation of snow and ice over the years. But this does not address the broader question of what causes ice ages. Actually, we need to address not only what causes ice ages, but also why there have been so few episodes of widespread glaciation in all of Earth history.

For more than a century, scientists have attempted to develop a comprehensive theory explaining all aspects of ice ages, but they have not yet been completely successful. One reason for their lack of success is that the climatic changes responsible for glaciation, the cyclic occurrence of glacial–interglacial episodes, and short-term events such as the Little Ice Age operate on vastly different time scales.

Only a few periods of glaciation are recognized in the geologic record, each separated from the others by long intervals of mild climate. Such long-term climatic changes probably result from slow geographic changes related to plate tectonic activity. Moving plates carry continents to high latitudes where glaciers exist, provided they receive enough precipitation as snow. Plate collisions, the subsequent uplift of vast areas far above sea level, and the changing atmospheric and oceanic circulation patterns caused by the changing shapes and positions of plates also contribute to long-term climate change.

The Milankovitch Theory

During the 1920s, the Serbian astronomer Milutin Milankovitch proposed that minor irregularities in

Figure 13.17 Development of Features Produced by Past Continental Glaciers

a. This retreating glacier once covered a larger area as indicated by the terminal moraine.

b. Moraines, eskers, drumlins, kettles, and outwash are all features found in areas once covered by continental glaciers.

© 2013 Cengage Learning

Earth's rotation and orbit are sufficient to alter the amount of solar radiation received at any given latitude and hence bring about climate changes. Now called the **Milankovitch theory**, this reasoning was initially ignored but has received renewed interest since the 1970s and is now widely accepted.

Milankovitch attributed the onset of the glacial episodes to variations in three aspects of Earth's orbit. The first is *orbital eccentricity*, which is the degree to which Earth's orbit around the sun changes over time (Figure 13.20a). When the orbit is nearly circular, both the Northern and Southern Hemispheres have similar contrasts between the seasons. However, if the orbit is more

Milankovitch theory An explanation for the cyclic variations in climate and the onset of ice ages as a result of irregularities in Earth's rotation and orbit.

Figure 13.18 Valley Train and Kettles

This deposit of sand and gravel is the valley train deposited by meltwaters from the Salmon Glacier in the Coast Mountains of British Columbia, Canada. Notice the circular, water-filled depressions; these are kettle lakes, or simply kettles if not filled with water. Also, notice in Figure 13.15a—which shows the terminus of the Salmon Glacier—that several blocks of ice are isolated from the glacier. Should these be partially or wholly covered in valley train deposits, they too will form kettles.

elliptical, hot summers and cold winters will occur in one hemisphere, whereas warm summers and cool winters will take place in the other hemisphere. Calculations indicate a roughly 100,000-year cycle between times of maximum eccentricity, which corresponds closely to the 20 warm–cold climatic cycles that took place during the Pleistocene.

Milankovitch also pointed out that the angle between Earth's axis and a line perpendicular to the plane of Earth's orbit shifts about 1.5 degrees from its current value of 23.5 degrees during a 41,000-year cycle (Figure 13.20b). Although changes in *axial tilt* have little effect on equatorial latitudes, they strongly affect the amount of solar radiation received at high latitudes and the duration of the dark period at and near Earth's poles. Coupled with the third aspect of Earth's orbit, precession of the equinoxes (Figure 13.20c), high latitudes might receive as much as 15% less solar radiation, certainly enough to affect glacial growth and melting.

Precession of the equinoxes, the last aspect of Earth's orbit that Milankovitch cited, refers to a change in the time of the equinoxes (Figure 13.20c). At present, the equinoxes take place on, or about, March 21 and September 21, when the Sun is directly over the equator. But as Earth rotates on its axis, it also wobbles as its axial tilt varies 1.5 degrees from its current value, thus changing the time of the equinoxes. Taken alone, the time of the equinoxes has little climatic effect, but changes in Earth's axial tilt also change the times of *aphelion* and *perihelion*, which are, respectively, when Earth is farthest from and closest to the Sun during its orbit (Figure 13.20c). Earth is now at perihelion, closest to the Sun, during Northern Hemisphere winters, but in about 11,000 years, perihelion will be in July. Accordingly, Earth will be at aphelion, farthest from the Sun, in January and have colder winters.

Continuous variations in Earth's orbit and axial tilt cause the amount of solar heat received at any latitude

Figure 13.19 Varves and a Dropstone in Glacial Deposits
These varves have a dropstone that was probably liberated from floating ice.

to vary slightly through time. The total heat received by the planet changes little, but according to Milankovitch, and now many scientists agree, these changes cause complex climatic variations and provided the triggering mechanism for the glacial–interglacial episodes of the Pleistocene.

Short-Term Climatic Events

Climatic events with durations of several centuries, such as the Little Ice Age, are too short to be accounted for by plate tectonics or Milankovitch cycles. Several hypotheses have been proposed to account for these short-term events, including variations in solar energy and volcanism.

Variations in solar energy could result from changes within the Sun itself or from anything that would reduce the amount of energy Earth receives from the Sun. The latter could result from the solar system passing through clouds of interstellar dust and gas or from substances in the atmosphere reflecting solar radiation back into space. Records kept during the past century indicate that during this time the amount of solar radiation has varied only slightly. Although variations in solar energy may influence short-term climatic events, such a correlation has not yet been demonstrated.

During large volcanic eruptions, tremendous amounts of ash and gases are spewed into the atmosphere, where they reflect incoming solar radiation and thus reduce atmospheric temperatures. Small droplets

of sulfur gases remain in the atmosphere for years and can have a significant effect on climate. Several large-scale volcanic events have occurred, such as the 1815 eruption of Tambora, and are known to have had climatic effects. However, no relationship between periods of volcanic activity and periods of glaciation has yet been established.

a Earth's orbit varies from nearly a circle (left) to an ellipse (right) and back again in about 100,000 years.

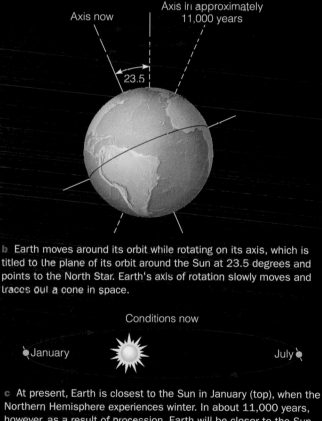

b Earth moves around its orbit while rotating on its axis, which is titled to the plane of its orbit around the Sun at 23.5 degrees and points to the North Star. Earth's axis of rotation slowly moves and traces out a cone in space.

c At present, Earth is closest to the Sun in January (top), when the Northern Hemisphere experiences winter. In about 11,000 years, however, as a result of procession, Earth will be closer to the Sun in July (bottom), when summer occurs in the Northern Hemisphere.

Figure 13.20 Milankovitch Theory
According to the Milankovitch Theory, minor irregularities in Earth's rotation and orbit may affect climatic changes.

14 The Work of Wind and Deserts

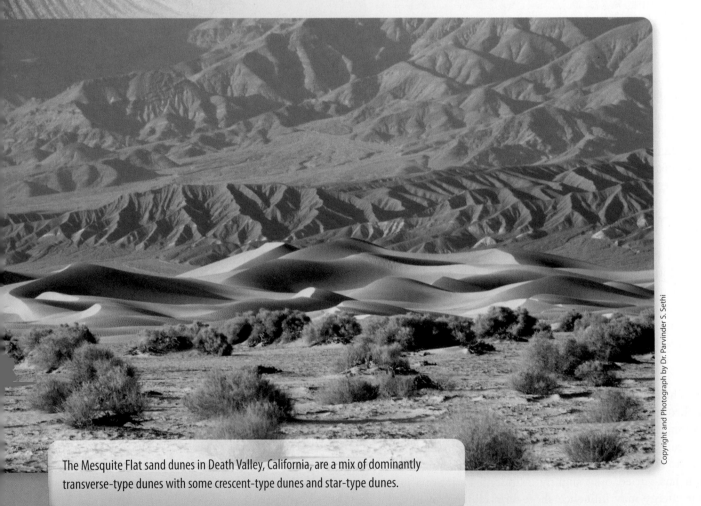

The Mesquite Flat sand dunes in Death Valley, California, are a mix of dominantly transverse-type dunes with some crescent-type dunes and star-type dunes.

Copyright and Photograph by Dr. Parvinder S. Sethi

Introduction

"**Because** of the relentless advance of deserts, **hundreds of thousands** of people have died of starvation or been **forced to migrate.**"

During the past several decades, deserts have been advancing across millions of acres of productive land, destroying rangeland, croplands, and even villages. Such expansion, estimated at 70,000 km² per year, has exacted a terrible toll in human suffering. Because of the relentless advance of deserts, hundreds of thousands of people have died of starvation or been forced to migrate as "environmental refugees" from their homelands to camps, where the majority are severely malnourished. This expansion of deserts into formerly productive lands is called **desertification** and is a major problem in many countries.

Most regions undergoing desertification lie along the margins of existing deserts, where a delicately balanced ecosystem serves as a buffer between the desert on one side and a more humid environment on the other. These regions have limited potential to adjust to increasing environmental pressures from natural causes as well as human activity. Ordinarily, desert regions expand and contract gradually in response to natural processes such as climatic change, but much of the recent desertification has been greatly accelerated by human activities.

desertification The expansion of deserts into formerly productive lands.

In many areas, the natural vegetation has been cleared as crop cultivation has expanded into increasingly drier desert fringes to support growing populations. Because grasses are the dominant natural vegetation in most desert fringe areas, raising livestock is a common economic activity. However, increasing numbers of livestock in many areas have greatly exceeded the land's ability to support them. Consequently, the vegetation cover that protects the soil has diminished, causing the soil to crumble and be stripped away by wind and water, which results in increased desertification and the many problems associated with it.

There are many important reasons to study deserts and the processes that are responsible for their formation. First, deserts cover large regions of Earth's surface. More than 40 percent of Australia is desert, and the Sahara occupies a vast part of northern Africa. Additionally, many desert regions are experiencing increases in population growth, such as the high desert area of southern California and parts of Nevada, as well as various locations in Arizona. Many of these places already have problems associated with increasing population and the strains it places on the environment, particularly the need for greater amounts of groundwater (see Chapter 12).

Furthermore, with the current debate about global warming, it is important to understand how desert processes operate and how global climate changes affect the various Earth systems and subsystems. Learning about the underlying causes of climate change by examining ancient desert regions may provide insight into the possible duration and severity of future climatic changes. This understanding can have important ramifications in decisions about whether burying nuclear waste in a desert, such as Yucca Mountain, Nevada, is as safe as some claim and is truly in our best interests as a society.

Learning Outcomes

After reading this unit, you should be able to do the following:

LO1 Discuss the role wind plays in transporting sediment

LO2 Explain the two processes of wind erosion

LO3 Identify the different types of wind deposits

LO4 Describe air-pressure belts and global wind patterns

LO5 Describe the distribution of deserts

LO6 Identify the various characteristics of deserts

LO7 Describe the different types of desert landforms

> By understanding how desertification operates, people can take steps to eliminate or reduce the destruction done, particularly in terms of human suffering.

LO1 SEDIMENT TRANSPORT BY WIND

Wind is a turbulent fluid and therefore transports sediment in much the same way as running water. Although wind typically flows at a greater velocity than water, it has a lower density and thus can carry only clay- and silt-size particles as *suspended load*. Sand and larger particles are moved along the ground as *bed load*.

Bed Load

Sediments that are too large or heavy to be carried in suspension by water or wind are moved as bed load either by *saltation* or by rolling and sliding. Saltation on land occurs when wind starts sand grains rolling and lifts and carries some grains short distances before they fall back to the surface. As the descending sand grains hit the surface, they strike other grains, causing them to bounce along (Figure 14.1). Wind-tunnel experiments show that once sand grains begin moving, they continue to move, even if the wind drops below the speed necessary to start them moving! This happens because once saltation begins, it sets off a chain reaction of collisions between sand grains that keeps the grains in constant motion.

> More than 6,000 years ago, the Sahara was a fertile savannah supporting a diverse fauna and flora, including humans. Then the climate changed, and the area became a desert. How did this happen? Will this region change back again in the future? These are some of the questions geoscientists hope to answer by studying deserts.

© Pixland/Jupiterimages

Saltating sand usually moves near the surface, and even when winds are strong, grains are rarely lifted higher than about a meter. If the winds are very strong, these wind-whipped grains can cause extensive abrasion. A car's paint can be removed by sandblasting in a short time, and its windshield will become completely frosted and translucent from pitting.

Suspended Load

Silt- and clay-sized particles constitute most of a wind's suspended load. Even though these particles are much smaller and lighter than sand-sized particles, wind usually starts the latter moving first. The reason for this phenomenon is that a very thin layer of motionless air lies next to the ground, where the small

Figure 14.1 Saltation

Most sand is moved near the ground surface by saltation. Sand grains are picked up by the wind and carried a short distance before falling back to the ground, where they usually hit other grains, causing them to bounce and move in the direction of the wind.

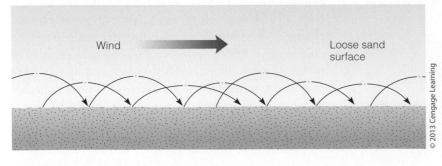

Wind

Loose sand surface

© 2013 Cengage Learning

silt and clay particles remain undisturbed. The larger sand grains, however, stick up into the turbulent air zone, where they can be moved. Unless the stationary air layer is disrupted, the silt and clay particles remain on the ground, providing a smooth surface.

This phenomenon can be observed on a dirt road on a windy day. Unless a vehicle travels over the road, little dust is raised even though it is windy. When a vehicle moves over the road, it breaks the calm boundary layer of air and disturbs the smooth layer of dust, which is picked up by the wind and forms a dust cloud in the vehicle's wake.

In a similar manner, when a sediment layer is disturbed, silt- and clay-sized particles are easily picked up and carried in suspension by the wind, creating clouds of dust or even dust storms. Once these fine particles are lifted into the atmosphere, they may be carried thousands of kilometers from their source.

LO2 WIND EROSION

Although wind action produces many distinctive erosional features and is an extremely efficient sorting agent, running water is responsible for most erosional landforms in arid regions despite the fact that stream channels are typically dry (Figure 14.2). Wind erodes material in two ways: *abrasion* and *deflation*.

Abrasion

Abrasion involves the impact of saltating sand grains on an object and is analogous to sandblasting. The effects of abrasion are usually minor, because sand, the most common agent of abrasion, is rarely carried more than a meter above the surface. Rather than creating major erosional features, wind abrasion typically modifies existing features by etching, pitting, smoothing, or polishing. Nonetheless, wind abrasion can produce many strange-looking and bizarre-shaped features.

Ventifacts are a common product of wind abrasion; these are stones whose surfaces have been polished, pitted, grooved, or faceted by the wind (Figure 14.3 and the Virtual Field Trip in this chapter). If the wind blows from different directions, or if the stone is moved, the ventifact will have multiple facets. Ventifacts are most common in deserts, yet they can form wherever stones are exposed to saltating sand grains—for example, on beaches in humid regions and some outwash plains, such as in New England.

Deflation

Another important mechanism of wind erosion is **deflation**, which is the removal of loose surface sediment by wind. Among the characteristic features of deflation in many arid and semiarid regions are *deflation*

> **abrasion** The process whereby rock is worn smooth by the impact of sediment transported by running water, glaciers, waves, or wind.
>
> **ventifact** A stone with a surface polished, pitted, grooved, or faceted by wind abrasion.
>
> **deflation** The removal of loose surface sediment by wind.

Figure 14.2 Erosion
Natural forces of wind and water have eroded the rocks in Monument Valley Navajo Tribal Park, Arizona, to create "mitten" buttes such as this one.

Copyright and Photograph by Dr. Parvinder S. Sethi

Figure 14.3 Ventifacts

Wind

a. A ventifact forms when windborne particles (1) abrade the surface of a rock, (2) forming a flat surface. If the rock is moved, (3) additional flat surfaces are formed.

b. Numerous ventifacts are visible in this photo, which also shows desert pavement in Death Valley, California. Desert pavement prevents further erosion and transport of a desert's surface materials by forming a protective layer of close-fitting, larger rocks.

Copyright and Photograph by Dr. Parvinder S. Sethi

desert pavement A surface mosaic of close-fitting pebbles, cobbles, and boulders found in many dry regions; it results from wind erosion of sand and smaller particles.

hollows, or *blowouts* (Figure 14.4). These shallow depressions of variable dimensions result from differential erosion of surface materials. Ranging in size from several kilometers in diameter and tens of meters deep, to small depressions only a few meters wide and less than a meter deep, deflation hollows are common in the southern Great Plains region of the United States.

In many dry regions, the removal of sand-sized and smaller particles by wind leaves a surface of pebbles, cobbles, and boulders. As the wind removes the fine-grained material from the surface, the effects of gravity and occasional heavy rain, and even the swelling of clay minerals, rearrange the remaining coarse particles into a mosaic of close-fitting rocks called **desert pavement** (Figures 14.3b and 14.5). Once desert pavement forms, it protects the underlying material from further deflation.

LO3 WIND DEPOSITS

Although wind is of minor importance as an erosional agent, it is responsible for impressive deposits, which are primarily of two types. The first, *dunes*, occur in several distinctive types, all of which consist of sand-sized particles that are usually deposited near their source. The second, *loess*, consists of layers of windblown silt and clay deposited over large areas downwind and commonly far from their source.

Copyright and Photograph by Dr. Parvinder S. Sethi

Figure 14.4 Deflation Hollow

A deflation hollow, shown here as the low area between two sand dunes in Death Valley, California, results when loose surface sediment is differentially removed by wind.

Figure 14.5 Desert Pavement

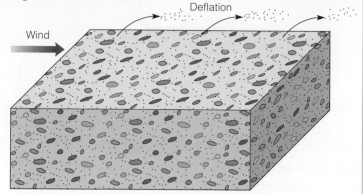

a. Fine-grained material is removed by wind,

b. leaving a concentration of larger particles that form desert pavement.

The Formation and Migration of Dunes

The most characteristic features in sand-covered regions are **dunes**, which are mounds or ridges of wind-deposited sand (Figure 14.6). Dunes form when wind flows over and around an obstruction, resulting in the deposition of sand grains, which accumulate and build up a deposit of sand. As they grow, these sand deposits become self-generating in that they form ever-larger wind barriers that further reduce the wind's velocity, resulting in more sand deposition and growth of the dune.

Most dunes have an asymmetrical profile, with a gentle windward slope and a steeper downwind or leeward slope that is inclined in the direction of the prevailing wind (Figure 14.7a). Sand grains move up the gentle windward slope by saltation and accumulate on the leeward side, forming an angle of 30 to 34 degrees from the horizontal, which is the angle of repose of dry sand. When this angle is exceeded by accumulating sand, the slope collapses, and the sand slides down the leeward slope, coming to rest at its base. As sand moves from a dune's windward side and periodically slides down its leeward slope, the dune slowly migrates in the direction of the prevailing wind (Figure 14.7b). When preserved in the geologic record, dunes help geologists determine the prevailing direction of ancient winds (Figure 14.8)

> **dune** A mound or ridge of wind-deposited sand.

Dune Types

Geologists recognize four major dune types (barchan, longitudinal, transverse, and parabolic), although intermediate forms also exist. The size, shape, and

Figure 14.6 Sand Dunes

Large sand dunes in Death Valley, California. The prevailing wind direction is from left to right, as indicated by the sand dunes in which the gentle windward side is on the left and the steeper leeward slope is on the right.

Copyright and Photograph by Dr. Parvinder S. Sethi

Figure 14.7 **Dune Migration**

Wind

Sand moves by saltation

Windward side

Leeward slope

a. Profile of a sand dune.

Wind

Direction of dune migration

b. Dunes migrate when sand moves up the windward side and slides down the leeward slope. Such movement of the sand grains produces a series of cross-beds that slope in the direction of wind movement.

© 2013 Cengage Learning

Copyright and Photograph by Dr. Parvinder S. Sethi

Figure 14.8 **Cross-Bedding**
Ancient cross-bedding in sandstone beds in Zion National Park, Utah, helps geologists determine the prevailing direction of the wind that formed these ancient sand dunes.

barchan dune A crescent-shaped sand dune with its tips pointing downwind.

longitudinal dune A long ridge of sand generally parallel to the direction of the prevailing wind.

transverse dune A ridge of sand with its long axis perpendicular to the wind direction.

arrangement of dunes result from the interaction of such factors as sand supply, the direction and velocity of the prevailing wind, and the amount of vegetation. Although dunes are usually found in deserts, they can also develop wherever sand is abundant, such as along the upper parts of many beaches.

Barchan dunes are crescent-shaped dunes whose tips point downwind (Figure 14.9). They form in areas that have a generally flat, dry surface with little vegetation, a limited supply of sand, and a nearly constant wind direction. Most barchans are small, with the largest reaching about 30 m high. Barchans are the most

mobile of the major dune types, moving at rates that can exceed 10 m per year.

Longitudinal dunes (also called *seif dunes*) are long, parallel ridges of sand aligned generally parallel to the direction of the prevailing winds; they form where the sand supply is somewhat limited (Figure 14.10). Longitudinal dunes result when winds converge from slightly different directions to produce the prevailing wind. They range in height from about 3 m to more than 100 m, and some stretch for more than 100 km. Longitudinal dunes are especially well developed in central Australia, where they cover nearly one-fourth of the continent. They also cover extensive areas in Saudi Arabia, Egypt, and Iran.

Transverse dunes form long ridges perpendicular to the prevailing wind direction in areas that have abundant sand and little or no vegetation (Figure 14.11).

Figure 14.9 Barchan Dunes

a. Barchan dunes form in areas that have a limited amount of sand, a nearly constant wind direction, and a generally flat, dry surface with little vegetation. The tips of barchan dunes point downward.

b. A ground-level view of several barchan dunes.

Figure 14.10 Longitudinal Dunes

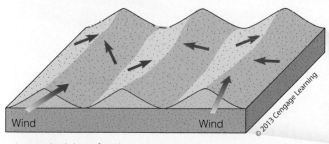

a. Longitudinal dunes form long, parallel ridges of sand aligned roughly parallel to the prevailing wind direction. They typically form where sand supplies are limited.

When viewed from the air, transverse dunes have a wavelike appearance and are therefore sometimes called *sand seas* (see the Virtual Field Trip in this chapter). The crests of transverse dunes can be as high as 200 m, and the dunes can be as wide as 3 km.

Parabolic dunes are most common in coastal areas with abundant sand, strong onshore winds, and a partial cover of vegetation (Figure 14.12). Although parabolic dunes have a crescent shape like barchan dunes, their tips point upwind. Parabolic dunes form when the vegetation cover is broken and deflation produces a deflation hollow, or blowout. As the wind transports the sand out of the depression, it builds up on the convex downwind dune crest. The central part of the dune is excavated by the wind while vegetation holds the ends and sides fairly well in place.

Loess

Wind-blown silt and clay deposits composed of angular quartz grains, feldspar, micas, and calcite are known as **loess**. The distribution of loess shows that it is derived from three main sources: deserts, Pleistocene glacial outwash deposits, and the floodplains of rivers in semiarid regions. Loess must be stabilized by moisture and vegetation in order to accumulate. Consequently, loess is not found in deserts, even though deserts provide much of its material. Because of its unconsolidated nature, loess is

> **parabolic dune** A crescent-shaped dune with its tips pointing upwind.
>
> **loess** Wind-blown deposits of silt and clay.

b. Longitudinal dunes, 15 m high, in the Gibson Desert, west central Australia. The bright blue areas between the dunes are shallow pools of rainwater, and the darkest patches are areas where the Aborigines have set fires to encourage the growth of spring grasses.

Figure 14.11 Transverse Dunes

Wind

a. Transverse dunes form long ridges of sand that are perpendicular to the prevailing wind direction in areas of little or no vegetation and abundant sand.

b. Transverse dunes, Great Sand Dunes National Monument, Colorado. The prevailing wind direction is from lower left to upper right.

Figure 14.12 Parabolic Dunes

Wind

a. Parabolic dunes typically form in coastal areas that have a partial cover of vegetation, a strong onshore wind, and abundant sand.

b. A parabolic dune developed along the Lake Michigan shoreline west of St. Ignace, Michigan.

easily eroded, and as a result, eroded loess areas are characterized by steep cliffs and rapid lateral and headward stream erosion (Figure 14.13).

Presently, loess deposits cover approximately 10 percent of Earth's land surface and 30 percent of the United States. The most extensive and thickest loess deposits are found in northeast China, where accumulations greater than 30 m thick are common. Loess-derived soils are some of the world's most fertile. It is therefore not surprising that the world's major grain-producing regions correspond to large loess deposits such as the North European Plain, Ukraine, and the Great Plains of North America.

Figure 14.13 Terraced Wheat Fields in the Loess Soil at Tangwa Village, China

Because of the unconsolidated nature of loess, many farmers live in hillside caves they carved from the loess.

© Lowel Georgia/CORBIS

LO4 AIR-PRESSURE BELTS AND GLOBAL WIND PATTERNS

To understand the work of wind and the distribution of deserts, we need to consider the global pattern of air-pressure belts and winds, which are responsible for Earth's atmospheric circulation patterns. Air pressure is the density of air exerted on its surroundings (i.e., its weight). When air is heated, it expands and rises, reducing its mass for a given volume and causing a decrease in air pressure. Conversely, when air is cooled, it contracts and air pressure increases. Therefore, those areas of Earth's surface that receive the most solar radiation, such as the equatorial regions, have low air pressure, whereas the colder areas, such as the polar regions, have high air pressure.

Air flows from high-pressure zones to low-pressure zones. If Earth did not rotate, winds would move in a straight line from one zone to another. Because Earth rotates, however, winds are deflected to the right of their direction of motion (clockwise) in the Northern Hemisphere and to the left of their direction of motion (counterclockwise) in the Southern Hemisphere. This deflection of air between latitudinal zones resulting from Earth's rotation is known as the **Coriolis effect**. The combination of latitudinal pressure differences and the Coriolis effect produces a worldwide pattern of east-west–oriented wind belts (Figure 14.14).

Earth's equatorial zone receives the most solar energy, which heats the surface air and causes it to rise. As the air rises, it cools and releases moisture that falls as rain in the equatorial region (Figure 14.14). The rising air is now much drier as it moves northward and southward toward each pole. By the time it reaches 20 to 30 degrees north and south latitudes, the air has become cooler and denser and begins to descend. Compression of the atmosphere warms the descending air mass and produces a warm, dry, high-pressure area, the perfect conditions for the formation of the low-latitude deserts of the Northern and Southern Hemispheres (Figure 14.15).

Coriolis effect The apparent deflection of a moving object from its anticipated course because of Earth's rotation. Winds and oceanic currents are deflected clockwise in the Northern Hemisphere and counterclockwise in the Southern Hemisphere.

© iStockphoto.com/Anton Zhukov

Many areas are dry because of their remoteness from moist maritime air and the presence of mountain ranges that produce a *rain-shadow desert*. When moist marine air moves inland and meets a mountain range, it is forced upward. As it rises, it cools, forming clouds and producing precipitation that falls on the windward side of the mountains. The air that descends on the leeward side of the mountain range is much warmer and drier, producing a rain-shadow desert.

LO5 THE DISTRIBUTION OF DESERTS

desert Any area that receives less than 25 cm of rain per year and that has a high evaporation rate.

Dry climates occur in the low and middle latitudes, where the potential loss of water by evaporation may exceed the yearly precipitation (Figure 14.15). Dry climates cover 30 percent of Earth's land surface and are subdivided into semiarid and arid regions. *Semiarid regions* receive more precipitation than arid regions, yet are moderately dry. Their soils are usually well developed and fertile and support a natural grass cover. *Arid regions*, generally described as **deserts**, are dry; they receive less than 25 cm of rain per year, have high evaporation rates, typically have poorly developed soils, and are mostly or completely devoid of vegetation.

The majority of the world's deserts are in the dry climates of the low and middle latitudes (Figure 14.15). The remaining dry climates of the world are found in the middle and high latitudes, mostly within continental interiors in the Northern Hemisphere (Figure 14.15).

LO6 CHARACTERISTICS OF DESERTS

To people who live in humid regions, deserts may seem stark and inhospitable. Instead of a landscape of rolling hills and gentle slopes with an almost continuous cover of vegetation, deserts are dry, have little vegetation, and consist of nearly continuous rock exposures, desert pavement, or sand dunes. Yet, despite the great contrast between deserts and more humid areas, the same geologic processes are at work, only operating under different climatic conditions.

Temperature, Precipitation, and Vegetation

The heat and dryness of deserts are well known. Many of the deserts of the low latitudes have average summer

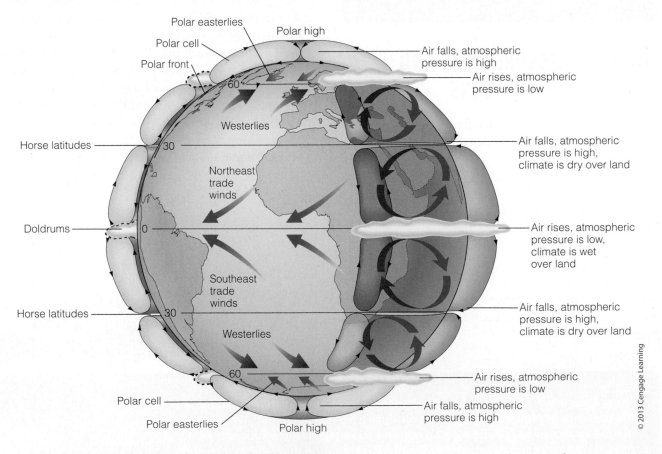

Figure 14.14 The General Circulation Pattern of Earth's Atmosphere
Air flows from high-pressure zones to low-pressure zones, and the resulting winds are deflected to the right of their direction of movement (clockwise) in the Northern Hemisphere, and to the left of their direction of movement (counterclockwise) in the Southern Hemisphere. This deflection of air between latitudinal zones resulting from Earth's rotation is known as the Coriolis effect.

Figure 14.15 The Distribution of Earth's Arid and Semiarid Regions

Semiarid regions receive more precipitation than arid regions, yet they are still moderately dry. Arid regions, generally described as deserts, are dry and receive less than 25 cm of rain per year. The majority of the world's deserts are located in the dry climates of the low and middle latitudes.

temperatures that range between 32°C and 38°C. It is not uncommon for some low-elevation inland deserts to record daytime highs of 46°C to 50°C for weeks at a time. During the winter months, when the Sun's angle is lower and there are fewer daylight hours, daytime temperatures average between 10°C and 18°C.

Although deserts are defined as regions that receive, on average, less than 25 cm of rain per year, the amount of rain that falls each year is unpredictable and unreliable. It is not uncommon for an area to receive more than an entire year's average rainfall in one cloudburst and then to receive little rain for several years. Thus, yearly rainfall averages can be misleading.

Deserts display a wide variety of vegetation (Figure 14.16). Although the driest deserts, or those with large areas of shifting sand, are almost devoid of vegetation, most deserts support at least a sparse plant cover that, when closely examined, reveals a surprising diversity of plants that have evolved the ability to live in the near absence of water.

Desert plants are widely spaced, typically small, and grow slowly. Their stems and leaves are usually hard and waxy to minimize water loss by evaporation and

to protect the plant from sand erosion. Most plants have a widespread shallow root system to absorb the dew that forms each morning in all but the driest deserts and to help anchor the plant in what little soil there may be. In extreme cases, many plants lie dormant during particularly dry years and spring to life after the first rain shower with a beautiful profusion of flowers.

Figure 14.16 Desert Vegetation

Desert vegetation is typically sparse, widely spaced, and characterized by slow growth rates. The vegetation shown here is in Death Valley, California.

Virtual Field Trip

Desert Environments

Ready to Go!

What better place to learn about desert processes and features than Death Valley National Park, California. Aptly named, Death Valley National Park is the hottest, driest, and lowest of the nation's national parks. With summer daytime temperatures often above 49°C, scant rainfall (average annual precipitation is less than 5 cm), and signs reminding visitors that sea level is 86 m above them, Death Valley, for most people, is a very inhospitable place. But for geologists, it is a wonderful location where the geologic processes and products of an arid environment can be studied up close and personal. In this virtual field trip, you, too, will get to experience what it is like to walk on a salt pan and see the results of wind erosion and deposition. However, you will get to do this from the comfort of your living room or wherever you are connected to the Internet.

The oldest rocks in Death Valley consist of 1.8 billion-year-old gneisses exposed in the Black Mountain range. During the Paleozoic Era, the area in what is now Death Valley was below a warm, shallow sea, where the sandstones, limestones, and dolostones now exposed in the Panamint and Funeral Mountains were deposited. Volcanic activity dominated the region during the Paleogene and Neogene periods, and the results of this activity can be seen in the beautiful colored strata in the Artist's Palette area of Artist Drive.

Death Valley came into its own about 3 million years ago when tensional forces associated with the formation of the Basin and Range Province (see Chapter 17) produced a series of valleys and mountain ranges. Death Valley is one such valley that formed between the Panamint Range on the west and the Black Mountains on the east.

It is now time to see all of the beauty and geology that Death Valley National Park, California, offers. So put on a hat, lather up with the sunscreen so you don't get sunburned, and bring plenty of water while we explore and learn about desert processes and features.

DeerPoint, UT. topographic map, courtesy of USGS.
All photographs: Copyright and Photograph by Dr. Parvinder S. Sethi

GOALS OF THE TRIP

Although there are lots of things to see and learn about deserts and how they form, we will limit ourselves in this virtual field trip to three main desert processes and features:

1. The effects of wind erosion.
2. The features of wind deposition.
3. Desert landforms.

FOLLOW-UP QUESTIONS

1. What type of dunes would probably form if there were far less sand at Mesquite Flat Dunes than there presently is?

2. What type of sedimentary structure forms as a result of dune formation? [Hint: see Figure 14.8 in this chapter and the Sedimentary Rocks Virtual Field Trip]. What can those sedimentary structures tell you about the environment at that location in the geologic past?

3. Why is it so difficult to travel across Death Valley? Imagine what the emigrants must have felt like after crossing the country, only to be confronted with salt pans, sand dunes, and unrelenting heat.

What to See When You Go

Interestingly, most people associate desert features with wind action. However, although wind produces many distinctive erosional features, running water is responsible for most erosional landforms in arid regions. Nonetheless, wind abrasion typically modifies existing features by etching, pitting, smoothing, and polishing. In the photo above, we can see two beautifully formed *ventifacts*, which are stones whose surfaces have been polished, pitted, grooved, or faceted by the wind. In addition, these ventifacts are sitting on what geologists call *desert pavement*, which results when wind removes the fine-grained material from the surface, leaving behind a close-fitting mosaic of coarse particles that forms a pavement that protects the underlying material from further erosion (See Figures 14.3 and 14.5 of this chapter).

Just as wind can erode and produce some unusual-looking structures, it also deposits loose material as it moves over a landscape. The most characteristic features in sand-covered regions are dunes, which are mounds or ridges of wind-deposited sand. Dunes form when wind flows over and around an obstruction, resulting in the deposition of sand grains. Most dunes have an asymmetrical profile, with a gentle windward slope and a steeper downwind slope

that is inclined in the direction of the wind. Geologists recognize four major dune types, which form from the interaction of such factors as sand supply, the direction and velocity of the prevailing wind, and the amount of vegetation.

Playa lakes form in low areas following rainstorms. When a playa lake evaporates, the dry lake bed is called a *playa* or *salt pan*. As seen in this panorama of Badwater Basin above, salt pans are appropriately named. Here, salt deposits and salt ridges cover the floor of this area. The photo also shows a close-up of a typical salt polygon in which salt ridges define the polygon and mudcracks are visible within it. Formed from the evaporation of briny water, some of these deposits were at one time commercially viable, such as the borates which were hauled out of Death Valley by 20-mule team wagons during the late 1800s.

At this location, numerous *transverse dunes* are visible. Transverse dunes form long, undulating ridges perpendicular to the prevailing wind direction in areas that have abundant sand and little or no vegetation. In the white oval on the Virtual Field Trip screen, click on the windward part of the sand dune. Can you tell what the prevailing direction of wind is in this photo?

Weathering and Soils

Mechanical weathering is dominant in desert regions. Daily temperature fluctuations and frost wedging are the primary forms of mechanical weathering (see Chapter 6). The breakdown of rocks by roots and from salt crystal growth is of minor importance. Some chemical weathering does occur, but its rate is greatly reduced by aridity and the scarcity of organic acids produced by the sparse vegetation. Most chemical weathering takes place during the winter months, when there is more precipitation, particularly in the mid-latitude deserts.

Desert soils, if developed, are usually thin and patchy because the limited rainfall and the resultant scarcity of vegetation reduce the efficiency of chemical weathering and hence soil formation. Furthermore, the sparseness of the vegetative cover enhances wind and water erosion of what little soil actually forms.

Mass Wasting, Streams, and Groundwater

When traveling through a desert, most people are impressed by such wind-formed features as moving sand, sand dunes, and sand and dust storms. They may also notice the dry washes and dry streambeds. Because of the lack of running water, most people would conclude that wind is the most important erosional agent in deserts. They would be wrong! Running water, even though it occurs infrequently, causes most of the erosion in deserts. The dry conditions and sparse vegetation characteristic of deserts enhance water erosion.

Most of a desert's average annual rainfall of 25 cm or less comes in brief, heavy, localized cloudbursts. During these times, considerable erosion takes place, because the ground cannot absorb all of the rainwater. With so little vegetation to hinder the flow of water, runoff is rapid, especially on moderately to steeply sloping surfaces, resulting in flash floods and sheet flows. Dry stream channels quickly fill with raging torrents of muddy water and mudflows, which carve out steep-sided gullies and overflow their banks. During these times, a tremendous amount of sediment is rapidly transported and deposited far downstream.

Most desert streams are poorly integrated and flow only intermittently. Many of them never reach the sea because, as the water table is usually far deeper than the channels of most streams, they cannot draw upon groundwater to replace water lost to evaporation and absorption into the ground. This type of drainage, in which a stream's load is deposited within the desert, is called *internal drainage* and is common in most arid regions.

Although most deserts have internal drainage, some deserts have permanent through-flowing streams, such as the Nile and Niger rivers in Africa, the Rio Grande and Colorado rivers in the southwestern United States, and the Indus River in Asia. These streams can flow through desert regions because (1) their headwaters are well outside the desert and (2) water is plentiful enough to offset losses resulting from evaporation and infiltration.

Wind

Although running water does most of the erosional work in deserts, wind can also be an effective geologic agent capable of producing a variety of distinctive erosional (Figure 14.2) and depositional features (Figures 14.9 through 14.13). Wind is effective in transporting and depositing unconsolidated sand-, silt-, and dust-sized particles. Contrary to popular belief, most deserts are not sand-covered wastelands, but rather vast areas of rock exposures and desert pavement (Figure 14.3). Sand-covered regions, or sandy deserts, constitute less than 25 percent of the world's deserts. The sand in these areas has accumulated primarily by the action of wind.

LO7 DESERT LANDFORMS

Because of differences in temperature, precipitation, and wind, as well as the underlying rocks and recent tectonic events, landforms in arid regions vary considerably. Running water, although infrequent in deserts, is responsible for producing and modifying many distinctive landforms found there.

After an infrequent and particularly intense rainstorm, excess water not absorbed by the ground may accumulate in low areas and form *playa lakes* (Figure 14.17a). These lakes are temporary, lasting from a few hours to several months. Most of them are shallow and have rapidly shifting boundaries as water flows in or leaves by evaporation and seepage into the ground. The water is often very saline.

When a playa lake evaporates, the dry lake bed is called a **playa** or *salt pan* and is characterized by mud cracks and precipitated salt crystals (Figure 14.17b and the Virtual Field Trip in this chapter). Salts in some

Figure 14.17 Playas and Playa Lakes

a. A playa lake formed after a rainstorm near Badwater, Death Valley National Park, California. Playa lakes are ephemeral features, lasting from a few hours to several months.

b. Salt deposits and salt ridges cover the floor of this playa in the Mojave Desert, California. Salt crystals and mud cracks are characteristic features of playas.

playas are thick enough to be mined commercially. For example, borates have been mined in Death Valley, California, for more than 100 years.

Another common feature of deserts, particularly in the Basin and Range Province of the western United States, are alluvial fans. **Alluvial fans** form when sediment-laden streams flowing out from the generally straight, steep mountain fronts deposit their load on the relatively flat desert floor. Once beyond the mountain front, where no valley walls confine streams, the sediment spreads out laterally, forming a gently sloping and poorly sorted fan-shaped sedimentary

alluvial fan A cone-shaped accumulation of mostly sand and gravel deposited where a stream flows from a mountain valley onto an adjacent lowland.

Figure 14.18 Alluvial Fan

A ground view of an alluvial fan, Death Valley, California. Alluvial fans form when sediment-laden streams flowing out from a mountain deposit their load on the desert floor, forming a gently sloping, fan-shaped, sedimentary deposit.

pediment An erosion surface of low relief gently sloping away from the base of a mountain range.

mesa A broad, flat-topped erosional remnant bounded on all sides by steep slopes.

butte An isolated, steep-sided, pinnacle-like hill formed when resistant cap rock is breached, allowing erosion of less resistant underlying rocks.

deposit (Figure 14.18). Although alluvial fans are similar in origin and shape to deltas (see Chapter 11), they are formed entirely on land.

Most mountains in desert regions, including those of the Basin and Range Province, rise abruptly from gently sloping surfaces called **pediments**. Pediments are erosional bedrock surfaces of low relief that slope gently away from mountain bases (Figure 14.19). Most pediments are covered by a thin layer of debris, or alluvial fans.

Other easily recognized erosional remnants common to arid and semiarid regions are mesas and buttes (Figure 14.20). A **mesa** is a broad, flat-topped erosional remnant bounded on all sides by steep slopes. Continued weathering and stream erosion form isolated pillar-like structures known as **buttes**. Buttes and mesas consist of relatively easily weathered sedimentary rocks capped by nearly horizontal, resistant rocks such as sandstone, limestone, or basalt. They form when the resistant rock layer is breached, which allows rapid erosion of the less resistant underlying sediment.

Figure 14.19 Pediment

Pediments are erosional bedrock surfaces formed by erosion along a mountain front.

A pediment north of Mesquite, Nevada.

Figure 14.20 Buttes

Left Mitten Butte and Right Mitten Butte in Monument Valley on the border of Arizona and Utah.

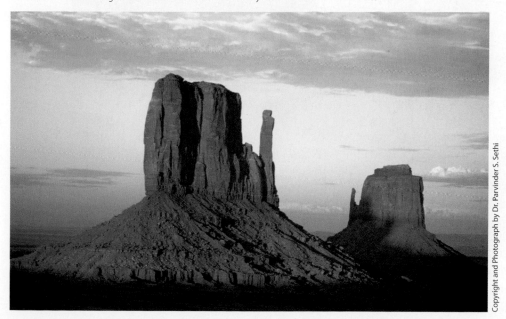

15 Oceans, Shorelines, and Shoreline Processes

Exposures of sandstone along the shoreline at Shore Acres State Park near Coos Bay, Oregon. This Eocene-age sandstone was deposited in a deltaic depositional environment. Notice that the sandstone layers dip at about 50 degrees, but the overlying layers (below the fence) are nearly horizontal, so this is a good example of an angular unconformity. During storms, waves impact the entire rock face, the top of which is more than 25 m above the sea.

Sue Monroe

Introduction

"Many centers of commerce and much of Earth's population are concentrated in a narrow band at or near shorelines."

In Chapter 2 we discussed several aspects of the seafloor such as oceanic ridges, deep-sea hydrothermal vents, and seamounts, as well as the nature of the continental margins. Our interest here is in the oceanic waters themselves and their impact on shorelines. You already know that the hydrosphere consists of all water on Earth, most of which (97.2%) is in the oceans. This vast interconnected body of saltwater covers 71% of Earth's surface, but parts of it are distinct enough for us to recognize the Pacific, Atlantic, Indian, and Arctic oceans (Figure 15.1). Seas, on the other hand, are marginal parts of oceans, as in the Sea of Japan and the Mediterranean Sea.

What lies beneath the ocean's surface is largely hidden from view, which accounts for why so many sensational stories have persisted for centuries. In about 350 B.C. the Greek Philosopher Plato claimed that a continent called Atlantis existed in the Atlantic Ocean west of the Strait of Gibraltar (Figure 15.2). Following the conquest of Atlantis by Athens, this vast continent supposedly sank, and now only "mud shallows" mark its former location. There is no geologic evidence that Atlantis ever existed, so why has this story persisted for so long? One reason is that stories of lost civilizations are popular, and another is that until fairly recently little was known about the deep seafloor.

In this enormous body of water we call oceans, wave energy is transferred through the water to shorelines, where it has a tremendous impact (see the chapter opening photo). Accordingly, understanding shoreline processes is important to oceanographers, geologists, and coastal engineers, as well as elected officials and city planners of coastal communities. Another important aspect of shorelines is rising sea level, because buildings and communities that were once inland are now in peril, must be protected or moved, or have already been destroyed. Furthermore, hurricanes expend much of their energy on shorelines, resulting in extensive coastal flooding, numerous fatalities, and widespread property damage.

We define a **shoreline** as the area of land in contact with the ocean or a lake, but we can expand this definition by noting that ocean shorelines include the land between low tide and the highest level on land affected by storm waves. Our main concern is with ocean shorelines, or seashores, but waves and nearshore currents are also effective in large lakes. However, even the largest lakes have insignificant tides.

The study of oceans and shorelines provides another excellent example of systems interactions—in this case, between the hydrosphere and the solid Earth. The atmosphere is also involved because energy is transferred from wind to water, thereby causing waves, which in turn generate nearshore currents. And, of course, the gravitational attraction of the Moon and Sun on the ocean waters is responsible for the rhythmic rise and fall of tides.

shoreline The area between mean low tide and the highest level on land affected by storm waves.

Learning Outcomes

After reading this unit, you should be able to do the following:

LO1 Understand the composition and circulation of seawater and describe the types of sediment on the seafloor.

LO2 Describe the behavior of tides, waves, and nearshore currents.

LO3 Understand the processes of shoreline erosion and deposition, and the concept of the nearshore sediment budget.

LO4 Identify the different types of coasts and their features.

LO5 Discuss the effects of storm waves and coastal flooding, and explain how coasts are managed as sea level rises.

LO6 Recognize the types of resources found in the oceans.

Figure 15.1 The Oceans

Map showing the Atlantic, Pacific, Indian, and Arctic oceans, and the ocean currents.

Warm water current
Cool water current

© 2013 Cengage Learning

LO1 SEAWATER, OCEANIC CIRCULATION, AND SEAFLOOR SEDIMENTS

salinity A measure of the dissolved solids in seawater, usually expressed in parts per thousand.

During its earliest history, Earth was probably hot, airless, and dry, but erupting volcanoes emitted several gases, the most abundant being water vapor, which accumulated in the atmosphere. As Earth cooled, the water vapor condensed, fell as rain, and began collecting on the surface. Oceans were present by at least 3.5 billion years ago, although their volumes and extent are unknown.

Seawater—Its Composition

Of the more than 70 chemical elements in solution in seawater, the most common ones are chloride ions (Cl^-) and sodium ions (Na^+), which together make up 85.6% of all dissolved substances and give seawater its most distinctive feature—its saltiness, or its **salinity**, a measure of the total quantity of dissolved solids. On average, 1 kg (1,000 g) of seawater contains 35 g of dissolved solids, or 35 parts per thousand, symbolized as $35\%_o$. In the open ocean, salinity varies from $32\%_o$ to $37\%_o$, although in some marginal seas, especially in dry, hot areas nearly isolated from the open ocean, values may exceed $40\%_o$.

Runoff from the continents, the source of most chemical elements in seawater, adds about 4 billion tons of dissolved solids to the oceans each year. Another source is *outgassing*, in which gases from within Earth are released into the oceans and the atmosphere by volcanoes at deep-sea hydrothermal vents (see Figure 2.7). In any case, the oceans have been salty for at least 1.5 billion years, but the salinity of seawater remains rather constant because continuous recycling of ions takes place; otherwise seawater would get saltier with time. Thus, seawater is in a state of dynamic equilibrium, meaning that additions are offset by losses. Ions are removed from seawater when evaporites such as rock salt (NaCl) and rock gypsum ($CaCO_4 \cdot 2H_2O$) are precipitated, when salt spray is blown onshore, when magnesium is used in dolomite and clay minerals, and when organisms use calcium or silica to construct their shells.

Figure 15.2 Atlantis

According to Plato, Atlantis was a continent west of the Pillars of Hercules, now called the Strait of Gibraltar. In this map from Anthanasium Kircher's *Mundus Subterraneous* (1664), north is toward the bottom of the map. The Strait of Gibraltar is the narrow area between Hispania (Spain) and Africa.

Based on the decreasing intensity of light with depth, scientists define two layers in the oceans: The upper layer, called the **photic zone**, is usually 100 m or less deep and receives enough light for organisms to photosynthesize. Below is the **aphotic zone**, where too little light is available for photosynthesis, and most organisms depend directly or indirectly on organic substances that "rain" down from the photic zone.

Oceanic Circulation

As wind blows over a water surface, some of its energy is transferred to the water, which generates surface currents and waves. Figure 15.1 shows the global surface-water current patterns averaged over a long time. Notice that surface currents in the Northern Hemisphere are deflected to the right (clockwise) of their direction of motion and to the left (counterclockwise) of their direction of motion in the Southern Hemisphere. This deflection is the **Coriolis effect** that results from Earth's rotation. The combination of wind and the Coriolis effect produces large-scale water circulation systems known as **gyres** between the 60 degree parallels in the Atlantic, Pacific, and Indian oceans (Figure 15.1). One of the best known ocean currents is the *Gulf Stream*, which is actually part of the much larger North Atlantic gyre.

Gyres are important as a world temperature control, because seawater near the equator absorbs huge amounts of heat and transports it to high latitudes in warm currents. Cold currents originate at high latitudes and flow toward the equator. Indeed, the reason that some rather northerly countries, such as Scotland, have mild climates is related to these currents. On the other hand, the current along the northern and central coasts of California is cold, which keeps that area much cooler than the inland parts of the state.

In addition to surface currents, horizontal circulation of water also takes place in the deep ocean basins because of differences in temperature and density of adjacent water masses; a water mass of greater density (colder or saltier) will displace and flow beneath a water mass of lesser density. Deep-ocean circulation affects about 90% of all ocean water, but because studying it is expensive and time-consuming, scientists know less about it than about other circulation patterns.

Vertical circulation occurs in the oceans when **upwelling** slowly transfers cold water from depth to the surface and when **downwelling** carries warm

Figure 15.3 The Evolution of Coral Reefs

The origin of a fringing reef, barrier reef, and atoll as the plate upon which the fringing reef is carried into deeper water.

Fringing reef   Barrier reef   Atoll

a.   b.   c.

© 2013 Cengage Learning

A fringing reef forms around an island in the tropics.

The island sinks as the oceanic plate on which it rides moves away from a spreading center. In this case, the island does not sink at a rate faster than coral organisms can build upward.

The island eventually disappears beneath the surface, but the coral remains at the surface as an atoll.

d. An underwater view of a coral reef in the Red Sea.

Courtesy of Carl Roessler

pelagic clay Brown or red deep-sea sediment composed of clay-sized particles.

ooze Deep-sea sediment composed mostly of shells of marine animals and plants.

reef A mound-like, wave-resistant structure composed of the skeletons of organisms.

water from the surface to depth. Upwelling is by far the most important process, because as it transfers water upward, it also carries nutrients, particularly nitrates and phosphate, into the photic zone. Here, high concentrations of plankton are sustained that in turn support other organisms. In fact, less than 1% of the ocean surface is in regions of upwelling, yet they support more than 50% by weight of all fishes. In addition, most of Earth's sedimentary rocks containing phosphate were deposited along continental margins where upwelling takes place.

Seafloor Sediments

Much of the sediment eroded from continents is deposited on the continental margins, but some, mostly silt-and clay-sized particles, is carried into the deep ocean basins. Most of this seafloor sediment is *pelagic*, meaning that it settled from suspension far from land. **Pelagic clay** is brown or red, and composed mostly of clay-sized particles, whereas **ooze** is made up of the tiny shells of marine organisms. If dominated by calcium carbonate ($CaCO_3$) skeletons it is *calcarious ooze*, or if mostly silica (SiO_2) skeletons it is *siliceous ooze*.

The term **reef** has many meanings, but here we are concerned with reefs defined as mound-like, wave-resistant structures composed of the shells of marine organisms (Figure 15.3). Although commonly called *coral reefs*, they actually have a solid framework of skeletons of corals and clams and encrusting organisms such as sponges and algae. Most reefs are found in tropical seas, where the water temperature does not fall below about 20°C, and rarely at depths of more than 50 m because many corals rely on symbiotic algae that must have enough sunlight for photosynthesis.

Most reefs are one of three basic types: fringing, barrier, and atoll. *Fringing reefs* are up to 1 km wide, solidly attached to a landmass, have a rough, table-like surface, and on their seaward side slope steeply down to the seafloor. *Barrier reefs* are similar except that a lagoon separates them from the mainland. The 2,000-km-long Great Barrier Reef of Australia is a good example. Circular to oval reefs surrounding a lagoon are *atolls*. They form around volcanic islands that subside below sea level as the plate they rest on moves into progressively deeper water. As subsidence occurs, the reef-building organisms grow upward so that the living part of the reef remains in shallow water (Figure 15.3).

LO2 SHORELINES AND SHORELINE PROCESSES

We have already defined an *ocean shoreline* (or *seashore*) as the land between low tide and the highest level on land affected by storm waves. How does a shoreline differ from a coast? Actually, the terms are commonly used interchangeably, but *coast* is a more inclusive term and encompasses the shoreline as well as an area of indefinite width seaward and landward of the shoreline (Figure 15.4). Our main concern here is with tides, nearshore currents, and waves, the processes that modify seashores.

In the marine realm, several biological, chemical, and physical processes are operating continuously. For example, organisms change the local chemistry of seawater and contribute their skeletons to nearshore sediments. However, the processes most important for modifying shorelines are purely physical ones, especially tides, waves, and nearshore currents.

Remember, a *shoreline* differs from a *coast*, because coast is a more inclusive term that includes the shoreline as well as an area of indefinite width both seaward and landward of the shoreline. For instance, in addition to the shoreline area, a coast includes nearshore sandbars and islands, and areas on land such as wind-blown sand dunes, marshes, and cliffs.

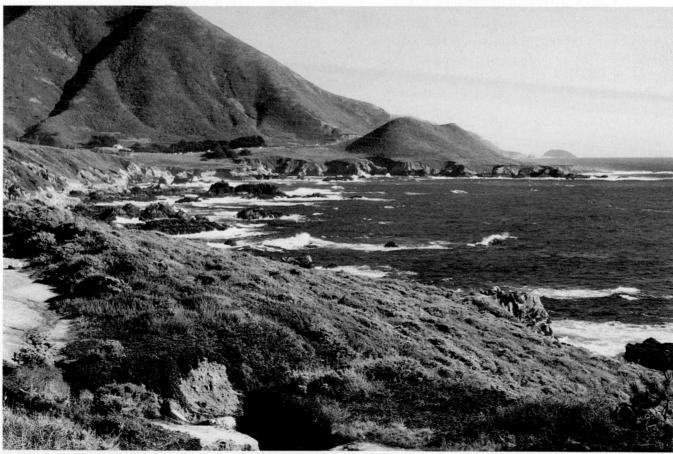

Figure 15.4 Shorelines and Coasts

The shoreline of this part of the U.S. Pacific Coast consists of the area from about where the waves break to the base of the sea cliffs. However, the coast extends farther seaward and also includes the sea cliffs and an area some distance inland.

Tides

The surface of the oceans rises and falls twice daily in response to the gravitational attraction of the Moon and Sun. These regular fluctuations in the ocean's surface, or **tides**, result in most seashores having two daily high tides and two low tides as sea level rises and falls from a few centimeters to more than 15 m (Figure 15.5). A complete tidal cycle includes a *flood tide* that progressively covers more and more of a nearshore area until high tide is reached, followed by an *ebb tide*, during which the nearshore area is once again exposed (Figure 15.5).

Both the Moon and the Sun have sufficient gravitational attraction to exert tide-generating forces strong enough to deform the solid body of Earth, but they have a much greater influence on the oceans. The Sun is 27 million times more massive than the Moon, but it is 390 times as far from Earth, and its tide-generating force is only 46% as strong as that of the Moon. Accordingly, the tides are dominated by the Moon, but the Sun plays an important role as well.

Jitloac/
Shutterstock.com

If we consider only the Moon acting on a spherical, water-covered Earth, its tide-generating forces produce two bulges on the ocean surface (Figure 15.6). One bulge points toward the Moon, because it is on the side of Earth where the Moon's gravitational attraction is greatest. The other bulge is on the opposite side of Earth; it points away from the Moon, because of centrifugal force due to Earth's rotation, and the Moon's gravitational attraction is less. These two bulges always point toward and away from the Moon (Figure 15.6a), so as Earth rotates and the Moon's position changes, an observer at a particular shoreline location experiences the rhythmic rise and fall of tides twice daily, but the heights of two successive high tides may vary depending on the Moon's inclination with respect to the equator.

The Moon revolves around Earth every 28 days, so its position with respect to any latitude changes slightly each day. That is, as the Moon moves in its orbit and Earth rotates on its axis, it takes the Moon 50 minutes longer each day to return to the same position it was in the previous day. Thus, an observer would experience a high tide at 1:00 P.M. on one day, for example, and at 1:50 P.M. on the following day.

When the Moon and Sun are aligned every two weeks, their forces added together generate *spring tides*, which are about 20% higher than average tides (Figure 15.6b). When the Moon and Sun are at right angles to each another, also at two-week intervals, the Sun's tide-generating force cancels some of the Moon's, and *neap tides*, which are about 20% lower than average, occur (Figure 15.6c).

James S. Monroe

a. Low tide.

Figure 15.5 Low and High Tides

Low tide (a) and high tide (b) in Turnagain Arm, part of Cook Inlet in Alaska. The tidal range here is about 10 m. Turnagain Arm is a huge fiord now being filled with sediment carried in by rivers. Notice the mudflats in (a).

b. High tide.

James S. Monroe

Figure 15.6 Tidal Bulges

The gravitational attraction of the Moon and Sun causes tides. The sizes of the tidal bulges are greatly exaggerated.

a. Tidal bulges if only the Moon caused them.

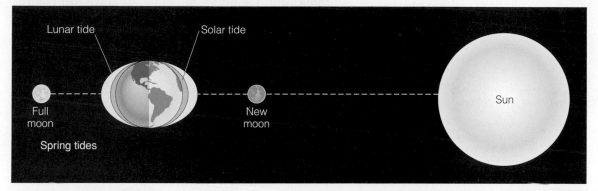

b. When the Moon is new or full, the solar and lunar tides reinforce one another, causing spring tides, the highest high tides and lowest low tides.

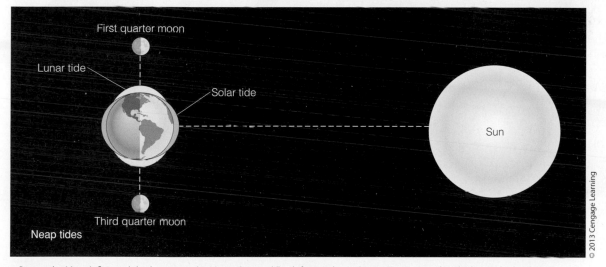

c. During the Moon's first and third quarters, the Moon, Sun, and Earth form right angles, causing neap tides, the lowest high tides and highest low tides.

© 2013 Cengage Learning

Tidal ranges are also affected by shoreline configuration. Broad, gently sloping continental shelves as in the Gulf of Mexico have low tidal ranges, whereas steep, irregular shorelines experience much greater rise and fall of tides. Tidal ranges are greatest in some narrow, funnel-shaped bays and inlets. The Bay of Fundy in Nova Scotia has a tidal range of 16.5 m, and ranges greater than 10 m occur in several other areas.

Tides have an important impact on shorelines, because the area of wave attack constantly shifts onshore and offshore as the tides rise and fall. Tidal currents themselves, however, have little modifying effect on shorelines, except in narrow passages where tidal current velocity is great enough to erode and transport sediment.

Waves

You can see **waves**, or oscillations of a water surface, on all bodies of water, but they are best developed in the oceans. In fact, waves are directly or indirectly responsible for most erosion, sediment transport,

> **wave** An undulation on the surface of a body of water, resulting in the water surface rising and falling.

and deposition along shorelines. Wave terminology is illustrated with a typical series of waves in Figure 15.7a. A *crest*, as you would expect, is the highest part of a wave, whereas the low area between crests is a *trough*. The distance from crest to crest (or trough to trough) is the *wavelength*, and the vertical distance from trough to crest is *wave height*. You can calculate the speed at which a wave advances, called *celerity* (C), by the formula

$$C = L/T$$

where L is wavelength and T is wave period—that is, the time it takes for two successive wave crests, or troughs, to pass a given point.

The speed of wave advance (C) is actually a measure of the velocity of the wave form rather than the speed of the molecules of water in a wave. When waves move across a water surface, the water moves in circular orbits but shows little or no net forward movement (Figure 15.7a). Only the wave form moves forward, and as it does, it transfers energy in the direction of wave movement.

Figure 15.7 Waves and Wave Terminology

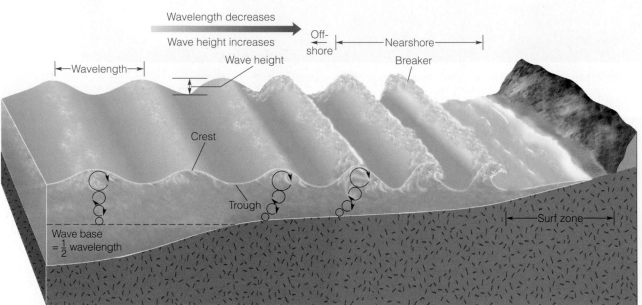

a. Waves and the terminology applied to them. Note that swells are disrupted when they encounter water shallower than wave base, and they eventually form breakers.

b. A plunging breaker on the north shore of Oahu, Hawaii.

c. A spilling breaker.

The diameters of the orbits that water follows in waves diminish rapidly with depth, and at a depth of about one-half wavelength (*L/2*), called **wave base**, they are essentially zero. Thus, at a depth exceeding wave base, the water and seafloor or lake floor are unaffected by surface waves (Figure 15.7a).

Wave Generation

Most geologic work on shorelines is accomplished by wind-generated waves, especially storm waves. When wind blows over water—that is, one fluid (air) moves over another fluid (water)—friction between the two transfers energy to the water, causing the water surface to oscillate.

In areas where waves are generated, such as beneath a storm center at sea, sharp-crested, irregular waves called *seas* develop. Seas have various heights and lengths, and one wave cannot be easily distinguished from another. But as seas move out from their area of generation, they are sorted into broad *swells* with rounded, long crests, and all are about the same size (Figure 15.7a).

The harder and longer the wind blows, the larger are the waves, but these are not the only factors that control wave size. High-velocity wind blowing over a small pond will never generate large waves regardless of how long it blows. In fact, waves on ponds and most lakes appear only while the wind is blowing. In contrast, the surface of the ocean is always in motion, and waves with heights of 34 m have been recorded during storms in the open sea.

The reason for the disparity between wave sizes on ponds and lakes and on the oceans is the **fetch**, which is the distance the wind blows over a continuous water surface. So, on ponds and lakes, fetch corresponds to their length or width, depending on wind direction. To produce waves of greater length and height, more energy must be transferred from wind to water; hence large waves form beneath large storms at sea.

Shallow-Water Waves and Breakers

Swells moving out from an area of wave generation lose little energy as they travel long distances across the ocean. In these deepwater swells, the water surface oscillates and water moves in circular orbits, but little net displacement of water takes place in the direction of wave travel (Figure 15.7a). Of course, wind blows some water from wave crests, thus forming whitecaps with foamy white crests, and surface currents transport water great distances, but deepwater waves accomplish little actual water movement. When these waves enter progressively shallower water, however, the wave shape changes, and water is displaced in the direction of wave advance.

wave base The depth corresponding to about one-half wavelength, below which water is unaffected by surface waves.

fetch The distance the wind blows over a continuous water surface.

Under some circumstances, two wave crests merge to form rogue waves that are three or four times higher than the average. These waves can rise unexpectedly out of an otherwise comparatively calm sea and threaten even the largest ships. During the last two decades, more than 200 supertankers and container ships have been lost at sea, many of them apparently hit by these huge waves. As recently as April 16, 2005, the Norwegian cruise ship Dawn was damaged by a 21-m-high rogue wave that flooded more than 60 cabins and injured four passengers.

Elena Elisseeva/Shutterstock.com

breaker A wave that steepens as it enters shallow water until its crest plunges forward.

wave refraction The bending of waves so that they move nearly parallel to the shoreline.

longshore current A current resulting from wave refraction found between the breaker zone and a beach that flows parallel to the shoreline.

Broad, undulating deepwater waves are transformed into sharp-crested waves as they enter shallow water. This transformation begins at a water depth corresponding to wave base—that is, one-half wavelength (Figure 15.7a). At this point, the waves "feel" the seafloor, and the orbital motion of water within the waves is disrupted. As waves continue moving shoreward, the speed of wave advance and wavelength decrease, but wave height increases. Thus, as they enter shallow water, waves become oversteepened as the wave crest advances faster than the wave form, and eventually the crest plunges forward as a **breaker** (Figure 15.7b). Breaking waves might be several times higher than their deepwater counterparts, and when they break, they expend their kinetic energy on the shoreline.

The waves just described are the classic *plunging breakers* (Figure 15.7b) that crash onto shorelines with steep offshore slopes, such as those on the north shore of Oahu in the Hawaiian Islands. In contrast, shorelines where the offshore slope is more gentle usually have *spilling breakers*, where the waves build up slowly and the wave's crest spills down the wave front (Figure 15.7c).

Nearshore Currents

The area extending seaward from the upper limit of the shoreline to just beyond the area of breaking waves is conveniently designated as the *nearshore zone*. Within the nearshore zone are the breaker zone and a surf zone, where water from breaking waves rushes forward and then flows seaward as backwash. The nearshore zone's width varies, depending on the length of approaching waves, because long waves break at a greater depth, and thus farther offshore, than do short waves. Incoming waves are responsible for two types of currents in the nearshore zone: *longshore currents* and *rip currents*.

Wave Refraction and Longshore Currents

Deepwater waves have long, continuous crests, but rarely are their crests parallel with the shoreline (Figure 15.8). In other words, they seldom approach a shoreline head-on, but rather at some angle. Thus, one part of a wave enters shallow water, where it encounters wave base and begins breaking before other parts of the same wave. As a wave begins to break, its velocity diminishes, but the part of the wave still in deep water races ahead until it too encounters wave base. The net effect of this oblique approach is that waves bend so that they more nearly parallel the shoreline, a phenomenon known as **wave refraction** (Figure 15.8).

Even though waves are refracted, they still usually strike the shoreline at some angle, causing the water between the breaker zone and the beach to flow parallel to the shoreline. These **longshore currents**, as they are called, are long and narrow and flow in the same general direction as the approaching waves. These currents are particularly important in transporting and depositing sediment in the nearshore zone.

Figure 15.8 Wave Refraction
Waves are refracted as they enter shallow water, but even so they usually come in at an angle to the shoreline.

Rip Currents Waves carry water into the nearshore zone, so there must be a mechanism for mass transfer of water back out to sea. One way in which water moves seaward from the nearshore zone is in **rip currents**, narrow surface currents that flow out to sea through the breaker zone (Figure 15.9). Surfers commonly take advantage of rip currents for an easy ride out beyond the breaker zone, but these currents pose a danger to inexperienced swimmers. Some rip currents flow at several kilometers per hour, so if a swimmer is caught in one, it is useless to try to swim directly back to shore. Instead, because rip currents are narrow and usually nearly perpendicular to the shore, one can swim parallel to the shoreline for a short distance and then turn shoreward with little difficulty.

Rip currents are circulating cells fed by longshore currents that increase in velocity from midway between each rip current (Figure 15.9). When waves approach a shoreline, the amount of water builds up until the excess moves out to sea through the breaker zone.

Rip currents commonly develop where wave heights are lower than in adjacent areas, and differences in wave height are controlled by variations in water depth. For instance, if waves move over a depression, the height of the waves over the depression tends to be less than in adjacent areas, forming the ideal environment for rip currents.

LO3 EROSION AND DEPOSITION ALONG SHORELINES

Erosion and deposition by waves and nearshore currents account for many interesting seashore features, such as sea stacks, arches, beaches, and spits, all of which are easy to recognize.

Figure 15.9 Rip Currents

a. Rip currents are fed on each side by currents moving parallel to the shoreline.

b. Suspended sediment, indicated by discolored water, is being carried seaward in these rip currents.

wave-cut platform A beveled surface that slopes gently seaward; formed by the erosion and retreat of a sea cliff.

marine terrace A wave-cut platform now above sea level.

headland Part of a shoreline commonly bounded by cliffs that extends out into the sea or a lake.

Erosion and Wave-Cut Platforms

On many seashores, erosion creates steep or vertical slopes known as *sea cliffs*. During storms, sea cliffs are pounded by waves (hydraulic action), worn by the impact of sand and gravel (abrasion) (Figure 15.10), and eroded by dissolution involving the solvent action of seawater.

Tremendous energy from waves is concentrated on the lower parts of sea cliffs and has the most effect on those composed of sediment or highly fractured rocks. In any case, the net effect is erosion of the sea cliff and the landward migration of the cliff face.

As sea cliffs are undercut by hydraulic action and abrasion at their bases, their upper parts are left unsupported and susceptible to mass wasting processes. Thus, sea cliffs retreat landward little by little, and as they do so, beveled surfaces called **wave-cut platforms** develop that slope gently seaward (Figure 15.11a). Broad wave-cut platforms are common in many areas, but invariably the water over them is shallow, because the abrasive planing action of waves is effective to a depth of only about 10 m (Figure 15.11b). The sediment eroded from sea cliffs is transported seaward and deposited as a *wave-built platform*, which is a seaward extension of the wave-cut platform (Figure 15.11a). If a wave cut platform is uplifted above sea level it is called a **marine terrace**.

Sea Caves, Arches, and Stacks

Sea cliffs do not retreat uniformly, because some shoreline materials are more resistant to erosion than others. **Headlands** are seaward-projecting parts of the shoreline that are eroded on both sides by wave refraction (Figure 15.12). *Sea caves* form on opposite sides of a headland, and if these caves join, they form a *sea arch*

Figure 15.10 Wave Erosion by Abrasion and Hydraulic Action

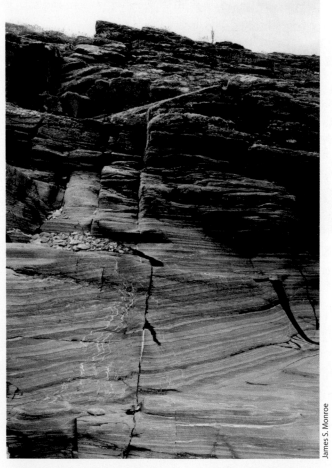

a. The rocks in the lower part of this image on a small island in the Irish Sea have been smoothed by abrasion, but the rocks higher up are out of the reach of waves.

b. Hydraulic action and abrasion have undercut these sea cliffs near Bodega Bay, California. Erosion was particularly intense during storms of February 1998. Attempts to stabilize the shoreline have failed and the homes are now gone, and others are now threatened.

Figure 15.11 Origin of a Wave-Cut Platform

a. Wave erosion causes a sea cliff to migrate landward, leaving a gently sloping surface, called a wave-cut platform. A wave-built platform originates by deposition at the seaward margin of the wave-cut platform.

b. This gently sloping surface along the coast of Oregon is a wave-cut platform. Notice the sea stacks rising above the platform.

(Figure 15.12b). Continued erosion causes the span of an arch to collapse, creating isolated *sea stacks* on wave-cut platforms (Figure 15.12c).

In the long run, shoreline processes tend to straighten an initially irregular shoreline. Wave refraction causes more wave energy to be expended on headlands and less on embayments. Thus, headlands erode, and some of the sediment formed by erosion is deposited in the embayments.

We mentioned that longshore currents are effective at transporting sediment. In fact, we can think of the area from the breaker zone to the upper limit of the surf zone as a "river" that flows along the shoreline. Unlike rivers on land, though, this shoreline river's direction of flow changes if waves approach from a different direction. Nevertheless, the analogy is apt, and just like rivers on land, a longshore current's capacity for transport varies with flow velocity and water depth.

Wave refraction and the resulting longshore currents are the primary agents of sediment transport and deposition on shorelines, but tides also play a role, because, as they rise and fall, the position of wave attack shifts onshore and offshore. Rip currents play no role in shoreline deposition, but they do transport fine-grained sediment (silt and clay) offshore through the breaker zone.

Beaches

By definition, a **beach** is a deposit of unconsolidated sediment extending landward from low tide to a change in topography, such as a line of sand dunes, a sea cliff, or the point where permanent vegetation begins. Typically, a beach has several component parts (Figure 15.13a), including a backshore that is usually dry, being covered by water only during storms or exceptionally high tides. The *backshore* consists of one or more *berms*, platforms composed of sediment deposited by waves; the berms are either nearly horizontal or slope gently landward. The sloping area below a berm exposed to wave swash is the beach face. The *beach face* is part of the *foreshore*, an area covered by water during high tide but exposed during low tide.

Depending on shoreline materials and wave intensity, beaches may be discontinuous, existing as only *pocket beaches* in protected areas such as embayments, or they may be continuous for long distances (Figure 15.13b, c). Some of the sediment on beaches is derived from weathering and wave erosion of the shoreline, but most of it is transported to the coast by streams and redistributed by

beach Any deposit of sediment extending landward from low tide to a change in topography or where permanent vegetation begins.

Figure 15.12 Erosion of a Headland

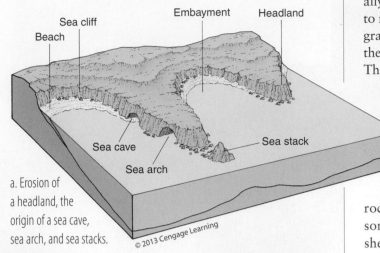

a. Erosion of a headland, the origin of a sea cave, sea arch, and sea stacks.

© 2013 Cengage Learning

b. A sea arch at Samuel H. Boardman State Park, Oregon

c. Sea stacks at Ruby Beach in Olympic National Park, Washington.

longshore currents. As we previously noted, waves usually strike beaches at some angle, causing the sand grains to move up the beach face at a similar angle; as the sand grains are carried seaward in the backwash, however, they move perpendicular to the long axis of the beach. Thus, individual sand grains move in a zigzag pattern in the direction of longshore currents. This movement is not restricted to the beach; it extends seaward to the outer edge of the breaker zone.

Quartz is the most common mineral in most beach sands, but there are some notable exceptions. For example, the black sand beaches of Hawaii are composed of basalt rock fragments or small grains of volcanic glass, and some Florida beaches are composed of the fragmented shells of marine organisms. In short, beaches are composed of whatever material is available.

In an attempt to widen a beach or prevent erosion, shoreline residents often build *groins*, structures that project seaward at right angles from the shoreline. Groins interrupt the flow of longshore currents, causing sand deposition on their upcurrent sides and widening of the beach at that location. However, erosion inevitably occurs on the downcurrent side of a groin.

Seasonal Changes in Beaches

The loose grains on beaches are constantly moved by waves, but the overall configuration of a beach remains unchanged as long as equilibrium conditions persist. We can think of the beach profile consisting of a berm or berms and a beach face, as a profile of equilibrium; that is, all parts of the beach are adjusted to the prevailing conditions of wave intensity, nearshore currents, and materials composing the beach (Figure 15.13).

Tides and longshore currents affect the configuration of beaches to some degree, but storm waves are by far the most important agent modifying their equilibrium profile. In many areas, beach profiles change with the seasons, so we recognize *summer beaches* and *winter beaches*, each of which is adjusted to the conditions prevailing at those times. Summer beaches are sand-covered and have a wide berm, a gently sloping beach face, and a smooth offshore profile. Winter beaches, in contrast, tend to be coarser grained and steeper; they have a small berm or none at all, and their offshore profiles reveal sandbars paralleling the shoreline.

Seasonal changes in beach profiles are related to changing wave intensity. During the winter, energetic

Figure 15.13 Beaches

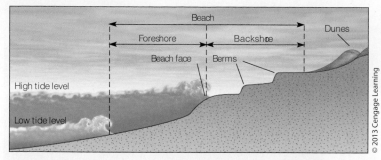

a. Diagram of a beach showing its component parts..

© 2013 Cengage Learning

b. The Grand Strand of South Carolina, shown here at Myrtle Beach, is 100 km of nearly continuous beach.

Michael Slear

c. A pocket beach near Seal Rock, Oregon.

James S. Monroe

storm waves erode the sand from beaches and transport it offshore, where it is stored in sandbars. The same sand that was eroded from a beach during the winter returns the next summer when it is driven onshore by more gentle swells. The volume of sand in the system remains more or less constant; it simply moves farther offshore or onshore depending on wave energy.

The terms *winter beach* and *summer beach*, though widely used, are somewhat misleading. A winter beach profile can develop at any time if there is a large storm, and likewise a summer beach profile can develop during a prolonged winter calm period.

Spits, Baymouth Bars, and Tombolos

Beaches are the most familiar depositional features of coasts, but spits, baymouth bars, and tombolos are also common. In fact, these features are simply continuations of a beach. A **spit**, for instance, is a fingerlike projection of a beach into a body of water such as a bay, and a **baymouth bar** is a spit that has grown until it completely closes off a bay from the open sea (Figure 15.14). Both are composed of sand, more rarely gravel, that was transported and deposited by longshore currents at the point where they weakened as they entered the deeper water of a bay's opening. Some spits are modified by waves so that their free ends are curved; they go by the name *hook* or *recurved spit* (Figure 15.14a).

> **spit** A fingerlike projection of a beach into a body of water such as a bay.
>
> **baymouth bar** A spit that has grown until it closes off a bay from the open sea or lake.

Figure 15.14 Spits, Baymouth Bars, and Tombolos

a. Spits form where longshore currents deposit sand in deeper water, as at the entrance to a bay. A baymouth bar is simply a spit that has grown until it extends across the mouth of a bay.

b. A spit at the mouth of the Russian River near Jenner, California.

c. Rodeo Beach north of San Francisco, California, is a baymouth bar.

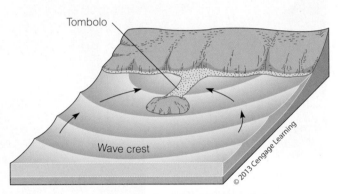

d. Wave refraction around an island and the origin of a tombolo.

tombolo A type of spit that extends out from the shoreline and connects the mainland with an island.

barrier island A long, narrow island of sand parallel to a shoreline but separated from the mainland by a lagoon.

A **tombolo** is a type of spit that extends out from the shoreline to an island (Figure 15.14d). A tombolo forms on the shoreward side of an island as wave refraction around the island creates converging currents that turn seaward and deposit a sandbar.

Spits and baymouth bars constitute a continuing problem where bays must be kept open for pleasure boating, commercial shipping, or both. Obviously, a bay closed off by a sandbar is of little use for either endeavor, so a bay must be regularly dredged or protected from deposition by longshore currents.

Barrier Islands

Long, narrow islands of sand lying a short distance offshore from the mainland are **barrier islands** (Figure 15.15). Everyone agrees that barrier islands form on gently sloping continental shelves where abundant sand is available and where both wave energy and the tidal range are low. For these reasons, many barrier islands are located along the United States' Atlantic and Gulf Coasts. But even though it is well known where barrier islands form, the details of their origin are still unresolved. According to one model, they formed as spits that became detached from land, whereas another model holds that they formed as beach ridges that subsequently subsided.

Most barrier islands are migrating landward as a result of erosion on their seaward sides and deposition on their landward sides. This is a natural part of barrier

Figure 15.15 Barrier Islands

View from space of the barrier islands along the Gulf Coast of Texas. Notice that a lagoon up to 20 km wide separates the long, narrow barrier islands from the mainland.

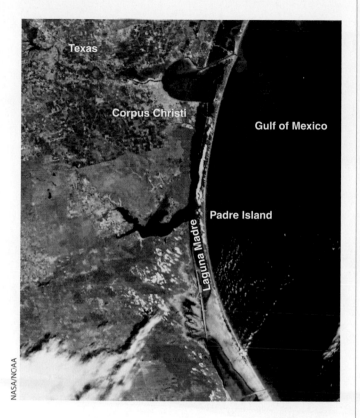

NASA/NOAA

island evolution, and it takes place rather slowly, but it takes place fast enough to cause many problems for island residents and communities.

The Nearshore Sediment Budget

We can think of the gains and losses of sediment in the nearshore zone in terms of a **nearshore sediment budget** (Figure 15.16). If a nearshore system has a balanced budget, sediment is supplied as fast as it is removed, and the volume of sediment remains more or less constant, although sand may shift offshore and onshore with the changing seasons. A *positive budget* means that gains exceed losses, whereas a *negative budget* means that losses exceed gains. If a negative budget prevails long enough, a nearshore system is depleted, and beaches may disappear.

Although there are a few exceptions, most sediment on beaches is transported to the shoreline by streams and then redistributed along the shoreline by longshore currents. Thus, longshore currents play a role in the nearshore sediment budget, because they continuously move sediment into and away from beach systems.

nearshore sediment budget The balance between additions and losses of sediment in the nearshore zone.

The primary ways that a nearshore system loses sediment are offshore transport, wind, and deposition in submarine canyons. Offshore transport mostly involves fine-grained sediment carried seaward, where it eventually settles in deeper water. Wind is an important process, because it removes sand from beaches and blows it inland, where it piles up as sand dunes.

If the heads of submarine canyons are near the shore, huge quantities of sand are funneled into them and deposited in deeper water. La Jolla and Scripps submarine canyons off the coast of southern California funnel an estimated 2 million m^3 of sand each year. In most areas, however, submarine canyons are too far offshore to interrupt the flow of sand in the nearshore zone.

It should be apparent from this discussion that if a nearshore system is in equilibrium, its incoming supply of sediment exactly offsets its losses. Such a delicate balance tends to continue unless the system is somehow disrupted. One change that affects this balance is the construction of dams across the streams that supply sand. Once dams have been built, all sediment from the upper reaches of the drainage systems is trapped in reservoirs and thus cannot reach the shoreline.

LO4 TYPES OF COASTS

Coasts are difficult to classify because of variations in the factors that control their development and variations in their composition and configuration. Rather than attempt to categorize all coasts, we shall simply note that two types of coasts have already been discussed: those dominated by deposition and those dominated by erosion. In addition, we will examine coasts in terms of their changing relationship to sea level. But note that although some coasts, such as those in southern California, are described as emergent (uplifted), these same coasts may be erosional as well. In other words, coasts commonly possess features that allow them to be classified in more than one way.

INPUT V⁺ Sediment added from erosion by longshore transport onto beach
OUTPUTS V⁻ Sediment carried down-coast from the beach by longshore transport
W⁻ Sediment blown inland by wind
O⁻ Sediment cascading down the submarine slope
STABLE BEACH: (V⁺) + (V⁻ + W⁻ + O⁻) = 0

River

Inland dunes

W⁻

Beach

V⁺

V⁻

O⁻

Submarine landslide

© 2013 Cengage Learning

Figure 15.16
The Nearshore Sediment Budget
The long-term sediment budget can be assessed by considering inputs versus outputs. If inputs and outputs are equal, the system is in a steady state, or equilibrium. If outputs exceed inputs, however, the beach has a negative budget and erosion occurs. Accretion takes place when the beach has a positive budget, with inputs exceeding outputs.

submergent coast A coast along which sea level rises with respect to the land or the land subsides.

emergent coast A coast where the land has risen with respect to sea level.

Depositional and Erosional Coasts

Depositional coasts, such as much of the Atlantic and Gulf coasts, are characterized by an abundance of detrital sediment and such depositional landforms as wide sandy beaches, deltas, and barrier islands (Figure 15.13b). In contrast, erosional coasts are steep and irregular and typically lack well-developed beaches, except in protected areas. They are further characterized by sea cliffs, wave-cut platforms, and sea stacks. Many of the coasts along the West Coast of North America fall into this category (Figure 15.13c).

Submergent and Emergent Coasts

If sea level rises with respect to the land or the land subsides, coastal regions are flooded and said to be **submergent coasts** or *drowned coasts* (Figure 15.17a). Much of the East Coast of North America from Maine southward through South Carolina was flooded during the rise in sea level following the Pleistocene Epoch, so it is extremely irregular. Recall that during the expansion of glaciers during the Pleistocene, sea level was as much as 130 m lower than at present, and that streams eroded their valleys more deeply and extended across continental shelves. When sea level rose, the lower ends of these valleys were drowned, forming *estuaries* such as Delaware and Chesapeake bays (Figure 15.17a). Estuaries are simply the seaward ends of river valleys where seawater and freshwater mix.

Emergent coasts are found where the land has risen with respect to sea level (Figure 15.17b). Emergence

Figure 15.17 Submergent and Emergent Coasts

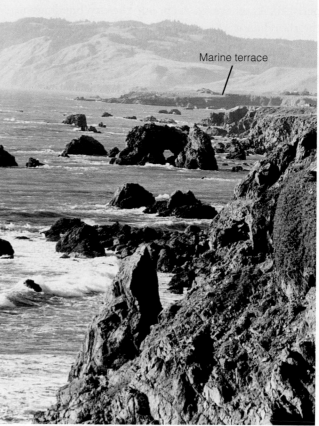

a. Submergent coasts tend to be extremely irregular, with estuaries such as Chesapeake and Delaware bays. They formed when the East Coast of the United States was flooded as sea level rose following the Pleistocene Epoch.

b. Emergent coasts tend to be steep and straighter than submergent coasts. Notice the several sea stacks and the sea arch. Also, a marine terrace is visible in the distance.

takes place when water is withdrawn from the oceans, as occurred during the Pleistocene expansion of glaciers. Presently, coasts are emerging as a result of isostasy or tectonism. In northeastern Canada and the Scandinavian countries, for instance, the coasts are irregular, because isostatic rebound is elevating formerly glaciated terrain from beneath the sea.

LO5 THE PERILS OF LIVING ALONG A SHORELINE

No matter where we live there are risks. Great Plains residents worry about violent storms, whereas those living near active faults or active volcanoes must be aware of these potential dangers. Likewise, living along a shoreline poses some risk, one of the most obvious being storm waves and coastal flooding, especially during hurricanes. Strong wind is dangerous too, but most fatalities during hurricanes result from coastal flooding.

Storm Waves and Coastal Flooding

Coastal flooding results when large waves are driven onshore by hurricanes and by heavy rainfall, as much as 60 cm in as little as 24 hours. In addition, as a hurricane moves over the ocean, low atmospheric pressure causes the ocean's surface to bulge upward as much as 0.5 m. When the eye of the storm reaches the shoreline, the bulge, coupled with wind-driven waves, piles up in a **storm surge** that may rise several meters above normal high tide and flood areas far inland.

In 1900, hurricane-driven waves surged inland, eventually covered the entire island that Galveston, Texas,

storm surge The surge of water onto a shoreline as a result of a bulge in the ocean's surface beneath the eye of a hurricane and wind-driven waves.

was on, and, as a result, between 6,000 and 8,000 people perished. As the waves swept into the city, buildings near the shoreline were battered to pieces and "great beams and railway ties were lifted by the [waves] and driven like battering rams into dwellings and business houses."[*] In an effort to protect the city from future storms, a huge seawall was constructed, and the entire city was elevated to the level of the top of the seawall (Figure 15.18).

The seawall has largely protected Galveston from more recent storm surges, but the same is not true for several other areas. Large-scale coastal flooding took place when Hurricane Isabel hit the Outer Banks of North Carolina in 2003. In 2004, Florida was hit by four hurricanes, causing widespread wind damage and coastal flooding. The U. S. Gulf Coast was hardest hit in 2005, first in August by Hurricane Katrina and then again in September by Hurricane Rita. When Hurricane Katrina roared ashore on August 29, 2005, high winds, a huge storm surge, and coastal flooding destroyed nearly everything in an area of 230,000 km². Gulfport and Biloxi in Mississippi were mostly leveled, but most of the public's attention focused on New Orleans, Louisiana (Figure 15.19).

When Hurricane Katrina came ashore, the levees initially held, but on the next day, some of the floodwalls were breached, and about 80% of the city was flooded. Because New Orleans is mostly below sea level, the floodwaters could not drain out naturally. In fact, the city has 22 pumping stations to remove water that builds up from normal rainstorms, but as the city flooded, the pumps were overwhelmed, and when the electricity failed, they were useless. All in all, Hurricane Katrina was the most expensive natural disaster in U.S. history; the property damages exceeded $100 billion, and more than 1,800 people died, mostly in Louisiana. Then, in September, Hurricane Rita hit the Gulf Coast, mostly in Texas, and caused about 120 more fatalities and nearly $12 billion in damages.

How Are Coastal Areas Managed as Sea Level Rises?

During the last century, sea level rose about 12 cm worldwide, and all indications are that it will continue to rise. The absolute rate of sea-level rise in a shoreline region depends on two factors. The first is the volume of water in the ocean basins, which is increasing as a result of glacial ice melting and the thermal expansion of near-surface seawater.

The second factor is the rate of uplift, or subsidence of a coastal area. In some areas, uplift is occurring fast enough that sea level is actually falling with respect to the land. In other areas, sea level is rising while the coastal region is simultaneously subsiding, resulting in a net change in sea level of as much as 30 cm per century. Perhaps such a "slow" rate of sea-level change seems insignificant; after all, it amounts to only a few millimeters per year. But in gently sloping coastal areas, as in the eastern United States from New Jersey southward, even a slight rise in sea level will eventually have widespread effects.

Many of the nearly 300 barrier islands along the East and Gulf Coasts of the United States are migrating landward as sea level rises (Figure 15.20). Landward migration of barrier islands would pose few problems if it were not for the numerous communities, resorts, and vacation homes located on them. Moreover, barrier islands are not the only threatened areas. For example, Louisiana's coastal wetlands—an important wildlife habitat and seafood-producing area—are currently being lost at a rate of about 90 km² per year. Much

[*] L. W. Bates, Jr., "Galveston—A City Built upon Sand," *Scientific American*, 95(1906), p. 64.

Figure 15.18 Seawall Construction

Construction on this seawall to protect Galveston, Texas, from storm waves began in 1902. Notice that the wall is curved to deflect waves upward.

Rosenberg Library, Galveston, Texas – Photo by Maurer

Figure 15.19 Hurricane Katrina, 2005

a. Although many areas were hit hard by Katrina, New Orleans was extensively flooded when the floodwalls built to protect the city from Lake Pontchartrain and the Mississippi River failed.

b. Destruction caused by the storm surge from Hurricane Katrina. This image shows a man in Biloxi, Mississippi, trying to find his house.

of this loss results from sediment compaction, but rising sea level exacerbates the problem.

Rising sea level also directly threatens many beaches upon which communities depend for revenue. The beach at Miami Beach, Florida, for instance, was disappearing at an alarming rate until the Army Corps of Engineers began replacing the eroded beach sand. The problem is even more serious in other countries. A rise in sea level of only 2 m would inundate large areas of the U.S. East and Gulf Coasts, but it would cover 20% of the entire country of Bangladesh. Other problems associated with rising sea level include increased coastal flooding during storms and saltwater incursions that may threaten groundwater supplies.

Armoring shorelines with *seawalls* and using *riprap* (piles of stones, Figure 15.21) help protect beachfront

Figure 15.20 Barrier Island Migration

a. A barrier island.

b. A barrier island migrates landward as sea level rises and storm waves carry sand from its seaward side into its lagoon.

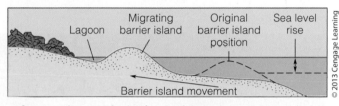

c. Over time, the entire island shifts toward the land.

structures, but both projects are initially expensive and during large storms are commonly damaged or destroyed. Seawalls do afford some protection and are seen in many coastal areas along the oceans and large lakes, but some states, including North and South Carolina, Rhode Island, Oregon, and Maine, no longer allow their construction.

Because we can do nothing to prevent sea level from rising, engineers, scientists, planners, and political leaders must examine what they can do to prevent, or minimize, the effects of shoreline erosion. At present, only a few viable options exist. One is to put strict controls on coastal development. North Carolina, for example, permits large structures to be built no closer to the shoreline than 60 times the annual erosion rate. Although a growing awareness of shoreline processes has resulted in similar legislation elsewhere, some states have virtually no restrictions on coastal development.

LO6 RESOURCES FROM THE OCEANS

Seawater contains many elements in solution, the most common being sodium (Na) and chlorine (Cl), which are extracted for salt by evaporation or by mining salt deposits. A large part of the world's magnesium comes

Figure 15.21 Riprap and Sand Berm
Riprap made up of large pieces of basalt was piled on this beach to protect a luxury hotel just to the left of this image.

Sue Monroe

from seawater, and several other elements and compounds are extracted from seawater.

The United States, by presidential proclamation issued in 1983, claims sovereign rights over an area designated the **Exclusive Economic Zone (EEZ)**, which extends seaward 200 nautical miles (371 km). Many other nations make similar claims. Resources within the U.S. EEZ include sand and gravel for construction, and about 34% of all domestic oil production comes from wells on the continental shelf. The *Deepwater Horizon* drill rig that sank in the Gulf of Mexico about 66 km off the Louisiana coast was a semi-submersible offshore rig that was equipped to operate in water 2,100 m deep. Unfortunately, on April 20, 2010, as it was drilling in 1,524 m of water, it exploded, killing 11 workers, beginning the largest oil spill in the history of the petroleum industry (Figure 15.22).

U.S. Coast Guard photo

Figure 15.22 *Deepwater Horizon* Oil Spill
The *Deepwater Horizon* drilling platform not long before it sank in the Gulf of Mexico on April 22, 2010.

Other resources within the EEZ include spherical objects called manganese nodules, which are composed of manganese and iron oxides, as well as copper, nickel, and cobalt. Another important resource found in shallow marine deposits is phosphate-rich sedimentary rock known as *phosphorite*; it is used for chemical fertilizers and animal feed supplements, as well as matches, metallurgy, preserved foods, and ceramics.

A potential resource within the EEZ is methane hydrate, consisting of single methane molecules bound up in networks formed by frozen water. It is also found in permafrost soils. Methane hydrate is stable at water depths of more than 500 m and near-freezing temperatures. According to one estimate, the carbon in these deposits is double that in all coal, oil, and natural gas. However, no one knows if methane hydrate can be effectively recovered and used as an energy source. Another factor that must be evaluated is that the volume of methane hydrate is about 3,000 times as much as in the atmosphere—and methane is 10 times more effective than carbon dioxide as a greenhouse gas.

In some coastal locations a dam may be built across an inlet so that it regulates the flow of water landward during flood tides and releases water seaward at ebb tide. In so doing, the moving water produces electricity by spinning turbines connected to generators, just as at hydroelectric power plants. Unfortunately, there are not many coastal locations suitable for tidal power plants. Furthermore, the initial cost of such installations is high, and they can have disastrous effects on the ecology of estuaries.

16 Geologic Time: Concepts and Principles

The Grand Canyon, Arizona. Major John Wesley Powell led two expeditions down the Colorado River and through the canyon in 1869 and 1871. He was struck by the seemingly limitless time represented by the rocks exposed in the canyon walls and by the recognition that these rock layers, like the pages in a book, contain the geologic history of this region.

Copyright and Photograph by Dr. Parvinder S. Sethi

Introduction

"Vast periods of time set geology apart from most of the other sciences."

I n 1869, Major John Wesley Powell, a Civil War veteran who lost his right arm in the battle of Shiloh, led a group of hardy explorers down the uncharted Colorado River through the Grand Canyon. With no maps or other information, Powell and his group ran the many rapids of the Colorado River in fragile wooden boats, hastily recording what they saw. Powell wrote in his diary that "all about me are interesting geologic records. The book is open and I read as I run."

Probably no one has contributed as much to the understanding of the Grand Canyon as Major Powell. In recognition of his contributions, the Powell Memorial was erected on the South Rim of the Grand Canyon in 1969 to commemorate the 100th anniversary of this history-making first expedition.

Most tourists today, like Powell and his fellow explorers in 1869, are astonished by the seemingly limitless time represented by the rocks exposed in the walls of the Grand Canyon (see the Virtual Field Trip in this

chapter). For most visitors, viewing a 1.5-km-deep cut into Earth's crust is the only encounter they'll ever have with the magnitude of geologic time. When standing on the rim and looking down into the Grand Canyon, we are really looking far back in time, more than 1 billion years—all the way back to the early history of our planet.

Vast periods of time set geology apart from most of the other sciences, and an appreciation of the immensity of geologic time is fundamental to understanding the physical and biological history of our planet. Besides gaining an appreciation of the immensity of geologic time, however, one of the most valuable lessons you will learn in this chapter is how to apply the various fundamental geologic principles to solve basic geologic problems. The logic used in applying these principles to interpret the geologic history of an area involves basic reasoning skills that can be transferred to, and used in, almost any profession or discipline.

relative dating The process of determining the age of an event as compared to other events; involves placing geologic events in their correct chronological order but does not involve the consideration of when the events occurred in number of years ago.

Learning Outcomes

After reading this unit, you should be able to do the following:

LO1 Explain how geologic time is measured

LO2 Review early concepts of geologic time and Earth's age

LO3 Understand James Hutton's contribution to the recognition of geologic time

LO4 Describe relative dating methods and apply them to decipher the geologic history of an area

LO5 Explain the process of correlating rock units

LO6 Explain absolute dating methods

LO7 Describe the development of the geologic time scale

LO8 Describe and understand the methods used to determine the chronology of climate change

LO1 HOW IS GEOLOGIC TIME MEASURED?

In some respects, time is defined by the methods used to measure it. Geologists use two different frames of reference when discussing geologic time: relative dating and absolute dating. **Relative dating** is placing geologic events

© iStockphoto.com/pixhook

in a sequential order as determined from their position in the geologic record. Relative dating will not tell us how long ago a particular event took place, only that one event preceded another.

The principles used to determine relative dating were discovered hundreds of years ago, and since then, they have been used to construct the *relative geologic time scale* (Figure 16.1). Furthermore, these principles are still widely used by geologists today.

Absolute dating provides specific dates for rock units or events expressed in years before the present. *Radiometric dating* is the most common method of obtaining absolute ages. Dates are calculated from the natural decay rates of various radioactive elements present in trace amounts in some rocks. It was not until the discovery of radioactivity near the end of the 19th century that absolute ages could be accurately applied to the relative geologic time scale.

Today, the geologic time scale is really a dual scale: a relative scale based on rock sequences with radiometric dates expressed as years before the present (Figure 16.1).

LO2 EARLY CONCEPTS OF GEOLOGIC TIME AND EARTH'S AGE

The concept of geologic time and its measurement have changed throughout human history. Some early Christian scholars and clerics tried to establish the date of creation by analyzing historical records and the genealogies found in Scripture.

Based on their analyses, they generally believed that Earth and all its features were no more than about 6,000 years old. Thus, the idea of a very young Earth provided the basis for most Western chronologies of Earth history prior to the 18th century.

During the 18th and 19th centuries, several attempts were made to determine Earth's age on the basis of scientific evidence rather than revelation. The French zoologist Georges Louis de Buffon (1707–1788) assumed that Earth gradually cooled to its present

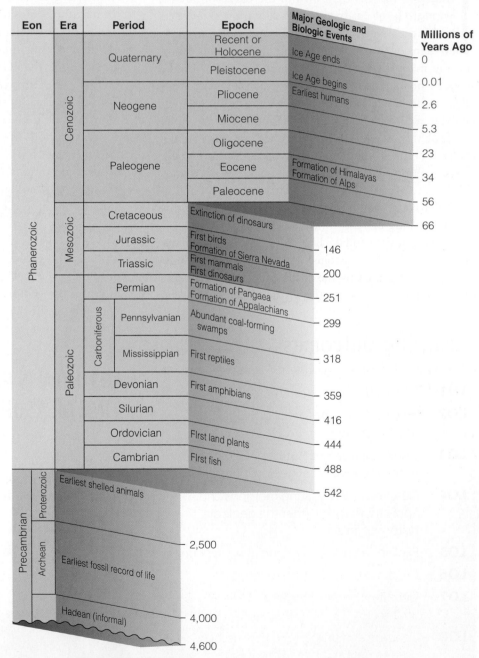

Figure 16.1 The Geologic Time Scale

Some of the major geologic and biologic events are indicated for the various eras, periods, and epochs.

DATES ARE FROM THE 2009 INTERNATIONAL STRATIGRAPHIC CHART, © 2009 BY THE INTERNATIONAL COMMISSION ON STRATIGRAPHY.

Early scholar studying the Bible

iStockphoto.com/rammekens

LO3 JAMES HUTTON AND THE RECOGNITION OF GEOLOGIC TIME

Many consider the Scottish geologist James Hutton (1726–1797) to be the father of modern geology. His detailed studies and observations of rock exposures and present-day geologic processes were instrumental in establishing the principle of *uniformitarianism* (see Chapter 1), the concept that the same processes seen today have oper-

James Hutton, father of modern geology

© North Wind Picture Archives/Alamy

ated over vast amounts of time. Because Hutton relied on known processes to account for Earth history, he concluded that Earth must be very old and wrote that "we find no vestige of a beginning, and no prospect of an end."

In 1830, Charles Lyell published a landmark book, *Principles of Geology*, in which he championed Hutton's concept of uniformitarianism. Instead of relying on catastrophic events to explain various Earth features, Lyell recognized that imperceptible changes brought about by present-day processes could, over long periods of time, have tremendous cumulative effects. Through his writings, Lyell firmly established uniformitarianism as the guiding principle of geology.

LO4 RELATIVE DATING METHODS

Before the development of radiometric dating techniques, geologists had no reliable means of absolute dating and therefore depended solely on relative dating methods. Recall that relative dating places events in sequential order but does not tell us how long ago an event took place.

Fundamental Principles of Relative Dating

The 17th century was an important time in the development of geology as a science because of the widely circulated writings of the Danish anatomist Nicolas Steno (1638–1686). Steno observed that when streams flood, they spread out across their floodplains and deposit layers of sediment that bury organisms dwelling on the floodplain. Subsequent floods produce new layers of sediments that are deposited or superposed over previous deposits. When lithified, these layers of sediment become sedimentary rock.

condition from a molten beginning. To simulate this history, he melted iron balls of various diameters and allowed them to cool to the surrounding temperature. By extrapolating their cooling rate to a ball the size of Earth, he determined that Earth was at least 75,000 years old. Although this age was much older than that derived from Scripture, it was still vastly younger than we now know our planet to be.

Other scholars were equally ingenious in attempting to calculate Earth's age. For example, if deposition rates could be determined for various sediments, geologists reasoned that they could calculate how long it would take to deposit any rock layer. They could then extrapolate how old Earth was from the total thickness of sedimentary rock in its crust. Rates of deposition vary, however, even for the same type of rock. Furthermore, it is impossible to estimate how much of a rock has been removed by erosion, or how much a rock sequence has been reduced by compaction. As a result of these variables, estimates of Earth's age ranged from younger than 1 million years to older than 2 billion years.

Besides trying to determine Earth's age, the naturalists of the 18th and 19th centuries were formulating some of the fundamental geologic principles that are used in deciphering Earth history. From the evidence preserved in the geologic record, it was clear to them that Earth is very old and that geologic processes have operated over long periods of time.

principle of superposition A principle holding that in a vertical sequence of undeformed sedimentary rocks, the relative ages of the rocks can be determined by their position in the sequence—oldest at the bottom, followed by successively younger layers.

principle of original horizontality According to this principle, sediments are deposited in horizontal or nearly horizontal layers.

principle of lateral continuity A principle holding that rock layers extend outward in all directions until they terminate.

principle of cross-cutting relationships A principle holding that an igneous intrusion or fault must be younger than the rocks it intrudes or cuts across.

principle of inclusions A principle holding that inclusions or fragments in a rock unit are older than the rock unit itself; for example, granite inclusions in sandstone are older than the sandstone.

Thus, in an undisturbed succession of sedimentary rock layers, the oldest layer is at the bottom and the youngest layer is at the top. This **principle of superposition** is the basis for relative-age determinations of strata and their contained fossils (Figure 16.2a; see also the Virtual Field Trip in Chapter 6 and this chapter).

Steno also observed that, because sedimentary particles settle from water under the influence of gravity, sediment is deposited in essentially horizontal layers, thus illustrating the **principle of original horizontality** (Figure 16.2a; see also the Virtual Field Trip in Chapter 6 and this chapter). Therefore, a sequence of sedimentary rock layers that is steeply inclined from the horizontal must have been tilted after deposition and lithification (Figure 16.2b).

Steno's third principle, the **principle of lateral continuity**, states that sediment extends laterally in all directions until it thins and pinches out, or terminates, against the edge of the depositional basin (Figure 16.2a; see also the Virtual Field Trip in this chapter).

James Hutton is credited with formulating the **principle of cross-cutting relationships**. Based on his detailed studies and observations of rock exposures in Scotland, Hutton recognized that an igneous intrusion or fault must be younger than the rocks it intrudes or displaces (Figure 16.3).

Another way to determine relative ages is by using the **principle of inclusions**. This principle holds that

Figure 16.2 The Principles of Original Horizontality, Superposition, and Lateral Continuity

Copyright and Photograph by Dr. Parvinder S. Sethi

a. The sedimentary rocks of Bryce Canyon National Park, Utah, were originally deposited horizontally in a variety of continental environments (principle of original horizontality). The oldest rocks are at the bottom of this highly dissected landscape, and the youngest rocks are at the top, forming the rims (principle of superposition). The exposed rock layers extend laterally in all directions for some distance (principle of lateral continuity).

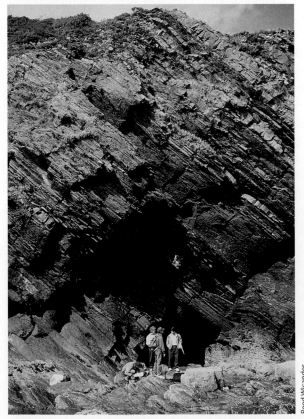

Reed Wicander

b. These shales and limestones of the Postolonnec Formation, at Postolonnec Beach, Crozon Peninsula, France, were originally deposited horizontally but have been significantly tilted since their formation.

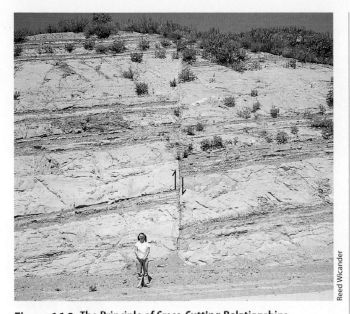

Figure 16.3 **The Principle of Cross-Cutting Relationships**
A small fault (arrows show direction of movement) cuts across, and thus displaces, tilted sedimentary beds along the Templin Highway in Castaic, California. The fault is therefore younger than the youngest beds that are displaced.

Reed Wicander

inclusions, or fragments of one rock contained within a layer of another, are older than the rock layer itself. The batholith shown in Figure 16.4a contains sandstone inclusions, and the sandstone unit shows the effects of baking. Accordingly, we conclude that the sandstone is older than the batholith. In Figure 16.4b, however,

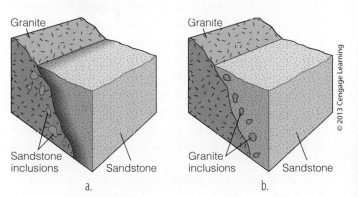

© 2013 Cengage Learning

Figure 16.4 **The Principle of Inclusions**

a. The batholith is younger than the sandstone because the sandstone has been baked at its contact with the granite and the granite contains sandstone inclusions.

b. Granite inclusions in the sandstone indicate that the batholith was the source of the sandstone and therefore is older.

c. Outcrop in northern Wisconsin showing basalt inclusions (dark gray) in granite (white). Accordingly, the basalt inclusions are older than the granite.

the sandstone contains granite rock pieces, indicating that the batholith was the source rock for the inclusions and is therefore older than the sandstone.

Fossils have been known for centuries, yet their utility in relative dating and geologic mapping was not fully appreciated until the early 19th century. William Smith (1769–1839), an English civil engineer involved in surveying and building canals in southern England, independently recognized the principle of superposition by reasoning that the fossils at the bottom of a sequence of strata are older than those at the top of the sequence. This recognition served as the basis for the **principle of fossil succession** or the *principle of faunal and floral succession*, as it is sometimes called (Figure 16.5).

principle of fossil succession A principle holding that fossils, and especially groups or assemblages of fossils, succeed one another through time in a regular and predictable order.

Unconformities

Our discussion so far has been concerned with *conformable strata*—sequences in which no depositional breaks of any consequence occur; that is, sedimentation was more or less continuous. A bedding plane between strata may represent a depositional break of anywhere from minutes to tens of years, but it is inconsequential in the context

Basalt inclusion

Granite

James S. Monroe

c.

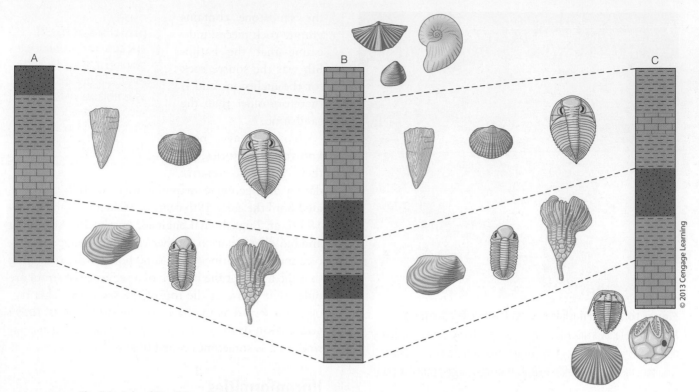

Figure 16.5 The Principle of Fossil Succession

This generalized diagram shows how geologists use the principle of fossil succession to identify strata of the same age in different areas. The rocks in the three sections encompassed by the dashed lines contain similar fossils and are therefore the same age. Note that the youngest rocks in this region are in section B, whereas the oldest rocks are in section C.

unconformity A break in the geologic record represented by an erosional surface separating younger strata from older rocks.

disconformity An unconformity in which the rock layers above and below are parallel.

angular unconformity An unconformity below which older rocks dip at a different angle (usually steeper) than overlying strata.

of geologic time. However, in some sequences of strata, surfaces known as **unconformities** may be present, representing times of nondeposition, erosion, or both. Unconformities encompass long periods of geologic time, perhaps millions or tens of millions of years. Accordingly, the geologic record is incomplete wherever an unconformity is present, just as a book with missing pages is incomplete, and the interval of geologic time not represented by strata is called a *hiatus* (Figure 16.6).

Geologists recognize three types of uncomformities (to view all three types of unconformities exposed

in the Grand Canyon, see the Virtual Field Trip in this chapter). A **disconformity** is a surface of erosion, or nondeposition, separating younger from older rocks, both of which are parallel with one another (Figure 16.7). Unless the erosional surface separating the older from the younger parallel beds is well defined, or distinct, the disconformity frequently resembles an ordinary bedding plane. Hence, many disconformities are difficult to recognize and must be identified on the basis of fossil assemblages.

An **angular unconformity** is an erosional surface on tilted or folded strata over which younger strata were deposited (Figure 16.8). The strata below the unconformable surface generally dip more steeply than those above, producing an angular relationship.

The most famous angular unconformity in the world is probably the one at Siccar Point, Scotland (Figure 16.8b). It was here that James Hutton realized that severe upheavals had tilted the lower rocks and formed mountains that were then worn away and covered by younger, flat-lying rocks. The erosional surface between the older, tilted rocks and the younger, flat-lying strata

Figure 16.6 The Development of a Hiatus and an Unconformity

a. Deposition began 12 million years ago (mya) and continued more or less uninterrupted until 4 mya.

b. Between 3 and 4 mya, an episode of erosion occurred. During that time some of the strata deposited earlier was eroded.

c. A hiatus of 3 million years thus exists between the older strata and the strata that formed during a renewed episode of deposition that began 3 mya.

d. The actual stratigraphic record as seen in an outcrop today. The unconformity is the surface separating the strata and represents a major break in our record of geologic time.

© 2013 Cengage Learning

meant that a significant gap existed in the geologic record. Although Hutton did not use the term *unconformity*, he was the first to understand and explain the significance of such discontinuities in the geologic record.

A **nonconformity** is the third type of unconformity. Here, an erosional surface cut into metamorphic or igneous rocks is covered by sedimentary rocks (Figure 16.9). This type of unconformity closely resembles an intrusive igneous contact with sedimentary rocks. The principle of inclusions (Figure 16.4) is helpful in determining whether the relationship between the underlying igneous rocks and the overlying sedimentary rocks is the result of an intrusion or erosion. In the case of an intrusion, the igneous rocks are younger, whereas in the case of erosion, the sedimentary rocks are younger. Being able to distinguish between a nonconformity and an intrusive contact is important because they represent different sequences of events.

Applying the Principles of Relative Dating

We can decipher the geologic history of the area represented by the block diagram in Figure 16.10 by applying the various relative dating principles just discussed. The methods and logic used in this example are the same as those applied by 19th-century geologists in constructing the geologic time scale.

According to the principles of superposition and original horizontality, beds A–G were deposited horizontally; then either they were tilted, faulted (H), and eroded, or after deposition they were faulted (H), tilted, and then eroded. Because the fault cuts beds A–G, it

nonconformity An unconformity in which stratified sedimentary rocks overlie an erosion surface cut into igneous or metamorphic rocks.

Figure 16.7 Formation of a Disconformity

Deposition

Uplift and erosion

Uplift and erosion

Deposition

Disconformity

© 2013 Cengage Learning

a. Formation of a disconformity.

Jurassic rocks

Mississippian rocks

James S. Monroe

b. Disconformity between Mississippian and Jurassic strata in Montana. The geologist at the upper left is sitting on Jurassic strata, and his right foot is resting on Mississippian rocks. This disconformity represents approximately 165 million years.

Figure 16.8 Formation of an Angular Unconformity

Deposition

Uplift and erosion

Erosion

Uplift and tilting

Deposition

Angular unconformity

© 2013 Cengage Learning

Marti Bryant Miller

a. Formation of an angular unconformity.

b. Angular unconformity at Siccar Point, Scotland. James Hutton first realized the significance of unconformities at this site in 1788.

Figure 16.9 Formation of a Nonconformity

Deposition

Uplift and erosion

Nonconformity

Uplift and erosion of overlying sediments

Intrusion of magma

© 2013 Cengage Learning

a. Formation of a nonconformity.

b. Nonconformity between Precambrian metamorphic rocks and the overlying Paleozoic sedimentary rocks in the Grand Canyon, Arizona.

Paleozoic sedimentary rocks

Nonconformity

Precambrian metamorphic rocks

Marti Bryant Miller

Sedimentary Rocks

Igneous Rocks

Figure 16.10 Block Diagram of a Hypothetical Area

A block diagram of a hypothetical area in which the various relative dating principles can be applied to determine its geologic history.

© 2013 Cengage Learning

must be younger than the beds according to the principle of cross-cutting relationships.

Beds J–L were then deposited horizontally over this erosional surface, producing an angular unconformity (I). Following deposition of these three beds, the entire sequence was intruded by a dike (M), which, according to the principle of cross-cutting relationships, must be younger than all of the rocks that it intrudes.

The entire area was then uplifted and eroded; next, beds P and Q were deposited, producing a disconformity (N) between beds L and P and a nonconformity (O) between the igneous intrusion M and the sedimentary bed P. We know that the relationship between igneous intrusion M and the overlying sedimentary bed P is a nonconformity because of the inclusions of M in P (principle of inclusions).

At this point, there are several possibilities for reconstructing the geologic history of this area. According to the principle of cross-cutting relationships, dike R must be younger than bed Q because it intrudes into it. It could have intruded anytime *after* bed Q was deposited; however, we cannot determine whether R was formed right after Q, right after S, or after T was

formed. For purposes of this history, we will say that it intruded after the deposition of bed Q.

Following the intrusion of dike R, lava S flowed over bed Q, followed by the deposition of bed T. Although the lava flow (S) is not a sedimentary unit, the principle of superposition still applies because it flowed onto the surface, just as sediments are deposited on Earth's surface.

We have established a relative chronology for the rocks and events of this area by using the principles of relative dating. Remember, however, that we have no way of knowing how many years ago these events occurred unless we can obtain radiometric dates for the igneous rocks. With these dates, we can establish the range of absolute ages between which the different sedimentary units were deposited and also determine how much time is represented by the unconformities.

LO5 CORRELATING ROCK UNITS

To decipher Earth history, geologists must demonstrate the time equivalency of rock units in different areas. This

Virtual Field Trip

Geologic Time

Ready to Go!

During this virtual field trip, you will be visiting Grand Canyon National Park, Arizona, one of the seven wonders of the world. Established as a national park in 1919, the Grand Canyon is visited by more than five million people every year! It is here that most people are first exposed to the idea of "deep time," or geologic time—that is, the concept of seemingly limitless amounts of time. When standing on the South Rim and looking down into the 1.5 km-deep canyon cut by the Colorado River, one can't help but be impressed with the magnitude of time represented by the rocks exposed in the canyon walls. It is here in the Grand Canyon that we will see examples of the various relative dating principles used to place geologic events in their correct sequential order, as well as examples of the three types of unconformities, all of which are discussed in this chapter. In addition to the Grand Canyon, we will also be visiting Capitol Reef National Park, Utah, where you will be able to see a variety of sedimentary features and examples of the various relative dating principles you have already read about.

It is now time to see actual examples of the concepts and geologic features discussed in your textbook. And, there is no better place to start this journey then the Grand Canyon! See you there.

DeerPoint, UT. topographic map, courtesy of USGS.
All photographs: Copyright and Photograph by
Dr. Parvinder S. Sethi

GOALS OF THE TRIP

In this field trip you will come face to face with such important concepts as

1. The various principles of relative dating.
2. Unconformities.
3. Correlation of rock units.

FOLLOW-UP QUESTIONS

1. Using the principles of relative dating and knowledge of the types of unconformities you have learned, reconstruct the geologic history of the rock units in the Grand Canyon as seen and numbered in the "Reconstructing Geologic Events" slide.

2. According to Figure 16.12, the rocks forming the rim of the Grand Canyon in Arizona include the Moenkopi Formation. The oldest rock unit in Capitol Reef National Park is the Moenkopi Formation. Discuss how you can use the various principles of relative dating to reconstruct a more complete geologic history of the Colorado Plateau from these two locations.

3. Why is the Grand Canyon such an ideal location to understand what is meant by "deep time," or geologic time?

What to See When You Go

Long before radioactivity was discovered and geologists could assign actual dates to rock units and geologic events, Earth scientists used a variety of relative dating methods (articulated as principles) to place rock sequences and geologic events in sequential order. In this spectacular sunset photo of the Grand Canyon, taken from Hopi Point, one can easily see the three interrelated principles of relative dating: the principles of original horizontality, superposition, and lateral continuity. Here, the rocks of the Grand Canyon extend in all directions (*principle of lateral continuity*), were deposited horizontally (*principle of original horizontality*), with the oldest rocks at the bottom of the canyon, and the youngest rocks forming the canyon rim (*principle of superposition*).

What to See When You Go

Recall from your reading in this chapter that *unconformities* are breaks in the geologic record represented by an erosional surface or a surface of nondeposition separating younger rocks from older rocks. There are three types of unconformities, all of which can be seen in the Grand Canyon. In the photo of the Great Unconformity shown above, sedimentary rocks of Cambrian age overlie metamorphic rocks of Precambrian age. This surface represents more than a billion years of geologic time for which we don't have a record at this location. It is an example of a *nonconformity*.

An excellent example of an *angular unconformity* can be observed at Lipan Point, a few miles from Zuni Point. Here, older rocks can be seen dipping at a different angle than the younger overlying rocks.

This photo shows the rocks of the Grand Canyon and illustrates a *disconformity*, that is, an unconformity between parallel rock layers, in this case the Muav Limestone (Cambrian age, Layer 3) and the overlying Redwall Limestone (Mississippian age, Layer 4). Note that the bedding plane between the Muav and Redwall Limestone looks just like the other bedding planes. We know this is a disconformity because the rocks have been dated by fossils (principle of fossil succession) and indicate that rocks representing the Ordovician through Devonian periods are missing here.

In order to decipher Earth history, geologists had to demonstrate time equivalency of rock units in different areas, a process known as *correlation*. This view of Capitol Reef National Park was photographed from the Scenic Drive. You can clearly see the red-colored sedimentary rocks at this location, and you should note that the reddish-brown layers near the bottom of the sequence are slightly tilted. This is due to the rocks being uplifted by tectonic forces. When surface exposures, such as here, are adequate, rocks can easily be traced laterally on the basis of similarity of rock types and position in a sequence.

process is known as **correlation**. If surface exposures are adequate, units may simply be traced laterally (principle of lateral continuity), even if occasional gaps exist (Figure 16.11). Other criteria used to correlate units are similarity of rock type, position in a sequence, and key beds (see the Virtual Field Trip in this chapter). *Key beds* are units, such as coal beds or volcanic ash layers, that are sufficiently distinctive to allow identification of the same unit in different areas (Figure 16.11).

Generally, no single location in a region has a geologic record of all events that occurred during its history; therefore, geologists must correlate from one area to another to determine the complete geologic history of the region. An excellent example is the history of the Colorado Plateau (Figure 16.12). This region provides a record of events occurring over approximately 2 billion years, but because of the forces of erosion, the entire record is not preserved at any single location. Within the walls of the Grand Canyon are rocks of the Precambrian and Paleozoic eras, whereas Paleozoic and Mesozoic Era rocks are found in Zion National Park, and Mesozoic and Cenozoic Era rocks are exposed in Bryce Canyon National Park (Figure 16.12). By correlating the uppermost rocks at one location with the lowermost equivalent rocks of another area, however, geologists can decipher the history of the entire region.

Although geologists match up rocks on the basis of similar rock type and superposition, correlation of this type can be done only in a limited area, where beds can be traced from one site to another. To correlate rock units over a large area, or to correlate age-equivalent units of different composition, fossils and the principle of fossil succession must be used.

Fossils are useful as relative time indicators because they are the remains of organisms that lived for a certain length of time during the geologic past. Fossils that are easily identified, are geographically widespread, and existed for a rather short interval of geologic time are particularly useful. Such fossils are **guide fossils**, or *index fossils* (Figure 16.13). The trilobite *Paradoxides* and the brachiopod *Atrypa* meet these criteria and are therefore good guide fossils. In contrast, the brachiopod *Lingula* is easily identified and widespread, but its long geologic range of Ordovician to Recent makes it of little use in correlation.

© 2013 Cengage Learning

Key bed
volcanic ash

Figure 16.11 Correlating Rock Units
In areas of adequate exposure, rock units can be traced laterally, even if occasional gaps exist, and correlated on the basis of similarity in rock type and position in a sequence. Rocks can also be correlated by a key bed—in this case, volcanic ash.

LO6 ABSOLUTE DATING METHODS

Although most of the isotopes of the 92 naturally occurring elements are stable, some are radioactive and spontaneously decay

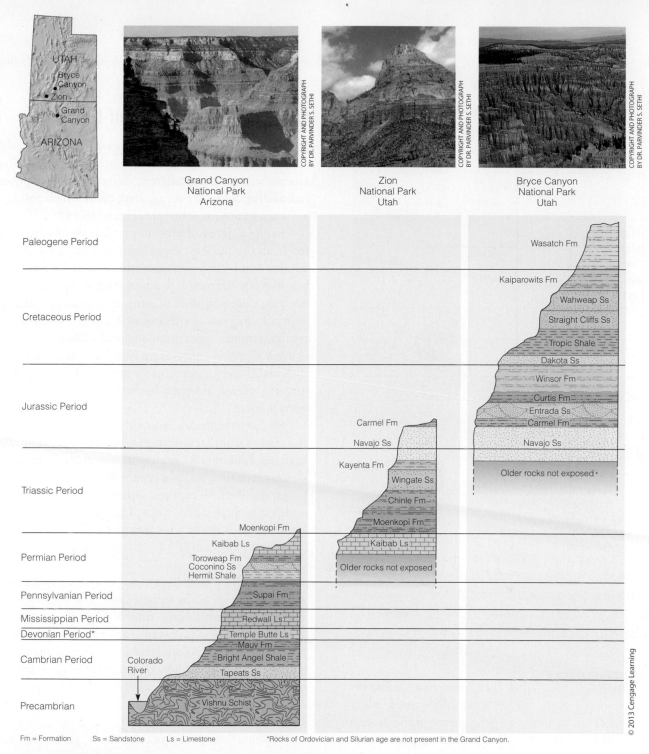

Grand Canyon
National Park
Arizona

Zion
National Park
Utah

Bryce Canyon
National Park
Utah

Fm = Formation   Ss = Sandstone   Ls = Limestone   *Rocks of Ordovician and Silurian age are not present in the Grand Canyon.

Figure 16.12 Correlation of Rock Units within the Colorado Plateau

At each location only a portion of the geologic record of the Colorado Plateau is exposed. By correlating the youngest rocks at one exposure with the oldest rocks at another exposure, geologists can determine the entire history of the region. For example, the rocks forming the rim of the Grand Canyon, Arizona, are the Kaibab Limestone and Moenkopi Formation—the youngest rocks exposed in the Grand Canyon. The Kaibab Limestone and Moenkopi Formation are the oldest rocks exposed in Zion National Park, Utah, and the youngest rocks are the Navajo Sandstone and Carmel Formation. The Navajo Sandstone and Carmel Formation are the oldest rocks exposed in Bryce Canyon National Park, Utah. By correlating the Kaibab Limestone and Moenkopi Formation between the Grand Canyon and Zion National Park, geologists have extended the geologic history from the Precambrian to the Jurassic. And by correlating the Navajo Sandstone and Carmel Formation between Zion and Bryce Canyon National Parks, geologists can extend the geologic history through the Paleogene Period. Thus, by correlating the rock exposures between these areas and applying the principle of superposition, geologists can reconstruct the geologic history of the region.

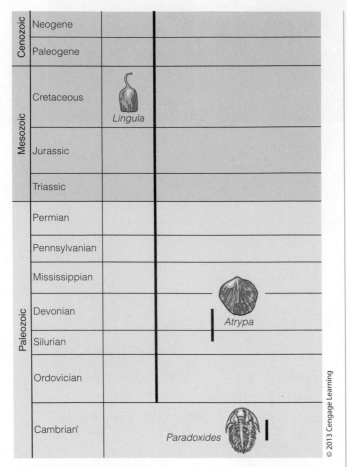

Figure 16.13 Guide Fossils

Comparison of the geologic ranges (heavy vertical lines) of three marine invertebrate animals. *Lingula* is of little use in correlation because it has such a long range. But *Atrypa* and *Paradoxides* are good guide fossils because both are widespread, easily identified, and have short geologic ranges. Thus, both can be used to correlate rock units that are widely separated and to establish the relative age of a rock that contains them.

radioactive decay
The spontaneous change of an atom to an atom of a different element by emission of a particle from its nucleus (alpha and beta decay) or by electron capture.

half-life The time necessary for half of the original number of radioactive atoms of an element to decay to a new, and more stable daughter product; for example, the half-life for potassium 40 is 1.3 billion years.

to other, more stable isotopes of elements, releasing energy in the process. The discovery in 1903 by Pierre and Marie Curie that radioactive decay produces heat meant that geologists finally had a mechanism for explaining Earth's internal heat that did not rely on residual cooling from a molten origin. Furthermore, geologists now had a powerful tool to date geologic

events accurately and to verify the long time periods postulated by Hutton and Lyell.

Radioactive Decay and Half-Lives

Radioactive decay is the process whereby an unstable atomic nucleus is spontaneously transformed into an atomic nucleus of a different element. Three types of radioactive decay are recognized, all of which result in a change of atomic structure (Figure 16.14).

In *alpha decay*, 2 protons and 2 neutrons are emitted from the nucleus, resulting in the loss of 2 atomic numbers and 4 atomic mass numbers. In *beta decay*, a fast-moving electron is emitted from a neutron in the nucleus, changing that neutron to a proton and consequently increasing the atomic number by 1, with no resultant atomic mass number change. *Electron capture* is when a proton captures an electron from an electron shell and thereby converts to a neutron, resulting in the loss of 1 atomic number, but not changing the atomic mass number.

Some elements undergo only one decay step in the conversion from an unstable form to a stable form. For example, rubidium 87 decays to strontium 87 by a single beta emission, and potassium 40 decays to argon 40 by a single electron capture. Other radioactive elements undergo several decay steps. Uranium 235 decays to lead 207 by 7 alpha and 6 beta steps, whereas uranium 238 decays to lead 206 by 8 alpha and 6 beta steps (Figure 16.15).

When discussing decay rates, it is convenient to refer to them in terms of half-lives. The **half-life** of a radioactive element is the time it takes for half of the atoms of the original unstable *parent element* to decay to atoms of a new, more stable *daughter element*. The half-life of a given radioactive element is constant and can be precisely measured. Half-lives of various radioactive elements range from less than a billionth of a second to 49 billion years.

Radioactive decay occurs at a geometric rate rather than a linear rate (Figure 16.16). Therefore, a graph of the decay rate produces a curve rather than a straight line (Figure 16.16b). For example, an element with 1,000,000 parent atoms will have 500,000 parent atoms and 500,000 daughter atoms after one half-life. After two half-lives, it will have 250,000 parent atoms (one-half of the previous parent atoms, which is equivalent to one-fourth of the original parent atoms) and 750,000 daughter atoms. After three half-lives, it will have 125,000 parent atoms (one-half of the previous parent atoms, or one-eighth of the original parent

Figure 16.14 Three Types of Radioactive Decay

Parent nucleus — Alpha particle — Daughter nucleus

Changes in atomic number and atomic mass number

Atomic number = −2
Atomic mass number = −4

● Proton ○ Neutron ○ Electron

a. Alpha decay, in which an unstable parent nucleus emits 2 protons and 2 neutrons.

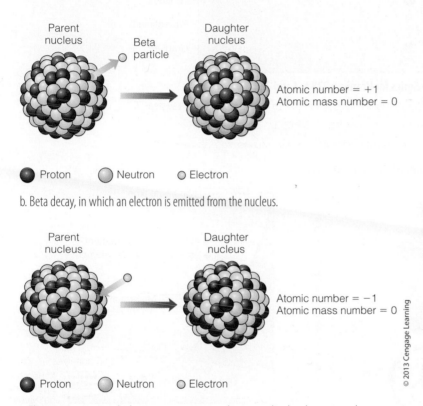

Parent nucleus — Beta particle — Daughter nucleus

Atomic number = +1
Atomic mass number = 0

● Proton ○ Neutron ○ Electron

b. Beta decay, in which an electron is emitted from the nucleus.

Parent nucleus — Daughter nucleus

Atomic number = −1
Atomic mass number = 0

● Proton ○ Neutron ○ Electron

c. Electron capture, in which a proton captures an electron and is thereby converted to a neutron.

© 2013 Cengage Learning

atoms) and 875,000 daughter atoms, and so on, until the number of parent atoms remaining is so few that they cannot be accurately measured by present-day instruments.

By measuring the parent–daughter ratio and knowing the half-life of the parent (which has been determined in the laboratory), geologists can calculate the age of a sample that contains the radioactive element. The parent–daughter ratio is usually determined by a *mass spectrometer*, an instrument that measures the proportions of atoms of different masses.

Sources of Uncertainty

The most accurate radiometric dates are obtained from igneous rocks. As magma cools and begins to crystallize, radioactive parent atoms are separated from previously formed daughter atoms. Because they are the right size, some radioactive parent atoms are incorporated into the crystal structure of certain minerals. The stable daughter atoms, though, are a different size from the radioactive parent atoms and consequently cannot fit into the crystal structure of the same mineral as the parent atoms. Therefore, a mineral crystallizing in cooling magma will contain radioactive parent atoms but no stable daughter atoms (Figure 16.17). Thus, the time that is being measured is the time of crystallization of the mineral that contains the radioactive atoms and *not* the time of formation of the radioactive atoms.

To obtain accurate radiometric dates, geologists must be sure that they are dealing with a *closed system*, meaning that neither parent nor daughter atoms have been added or removed from the system since crystallization, and that the ratio between them results only from radioactive decay. Otherwise, an inaccurate date will result. If daughter atoms have leaked out of the mineral being analyzed, the calculated age will be too young; if parent atoms have been removed, the calculated age will be too old.

Leakage may take place if the rock is heated or subjected to intense pressure as can sometimes occur during metamorphism. If this happens, some of the parent or daughter atoms may be driven from the mineral being analyzed, resulting in an inaccurate age determination. If the daughter product was completely removed, then one would be measuring the time since metamorphism (a useful measurement itself) and not the time since crystallization of the mineral (Figure 16.18).

Because heat and pressure affect the parent–daughter ratio, metamorphic rocks are difficult to date accurately. Remember that although the resulting parent–daughter ratio of the sample being analyzed may have been affected by heat, the decay rate of the parent element remains constant, regardless of any physical or chemical changes.

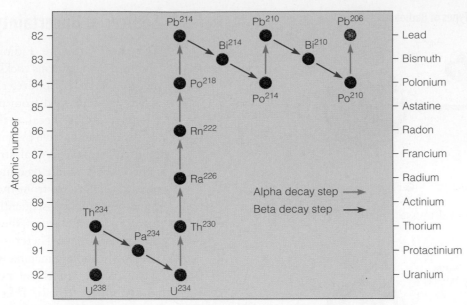

Figure 16.15 Radioactive Decay Series for Uranium 238 to Lead 206
Radioactive uranium 238 decays to its stable daughter product, lead 206, by 8 alpha and 6 beta decay steps. A number of different isotopes are produced as intermediate steps in the decay series.

BASED ON DATA FROM S. M. RICHARDSON AND H. Y. MCSWEEN, JR., *GEOCHEMISTRY—PATHWAYS AND PROCESSES*, PRENTICE-HALL.

Figure 16.16 Uniform, Linear Change Compared to Geometric Radioactive Decay

a. Uniform, linear change is characteristic of many familiar processes. In this example, water is being added to a glass at a constant rate.

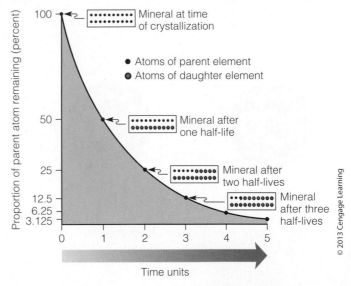

b. A geometric radioactive decay curve, in which each time unit represents one half-life, and each half-life is the time it takes for half of the parent element to decay to the daughter element.

Long-Lived Radioactive Isotope Pairs

Table 16.1 shows the five common, long-lived parent–daughter isotope pairs used in radiometric dating. Long-lived pairs have half-lives of millions or billions of years. All of these were present when Earth formed and are still present in measurable quantities. Other shorter-lived radioactive isotope pairs have decayed to the point that only small quantities near the limit of detection remain.

The most commonly used isotope pairs are the uranium–lead and thorium–lead series, which are used principally to date ancient igneous intrusives, lunar samples, and some meteorites. The rubidium–strontium pair is also used for very old samples and has been effective in dating the oldest rocks on Earth, as well as meteorites.

Figure 16.17 Crystallization of Magma Containing Radioactive Parent and Stable Daughter Atoms

Magma

Mineral crystallizing from magma

Igneous rock

● Radioactive parent atoms
● Stable daughter atoms

a. Magma contains both radioactive parent atoms and stable daughter atoms. The radioactive parent atoms are larger than the stable daughter atoms.

b. As magma cools and begins to crystallize, some of the radioactive atoms are incorporated into certain minerals because they are the right size and can fit into the crystal structure. In this example, only the larger radioactive parent atoms fit into the crystal structure. Therefore, at the time of crystallization, minerals in which the radioactive parent atoms can fit into the crystal structure will contain 100% radioactive parent atoms and 0% stable daughter atoms.

c. After one half-life, 50% of the radioactive parent atoms will have decayed to stable daughter atoms such that those minerals that had radioactive parent atoms in their crystal structure will now have 50% radioactive parent atoms and 50% stable daughter atoms.

© 2013 Cengage Learning

Figure 16.18 Effects of Metamorphism on Radiometric Dating

The effect of metamorphism in driving out daughter atoms from a mineral that crystallized 700 million years ago (mya). The mineral is shown immediately after crystallization (a), then at 400 million years (b), when some of the parent atoms had decayed to daughter atoms. Metamorphism at 350 mya (c) drives the daughter atoms out of the mineral into the surrounding rock. If the rock has remained a closed chemical system throughout its history, dating the mineral today (d) yields the time of metamorphism, whereas dating the whole rock provides the time of its crystallization, 700 million years ago.

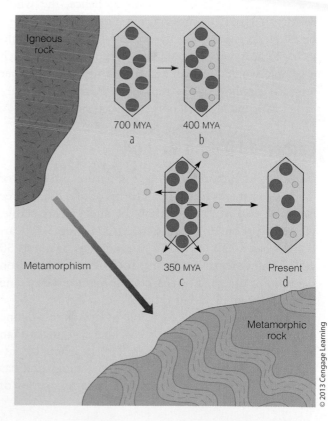

Igneous rock

700 MYA
a

400 MYA
b

Metamorphism

350 MYA
c

Present
d

Metamorphic rock

© 2013 Cengage Learning

TABLE 16.1

Five of the Principal Long-Lived Radioactive Isotope Pairs Used in Radiometric Dating

ISOTOPES		HALF-LIFE OF PARENT (YEARS)	EFFECTIVE DATING RANGE (YEARS)	MINERALS AND ROCKS THAT CAN BE DATED
PARENT	DAUGHTER			
Uranium 238	Lead 206	4.5 billion	10 million to 4.6 billion	Zircon Uraninite
Uranium 235	Lead 207	704 million		
Thorium 232	Lead 208	14 billion		
Rubidium 87	Strontium 87	48.8 billion	10 million to 4.6 billion	Muscovite Biotite Potassium feldspar Whole metamorphic or igneous rock
Potassium 40	Argon 40	1.3 billion	100,000 to 4.6 billion	Glauconite Hornblende Muscovite Whole volcanic rock Biotite

carbon-14 dating technique Absolute dating technique relying on the ratio of C^{14} to C^{12} in an organic substance; useful back to about 70,000 years ago.

The potassium–argon method is typically used for dating fine-grained volcanic rocks from which individual crystals cannot be separated; hence, the whole rock is analyzed. Because argon is a gas, great care must be taken to ensure that the sample has not been subjected to heat, which would allow argon to escape; such a sample would yield an age that is too young. Other long-lived radioactive isotope pairs exist, but they are rather rare and are used only in special situations.

Carbon-14 Dating Method

Carbon is an important element in nature and is one of the basic elements found in all forms of life. It has three isotopes; two of these, carbon 12 and 13, are stable, whereas carbon 14 is radioactive (see Figure 3.3). Carbon 14 has a half-life of 5,730 years plus or minus 30 years. The **carbon-14 dating technique** is based on the ratio of carbon 14 to carbon 12 and is used to date formerly living material.

The short half-life of carbon 14 makes this dating technique practical only for specimens younger than about 70,000 years. Consequently, the carbon-14 dating method is especially useful in archaeology and has greatly helped unravel the events of the latter portion of the Pleistocene Epoch.

Carbon 14 is constantly formed in the upper atmosphere when cosmic rays, which are high-energy particles (mostly protons), strike the atoms of upper-atmospheric gases, splitting their nuclei into protons and neutrons. When a neutron strikes the nucleus of a nitrogen atom (atomic number 7, atomic mass number 14), it may be absorbed into the nucleus and a proton emitted. Thus, the atomic number of the atom decreases by 1, whereas the atomic mass number stays the same. Because the atomic number has changed, a new element, carbon 14 (atomic number 6, atomic mass number 14), is formed. The newly formed carbon 14 is rapidly assimilated into the carbon cycle and, along with carbon 12 and 13, is absorbed in a nearly constant ratio by all living organisms (Figure 16.18). When an organism dies, however, carbon 14 is not replenished, and the ratio of carbon 14 to carbon 12 decreases as carbon 14 decays back to nitrogen by a single beta decay step (Figure 16.19).

Currently, the ratio of carbon 14 to carbon 12 is remarkably constant both in the atmosphere and in living organisms. There is good evidence, however, that the production of carbon 14, and thus the ratio of carbon 14 to carbon 12, has varied somewhat during the past several thousand years. This was determined by comparing ages established by carbon-14 dating of wood samples with ages established by counting annual tree rings in the same samples. As a result, carbon-14 ages have been corrected to reflect such variations in the past.

Figure 16.19 Carbon-14 Dating Method

The carbon cycle showing the formation of carbon 14 in the upper atmosphere, its dispersal and incorporation into the tissues of all living organisms, and its decay back to nitrogen 14 by beta decay.

LO7 DEVELOPMENT OF THE GEOLOGIC TIME SCALE

The geologic time scale is a hierarchical scale in which the 4.6-billion-year history of Earth is divided into time units of varying duration (Figure 16.1). It did not result from the work of any one individual, but rather evolved, primarily during the 19th century, through the efforts of many people.

By applying relative dating methods to rock outcrops, geologists in England and western Europe defined the major geologic time units without the benefit of radiometric dating techniques. Using the principles of superposition and fossil succession, they correlated various rock exposures and pieced together a composite geologic section. This composite section is, in effect, a relative time scale because the rocks are arranged in their correct sequential order.

By the beginning of the 20th century, geologists had developed a relative geologic time scale but did not yet have any absolute dates for the various time–unit boundaries. Following the discovery of radioactivity near the end of the 19th century, radiometric dates were added to the relative geologic time scale (Figure 16.1).

Because sedimentary rocks, with rare exceptions, cannot be radiometrically dated, geologists have had to rely on interbedded volcanic rocks and igneous intrusions to apply absolute dates to the boundaries of the various subdivisions of the geologic time scale (Figure 16.20).

An ashfall or lava flow provides an excellent marker bed that is a time-equivalent surface, supplying a minimum age for the sedimentary rocks below and a maximum age for the rocks above. Ashfalls are particularly useful because they may fall over both marine and nonmarine sedimentary environments and therefore can provide a connection between these different environments.

Figure 16.20 Determining Absolute Dates for Sedimentary Rocks
The absolute ages of sedimentary rocks can be determined by dating associated igneous rocks. In a and b, sedimentary rocks are bracketed by rock bodies for which absolute ages have been determined.

150 to 180 MYA

Nonconformity

Dike

Batholith 180 MYA

a.

150 MYA

> 600 to < 675 MYA

> 675–750 MYA

Lava flow (600 MYA)

Ash fall (675 MYA)

Sedimentary rocks regionally metamorphosed (750 MYA)

b.

© 2013 Cengage Learning

Thousands of absolute ages are now known for sedimentary rocks of known relative ages, and these absolute dates have been added to the relative time scale. In this way, geologists have been able to determine both the absolute ages of the various geologic periods and their durations (Figure 16.1). In fact, the dates for the era, period, and epoch boundaries of the geologic time scale are still being refined as more accurate dating methods are developed and new exposures dated. The ages shown in Figures 1.17 and 16.1 are the most recently published (2009).

LO8 GEOLOGIC TIME AND CLIMATE CHANGE

Given the debate concerning global warming and its possible implications, it is extremely important to be able to reconstruct past climatic regimes as accurately as possible. To model how Earth's climate system has responded to changes in the past and to use that information for simulations of future climate scenarios, geologists must have a geologic calendar that is as precise and accurate as possible. The ability to accurately determine when past climate changes occurred helps geologists correlate these changes with regional and global geologic events to see if there are any possible connections.

One interesting method that is becoming more common in reconstructing past climates is to analyze stalagmites from caves. Recall that stalagmites are icicle-shaped structures rising from a cave floor and formed of calcium carbonate precipitated from evaporating water (see Chapter 12). A stalagmite therefore records a layered history because each newly precipitated layer of calcium carbonate is younger than the previously precipitated layer (Figure 16.21).

Thus, a stalagmite's layers are oldest in the center at its base and progressively younger as they move outward (principle of superposition). Using techniques based on ratios of uranium 234 to thorium 230, and reliable back to approximately 500,000 years, geologists can achieve very precise radiometric dates on individual layers of a stalagmite.

A study of stalagmites from Crevice Cave in Missouri revealed a history of climatic and vegetation change in the mid-continent region of the United States during the interval between 75,000 and 25,000 years ago. Dates obtained from the Crevice Cave stalagmites were correlated with major changes in vegetation and average temperature fluctuations obtained from carbon 13 and oxygen 18 isotope profiles to reconstruct a detailed picture of climate changes during this time period.

As a result, precise dating techniques in stalagmite studies provide an accurate chronology that allows geologist to model climate systems of the past and perhaps to determine what causes global climate changes and their duration. By analyzing past environmental and climate changes and their duration, geologists hope they can use these data, sometime in the near future, to predict and possibly modify regional climate changes.

Figure 16.21 Stalagmites and Climate Change

Stalactites

Stalagmite

1 Newly forming layers of calcite in a stalagmite contain U^{234} (substituting for calcium).

2 The inside of a stalagmite is layered like an onion, showing its incremental growth.

3 U^{234} decays to Th^{230} at a predictable, measurable rate.

$\dfrac{U^{234}}{Th^{230}}$

Time

Time of layer formation

4 The age of each layer can be determined by measuring its ratio of U^{234}/Th^{230}. Layer 5 (to the left) is older than layer 2. It has a lower U^{234}/Th^{230} ratio.

O^{18}/O^{16} in water

Temperature

a. Stalagmites are icicle-shaped structures rising from the floor of a cave that are formed by the precipitation of calcium carbonate from evaporating water. A stalagmite is thus layered, with the oldest layer in the center and the youngest layers on the outside. Uranium 234 frequently substitutes for the calcium ion in the calcium carbonate of the stalagmite. Uranium 234 decays to thorium 230 at a predictable and measurable rate. Therefore, the age of each layer of the stalagmite can be dated by measuring its ratio of uranium 234 to thorium 230.

b. There are two isotopes of oxygen, a light one, oxygen 16, and a heavy one, oxygen 18. Because the oxygen 16 isotope is lighter than the oxygen 18 isotope, it vaporizes more readily than the oxygen 18 isotope when water evaporates. Therefore, as the climate becomes warmer, evaporation increases, and the O^{18}/O^{16} ratio becomes higher in the remaining water. Water in the form of rain or snow percolates into the ground and becomes trapped in the pores between the calcite forming the stalagmites.

1 The layers in a core or slice of a stalagmite are dated radiometrically.

2 The pore water is analyzed in each layer for O^{18}/O^{16} and species of plants (from the pollen).

3 A record of climatic change is put together for the area of the caves.

Climate conditions

Warmer, drier

Present

Cooler, wetter

Time ⟶   (today)

c. The layers of a stalagmite can be dated by measuring the U^{234}/Th^{230} ratio, and the O^{18}/O^{16} ratio determined for the pore water trapped in each layer. Thus, a detailed record of climatic change for the area can be determined by correlating the climate of the area as determined by the O^{18}/O^{16} ratio to the time period determined by the U^{234}/Th^{230} ratio.

© 2013 Cengage Learning

17 | Earth History

Outcrop of cross-bedded Upper Cambrian sandstone in the Dells area of Wisconsin. These beds were deposited by the transgressing Sauk Sea and are part of the Sauk sequence. They have subsequently been exposed by erosion of the Wisconsin River.

Sue Monroe

Introduction

"Imagine a barren, lifeless, waterless, hot **planet with a poisonous** atmosphere . . . what Earth was like **shortly after it** formed."

Imagine a barren, lifeless, waterless, hot planet with a poisonous atmosphere. Cosmic radiation is intense, meteorites and comets crash to the ground, and volcanoes erupt almost continuously. Storms form in the turbulent atmosphere, lightning flashes much of the time, but no rain falls, because all water is in the form of vapor due to the high temperature, day and night. And because the atmosphere has no free oxygen, nothing burns, yet pools and streams of molten rock radiate a continuous red glow. This may seem like a description from a science fiction novel, but it is probably a reasonably accurate account of what Earth was like shortly after it formed (Figure 17.1a).

Our concept of time is totally inadequate to comprehend the magnitude of geologic time, or what some call *deep time*. After all, we have no frame of reference for millions or billions of years, so the 4.6-billion-year-existence of Earth is more than we can grasp. Consider

Learning Outcomes

After reading this unit, you should be able to do the following:

LO1 Describe Earth's Precambrian history

LO2 Describe the Paleozoic geography of Earth

LO3 Describe the Paleozoic evolution of North America

LO4 Understand and explain the history of Paleozoic mobile belts

LO5 Explain the role of microplates in Paleozoic geology

LO6 Describe the stages in the breakup of Pangaea

LO7 Explain the Mesozoic history of North America

LO8 Explain Cenozoic Earth history

LO9 Describe the geologic history of the North American Cordillera

this: Suppose that 1 second equals 1 year. If so, and you were to count to 4.6 billion, the task would take you and your descendants nearly 146 years! In fact, the time we designate as Precambrian alone constitutes about 88% of all geologic time.

This chapter has the ambitious task of sketching the history of Earth from its original state to its current and very different conditions. In Chapter 1, we introduced the concept of a *system* as a combination of related parts that operate in an organized fashion, and we gave examples of interactions in discussing volcanism, plate tectonics, running water, and glaciation. After Earth formed some 4.6 billion years ago, its systems became operative, although not all at the same time or in their present form. For instance, Earth did not differentiate into a core and mantle until millions of years after it initially formed, and we know of no crust older than 4.0 billion years. However, once Earth did differentiate into layers, internal heat drove plate movements and the crust began evolving—as it continues to do.

LO1 PRECAMBRIAN EARTH HISTORY

Precambrian is a widely used term that refers to both time and rocks. As a time term, it includes all geologic time from Earth's origin 4.6 billion years ago to the beginning of the Phanerozoic Eon 542 million years ago (see Figure 17.1b). The term also refers to all rocks lying below rocks of the Cambrian System. Unfortunately, no rocks are known for the first 600 million years of geologic time, so our geologic record begins about 4.0 billion years ago, with the oldest known rocks on Earth, the Acasta Gneiss of Canada (see Figure 7.1). The geologic record we do have for the Precambrian, particularly its older part, is difficult to decipher, because many of these ancient rocks (1) have been metamorphosed and complexly deformed; (2) in many areas, they are deeply buried beneath younger rocks; and (3) they contain few fossils useful for determining relative ages.

Geologists divide the Precambrian into three parts: the Hadean (an informal term) (4.6 to 4.0 billion years ago), the Archean Eon (4.0 to 2.5 billion years ago), and the Proterozoic Eon (2.5 billion to 542 million years ago). The eons are subdivided by using prefixes such

Figure 17.1 The Precambrian Earth and Precambrian Time Scale

a. Earth as it may have appeared soon after it formed. No rocks are known from this time, but geologists can make reasonable inferences about the nature of the newly formed planet.

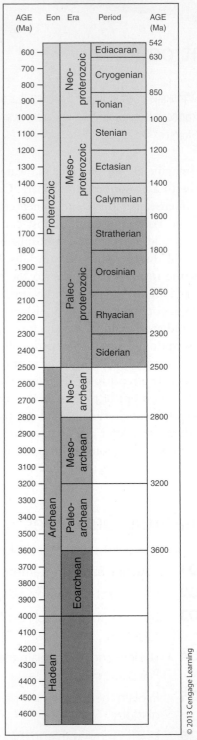

> **shield** A vast area of exposed ancient rocks on a continent; the exposed part of a craton.

as *paleo* (old or ancient), *neo* (new or recent), and so on (Figure 17.1b). So the Precambrian lasted for more than 4.0 billion years, and Earth has existed for 4.6 billion years. Our task here is to cover this lengthy interval in only a few pages.

The Origin and Evolution of Continents

Rocks 3.8 to 4.0 billion years old that are thought to represent continental crust are known from several areas, including Minnesota, Greenland, Canada, and South Africa. Furthermore, these rocks are metamorphic, so they were altered from even older rocks.

According to one model for the origin of continents, the oldest crust was thin, unstable, and composed of ultramafic igneous rock. At first, this early crust was disrupted by rising basaltic magma at ridges and was consumed at subduction zones. A second stage in crustal evolution began when partial melting of earlier-formed basaltic crust resulted in the formation of andesitic island arcs, and partial melting of lower crustal andesites yielded granitic magmas that were emplaced in the crust. As plutons were emplaced in these island arcs, they became more like continental crust. By approximately 4.0 billion years ago, plate movements accompanied by subduction and collisions of island arcs had formed several granitic continental nuclei (Figure 17.2).

Shields, Platforms, and Cratons

Each continent has a vast area of exposed Precambrian rocks known as a **shield**. Extending outward

b. This most recent version of the geologic time scale was published by the International Commission on Stratigraphy (ICS) in 2009. See Figures 1.17 and 16.1 for the complete time scale. Notice the use of the prefixes *eo* (early or dawn), *paleo* (old or ancient), *meso* (middle), and *neo* (new or recent). The age columns on the left and right sides of the time scale are in hundreds and thousands of millions of years (1,800 million years = 18 billion years, for example).

Figure 17.2 Three Stages in the Origin of Granitic Continental Crust

Andesitic island arcs formed by the partial melting of basaltic oceanic crust are intruded by granitic magmas. As a result of plate movements, island arcs collide and form larger units, or cratons.

a. Two island arcs on separate plates move toward each other.

b. The island arcs shown in (a) collide, forming a small craton, and another island arc approaches this craton.

c. The island arc shown in (b) collides with the craton.

© 2013 Cengage Learning

platform The broad area extending from a shield, but covered by younger rocks; a platform and shield form a craton.

craton The relatively stable part of a continent; consists of a shield and a buried extension of a shield known as a platform; the ancient nucleus of a continent.

Canadian shield The exposed part of the North American craton; mostly in Canada but also in Minnesota, Wisconsin, Michigan, and New York.

greenstone belt A linear or podlike association of igneous and sedimentary rocks; typically synclinal and consists of lower and middle volcanic units and an upper sedimentary rock unit.

from shields are **platforms**, consisting of Precambrian rocks beneath more recent rocks. A shield and platform collectively form a **craton**, which we can think of as the stable core or nucleus of a continent (Figure 17.3). Many of the rocks within cratons have been strongly deformed, intruded by plutons, and altered by metamorphism, but they have experienced little or no deformation since the end of the Precambrian. Their stability since that time contrasts sharply with their Precambrian history of orogenic activity.

The **Canadian shield**, the exposed part of the North American craton, is a vast area of subdued topography, lakes, and exposed Archean and Proterozoic rocks, thinly covered in places by Pleistocene glacial deposits (Figure 17.3). Beyond the Canadian shield, exposures of Precambrian rocks are limited to areas of deep erosion, such as the Grand Canyon, and areas of orogeny, such as the Appalachian and Rocky Mountains.

Geologists have delineated several smaller units within the Canadian shield, each of which is recognized by radiometric ages and trends of geologic structures.

These smaller units, as well as others making up the platform, are the subunits that constitute the North American craton. Each smaller unit was likely an independent minicontinent that later assembled into a large craton. The amalgamation of these units took place along deformation belts during the Paleoproterozoic Eon.

Archean Earth History

By far, the most common rocks of Archean age are complexes consisting of granite and gneiss, as well as subordinate, but reasonably common rock successions, known as **greenstone belts**. Greenstone belts make up only about 10 percent of Archean rocks, and yet they are important in unraveling some of the complexities of Archean tectonic events.

An ideal greenstone belt has three major rock units: The lower and middle units are mostly volcanic rocks, and the upper unit is sedimentary. Low-grade metamorphism and the origin of the minerals chlorite, epidote, and actinolite give the volcanic rocks a greenish color.

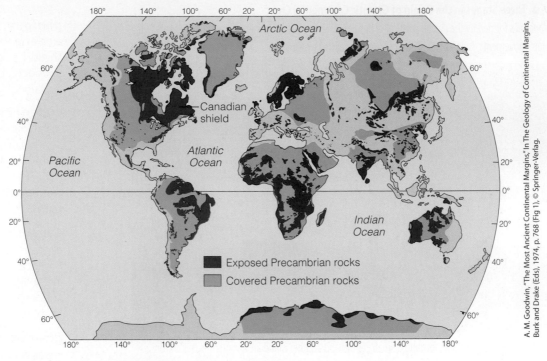

Figure 17.3 The Distribution of Precambrian Rocks
Areas of exposed Precambrian rocks constitute the shields, whereas the platforms consist of buried Precambrian rocks. A shield and its adjoining platform make up a craton.

A. M. Goodwin, "The Most Ancient Continental Margins," In The Geology of Continental Margins, Burk and Drake (Eds), 1974, p. 768 (Fig 1), © Springer-Verlag.

Laurentia A Proterozoic continent composed mostly of North America, Greenland, and parts of Scotland and Scandinavia.

Most greenstone belts have a synclinal structure and have been intruded by granitic plutons and complexly folded and cut by thrust faults (Figure 17.4a).

A currently widely accepted model for the origin of greenstone belts holds that they develop in *back-arc basins* that first open and then close. An early stage of extension takes place when the back-arc basin opens, during which time volcanism and sedimentation take place, and finally an episode of compression occurs as it closes (Figure 17.5). During closure, the rocks are intruded by plutons and metamorphosed, and the greenstone belt takes on a synclinal form as it is folded and faulted.

Undoubtedly, the present style of plate tectonics, involving the opening and closing of ocean basins, has been the primary agent of Earth evolution since at least the Paleoproterozoic. But many geologists are now convinced that some form of plate tectonics was operating during the Archean as well. However, Earth's radiogenic heat production has diminished through time, so during the Archean, when more heat was available, seafloor spreading and plate movements took place more rapidly.

Nevertheless, by the Paleoproterozoic and since then, a plate tectonic style like that of the present has operated.

Proterozoic Earth History

The origin of greenstone belts and granite-gneiss complexes continued into the Proterozoic but at a considerably reduced rate. Whereas most Archean rocks have been metamorphosed, many Proterozoic rocks have been little altered. Furthermore, the Proterozoic was a time of deposition of *banded iron formations*, consisting of alternating layers of iron minerals and chert; *red beds*, sandstones, siltstones, and shales with iron oxide cement; and sedimentary rocks indicating two episodes of widespread glaciation. Finally, widespread assemblages of sandstone, carbonates, and shale deposited on passive continental margins were common (Figure 17.6).

In the preceding section, we noted that Archean crust assembled through a series of island-arc and mini-continent collisions, providing the nuclei around which Proterozoic continental crust accreted. One large landmass so formed, called **Laurentia**, consisted mostly of North America and Greenland, parts of northwestern Scotland, and perhaps parts of the Baltic Shield of Scandinavia.

Figure 17.4 Greenstone Belts

b. Pillow lava of the Ishpeming greenstone belt in Michigan.

Greenstone belt succession

■ Granitic intrusives

▨ Upper sedimentary unit: sandstones and shales most common

■ Middle volcanic unit: mainly basalt

▨ Lower volcanic unit: mainly peridotite and basalt

■ Granite–gneiss complex

a. Two adjacent greenstone belts show their synclinal structure. The two lower units in greenstone belts are mostly volcanic rocks, whereas the upper unit is sedimentary.

c. Gneiss from a granite-gneiss complex in Ontario, Canada.

Figure 17.5 Origin of a Greenstone Belt in a Back-Arc Marginal Basin

a. Basalt lavas and sediment derived from the continent and island arc fill the back-arc marginal basin.

b. Closure of the back-arc marginal basin causes compression and deformation. The evolving greenstone belt is deformed into a syncline-like structure into which granitic magmas intrude.

Figure 17.6 Paleoproterozoic and Neoproterozoic Sedimentary Rocks

Many of the sedimentary rocks of Proterozoic-age are parts of sandstone-carbonate-shale assemblages that were deposited on passive continental margins.

a. Outcrop of the Paleoproterozoic Mesnard Quartzite in Michigan. The crests of the ripple marks point toward the observer.

b. The Paleoproterozoic Kona Dolomite of Michigan. The bulbous structures are stromatolites that were produced from the activities of cyanobacteria (blue-green algae).

c. This billion-year-old (Neoproterozoic) outcrop of sandstone and mudstone in Glacier National Park in Montana has been only slightly altered by metamorphism.

orogen A linear part of Earth's crust that was, or is, being deformed during an orogeny.

The first major episode of Laurentia's crustal evolution took place during the Paleoproterozoic, between 2.0 and 1.8 billion years ago (bya). Several major **orogens**—zones of deformed rocks—developed, many of which were metamorphosed and intruded by plutons. So this was a time of continental accretion during which collisions between Archean-age crust formed a large craton (Figure 17.7a). Following the Paleoproterozoic amalgamation of cra-

tons, considerable accretion took place between 1.8 and 1.6 bya in what is now the southwestern and central United States, as successively younger belts were sutured to the craton (Figure 17.7b).

Between 1.6 and 1.3 bya, extensive igneous activity—particularly the emplacement of granitic plutons and eruptions of rhyolite and ash flows—occurred that was unrelated to orogenic activity. According to one hypothesis, these rocks resulted from large-scale upwelling of magma beneath a supercontinent.

Another important event in the evolution of Laurentia, the *Grenville orogeny* in the eastern United

Figure 17.7 Proterozoic Evolution of Laurentia

These three illustrations show the overall trends in the Proterozoic evolution of Laurentia, but they do not show many of the details of this long, complex episode in Earth history.

a. During the Paleoproterozoic, Archean cratons were sutured along deformation belts called orogens.

b. Laurentia grew along its southeastern margin by accretion of the Yavapai and Mazatzal provinces.

© 2013 Cengage Learning

States and Canada, took place between 1.3 and 1.0 bya (Neoproterozoic) (Figure 17.7c). Grenville deformation represents the final episode of Proterozoic continental accretion of Laurentia. Contemporaneous with Grenville deformation was an episode of rifting in Laurentia, resulting in the origin of the *Midcontinent Rift*. The central part of the rift is filled with hundreds of overlapping basaltic lava flows and sedimentary rocks forming a pile several kilometers thick.

LO2 THE PALEOZOIC GEOGRAPHY OF EARTH

By the beginning of the Paleozoic Era, six major continents were present. Besides these large landmasses, geologists have also identified numerous microcontinents, such as *Avalonia* (composed of parts of present-day Belgium, northern France, England, Wales, Ireland, the Maritime Provinces and Newfoundland of Canada, as well as parts of the New England area of the United States), and various island

■ 1.3–1.0 billion	} Mesoproterozoic
□ 1.55–1.35 billion	
■ 1.69–1.65 billion	} Paleoproterozoic
□ 1.76–1.72 billion	
■ 1.9–1.8 billion	
■ 2.0–1.8 billion	} Archean
□ >2.5 billion	

© 2013 Cengage Learning

c. The last episodes in the Proterozoic accretion of Laurentia involved the origin of the Granite-Rhyolite province and the Grenville-Llano province.

arcs associated with microplates. The six major Paleozoic continents are *Baltica* (Russia west of the Ural Mountains and the major part of northern Europe), *China* (a complex area consisting of at least three Paleozoic continents that were not widely separated and are here considered to include China, Indochina, and the Malay Peninsula), *Gondwana* (Africa, Antarctica, Australia, Florida, India, Madagascar, and parts of the Middle East and southern Europe), *Kazakhstania* (a triangular continent centered on Kazakhstan but considered by some to be an extension of the Paleozoic Siberian continent), *Laurentia* (most of present North America, Greenland, northwestern Ireland, and Scotland), and *Siberia* (Russia east of the Ural Mountains and Asia north of Kazakhstan and south of Mongolia). The paleogeographic reconstructions that follow are based on the methods used to determine and interpret location, geographic features, and environmental conditions on the paleocontinents.

In contrast to today's global geography, the Cambrian world consisted of these six continents dispersed around the globe at low tropical latitudes (Figure 17.8a). Water circulated freely among ocean basins, and the polar regions were apparently ice free. By the Late Cambrian, shallow seas had covered large areas of Laurentia, Baltica, Siberia, Kazakhstania, and China, whereas highlands were present in northeastern Gondwana, eastern Siberia, and central Kazakhstania.

During the Ordovician and Silurian periods, plate movement played a major role in the changing global geography. Gondwana moved southward during the Ordovician and began to cross the South Pole, as indicated by Upper Ordovician glacial deposits found today in the Sahara Desert (Figure 17.8b). During the Early Ordovician, the microcontinent Avalonia separated from Gondwana and began moving northeastward, where it would finally collide with Baltica during the Late Ordovician—Early Silurian. In contrast to the passive continental margin Laurentia exhibited during the Cambrian, an active convergent plate boundary formed along its eastern margin during the Ordovician, as indicated by the Late Ordovician *Taconic orogeny* that occurred in New England.

During the Silurian, Baltica, along with the newly attached Avalonia, moved northwestward relative to Laurentia and collided with it to form the larger continent of *Laurasia*. This collision, which closed the northern *Iapetus Ocean*, is marked by the *Caledonian orogeny*. After this orogeny, the southern part of the Iapetus Ocean still remained open between Laurentia and Avalonia–Baltica. Siberia and Kazakhstania moved from a southern equatorial position during the Cambrian to north temperate latitudes by the end of the Silurian Period.

During the Devonian, as the southern Iapetus Ocean narrowed between Laurasia and Gondwana, mountain building continued along the eastern margin of Laurasia with the *Acadian orogeny* (Figure 17.9a). Erosion of the ensuing highlands spread vast amounts of reddish fluvial sediments over large areas of northern Europe and eastern North America.

Other Devonian tectonic events, probably related to the collision of Laurentia and Baltica, include the Cordilleran *Antler orogeny* and the change from a passive continental margin to an active convergent plate boundary in the Uralian mobile belt of eastern Baltica. The distribution of reefs, evaporites, and red beds, as well as the existence of similar floras throughout the world, suggest a rather uniform global climate during the Devonian Period.

During the Carboniferous Period, southern Gondwana moved over the South Pole, resulting in extensive continental glaciation. The advance and retreat of these glaciers produced global changes in sea level that affected sedimentation patterns on the cratons. As Gondwana continued moving northward, it collided with Laurasia during the Early Carboniferous and continued suturing with it during the rest of the Carboniferous. The final phase of collision between Gondwana and Laurasia is indicated by the Ouachita Mountains of Oklahoma and Arkansas, formed by thrusting during the Late Carboniferous and Early Permian.

Elsewhere, Siberia collided with Kazakhstania and moved toward the Uralian margin of Laurasia (Baltica), colliding with it during the Early Permian. By the end of the Carboniferous, the various continental landmasses were fairly close together as Pangaea began taking shape.

The assemblage of Pangaea concluded during the Permian with the completion of many of the continental collisions that began during the Carboniferous (Figure 17.9b). An enormous single ocean, *Panthalassa*, surrounded Pangaea and spanned Earth from pole to pole. Waters of this ocean probably circulated more freely than at present, resulting in more equable water temperatures.

The formation of a single landmass had climatic consequences for the terrestrial environment as well. Terrestrial Permian sediments indicate that arid and semiarid conditions were widespread over Pangaea. The mountain ranges produced by the *Hercynian*, *Alleghenian*, and *Ouachita orogenies* were high enough to create rainshadows that blocked the moist, subtropical, easterly winds—much as the southern Andes Mountains do in western South America today. This produced very dry conditions in North America and Europe, as indicated by the extensive Permian red beds and evaporites found in western North America, central Europe, and parts of Russia. Permian coals, indicative of abundant rainfall, were

Figure 17.8 Paleogeography for the Late Cambrian and Ordovician Periods

REPRINTED BY PERMISSION OF RONALD BLAKEY, NORTHERN ARIZONA UNIVERSITY.

a. Late Cambrian Period.

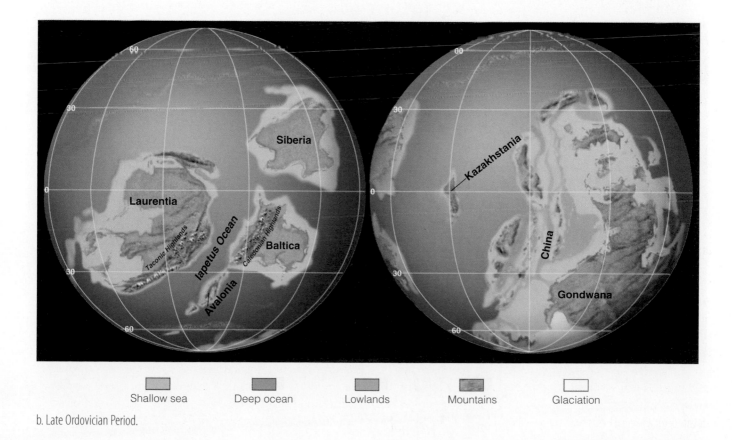

b. Late Ordovician Period.

Figure 17.9 Paleogeography for the Late Devonian and Permian Periods

REPRINTED BY PERMISSION OF RONALD BLAKEY, NORTHERN ARIZONA UNIVERSITY.

Shallow sea   Deep ocean   Lowlands   Mountains

a. Late Devonian Period.

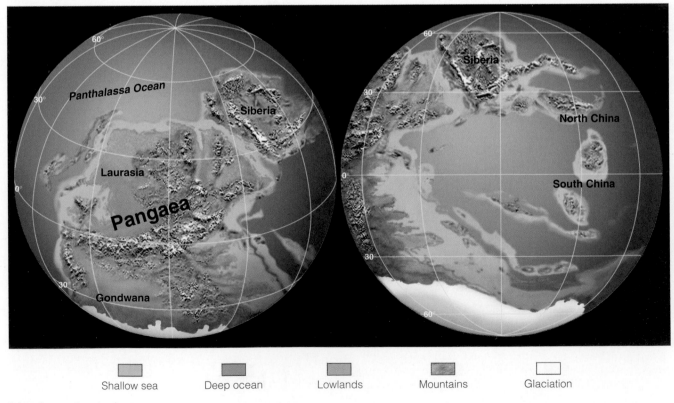

Shallow sea   Deep ocean   Lowlands   Mountains   Glaciation

b. Late Permian Period.

mostly limited to the northern temperate belts (latitude 40 to 60 degrees north), whereas the last remnants of the Carboniferous ice sheets continued their recession.

LO3 THE PALEOZOIC EVOLUTION OF NORTH AMERICA

We divide the Paleozoic history of the North American craton into two parts, the first dealing with the relatively stable continental interior over which shallow seas advanced (transgressed) and retreated (regressed), and the second with the mobile belts where mountain building occurred.

Geologists commonly divide the sedimentary record of North America into six cratonic sequences. A *cratonic sequence* is a major transgressive-regressive cycle bounded by craton-wide unconformities. The transgressive phase is usually well-preserved, whereas the regressive phase of each sequence is marked by an unconformity.

The Sauk Sequence

Rocks of the **Sauk sequence** (Late Neoproterozoic–Early Ordovician) record the first major transgression onto the North American craton. During the Neoproterozoic and Early Cambrian, deposition of marine sediments was limited to the passive shelf areas of the Appalachian and Cordilleran borders of the craton. The craton itself was above sea level and experiencing weathering and erosion. Because North America was located in a tropical climate at this time (Figure 17.8a), and because there is no evidence of any terrestrial vegetation, we can conclude that weathering and erosion of the exposed Precambrian basement rocks must have proceeded rapidly.

During the Middle Cambrian, the transgressive phase of the Sauk began with shallow seas encroaching over the craton. By the Late Cambrian, the Sauk Sea had covered most of North America, leaving only a portion of the Canadian shield and a few large islands above sea level (Figure 17.10). These islands, collectively named the *Transcontinental Arch*, extended from New Mexico to Minnesota and the Lake Superior region.

The Tippecanoe Sequence

As the Sauk Sea regressed from the craton during the Early Ordovician, a landscape of low relief emerged. The exposed rocks were predominantly limestones and dolostones that were deeply eroded because North America was still located in a tropical environment (Figure 17.8b).

The resulting craton-wide unconformity marks the boundary between the Sauk and Tippecanoe sequences.

Like the Sauk sequence, deposition of the **Tippecanoe sequence** (Middle Ordovician–Early Devonian) began with a major transgression onto the craton. This transgressing sea deposited clean, well-sorted quartz sands over most of the craton. The Tippecanoe basal sandstones were followed by widespread carbonate deposition. The limestones were formed by calcium carbonate–secreting marine organisms such as corals and brachiopods.

As the Tippecanoe Sea gradually regressed from the craton during the Late Silurian, precipitation of evaporite minerals took place in the Appalachian, Ohio, and Michigan basins (Figure 17.11). In the Michigan Basin alone, approximately 1,500 m of sediments were deposited, nearly half of which are halite and anhydrite.

By the Early Devonian, the regressing Tippecanoe Sea had retreated to the craton margin, exposing an extensive lowland topography. During this regression, marine deposition was initially restricted to a few interconnected cratonic basins and finally, by the end of the Tippecanoe, to only the margins surrounding the craton.

As the Tippecanoe Sea regressed during the Early Devonian, the craton experienced mild deformation, resulting in the formation of many domes, arches, and basins. These structures were mostly eroded during the time the craton was exposed, and deposits from the ensuing and encroaching Kaskaskia Sea eventually covered them.

The Kaskaskia Sequence

The boundary between the Tippecanoe sequence and the overlying **Kaskaskia sequence** (Middle Devonian–Late

Sauk sequence A widespread association of sedimentary rocks bounded above and below by unconformities; deposited during a Late Neoproterozoic to Early Ordovician transgressive-regressive cycle of the Sauk Sea.

Tippecanoe sequence A widespread body of sedimentary rocks bounded above and below by unconformities; deposited during a Middle Ordovician to Early Devonian transgressive-regressive cycle of the Tippecanoe Sea.

Kaskaskia sequence A widespread sequence of Middle Devonian to Upper Mississippian sedimentary rocks bounded above and below by unconformities; deposited during a transgressive-regressive cycle of the Kaskaskia Sea.

Figure 17.10 Paleogeography of North America During the Cambrian Period

Note the position of the Cambrian paleoequator. During this time, North America straddled the equator, as indicated in Figure 17.8a.

Islands
Land
Shallow sea
Deep ocean

Cordilleran mobile belt

Muddy bottom

Carbonate bottom

Sandy bottom

Canadian Shield

Sandy bottom

Carbonate bottom

Ouachita - Appalachian mobile belt

Muddy bottom

Transcontinental Arch

Paleoequator

© 2013 Cengage Learning

Mississippian) is marked by a major unconformity. As the Kaskaskia Sea transgressed over the low-relief landscape of the craton, most of the basal beds consisted of clean, well-sorted, quartz sandstones.

Except for widespread Upper Devonian and Lower Mississippian black shales, the majority of Kaskaskian rocks are carbonates, including reefs, and associated evaporite deposits. In many other parts of the world, such as southern England, Belgium, central Europe, Australia, and Russia, the Middle and early Late Devonian epochs were times of major reef building.

During the Late Mississippian regression of the Kaskaskia Sea from the craton, carbonate deposition was replaced by vast quantities of detrital sediments. Before the end of the Mississippian, the Kaskaskia Sea had retreated to the craton margin, once again exposing the craton to widespread weathering and erosion, resulting in a craton-wide unconformity at the end of the Kaskaskia sequence.

Figure 17.11 Paleogeography of North America During the Silurian Period
Note the development of reefs in the Michigan, Ohio, and Indiana-Illinois-Kentucky areas.

Land
Mountains
Barrier reefs
Evaporites
Shallow sea
Deep ocean

Cordilleran mobile belt

Muddy bottom

Craton

Carbonate bottom

Paleoequator

Michigan Basin

Barrier reefs

Muddy bottom

Alluvial plain

Taconic Highlands

Appalachian mobile belt

Ouachita

Muddy bottom

© 2013 Cengage Learning

The Absaroka Sequence

The extensive unconformity separating the Kaskaskia and Absaroka sequences essentially divides the strata into the North American Mississippian and Pennsylvanian systems. These two systems are closely equivalent to the European Lower and Upper Carboniferous. The rocks of the **Absaroka sequence** (Late Mississippian–Early Jurassic) not only differ from those of the Kaskaskia sequence but also result from different tectonic regimes.

One characteristic feature of Pennsylvanian rocks is their repetitive pattern of alternating marine and nonmarine strata known as **cyclothems** (Figure 17.12). They result from repeated alternations of marine and nonmarine environments, usually in areas of

Absaroka sequence
A widespread association of sedimentary rocks bounded above and below by unconformities; deposited during a Late Mississippian to Early Jurassic transgressive-regressive cycle of the Absaroka Sea.

cyclothem A vertical sequence of cyclically repeating sedimentary rocks resulting from alternating periods of marine and nonmarine deposition; commonly containing a coal bed.

low relief. Although seemingly simple, cyclothems reflect a delicate interplay between nonmarine deltaic and shallow marine interdeltaic and shelf environments.

Cyclothems represent transgressive and regressive sequences, with an erosional surface separating one cyclothem from another. Thus, an idealized cyclothem passes upward from fluvial–deltaic deposits, through coals, to detrital shallow-water marine sediments, and finally to limestones typical of an open marine environment (Figure 17.12a).

Such repetitious sedimentation over a widespread area requires an explanation. The hypothesis currently favored by many geologists is a rise and fall of sea level related to advances and retreats of Gondwanan continental glaciers. When the Gondwanan ice sheets advanced, sea level dropped; when they melted, sea level rose. Late Paleozoic cyclothem activity on all of the cratons closely corresponds to Gondwanan glacial–interglacial cycles.

During the Pennsylvanian, the area of greatest deformation occurred in the southwestern part

Figure 17.12 Columnar Section of a Complete Cyclothem

a. Columnar section of a complete cyclothem.

Disconformity
Brackish and
    nonmarine shales
Marine shales
Algal limestones with
    nearshore and brackish
    water invertebrate fossils
Limestones with offshore
    invertebrate fossils
Limestones and shale with
    offshore
    invertebrate fossils

Marine shales with nearshore
    invertebrate fossils

Coal
Underclay
Nonmarine shales and
    sandstones
Nonmarine sandstones
    Disconformity

b. Pennsylvanian coal bed, West Virginia.

Courtesy of Wayne E. Moore

c. Reconstruction of the environment of a Pennsylvanian coal-forming swamp.

© 2013 Cengage Learning

of the North American craton, where a series of fault-bounded uplifted blocks formed the *Ancestral Rockies* (Figure 17.13). These mountain ranges had diverse geologic histories and were not all elevated at the same time. Uplift of these mountains, some of which were elevated more than 2 km along near-vertical faults, resulted in the erosion of the overlying Paleozoic sediments and exposure of the Precambrian igneous and metamorphic basement rocks (Figure 17.14). As the mountains eroded, tremendous quantities of coarse, red sediments were deposited in the surrounding basins, where they are preserved in many areas, such as the spectacular Garden of the Gods, Colorado.

While the various intracratonic basins were filling with sediment during the Late Pennsylvanian, the Absaroka Sea slowly began retreating from the craton. During the Middle and Late Permian, the Absaroka Sea was restricted to west Texas and southern New Mexico, forming an interrelated complex of lagoonal, reef, and open-shelf environments (Figure 17.15). By the end of the Permian Period, the Absaroka Sea had

Figure 17.13 Paleogeography of North America During the Pennsylvanian Period

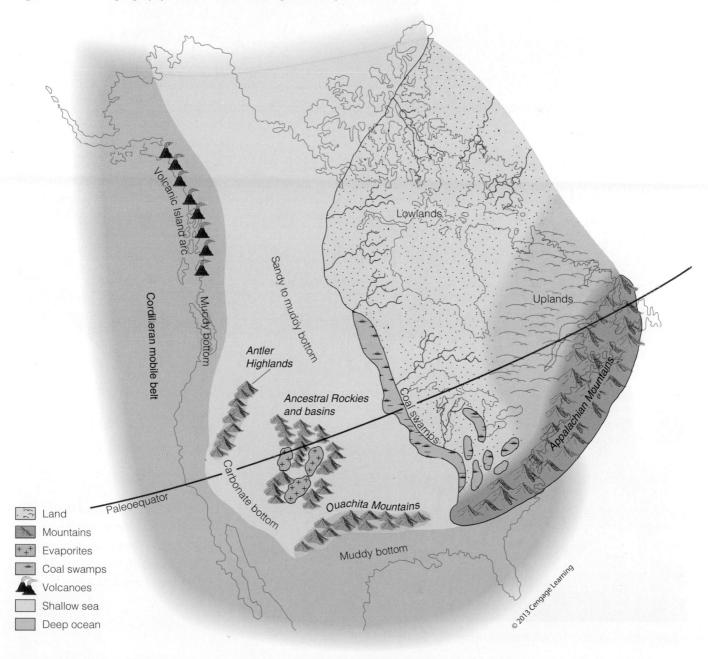

Land
Mountains
Evaporites
Coal swamps
Volcanoes
Shallow sea
Deep ocean

Volcanic Island arc
Cordilleran mobile belt
Muddy bottom
Sandy to muddy bottom
Antler Highlands
Ancestral Rockies and basins
Carbonate bottom
Paleoequator
Ouachita Mountains
Muddy bottom
Coal swamps
Lowlands
Uplands
Appalachian Mountains

© 2013 Cengage Learning

Figure 17.14 The Ancestral Rockies

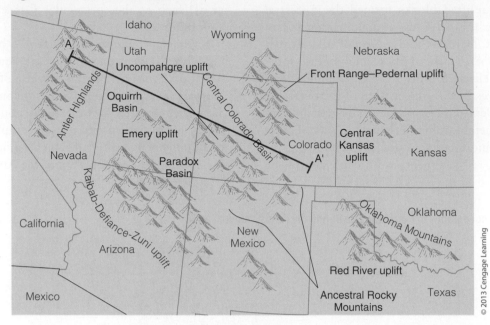

a. Location of the principal Pennsylvanian highland areas and basins of the southwestern part of the craton.

b. Block diagram of the Ancestral Rockies, elevated by faulting during the Pennsylvanian Period. Erosion of these mountains produced coarse, red sediments deposited in the basins adjacent to the Ancestral Rockies.

c. Garden of the Gods, Storm Sky View from Near Hidden Inn, Colorado Springs, Colorado.

retreated from the craton, exposing continental red beds that had been deposited over most of the southwestern and eastern region.

LO4 THE HISTORY OF THE PALEOZOIC MOBILE BELTS

Having examined the Paleozoic history of the craton, we now turn to the Paleozoic orogenic activity in the **mobile belts** (elongated areas of mountain-building activity along the margins of continents). The mountain building occurring during the Paleozoic Era had a profound influence on the climate and sedimentary history of the craton. In addition, it was part of the global tectonic regime that sutured the continents together, forming Pangaea by the end of the Paleozoic.

mobile belt An elongated area of deformation indicated by folds and faults; generally adjacent to a craton.

The Appalachian Mobile Belt

Throughout Sauk time (Neoproterozoic–Early Ordovician), the Appalachian region was a broad, passive

Figure 17.15 Paleogeography of North America During the Permian Period

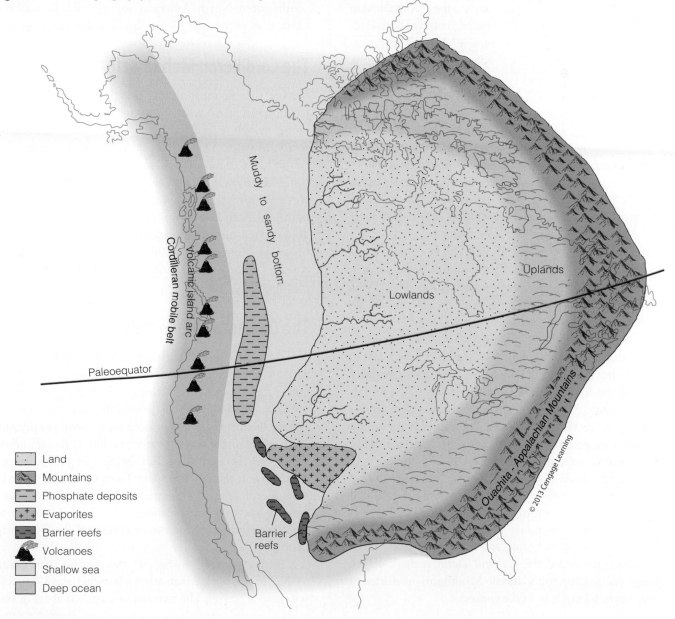

Land

Mountains

Phosphate deposits

Evaporites

Barrier reefs

Volcanoes

Shallow sea

Deep ocean

Taconic orogeny An Ordovician episode of mountain building that resulted in the deformation of the Appalachian mobile belt.

Acadian orogeny An episode of Devonian deformation in the northern Appalachian mobile belt resulting from the collision of Baltica with Laurentia.

Hercynian-Alleghenian orogeny Pennsylvanian to Permian orogenic event during which the Appalachian mobile belt in eastern North America and the Hercynian mobile belt of southern Europe were deformed.

continental margin. Sedimentation was closely balanced by subsidence as extensive carbonate deposits succeeded thick, shallow marine sands. During this time, movement along a divergent plate boundary was widening the Iapetus Ocean (Figure 17.16a).

Beginning with the subduction of the Iapetus plate beneath Laurentia (an oceanic–continental convergent plate boundary), the Appalachian mobile belt was born (Figure 17.16b). The resulting **Taconic orogeny**, named after the present-day Taconic Mountains of eastern New York, central Massachusetts, and Vermont, was the first of several orogenies to affect the Appalachian region.

A large *clastic wedge* (an extensive accumulation of mostly detrital sediments deposited adjacent to an uplifted area) formed in the shallow seas to the west of the Taconic orogeny. These deposits are thickest and coarsest nearest the highland area and become thinner and finer-grained away from the source area, eventually grading into carbonates on the craton. The clastic wedge resulting from the erosion of the Taconic Highlands is referred to as the *Queenston Delta*.

The second Paleozoic orogeny to affect Laurentia began during the Late Silurian and concluded at the end of the Devonian Period. The **Acadian orogeny** affected the Appalachian mobile belt from Newfoundland to Pennsylvania as sedimentary rocks were folded and thrust against the craton. As with the preceding Taconic orogeny, the Acadian orogeny occurred along an oceanic–continental convergent plate boundary. As the northern Iapetus Ocean continued to close during the Devonian, the plate carrying Baltica finally collided with Laurentia, forming a continental–continental convergent plate boundary along the zone of collision (Figure 17.16c). Weathering and erosion of the Acadian Highlands produced the *Catskill Delta*, a thick clastic wedge named for the Catskill Mountains in northern New York, where it is well exposed.

The Taconic and Acadian orogenies were part of the same major orogenic event related to the closing of the Iapetus Ocean. This event began with an oceanic–continental convergent plate boundary during the Taconic orogeny and culminated with a continental–continental convergent plate boundary during the Acadian orogeny as Laurentia and Baltica became sutured (Figure 17.16). After this, the **Hercynian-Alleghenian orogeny** began, followed by orogenic activity in the Ouachita mobile belt.

The Hercynian mobile belt of southern Europe and the Appalachian and Ouachita mobile belts of North America mark the zone along which Europe (part of Laurasia) collided with Gondwana. While Gondwana and southern Laurasia collided during the Pennsylvanian and Permian Periods, eastern Laurasia (Europe and southeastern North America) joined with Gondwana (Africa) as part of the Hercynian-Alleghenian orogeny.

The Cordilleran Mobile Belt

During the Neoproterozoic and Early Paleozoic, the Cordilleran area was a passive continental margin along which extensive continental shelf sediments were deposited. Beginning in the Middle Paleozoic, an island arc formed off the western margin of the craton. This eastward-moving island arc collided with the western border of the craton during the Late Devonian and Early Mississippian, resulting in a highland area termed the Antler Highlands (Figure 17.13). This orogenic event, the *Antler orogeny*, was the first in a series of orogenic events to affect the Cordilleran mobile belt.

The Ouachita Mobile Belt

The Ouachita mobile belt extends for approximately 2,100 km from the subsurface of Mississippi to the Marathon region of Texas. During the Late Neoproterozoic to Early Mississippian, shallow-water detrital and carbonate sediments were deposited on a broad continental shelf, while in the deeper-water portion of the adjoining mobile belt, bedded cherts and shales were accumulating. Beginning in the Mississippian Period, the rate of sedimentation increased dramatically as the region changed from a passive continental margin to an active convergent plate boundary, marking the beginning of the **Ouachita orogeny**.

Approximately 80% of the former Ouachita mobile belt is buried beneath a Mesozoic and Cenozoic sedimentary cover. The two major exposed areas in this

Figure 17.16 Evolution of the Appalachian Mobile Belt

Appalachian carbonate platform

Laurentia (North America)

Iapetus Ocean

Spreading ridge

a. During the Neoproterozoic to the Early Ordovician, the Iapetus Ocean was opening along a divergent plate boundary. Both the east coast of Laurentia and the west coast of Baltica were passive continental margins where large carbonate platforms existed.

Passive margin

Baltica (Europe)

Taconic Highlands

Oceanic–continental plate boundary

Narrowing Iapetus Ocean

Caledonian Highlands

b. Beginning in the Middle Ordovician, the passive margins of Laurentia and Baltica became oceanic–continental convergent plate boundaries, resulting in orogenic activity.

Queenston Delta clastic wedge

Laurentia

Laurasia

Baltica

Catskill Delta clastic wedge

Shallow sea

Acadian Caledonian Highlands

Shallow sea

c. By the Late Paleozoic, Laurentia and Baltica collided along a continental–continental convergent plate boundary, forming a larger continental landmass, Laurasia.

Suture zone

Old Red Sandstone clastic wedge

Continental crust

© 2013 Cengage Learning

region are the Ouachita Mountains of Oklahoma and Arkansas, and the Marathon Mountains of Texas.

Thrusting of sediments continued throughout the Pennsylvanian and Early Permian, driven by the compressive forces generated along the zone of subduction as Gondwana collided with Laurasia. The collision of Gondwana and Laurasia is marked by the formation of a large mountain range, most of which eroded during the Mesozoic Era. Only the rejuvenated Ouachita and Marathon Mountains remain of this once lofty mountain range.

LO5 THE ROLE OF MICROPLATES

It is becoming increasingly clear that accretion along the continental margins is more complicated than the somewhat simple, large-scale plate interactions that we have described. Geologists now recognize that numerous microplates, such as Avalonia (Figure 17.8b), existed during the Paleozoic and were involved in the orogenic events that occurred during that time.

A careful examination of Paleozoic global paleogeographic maps shows numerous microplates, and their location and role during the

formation of Pangaea must be taken into account. Thus, although the basic history of the formation of Pangaea during the Paleozoic remains the same, geologists now realize that microplates also played an important role in the formation of Pangaea. Furthermore, they help explain some previously anomalous geologic and paleontologic situations.

LO6 THE BREAKUP OF PANGAEA

Just as the formation of Pangaea influenced geologic and biologic events during the Paleozoic, the breakup of this supercontinent profoundly affected geologic and biologic events during the Mesozoic. The movement of continents affected global climatic and oceanic regimes as well as the climates of individual continents.

Geologic, paleontologic, and paleomagnetic data indicate that the breakup of Pangaea took place in four general stages. The first stage involved rifting between Laurasia and Gondwana during the Late Triassic. By the end of the Triassic, the newly formed and expanding Atlantic Ocean separated North America from Africa (Figure 17.17a). This was followed by the rifting of North America from South America sometime during the Late Triassic and Early Jurassic.

Figure 17.17 Paleogeography of the World During the Mesozoic Era
REPRINTED BY PERMISSION OF RONALD BLAKEY, NORTHERN ARIZONA UNIVERSITY.

a. Triassic Period.

b. Jurassic Period.

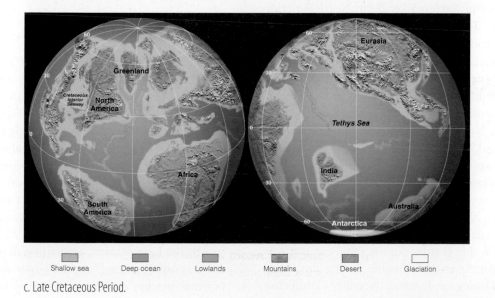

c. Late Cretaceous Period.

The second stage in Pangaea's breakup involved rifting and movement of the various Gondwana continents during the Late Triassic and Jurassic periods. As early as the Late Triassic, Antarctica and Australia, which remained sutured together, began separating from South America and Africa while India began drifting away from the Gondwana continent and moved northward (Figures 17.17a, b).

The third stage of breakup began during the Late Jurassic, when South America and Africa began to separate (Figure 17.17b). During this stage, the eastern end of the Tethys Sea began closing as a result of the clockwise rotation of Laurasia and the northward movement of Africa. This narrow Late Jurassic and Cretaceous seaway between Africa and Europe was the forerunner of the present Mediterranean Sea.

By the end of the Cretaceous, Australia and Antarctica had detached from each other, and India had moved into low southern latitudes and was nearly to the equator. South America and Africa were widely separated, and Greenland was essentially an independent landmass with only a shallow sea between it and North America and Europe (Figure 17.17c).

The final stage in Pangaea's breakup occurred during the Cenozoic. During this time, Australia continued moving northward, and Greenland was completely separated from Europe and North America and formed a separate landmass.

LO7 THE MESOZOIC HISTORY OF NORTH AMERICA

The beginning of the Mesozoic Era was essentially the same in terms of mountain building and sedimentation as the preceding Permian Period in North America (Figure 17.18). Terrestrial sedimentation continued over much of the craton, while block faulting and igneous activity began in the Appalachian region as North America and Africa began separating (Figure 17.17). The newly forming Gulf of Mexico was the site of extensive evaporite deposition during the Late Triassic and Jurassic as North America separated from South America (Figure 17.19).

A global rise in sea level during the Cretaceous resulted in worldwide transgressions onto the continents such that marine deposition was continuous over much of western North America (Figure 17.20).

A volcanic island arc system that formed off the western edge of the craton during the Permian was sutured to North America sometime during the Permian or Triassic. During the Jurassic, the entire Cordilleran area was involved in a series of major mountain-building episodes, resulting in the formation of the Sierra Nevada, the Rocky Mountains, and other lesser mountain ranges.

Eastern Coastal Region

During the Early and Middle Triassic, coarse detrital sediments derived from the erosion of the recently uplifted Appalachians (*Alleghenian orogeny*) filled various intermontane basins and spread over the surrounding areas. As weathering and erosion continued during the Mesozoic, this once lofty mountain system was reduced to a low-lying plain.

During the Late Triassic, the first stage in the breakup of Pangaea began with North America separating from Africa (Figure 17.17a). Fault-block basins developed in response to upwelling magma beneath Pangaea in a zone stretching from present-day Nova Scotia to North Carolina (Figure 17.21). Erosion of the adjacent fault-block mountains filled the adjacent basins with great quantities (up to 6,000 m) of poorly sorted, red, nonmarine detrital sediments.

As the Atlantic Ocean grew, rifting ceased along the eastern margin of North America, and this once active plate margin became a passive, trailing continental margin. The fault-block mountains produced by the rifting continued to erode during the Jurassic and Early Cretaceous until only an area of low-relief remained. Sediments from the erosion contributed to the growing eastern continental shelf. During the Cretaceous Period, the Appalachian region was re-elevated and once again shed sediments onto the continental shelf, forming a gently dipping, seaward-thickening wedge of rocks up to 3,000 m thick. These rocks are currently exposed in a belt extending from Long Island, New York, to Georgia.

Gulf Coastal Region

The Gulf Coastal region was above sea level until the Late Triassic. As North America separated from South America during the Late Triassic and Early Jurassic, the Gulf of Mexico began to form (Figure 17.19). With oceanic waters flowing into this newly formed, shallow, restricted basin, conditions were ideal for evaporite formation. More than 1,000 m of evaporites were precipitated at this time, and most geologists think that these Jurassic evaporites are the source of the Cenozoic salt domes found today in the Gulf of Mexico and southern Louisiana.

By the Late Jurassic, circulation in the Gulf of Mexico was less restricted, and evaporite deposition ended. Normal marine conditions returned to the area with

Figure 17.18 Paleogeography of North America During the Triassic Period

Land

Mountains

Fault-block basins

Volcanoes

Shallow sea

Deep ocean

Cordilleran mobile belt

Volcanic island arc

Muddy bottom

Sandy and muddy bottom

Red- colored coastal plains

Lowlands

Uplands

Appalachian Mountains and fault-block basins

Paleoequator

© 2013 Cengage Learning

alternating transgressing and regressing seas, resulting in sediments that were in turn covered and buried by thousands of meters of Cretaceous and Cenozoic sediments.

During the Cretaceous, the Gulf Coastal region, like the rest of the continental margin, was flooded by northward-transgressing seas forming a wide seaway that extended from the Arctic Ocean to the Gulf of Mexico (Figure 17.20).

Western Region

During the Permian, an island arc and ocean basin formed off the western North American craton (Figure 17.15), followed by subduction of an oceanic plate beneath the island arc and the thrusting of oceanic and island arc rocks eastward against the craton margin. This event, similar to the preceding Antler orogeny and known as the *Sonoma orogeny*, occurred at or near the Permian–Triassic boundary.

Following the Late Paleozoic–Early Mesozoic destruction of the volcanic island arc during the Sonoma orogeny, the western margin of North America became an oceanic–continental convergent plate boundary. During the Late Triassic, a steeply dipping subduction zone developed along the western margin of North America in response to the westward movement of North America over the Pacific plate. This newly created oceanic–continental plate boundary controlled Cordilleran

Figure 17.19 Paleogeography of North America During the Jurassic Period

Cordilleran mobile belt

Volcanic Island arc

Nevadan
mountain
building

Lowlands

Uplands

Appalachian Mountains

Sundance
Sea

© 2013 Cengage Learning

Newly forming
Gulf of Mexico

Land
Mountains
Volcanoes
Shallow sea
Deep ocean

Nevadan orogeny
Late Jurassic to
Cretaceous deformation
that strongly affected the
western part of North
America.

tectonics for the rest of the Mesozoic Era and for most of the Cenozoic Era; this subduction zone marks the beginning of the modern circum-Pacific orogenic system.

The general term *Cordilleran orogeny* is applied to the mountain-building activity that began during the Jurassic and continued into the Cenozoic (Figure 17.22). The Cordilleran orogeny consisted of a series of individually named, but interrelated, mountain-

building events that occurred in different regions at different times. Most of this Cordilleran orogenic activity is related to the continued westward movement of the North American plate as it overrode the Farallon plate, and its history is highly complex.

The first phase of the Cordilleran orogeny, the Nevadan orogeny, began during the Late Jurassic and continued into the Cretaceous as large volumes of granitic magma were generated at depth beneath the western edge of North America. These granitic masses ascended as huge batholiths that are now recognized as the Sierra Nevada, southern California, Idaho, and Coast Range batholiths (Figure 17.23).

Figure 17.20 Paleogeography of North America During the Cretaceous Period

Sevier and Laramide mountain building

Cordilleran mobile belt

Cretaceous Interior Seaway

Alluvial plain

Lowlands

Appalachian Mountains

Carbonate bottom

© 2013 Cengage Learning

Land
Mountains
Shallow sea
Deep ocean

The second phase of the Cordilleran orogeny, the **Sevier orogeny**, was mostly a Cretaceous event, even though it began in the Late Jurassic and is associated with the tectonic activity of the earlier Nevadan orogeny (Figure 17.22). Subduction of the Farallon plate beneath the North American plate continued during this time, resulting in numerous overlapping, low-angle thrust faults in which blocks of older rocks were thrust eastward on top of younger rocks. This deformation produced generally north-south–trending mountain ranges stretching from Montana to western Canada.

During the Late Cretaceous to Early Cenozoic, the final pulse of the Cordilleran orogeny occurred (Figure 17.22). The **Laramide orogeny** developed east of the Sevier orogenic belt in the present-day Rocky Mountain areas of New Mexico, Colorado, and Wyoming. Most of the features of the present-day Rocky Mountains resulted from the Cenozoic phase of the Laramide orogeny, and for that reason, it is discussed later in this chapter.

Sevier orogeny
Cretaceous deformation that affected the continental shelf and slope areas of the Cordilleran mobile belt.

Laramide orogeny A Late Cretaceous to Early Cenozoic episode of deformation in the area of the present-day Rocky Mountains.

Figure 17.21 North America Triassic Fault-Block Basins

a. Areas where Triassic fault-block basin deposits crop out in eastern North America.

b. After the Appalachians were eroded into a low-lying plain by the Middle Triassic, fault-block basins formed as a result of Late Triassic rifting between North America and Africa.

c. These valleys accumulated tremendous thickness of sediments and were themselves broken by a complex of normal faults during rifting.

d. Palisades of the Hudson River. This sill was one of many intruded into the Newark sediments during the Late Triassic rifting that marked the separation of North America from Africa.

Concurrent with the tectonism in the Cordilleran mobile belt, Early Triassic sedimentation on the western continental shelf consisted of shallow-water marine sandstones, shales, and limestones. During the Middle and Late Triassic, the western shallow seas regressed farther west, exposing large areas of former seafloor to erosion. Marginal marine and nonmarine Triassic rocks, particularly red beds, contribute to the spectacular and colorful scenery of the region.

These rocks represent a variety of depositional environments. The Upper Triassic *Chinle Formation*, for example, is widely exposed throughout the Colorado Plateau and is probably most famous for its petrified wood, spectacularly exposed in Petrified Forest National Park, Arizona. Although best known for its petrified wood, the Chinle Formation has also yielded fossils of amphibians and various reptiles, including small dinosaurs.

Early Jurassic deposits in a large part of the western region consist mostly of clean, cross-bedded sand-stones indicative of wind-blown deposits. The thickest and most prominent of these is the *Navajo Sandstone*, a widespread cross-bedded sandstone that accumulated in a coastal dune environment along the southwestern margin of the craton. The sandstone's most distinctive feature is its large-scale cross-beds, some of them more than 25 m high.

Marine conditions returned to the region during the Middle Jurassic when a seaway called the *Sundance Sea* twice flooded the interior of western North America (Figure 17.19). The resulting deposits were largely derived from erosion of the tectonic highlands to the west that paralleled the shoreline. These highlands resulted from intrusive igneous activity and associated volcanism that began during the Triassic.

During the Late Jurassic, a mountain chain formed in Nevada, Utah, and Idaho as a result of the deformation produced by the Nevadan orogeny. As the mountain chain grew and shed sediments eastward, the

Figure 17.22 Mesozoic Cordilleran Orogenies
Mesozoic orogenies occurring in the Cordilleran mobile belt.

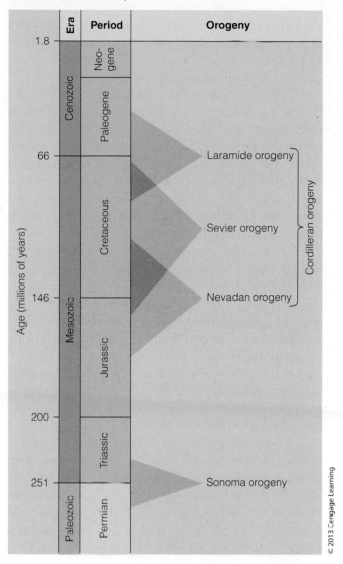

© 2013 Cengage Learning

Figure 17.23 Cordilleran Batholiths
Location of the Jurassic and Cretaceous batholiths in western North America.

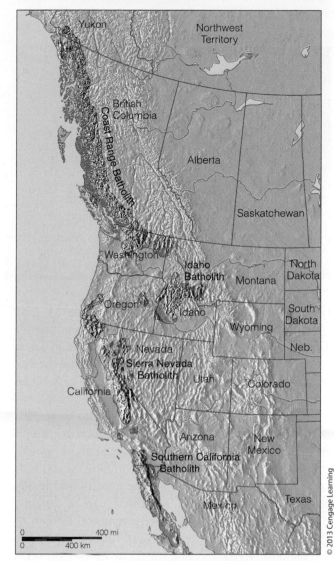

© 2013 Cengage Learning

Sundance Sea began retreating northward. A large part of the area formerly occupied by the Sundance Sea was then covered by multicolored detrital sediments that comprise the *Morrison Formation*, which contains the world's richest assemblage of Jurassic dinosaur remains (Figure 17.24).

Shortly before the end of the Early Cretaceous, Arctic waters spread southward over the craton, forming a large inland sea in the Cordilleran area. By the beginning of the Late Cretaceous, this incursion joined the northward-transgressing waters from the Gulf area to create an enormous *Cretaceous Interior Seaway* that occupied the area east of the Sevier orogenic belt (Figure 17.20). Extending from the Gulf of Mexico to the Arctic Ocean and more than 1,500 km wide at its maximum extent, this seaway effectively divided North America into two large landmasses until just before the end of the Late Cretaceous.

As the Mesozoic Era ended, the Cretaceous Interior Seaway withdrew from the craton. During this regression, marine waters retreated to the north and south, and marginal marine and continental deposition formed widespread coal-bearing deposits on the coastal plain.

LO8 CENOZOIC EARTH HISTORY

At 66 million years long, the Cenozoic Era is comparatively brief, constituting only 1.4 percent of all geologic

Figure 17.24 Dinosaur Bones in Bas Relief

The north wall of the visitors' center at Dinosaur National Monument, showing dinosaur bones in bas relief, just as they were deposited 140 million years ago in the Morrison Formation.

Reed Wicander

time. Nevertheless, 66 million years is an extremely long time, certainly long enough for significant evolution of Earth and its biota. Furthermore, Cenozoic rocks are at or near the surface and have been little altered, thereby making access to and interpretation of them easier than for rocks of previous eras (Figure 17.25).

Geologists divide the Cenozoic Era into three periods. The Paleogene Period (66 to 23 million years ago) includes the Paleocene, Eocene, and Oligocene epochs, the Neogene Period (23 to 2.6 million years ago) includes the Miocene and Pliocene epochs, and finally the Quaternary Period (2.6 million years ago to the present), which includes the Pleistocene and Recent (or Holocene) epochs (see Figure 16.1). Although you may see the term Tertiary Period for 66 to 1.8 million years ago, it is no longer recommended.

Many of Earth's features have long histories, but the present distribution of land and sea and the landforms of the continents developed recently in the context of geologic time. For instance, the Appalachian Mountain region began its evolution during the Precambrian, but its present expression is largely the product of Cenozoic uplift and erosion. Likewise, distinctive landforms such as glacial valleys, badlands topography, and volcanoes of our national parks developed during the past few thousand years to several millions of years.

Cenozoic Plate Tectonics and Orogeny

The Late Triassic fragmentation of Pangaea (Figure 17.17a) began an episode of plate motions that continues even now. As a result, Cenozoic orogenic activity was concentrated in two major belts; the *Alpine-Himalayan belt*, which extends from southern Europe and north Africa eastward through the Middle East and Asia; and the *circum-Pacific belt*, which includes orogens along the west coasts of the Americas as well as the eastern margin of Asia and the islands north of Australia and New Zealand.

Within the Alpine-Himalayan orogenic belt, the *Alpine orogeny* began during the Mesozoic, but major deformation also occurred from the Eocene to Late Miocene as the African and Arabian plates moved northward against Eurasia. Deformation resulting from plate convergence formed mountains between Spain and France, the Alps of mainland Europe, and the mountains of Italy and North Africa. This orogenic belt remains geologically active.

Farther east in the Alpine-Himalayan orogenic belt, the *Himalayan orogeny* resulted from the collision of India with Asia (see Figure 9.18). Sometime during the Eocene, India's northward drift rate decreased abruptly, indicating the probable time of collision. In any event, an orogeny resulted during which two continental plates became sutured, which is why the present-day Himalayas are far inland rather than at a continental margin.

Plate subduction in the circum-Pacific orogenic belt took place throughout the Cenozoic, giving rise to orogenies in the Aleutians, the Philippines, and Japan, and along the west coasts of North, Central, and South America. The Andes Mountains in western South America, for example, formed as a result of convergence of the Nazca and South American plates (see Figure 9.17). Spreading at the East Pacific Rise and subduction of the Cocos and Nazca plates beneath Central and South America, respectively, account for continuing orogenic activity in these regions.

Earth's systems continue to interact, accounting for a continuously evolving planet.

© Istockphoto.com/Ugur Evirgen

Figure 17.25 Cenozoic Sedimentary Rocks

Rocks of Cenozoic-age are at or near the surface and have been little altered by metamorphism, making them easier to interpret than rocks of the previous geologic eras. This is an exposure of the Brule Formation of Oligocene-age in Badlands National Park in South Dakota, which was deposited mostly in stream channels and on floodplains. Notice the sharp angular slopes and ridges and numerous ravines, which are typical of badlands topography.

LO9 THE NORTH AMERICAN CORDILLERA

Part of the circum-Pacific orogenic belt is the *North American Cordillera*, a complex mountainous region in western North America extending from Alaska into central Mexico (Figure 17.26). It has a long, complex geologic history involving accretion of island arcs along the continental margin, orogeny at an oceanic–continental boundary, vast outpourings of basaltic lavas, and block faulting. The most recent episode of large-scale deformation was the *Laramide orogeny*, which began during the Late Cretaceous, 85 to 90 million years ago. Like many other orogenies, it took place along an oceanic–continental boundary but was much farther inland than is typical (Figure 17.27). Recall that the Laramide orogeny was the final phase in a long episode of deformation referred to as the *Cordilleran orogeny*, which we discussed on page 379 in this chapter.

The Laramide orogeny ceased about 40 million years ago, but since that time the mountain ranges formed during the orogeny were eroded, and the valleys between the ranges filled with sediments. Many of the ranges were nearly buried in their own erosional debris, and their present-day elevations are the result of renewed uplift.

In other parts of the Cordillera, the Colorado Plateau was uplifted far above sea level, but the rocks were little deformed (Figure 17.28a). In the Basin and Range Province, block faulting began during the Middle Cenozoic and continues to the present. At its western edge, the province is bounded by a large escarpment that forms the east face of the Sierra Nevada (Figure 17.28b).

In the Pacific Northwest, an area of about 200,000 km², mostly in Washington, is covered by the Cenozoic Columbia River basalts (see Figure 5.13). Issuing from long fissures, these flows overlapped to produce an aggregate thickness of about 1,000 m. Widespread

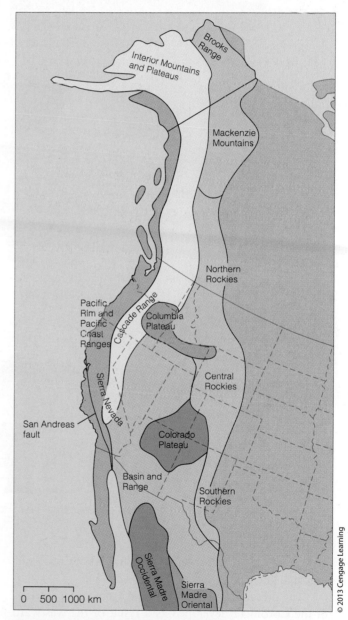

Figure 17.26 The North American Cordillera

The North American Cordillera is a complex mountainous region extending from Alaska into central Mexico. It consists of the elements shown here.

Figure 17.27 The Laramide Orogeny

The Laramide orogeny resulted when the Farallon plate was subducted beneath North America during the Late Cretaceous Period to Eocene Epoch.

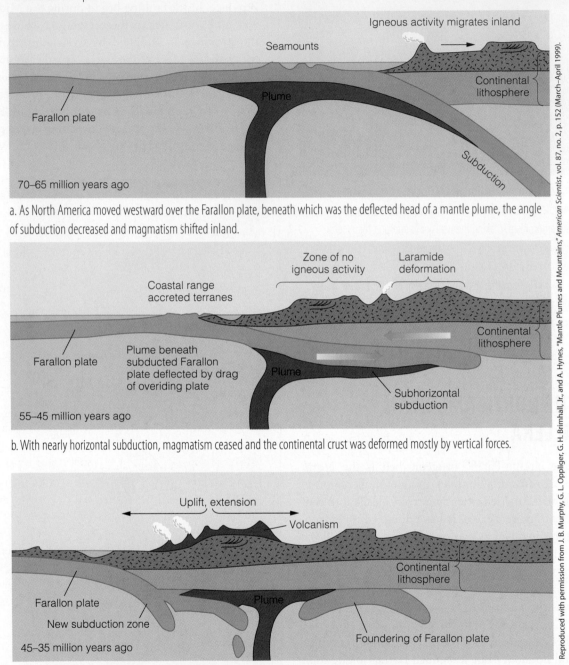

a. As North America moved westward over the Farallon plate, beneath which was the deflected head of a mantle plume, the angle of subduction decreased and magmatism shifted inland.

b. With nearly horizontal subduction, magmatism ceased and the continental crust was deformed mostly by vertical forces.

c. Disruption of the oceanic plate by the mantle plume marked the onset of renewed volcanism.

Reproduced with permission from J. B. Murphy, G. L. Oppliger, G. H. Brimhall, Jr., and A. Hynes, "Mantle Plumes and Mountains," *American Scientist*, vol. 87, no. 2, p. 152 (March–April 1999).

volcanism also took place in Oregon, Idaho, California, Arizona, and New Mexico (Figure 17.28c).

The present-day elements of the Pacific Coast section of the Cordillera developed as a result of the westward drift of North America, the partial consumption of the oceanic Farallon plate, and the collision of North America with the Pacific-Farallon Ridge (Figure 17.29). During the Early Cenozoic, the entire Pacific Coast was bounded by a subduction zone that stretched from Mexico to Alaska. Most of the Farallon plate was consumed at this subduction zone, and now only two small remnants exist: the Juan de Fuca and Cocos plates (Figure 17.29). The continuing subduction of these small plates accounts for seismicity and volcanism in the Cascade Range of the Pacific Northwest and Central America, respectively. Westward drift of the North American plate also resulted in its collision with the Pacific-Farallon Ridge and the origin of the Queen Charlotte and San Andreas transform faults (Figure 17.29).

Figure 17.28 The Colorado Plateau, Basin and Range Province, and Cenozoic Lava Flows

a. The Claron Formation in Bryce Canyon National Park in Utah is one of many exposed in the Colorado Plateau. This formation of gravel, sand, and mud was deposited mostly by streams during the Paleocene and Eocene epochs.

b. The Sierra Nevada at the western margin of the Basin and Range province has risen along normal faults, so that it is more than 3,000 m above the valley to the east.

c. Basalt lava flows of the Snake River Plain at Malad Gorge State Park in Idaho.

The Continental Interior and the Gulf Coastal Plain

The vast, shallow seas that had invaded the continents during the previous eras were largely absent during the Cenozoic. The notable exception was the Zuni Sea, which occupied a large area in the continental interior during part of the Cenozoic. Sediments derived from the Laramide highlands to the west and southwest were transported eastward and deposited in a variety of continental, transitional, and marine environments (Figure 17.25).

The Gulf Coast sedimentation pattern was established during the Jurassic and persisted through the Cenozoic. Much of the sediment deposited on the Gulf Coastal Plain was detrital, but in the Florida section of the coastal plain and the Gulf Coast of Mexico, significant carbonate deposition occurred. A carbonate platform was established in Florida during the Cretaceous, and carbonate deposition continues at the present.

Eastern North America

The eastern seaboard has been a passive continental margin since Late Triassic rifting separated North America from North Africa and Europe. The present distinctive topography of the Appalachian Mountains is the product of Cenozoic uplift and erosion. By the end of the Mesozoic, the Appalachian Mountains had been eroded to a plain. Cenozoic uplift rejuvenated the streams, which responded by renewed downcutting. As the streams eroded downward, they were superposed on resistant strata and cut large canyons across these strata. The distinctive topography of the Valley and Ridge Province is the product of Cenozoic erosion and preexisting geologic structures. It consists of northeast-southwest–trending ridges of resistant upturned strata and intervening valleys eroded into less-resistant strata.

Pleistocene Glaciation

We know today that the Pleistocene (Ice Age) began 2.6 million years ago and ended about 10,000 years ago. During this time, several intervals of widespread continental glaciation took place, especially in the Northern Hemisphere, each separated by warmer interglacial periods. In addition, valley glaciers were more common at lower elevations and latitudes, and many extended much farther than they do today.

As you would expect, the climatic effects responsible for Pleistocene glaciers were worldwide. Nevertheless, Earth was not as frigid as it is portrayed in movies

Figure 17.29 **Three Stages in Westward Drift of North America**

Three stages (a), (b), and (c) in the westward drift of North America and its collision with the Pacific–Farallon Ridge. As the North American plate overrode the ridge, its margin became bounded by transform faults rather than a subduction zone.

REPRINTED WITH PERMISSION FROM W. R. DICKINSON, "CENOZOIC PLATE TECTONIC SETTING OF THE CORDILLERAN REGION IN THE WESTERN UNITED STATES," *CENOZOIC PALEOGEOGRAPHY OF THE WESTERN UNITED STATES*, PACIFIC COAST SYMPOSIUM 3, 1979, P. 2 (FIG. 1).

and cartoons, nor was the onset of the climatic conditions leading to glaciation rapid. Indeed, evidence from several types of investigations indicates that the climate cooled gradually from the Eocene through the Pleistocene. Furthermore, evidence from seafloor sediments shows that 20 major warm–cold cycles have occurred during the last 2 million years.

Geologists know that, at their greatest extent, Pleistocene glaciers covered about three times as much of Earth's surface as they do now (Figure 17.30a). Like the vast ice sheets now in Greenland and Antarctica, they were probably 3 km thick. Geologists have identified four major Pleistocene glacial episodes that took place in North America—the *Nebraskan, Kansan, Illinoian,* and *Wisconsinan* (Figure 17.30b)—each named for the state in which the most southerly glacial deposits are well exposed. The three interglacial stages are named for localities of well-exposed soils and other deposits (Figure 17.28b).

Although the Pleistocene ended 10,000 years ago, we see the effects of glaciation in many parts of North America. Indeed, valley glaciers shaped and continue to modify mountains in Canada, Alaska, and several western states, and as a result have produced the majestic scenery seen in many areas (Figure 17.31a). And even though we have emphasized glaciers in this section, the

Pleistocene was also a time of continuing tectonism and volcanism (Figure 17.31b). In short, Earth remains a dynamic and evolving planet.

Continued global warming will result in a rise in mean sea level, as icecaps and glaciers melt and contribute their water to the world's ocean. Analysis of continued glacial ice loss in Antarctica and a reduction in Arctic sea ice in recent years indicate that global warming is occurring. However, what the long-term effects of this warming trend will be, and its duration, are still being debated.

Figure 17.30 Maximum Extent of Pleistocene Glaciers in North America

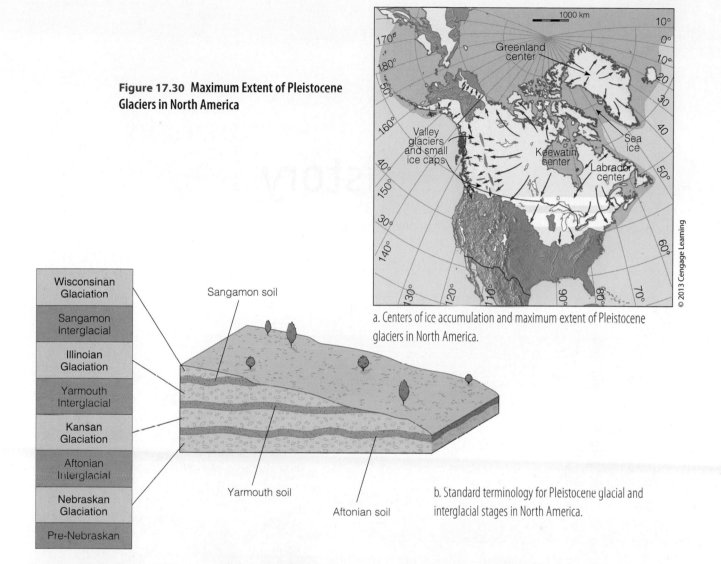

a. Centers of ice accumulation and maximum extent of Pleistocene glaciers in North America.

Wisconsinan Glaciation
Sangamon Interglacial
Illinoian Glaciation
Yarmouth Interglacial
Kansan Glaciation
Aftonian Interglacial
Nebraskan Glaciation
Pre-Nebraskan

Sangamon soil

Yarmouth soil

Aftonian soil

b. Standard terminology for Pleistocene glacial and interglacial stages in North America.

Figure 17.31 Ongoing Geologic Activity

Earth remains an active planet on which interactions among its systems continue to bring about change.

a. Valley glaciers continue to erode the Chugach Mountains in Alaska. The sharp, angular peaks and ridges, and broad, smooth valleys are typical of areas eroded by valley glaciers.

b. This 115-m-high cinder cone, called High Hole Crater, in northern California, lies on the flank of a huge shield volcano. The aa lava flow in the foreground is 1,100 years old.

18 Life History

In this scene from the Late Cretaceous, *Ankylosaurus* is defending itself from the large predator *Tyrannosaurus*. *Ankylosaurus* was the most heavily armored dinosaur, and it had a large club at the end of its tail that was almost certainly used for defense. This animal measured 8 to 10 m long and weighed about 4.5 metric tons. At 13 m long and weighing 5 metric tons, *Tyrannosaurus* was one of the largest carnivorous dinosaurs.

© Eivind Bovor

Introduction

"**We** now know that extinction is the rule, **not the exception.**"

Scientists have been seriously investigating the history of life on Earth for two centuries. Without science, our only knowledge of how the animals and plants around us changed would be from written human records. Such records are interesting, but they simply do not go far enough back in time to help us understand anything more than the current diversity of organisms and a few very recent extinctions.

By looking at the geologic record, scientists have clearly established the outline of Earth history going back some 4 billion years. Although there is much we still do not know about life history, we have learned a great deal. For example, we now know that extinction is the rule, not the exception. We are sure that more than 99 percent of all species that have ever existed in our seas and on land are now extinct. But we have also learned that the variety and complexity of organisms have increased after episodes of mass extinction time and time again.

This chapter has the ambitious task of summarizing what we know about the many and varied creatures that have inhabited our planet almost since Earth began. The evidence we rely on is the **fossils** that have endured through millions, even billions, of years. The quality of the *fossil record* varies considerably, depending on the types of organisms existing at a particular time and the environment in which they lived.

The preservation of any one organism as a fossil is rare, but fossils are nevertheless quite common. This apparent contradiction is easily explained when you consider that so many billions of organisms have existed during so many millions of years that if only a tiny fraction was preserved, the total number of fossils is phenomenal. In fact, fossils of many **vertebrate** animals (those with a segmented vertebral column), such as dinosaurs, are much more common than most people realize. In general, the fossil record does give a good overview of life history.

Learning Outcomes

After reading this unit, you should be able to do the following:

LO1 Discuss the evolution of life during the Precambrian

LO2 Review life during the Paleozoic Era

LO3 Discuss life of the Mesozoic Era

LO4 Review life during the Cenozoic Era

---

fossil The remains or traces of once-living organisms.

vertebrate Any animal that has a segmented vertebral column; includes fish, amphibians, reptiles, birds, and mammals.

stromatolite A biogenic sedimentary structure, especially in limestone, produced by the entrapment of sediment grains on sticky mats of photosynthesizing bacteria.

LO1 PRECAMBRIAN LIFE HISTORY

Scientists had long assumed that the fossils so common in Paleozoic rocks must have had a long earlier history, but little was known about these organisms. A few enigmatic Precambrian fossils had been reported, but they were mostly dismissed as inorganic structures rather than traces of organisms. In fact, the Precambrian was once called the *Azoic*, meaning "devoid of life."

Then, during the early 1900s, Charles Walcott proposed that layered, moundlike structures from Proterozoic rocks in Ontario, Canada, were ancient reefs constructed by algae. However, paleontologists did not demonstrate until 1954 that they were actually the products of organic activity. These structures, now called **stromatolites**, still form in a few environments such as Shark Bay, Australia, when sediment is trapped on sticky mats of photosynthesizing bacteria, commonly called *blue-green algae* (Figure 18.1). We now know that stromatolites are common in some Proterozoic-age rocks, but the oldest ones are in 3.3- to 3.5-billion-year-old (Paleoarchean) rocks in Australia.

The importance of stromatolites in Earth history should not be overlooked. The atmosphere of the planet

Figure 18.1 Stromatolites

James S. Monroe

Phil Playford, Geological Survey of Western Australia

a. Stromatolites from the Neoproterozoic in Glacier National Park, Montana. Note the gently curved structures or layers in the rock.

b. Present-day stromatolites displaying their pillow-shaped growth in Shark Bay, Australia, one of few places where they still live.

prokaryotic cell A cell that lacks a nucleus and organelles such as mitochondria and plastids; the cells of bacteria and cyanobacteria (blue-green algae).

eukaryotic cell A cell with an internal membrane-bounded nucleus that contains chromosomes and other internal structures; possessed by all organisms except bacteria.

during the Paleoarchean had no free oxygen (no oxygen gas), but stromatolites, like plants, produce oxygen as a by-product of photosynthesis. Slowly but surely, as the Archean and especially the Proterozoic unfolded, increasing amounts of oxygen accumulated in Earth's atmosphere.

Paleoarchean stromatolites are among the earliest fossils on Earth, but indirect evidence that life existed before stromatolites comes from 3.8-billion-year-old rocks in Greenland. These rocks contain small carbon spheres that may be of organic origin, but the evidence is not conclusive. All known fossils from Archean and Paleoproterozoic-age rocks are of single-celled bacteria lacking a cell nucleus and reproduced asexually, much as bacteria do today. Such cells are termed **prokaryotic cells** as opposed to **eukaryotic cells**, the latter having a cell nucleus and other internal structures not present in prokaryotic cells, and most reproduce sexually.

The origin of eukaryotic cells is one of the most important events in life history. No one doubts they were present by the Mesoproterozoic, and some evidence indicates that they first evolved as long as 1.4 billion years ago. A currently popular theory (supported by evidence mostly from living organisms) holds that

two or more prokaryotic cells entered into a beneficial symbiotic relationship, and the symbionts became increasingly interdependent until the unit could exist only as a whole.

The first eukaryotes (organisms composed of eukaryotic cells) were still single-celled, but by Paleoproterozoic or Mesoproterozoic time, the first multicelled organisms had made their appearance. Carbonaceous impressions of what appear to be multicelled algae are known from several areas, but the first multicelled animal fossils are found in Neoproterozoic-age rocks. Some of the oldest so far reported are from the 545- to 600-million-year-old Ediacaran fauna of Australia, consisting of multicelled animals now known from all continents except Antarctica. Some investigators think these fossils represent jellyfish, sea pens, segmented worms, and arthropods (Figure 18.2). One wormlike fossil has even been cited as a possible ancestor of the trilobites that were so common during the Early Paleozoic (Figure 18.2b). Other researchers disagree; they think these animals represent an early evolutionary development distinct from the ancestry of any present-day animals.

The nature of the Ediacaran fossils is debated, but all scientists agree that complex, multicelled animals were not only present by the Neoproterozoic but also widely distributed (Figure 18.2c). One good example is the fossil *Kimberella*, from Russia, which was perhaps a sluglike creature. In addition, tracks and trails provide compelling evidence of complex animals such as worms.

Although more fossils are being found, the Proterozoic fossil record is still not very good, because the animals then present lacked durable skeletons. However,

Figure 18.2 The Ediacaran Fauna of Australia (a and b) and Ediacaran-Like Fossils from Newfoundland (c)

a. The affinities of *Tribachidium* remain uncertain. It may be either a primitive echinoderm or a cnidarian.

b. *Spriggina* was originally thought to be a segmented worm (annelid), but now it appears more closely related to arthropods, possibly even an ancestor of trilobites.

c. These fossils have not been given scientific names. The most obvious ones are informally called *spindles*, whereas the one toward the lower center is a *feather duster*.

minute scraps of shell-like material and spicules, presumably from sponges, in Neoproterozoic rocks indicate that hard skeletal elements were present by then. Nevertheless, animals with durable skeletons of chitin (a complex organic substance), silica (SiO_2), and calcium carbonate ($CaCO_3$) were not abundant until the beginning of the Paleozoic Era.

LO2 PALEOZOIC LIFE HISTORY

At the beginning of the Paleozoic Era, animals with skeletons appeared rather abruptly in the fossil record. In fact, their appearance is described as an explosive development of new types of animals and is referred to as the "Cambrian explosion" by most scientists. This sudden appearance of new animals in the fossil record is rapid, however, only in the context of geologic time, having taken place over millions of years during the Early Cambrian Period.

The earliest of these animals with skeletons were *invertebrates*—that is, animals that lack a segmented vertebral column. The major invertebrate groups and their geologic ranges are given in Table 18.1. Rather than focusing on the history of each invertebrate group here, we will survey the evolution of the Paleozoic marine invertebrate communities through time, concentrating on the major features and changes that took place.

Marine Invertebrates

Although almost all of the major invertebrate phyla evolved during the Cambrian Period (Table 18.1), many were represented by only a few species. Whereas trace fossils are common and echinoderms diverse, trilobites, brachiopods, and archaeocyathids (bottom-dwelling organisms that constructed reeflike structures and lived only during the Cambrian) comprised the majority of Cambrian skeletonized life. It is important to remember,

TABLE 18.1

The Major Invertebrate Groups and Their Geologic Ranges

GROUP	RANGE	GROUP	RANGE
Phylum Protozoa	Cambrian–Recent	**Phylum Mollusca**	Cambrian–Recent
Class Sarcodina	Cambrian–Recent	Class Monoplacophora	Cambrian–Recent
Order Foraminifera	Cambrian–Recent	Class Gastropoda	Cambrian–Recent
Order Radiolaria	Cambrian–Recent	Class Bivalvia	Cambrian–Recent
Phylum Porifera	Cambrian–Recent	Class Cephalopoda	Cambrian–Recent
Class Demospongea	Cambrian–Recent	**Phylum Annelida**	Precambrian–Recent
Order Stromatoporoida	Cambrian–Oligocene	**Phylum Arthropoda**	Cambrian–Recent
Phylum Archaeocyatha	Cambrian	Class Trilobita	Cambrian–Permian
Phylum Cnidaria	Cambrian–Recent	Class Crustacea	Cambrian–Recent
Class Anthozoa	Ordovician–Recent	Class Insecta	Silurian–Recent
Order Tabulata	Ordovician–Permian	**Phylum Echinodermata**	Cambrian–Recent
Order Rugosa	Ordovician–Permian	Class Blastoidea	Ordovician–Permian
Order Scleractinia	Triassic–Recent	Class Crinoidea	Cambrian–Recent
Phylum Bryozoa	Ordovician–Recent	Class Echinoidea	Ordovician–Recent
Phylum Brachiopoda	Cambrian–Recent	Class Asteroidea	Ordovician–Recent
Class Inarticulata	Cambrian–Recent	**Phylum Hemichordata**	Cambrian–Recent
Class Articulata	Cambrian–Recent	Class Graptolithina	Cambrian–Mississippian

however, that the fossil record is biased toward organisms with durable skeletons and that we generally know little about the soft-bodied organisms of that time. At the end of the Cambrian Period, trilobites suffered mass extinctions, and even though they persisted until the end of the Paleozoic Era, their numbers were considerably diminished.

A major transgression that began during the Middle Ordovician (Tippecanoe sequence) resulted in the most widespread inundation of the craton in Earth history. This vast, shallow sea, which was uniformly warm during this time, opened numerous new marine habitats that were soon filled by a variety of organisms such that the Ordovician is characterized by a dramatic increase in the diversity of the total shelly fauna and is known to scientists as the "Great Ordovician Biodiversity Event" (Figure 18.3). The end of the Ordovician, however, was a time of mass extinctions in the marine realm. More than 100 families of marine invertebrates became extinct, and many geologists think these extinctions were the result of extensive glaciation that occurred in Gondwana at the end of the Ordovician Period (see Chapter 17).

The mass extinction at the end of the Ordovician Period was fol-lowed by rediversification and recovery of many of the decimated groups. In fact, the Silurian and Devonian were times of major reef building in which organic reef builders diversified in new ways, building massive reefs larger than any produced during the Cambrian or Ordovician.

Near the end of the Devonian, another mass extinction occurred that resulted in a worldwide near-total collapse of the massive reef communities. On land, however, the *seedless vascular plants* (which will be discussed shortly) were seemingly unaffected. Thus, extinctions at

Figure 18.3 Middle Ordovician Marine Community
Reconstruction of a Middle Ordovician seafloor fauna. Cephalopods, crinoids, colonial corals, bryozoans, trilobites, and brachiopods are shown.

this time were most extensive among marine life, particularly in the reef and pelagic communities.

The Carboniferous invertebrate marine community responded to the Late Devonian extinctions in much the same way that the Silurian invertebrate marine community responded to the Late Ordovician extinctions—that is, by renewed adaptive radiation and rediversification. However, large organic reefs like those existing earlier in the Paleozoic virtually disappeared and were replaced by small patch reefs that flourished during the Late Paleozoic.

The Permian invertebrate marine faunas resembled those of the Carboniferous (Figure 18.4). However, they were not as widely distributed because of the restricted size of the shallow seas on the cratons and the reduced shelf space along the continental margins (see Figure 17.9b).

The Permian Mass Extinction

The greatest recorded mass extinction to affect Earth's biota occurred at the end of the Permian Period (Figure 18.5). By the time the Permian ended, roughly 50 percent of all marine invertebrate families and about 90 percent of all marine invertebrate species became extinct. In addition, more than 65 percent of all amphibians and reptiles as well as nearly 33 percent of insects on land also became extinct.

1	Late Ordovician	(− 12%)
2	Late Devonian	(− 14%)
3	Late Permian	(− 52%)
4	Late Triassic	(− 12%)
5	Late Cretaceous	(− 11%)

Geologic time (10^6 years)

Figure 18.5 Phanerozoic Diversity of Marine Invertebrate and Vertebrate Families

REPRODUCED WITH PERMISSION FROM "MASS EXTINCTION IN THE MARINE FOSSIL RECORD," BY D.M. RAUP AND J.J. SEPKOSKI, *SCIENCE*, VOL. 215, P. 1502 (FIG. 2). © 1982 AMERICAN ASSOCIATION FOR THE ADVANCEMENT OF SCIENCE.

What caused such a crisis for both marine and land-dwelling organisms? Various hypotheses have been proposed, but no completely satisfactory answer has yet been found. Currently, many scientists think an episode of deep-sea anoxia and increased oceanic CO_2 levels resulted in a highly stratified ocean during the Late Permian. In other words, there was very little, if any, circulation of oxygen-rich surface waters into the deep ocean. During this time, stagnant waters also covered the shelf regions, thus affecting the shallow marine fauna.

During the Late Permian, widespread volcanic and continental fissure eruptions, such as the Siberian Traps, where lava flows covered more than 2 million km², were also taking place. These eruptions released not only large amounts of additional carbon dioxide into the atmosphere, but also high amounts of fluorine and chlorine, which could have damaged the ozone layer and helped contribute to increased climatic instability and ecologic collapse.

By the end of the Permian, a near-collapse of both the marine and terrestrial ecosystem had occurred. Although the ultimate cause of such devastation is still being debated and investigated, it is safe to say that it was probably a combination of interconnected and related geologic and biologic events.

Vertebrates

Besides the numerous invertebrate groups, vertebrates (animals with backbones) also evolved and diversified during the Paleozoic. Remains of the most primitive

The American Museum of Natural History, #K10269

Figure 18.4 Permian Patch-Reef Marine Community
Reconstruction of a Permian patch-reef community from the Glass Mountains of West Texas. Shown are algae, brachiopods, cephalopods, sponges, and corals.

vertebrates, the fish, are found in Upper Cambrian marine rocks in Wyoming. These fish, referred to as *ostracoderms*, were jawless, had poorly developed fins, had an external covering of bony armor, and lived from the Early Cambrian to Late Devonian (Figure 18.6a).

The evolution of jaws was a major evolutionary advance among primitive vertebrates. Although their jawless ancestors could only feed on detritus, jawed fish could chew food and become active predators, thus opening many new ecological niches.

The fossil remains of the first jawed fish are found in Lower Silurian rocks and belong to the *acanthodians*, a group of enigmatic fish characterized by large spines, scales covering much of the body, jaws, teeth, and reduced body armor (Figure 18.6c). Although the relationship of acanthodians to other fish is not yet well established, many scientists think that the acanthodians included the probable ancestors of the present-day bony and cartilaginous fish groups.

The other jawed fish, the *placoderms* (heavily armored jawed fish), evolved during the Silurian. The placoderms exhibited considerable variety, including

Currently, many scientists think that a large-scale marine regression, coupled with climatic changes in the form of global warming (caused by an increase in carbon dioxide levels), may have been responsible for the mass extinctions recorded in the fossil record.

small bottom-dwellers (Figure 18.6b), as well as some of the largest marine predators that ever existed.

The bony fish first evolved during the Devonian and are the most varied and numerous of all the fishes (Figure 18.6d). Among the bony fish, one group, known as *lobe-finned fish*, was particularly important, because this group included the probable ancestor of the amphibians, the first land-dwelling vertebrate animals (Figure 18.7). In fact, the similarity between the group of lobe-finned fish known as *crosspterygians* and the earliest amphibians is striking and one of the most

Figure 18.6 Recreation of a Devonian Seafloor

Recreation of a Devonian seafloor showing (a) an ostracoderm, (b) a placoderm, (c) an acanthodian, and (d) a bony fish.

© 2013 Cengage Learning

Figure 18.7 Late Devonian Landscape

Shown is *Ichthyostega*, both in the water and on land. *Ichthyostega* was an amphibian that grew to a length of about 1 m. The flora of the time was diverse, consisting of a variety of small and large seedless vascular plants.

widely cited examples of a transition from one major group to another. However, recent discoveries of older lobe-finned fish and newly published findings of fish ancestors of amphibians are filling in the gaps in the evolution from fish to amphibians.

Although amphibians were the first vertebrates to live on land, having evolved by the Devonian, they were not the first land-living organisms. Land plants, which probably evolved from green algae, first appeared during the Ordovician. Furthermore, insects, millipedes, spiders, and even snails invaded the land before amphibians.

The transition from water to land required animals to surmount several barriers. The most critical were drying out, reproduction, the effects of gravity, and the extraction of oxygen from the atmosphere by lungs rather than from water by gills. These problems were partly solved by some of the lobe-finned fish; they already had a backbone and limbs that could be used for support and walking on land, and they had lungs to extract oxygen from the atmosphere.

Amphibians were limited in colonizing the land, however, because they had to return to water to lay their gelatinous eggs and because they never completely solved the problem of drying out. In contrast, the reptiles evolved skin or scales to preserve their internal moisture. They also evolved an egg in which the developing embryo is surrounded by a liquid-filled sac and provided with both a food and a waste sac. The evolution of such an egg allowed vertebrates to colonize all parts of the land because they no longer had to return to the water as part of their reproductive cycle.

The oldest known reptiles evolved during the Mississippian Period and were small, agile animals that continued to diversify during the Pennsylvanian Period. One of the descendant groups of these early reptiles was the *pelycosaurs*, or finback reptiles, which became the dominant reptile group by the Permian (Figure 18.8).

The pelycosaurs became extinct during the Permian and were succeeded by the *therapsids*, mammal-like reptiles that evolved from the carnivorous pelycosaur lineage and rapidly diversified into herbivorous and carnivorous lineages. Among the therapsids, one group was the most mammal-like of all, and by Late Triassic time, true mammals had evolved from them.

As the Paleozoic Era came to an end, the therapsids constituted about 90 percent of the known reptile genera and occupied a wide range of ecological niches.

Edaphosaurus

Dimetrodon

Figure 18.8 Pelycosaurs

Most pelycosaurs, or finback reptiles, had a characteristic sail on their back. One hypothesis explains the sail as a type of thermoregulatory device. Other hypotheses are that it was a sexual display or a device to make the reptile look more intimidating. Shown here are the carnivore *Dimetrodon* and the herbivore *Edaphosaurus*.

The mass extinction that decimated the marine fauna at the close of the Paleozoic had an equally great effect on the terrestrial population. By the end of the Permian, about 90% of all marine invertebrate species were extinct, compared with more than two-thirds of all amphibians and reptiles. Plants, in contrast, apparently did not experience as great a turnover as invertebrates and animals.

Plants

Plants encountered many of the same problems animals faced when they made the transition to land: drying out, the effects of gravity, and reproduction. Plants adapted by evolving a variety of structural features that allowed them to invade the land during the Ordovician (Table 18.2).

The most common and widespread land plants are vascular plants, which have a tissue system of specialized cells for the movement of water and nutrients. Nonvascular plants, such as mosses and fungi, lack these specialized cells and are typically small and usually live in low, moist areas. Nonvascular plants were probably the first to make the transition to land, but their fossil record is poor.

The earliest known vascular land plants are small, leafless, Y-shaped stems from the Middle Silurian of Wales and Ireland. They are known as **seedless vascular plants**, because they did not produce seeds, nor did they have a true root system. Although these plants lived on land, they never completely solved the problem of drying out and were thus restricted to moist areas. Even their living descendants, such as ferns, are usually found in moist areas. During the Pennsylvanian Period, the seedless vascular plants became abundant and diverse because of the widespread coal-forming swamps (see Figure 17.12) that were ideally suited to their lifestyle (Figure 18.9).

Along with the evolution of diverse seedless vascular plants, another significant floral event occurred during the Devonian. The evolution of the seed at this time liberated vascular plants from their dependence on moist conditions and allowed them to spread over all parts of the land. The first to do so were the flowerless seed plants, or **gymnosperms**, which include the living

TABLE 18.2 Major Events in the Evolution of Land Plants

© 2013 Cengage Learning

Figure 18.9 Pennsylvanian Coal Swamp
Reconstruction of a Pennsylvanian coal swamp with its characteristic vegetation of seedless vascular plants.

cycads, conifers, and ginkgoes. Whereas the seedless vascular plants dominated the flora of the Pennsylvanian coal-forming swamps, the gymnosperms made up an important element of the Late Paleozoic flora, particularly in the nonswampy areas.

LO3 MESOZOIC LIFE HISTORY

The Mesozoic Era is designated as the "Age of Reptiles," alluding to the fact that reptiles were the most common land-dwelling vertebrate animals. Although dinosaurs and their relatives evoke considerable interest, many other groups of organisms were not only present, but also quite common. Many invertebrates flourished in the seas as well as on land, and vertebrates other than reptiles proliferated. For instance, mammals evolved from mammal-like reptiles during the Triassic, and birds probably evolved from small carnivorous dinosaurs during the Jurassic. Another wave of extinctions took place at the end of the Mesozoic.

Marine Invertebrates

Following the Permian mass extinctions, the Mesozoic was a time when marine invertebrates repopulated the seas. Among the mollusks, the clams, oysters, and snails became increasingly diverse and abundant, and the cephalopods were among the most important Mesozoic invertebrate groups. In contrast, the brachiopods never completely recovered from their near extinction and have remained a minor invertebrate group ever since. In areas of warm, clear, shallow marine waters, corals again proliferated, but these corals were of a new and more familiar type.

Single-celled animals known as *foraminifera* (Table 18.1) were also important, and they diversified tremendously during the Jurassic and Cretaceous periods. Floating, or planktonic, forms in particular became extremely common, but many of them became extinct at the end of the Mesozoic, and only a few types survived into the Cenozoic.

The Diversification of Reptiles

Reptile diversification began during the Mississippian Period with the evolution of the first animals capable of laying eggs on land. This ability to lay eggs on land, along with their skin or scales to preserve internal moisture, truly freed them from an aquatic environment. From this basic stock of so-called *stem reptiles*, all other reptiles, as well as birds and mammals, evolved.

archosaur One of a group of animals including dinosaurs, flying reptiles (pterosaurs), crocodiles, and birds.

dinosaur Any of the Mesozoic reptiles that belong to the groups designated as ornithischians and saurischians.

Saurischia An order of dinosaurs characterized by a lizardlike pelvis; includes theropods, prosauropods, and sauropods.

Ornithischia One of the two orders of dinosaurs, characterized by a birdlike pelvis; includes ornithopods, stegosaurs, ankylosaurs, pachycephalosaurs, and ceratopsians.

bipedal Walking on two legs as a means of locomotion, as in humans, birds, and some dinosaurs.

quadrupedal Walking on four legs as a means of locomotion.

Archosaurs and the Origin of Dinosaurs

Reptiles known as **archosaurs** (*archo*, "ruling," and *sauros*, "lizard") include crocodiles, pterosaurs (flying reptiles), dinosaurs, and birds. Including such diverse animals in a single group implies that they share a common ancestor, and indeed several characteristics unite them. For instance, all have teeth set in individual sockets, except today's birds, but the earliest birds had this feature.

All **dinosaurs** possess several shared characteristics yet differ enough for us to recognize two distinct orders: **Saurischia** and **Ornithischia**. Each order has a distinctive pelvic structure. Saurischian dinosaurs have a lizardlike pelvis and are thus called *lizard-hipped dinosaurs*, whereas ornithischians have a birdlike pelvis and are called *bird-hipped dinosaurs* (Figure 18.10). It is now clear that they had a common ancestor much like archosaurs, known from Middle Triassic rocks in Argentina (Figure 18.11a). These dinosaur ancestors were small (less than 1 m long), long-legged carnivores that walked and ran on their hind limbs, so they were **bipedal**, as opposed to **quadrupedal** animals, which move on all four limbs.

© iStockphoto.com/Robert Blandchard

Dinosaurs

The term *dinosaur* was proposed by Sir Richard Owen in 1842 to mean "fearfully great lizard," although now "fearfully" has come to mean "terrible"—thus the characterization of dinosaurs as "terrible lizards." But of course they were not terrible, or at least no more so than animals living today, and they were not lizards. Nevertheless, dinosaurs more than any other kind of animal have inspired awe and thoroughly captured the public imagination (Figure 18.11). Their popularization in cartoons, books, and movies has unfortunately often been inaccurate and contributed to misunderstandings. For instance, many people think that all dinosaurs were large. It is true that *many* were large, but, actually, dinosaurs varied from giants weighing several tens of metric tons to others no larger than a chicken.

Although various media now portray dinosaurs as more active animals, the misconception that they were lethargic, dimwitted beasts persists. Evidence now available indicates that some were quite active and perhaps even warm-blooded. It also appears that some species cared for their young long after hatching, a behavioral characteristic most often found in birds and mammals. Many questions remain unanswered about dinosaurs, but their fossils and the rocks that contain them are revealing more and more about their evolutionary relationships and behavior.

Among the saurischian dinosaurs, scientists define two suborders: theropods and sauropods. All *theropods* were bipedal carnivores that ranged in size from tiny *Compsognathus* to comparative giants such as *Tyrannosaurus* (see the chapter opening photo) and similar but even larger species. Some of the smaller theropods such as *Velociraptor* and its relative *Deinonychus*, with a large, sickle-shaped claw on each hind foot, likely used these claws in a slashing type of attack (Figure 18.11b). Some remarkable discoveries beginning in 1996 by Chinese paleontologists have yielded several species of small theropods with feathers.

Included among the sauropods are the giant, quadrupedal herbivores such as *Apatosaurus*, *Diplodocus*, and *Brachiosaurus*, the largest land animals of any kind. According to one estimate, *Brachiosaurus* weighed more than 75 metric tons, and partial remains indicate that even larger sauropods existed.

The diversity of ornithischians is evident from the five distinct suborders recognized: ornithopods, pachycephalosaurs, ankylosaurs, stegosaurs, and ceratopsians (Figure 18.10). All ornithischians were herbivores, but some were bipeds, whereas others were quadrupeds.

Figure 18.10 The Relationships Among Dinosaurs

Pelvises of ornithischian and saurischian dinosaurs are shown for comparison. All of the dinosaurs shown here were herbivores, except for the theropods. Note that bipedal and quadrupedal dinosaurs are found among both ornithischians and saurischians.

In fact, although ornithopods such as the well-known duck-billed dinosaurs were primarily bipeds, their well-developed forelimbs allowed them to walk on all fours as well (Figure 18.11c). Also included among the ornithischians were the heavily armored ankylosaurs (see the chapter opening photo); the stegosaurs, with plates on their backs as well as spikes on their tails for defense; the horned ceratopsians; and the peculiar, domeheaded pachycephalosaurs (Figure 18.10).

Warm-Blooded Dinosaurs?

Were dinosaurs *endotherms* (warm-blooded) like today's mammals and birds, or were they *ectotherms* (cold-blooded) like all of today's reptiles? Almost everyone now agrees that some compelling evidence exists for dinosaur endothermy.

Bones of endotherms typically have numerous passageways that, when the animals are alive, contain blood vessels, but considerably fewer passageways are present in bones of ectotherms. Proponents of dinosaur endothermy note that dinosaur bones are more similar to those of living endotherms. Crocodiles and turtles have this so-called endothermic bone, yet they are ectotherms, and some small mammals have bones more typical of ectotherms. It may be that bone structure is related more to body size and growth patterns than to endothermy.

Endotherms must eat more than comparable-size ectotherms, because their metabolic rates are so much higher. Consequently, endothermic predators require

Dinosaurs roamed Earth for more than 140 million years.

pterosaur Any of the Mesozoic flying reptiles.

large prey populations and thus constitute a much smaller proportion of the total animal population than their prey, usually only a few percent. Where data are sufficient to allow an estimate, dinosaur predators made up 3 percent to 5 percent of the total population. Nevertheless, uncer-

Figure 18.11 Dinosaurs

Many types and sizes of dinosaurs existed during the Mesozoic Era, but only a few are shown here.

a. This small, carnivorous biped, *Eoraptor*, was one of the first dinosaurs. These animals, or ones very much like them, are the probable ancestors of all later dinosaurs. *Eoraptor* lived during the Late Triassic (about 230 million years ago) in what is now Argentina. It was only about a meter long and likely weighed no more than 10 kg.

b. Lifelike restoration of *Deinonychus*. It was 3 m long and may have weighed 80 kg. This dinosaur had a huge curved claw on each back foot.

tainties in the data make this argument less than convincing for many paleontologists.

Endothermy seems to be a prerequisite for having a large brain, because a complex nervous system needs a rather constant body temperature. Some dinosaurs did have large brains compared to body size, especially the small- and medium-sized carnivores, but many did not. So brain size for some dinosaurs may be a convincing argument, but even more compelling evidence for theropod endothermy comes from their probable relationship to birds, and the recent discoveries in China of theropods with feathers or a featherlike outer covering. Today, only endotherms have hair, fur, or feathers for insulation.

Good arguments for endothermy exist for several types of dinosaurs, although the large sauropods were probably not endothermic, but nevertheless were capable of maintaining a rather constant body temperature. Large animals heat up and cool down more slowly than smaller ones, because they have a small surface area compared to their volume. With their comparatively smaller surface area for heat loss, sauropods probably retained heat more effectively than their smaller relatives.

In general, a fairly good case can be made for endothermy in many theropods and in some ornithopods. Nevertheless, disagreement exists, and for some dinosaurs the question is still open.

Flying Reptiles

Paleozoic insects were the first animals capable of flight, but the first among vertebrates were the **pterosaurs**, or flying reptiles. They were common in the skies from the Late Triassic until their extinction at the end of the Cretaceous (Figure 18.12). Adaptations for flight include a wing membrane sup-

c. The ornithopod dinosaurs were abundant and varied. *Maiasaura* nested in colonies and tended its young long after hatching.

ported by an elongated fourth finger; light, hollow bones; and development of those parts of the brain associated with muscular coordination and sight. The fact that at least one pterosaur species had a coat of hair or hairlike feathers suggests that it, and perhaps all pterosaurs, were endotherms.

Figure 18.12 Flying Reptiles (Pterosaurs)

a. *Pterodactylus* from the Late Jurassic is a good example of the earlier pterosaurs. It has a long tail, numerous teeth, and a wingspan of about 1.5 m.

b. Later pterosaurs such as *Pteranodon* of the Late Cretaceous were much larger, tailless, and had toothless beaks. *Pteranodon's* wingspan was about 6 m.

Most pterosaurs were no bigger than today's sparrows, robins, and crows. However, a few species had wingspans of several meters, and one Cretaceous pterosaur found in Texas had a wingspan of at least 12 m! The comparatively large pterosaurs probably took advantage of thermal updrafts to stay airborne, mostly by soaring but occasionally by flapping their wings for maneuvering. Smaller pterosaurs probably stayed aloft by vigorously flapping their wings, just as present-day small birds do.

Marine Reptiles
The most familiar Mesozoic marine reptiles are the rather porpoiselike **ichthyosaurs** (Figure 18.13a). Most of these fully aquatic animals were about 3 m long, but one species reached about 12 m. All ichthyosaurs had a streamlined body, a

ichthyosaur Any of the porpoiselike, Mesozoic marine reptiles.

Figure 18.13 Mesozoic Marine Reptiles

a. Icthyosaur.

b. A long-necked piesiosaur.

powerful tail for propulsion, and flipperlike forelimbs for maneuvering.

Their numerous sharp teeth indicate they were fish eaters, although they undoubtedly preyed on other marine organisms as well. Ichthyosaurs were so completely aquatic that it is doubtful they could come onto land, so females probably retained eggs within their bodies and gave birth to live young. A few fossils with small ichthyosaurs within the appropriate part of the body cavity support this interpretation.

Another well-known group, the **plesiosaurs**, belonged to one of two subgroups: short-necked and long-necked (Figure 18.13b). Most were modest-sized animals 3.6 to 6 m long, but one species found in Antarctica measures 15 m. Short-necked plesiosaurs may have been bottom feeders, but their long-necked cousins probably used their necks in a snakelike fashion to capture fish with their numerous sharp teeth. These animals probably came ashore to lay their eggs.

Birds

Several fossils with feather impressions have been discovered in the Jurassic Solnhofen Limestone of Germany, but in almost every other known physical feature these fossils are most similar to small theropods. This birdlike creature known as *Archaeopteryx* retained dinosaurlike teeth, tail, brain size, and hind-limb structure but had feathers and a wishbone, characteristics typical of birds (Figure 18.14).

Most paleontologists now think that some kind of small theropod was the ancestor of birds. Even the wishbone, consisting of fused clavicles so typical of birds, is found in a number of theropods, and recent discoveries of theropods in China with some kind of feathery covering are further evidence of this relationship.

Two more Mesozoic fossils shed more light on bird evolution. One specimen, from China, is slightly younger than *Archaeopteryx* and has both primitive and advanced features. For instance, it retains abdominal ribs similar to those of *Archaeopteryx* and theropods, but it has a reduced tail more typical of present-day birds. Another Mesozoic bird from Spain is also a mix of primitive and advanced characteristics.

Archaeopteryx's fossil record is not good enough to resolve whether it is the ancestor of today's birds or

an animal that died out without leaving descendants. Of course, that in no way diminishes the fact that it shows both reptile and bird characteristics. However, some scientists claim that fossils of two crow-sized individuals known as *Protoavis* are an even earlier bird than *Archaeopteryx*. These Late Triassic fossils have hollow bones and the breastbone structure of birds, but because no feather impressions were found, many paleontologists think these are specimens of small theropod dinosaurs.

Mammals

In a previous section, we briefly mentioned therapsids, or advanced mammal-like reptiles. One particular group of therapsids known as **cynodonts** was the most mammal-like of all and gave rise to mammals during the Late Triassic. This transition is especially well documented by fossils and is so gradual that classification of some fossils as either reptile or mammal is difficult.

In fact, the first mammals retained several reptilian characteristics but had mammalian features as well. A good example is their lower jaw. In typical mammals, the lower jaw is a single bone, whereas in reptiles it consists of several bones. The first mammals retained more than one bone in the lower jaw but had teeth and a jaw–skull joint characteristic of mammals. In fact, the quadrate and articular bones that formed the jaw–skull joint in reptiles were modified and became the incus and amalleus, two small bones in the mammalian middle ear. Another typical mammalian feature is occlusion, which means that the chewing teeth meet surface to surface to allow grinding, a feature also found in some advanced cynodonts.

In short, some mammalian features evolved more rapidly than others, thereby accounting for animals that have characteristics of both reptiles and mammals. Even though mammals appeared at the same time as dinosaurs, their diversity remained low, and all of them were small animals during the rest of the Mesozoic Era (Figure 18.14).

Mammals today belong to one of three groups: The *monotremes*, or egg-laying mammals; the *marsupial mammals*, or pouched animals; and the *placental mammals*, which includes most mammals with which we are familiar. The history of the monotremes is uncertain, but the marsupials and placentals diverged from a common ancestor during the Late Cretaceous (Figure 18.15). We discuss all three of these mammal groups later, under Learning Objective 4, "Cenozoic Life History."

Figure 18.14 Fossils of *Archaeopteryx* from the Jurassic-age Solnhofen Limestone of Germany

Notice the feather impressions on the wings and the long tail. This animal had feathers and a wishbone, making it a bird, but in most details of its anatomy, it resembled small theropod dinosaurs. For example, it had reptilian teeth, claws on its wings, and a long tail—none of which are found on birds today.

Tom McHugh/Photo Researchers, Inc.

Plants

Triassic and Jurassic land plant communities were composed of seedless vascular plants and gymnosperms. Among the gymnosperms, the conifers continued to diversify, and cycads that superficially resemble palms made their appearance. Both seedless vascular plants and gymnosperms are well represented in the present-day flora, but neither group is as abundant as it once was.

The long dominance of seedless plants and gymnosperms ended during the Early Cretaceous, when many were replaced by **angiosperms**, or flowering plants. Recent studies have identified both fossil and living gymnosperms that show close relationships to angiosperms. Since they first evolved, angiosperms have adapted to nearly every terrestrial habitat from mountains to deserts. Some have even adapted to shallow coastal waters. Their reproduction, involving flowers to attract animal pollinators and the evolution of enclosed seeds, largely accounts for their success. They now account for about 96% of all vascular plant species.

Cretaceous Mass Extinctions

The mass extinctions at the close of the Mesozoic were second in magnitude only to those at the end of the Paleozoic. Casualties of the Mesozoic extinctions include dinosaurs, flying reptiles, marine reptiles, and several kinds of marine creatures such as ammonites.

A hypothesis for these extinctions that has become popular since 1980 is based on a discovery at the Cretaceous–Paleogene boundary in Italy—a clay layer 2.5 cm thick with an abnormally high concentration of the platinum-group element iridium. Since this discovery, high iridium concentrations have been identified at many other Cretaceous–Paleogene boundary sites. The significance of this discovery lies in the fact that iridium is rare in crustal rocks but occurs in much higher concentrations in some meteorites. Several investigators proposed a meteorite impact to explain this iridium anomaly and further postulated that the impact of a large meteorite, perhaps 10 km in diameter, set in motion a chain of events that led to extinctions.

The meteorite-impact scenario goes something like this: On impact, about 60 times the mass of the meteorite was blasted from Earth's crust high into the atmosphere, and the heat generated at impact started raging fires that added more particulate matter to the atmosphere. Sunlight was blocked for several months, causing a temporary cessation of photosynthesis; food chains collapsed and extinctions followed. In addition, with sunlight greatly diminished, Earth's surface temperatures were drastically reduced and could have added to the biologic stress.

Some now claim that a probable impact site has been found centered on the town of Chicxulub on the Yucatán Peninsula of Mexico. The structure is about 180 km in diameter and lies beneath layers of Cenozoic sedimentary rock.

Most geologists now concede that a large meteorite impact occurred, but we also know that vast outpourings of lava were taking place in what is now India. Perhaps these eruptions brought about detrimental atmospheric changes. Furthermore, the vast, shallow seas that covered large parts of the continents had mostly withdrawn by the end of the Cretaceous, and the mild, equable Mesozoic climates became harsher and more seasonal by the end of that era.

LO4 CENOZOIC LIFE HISTORY

Even though we emphasize the evolution of mammals in this section, you should be aware of other important life events. Flowering plants continued to dominate land plant communities, the present-day groups of birds evolved during the Early Paleogene, and some marine invertebrates continued to diversify, eventually giving rise to today's marine fauna.

Mammals coexisted with dinosaurs for more than 100 million years, yet their Mesozoic fossil record indicates that they were not abundant, diverse, or very large. The largest one known was 1 m long and weighed 12 to 14 kg. Extinctions at the end of the Mesozoic eliminated dinosaurs and some of their relatives, thereby creating the adaptive opportunities that mammals quickly exploited. The "Age of Mammals," as the Cenozoic Era is commonly called, had begun.

Marine Invertebrates and Phytoplankton

Survivors of the Mesozoic extinctions populated the Cenozoic marine ecosystem. Especially abundant among the invertebrates were the foraminifera, radiolarians, corals, bryozoans, mollusks, and echinoids. Only a few species of phytoplankton survived into the Cenozoic, but those that did flourished and diversified. Marine and freshwater plants called *diatoms*, with silica skeletons, were particularly abundant.

Foraminifera were also abundant, and their shells accumulated to form thick limestones, some of which the ancient Egyptians used to construct the Sphinx and the Pyramids. Corals once again became the dominant reef builders, and just as during the Mesozoic, bivalves and gastropods were major components of the marine invertebrate communities.

Diversification of Mammals

Among living mammals, **monotremes**, such as the platypus, lay eggs, whereas marsupials and placentals give birth to live young. **Marsupial mammals** are born in an immature, almost embryonic condition, and then develop further in their mother's pouch. **Placental mammals** have no shelled egg, but they have developed a placenta within the uterus across which nutrients and oxygen are carried from the mother to the developing embryo. As a result, placental mammals are much more fully developed than marsupials before birth.

A measure of the success of placental mammals is that more than 90% of all mammals, fossil and living, are placental. Judging from the fossil record, monotremes have never been very common; the only living ones are platypuses and spiny anteaters of the Australian region. Marsupials have been more successful in terms of number of species and geographic distribution, but even they have been largely restricted to South America and the Australian region.

As we have already noted, mammals first evolved during the Triassic, and by the Late Cretaceous the marsupials and placentals diverged from a common ancestor (Figure 18.15). Then, following the Late Cretaceous extinctions, mammals began a major adaptive radiation that continued throughout the Cenozoic Era. Several groups of Paleocene mammals are *archaic*, meaning that they were holdovers from the Mesozoic Era or that they did not give rise to any of today's mammals (Figure 18.16a). But also among these mammals were the first rodents, rabbits, primates, carnivores, and hoofed mammals. However, even these had not yet become clearly differentiated from their ancestors, and the differences between herbivores and carnivores were slight. Most were small; large mammals were not present until the Late Paleocene, and the first giant terrestrial mammals did not appear until the Eocene (Figure 18.16b).

Diversification continued during the Eocene, when several more types of mammals appeared, but if we could go back and visit this time, we probably would not recognize many of these animals. Some would be vaguely familiar, but the ancestors of horses, camels, rhinoceroses, and elephants would bear little resemblance to their living descendants. By Oligocene time, all of the orders of existing mammals were present, but diversification continued as more familiar families and genera appeared. Miocene and Pliocene mammals were mostly mammals that we could readily identify, although a few unusual types still existed (Figure 18.17).

Cenozoic Mammals

Mammals arose from mammal-like reptiles known as *cynodonts* during the Late Triassic. Following the Mesozoic extinctions, they began an adaptive radiation

Figure 18.15 Mesozoic Mammals

Both of these fossils come from Early Cretaceous-age rocks in China.

a. Restoration of the oldest known marsupial mammal, *Sinodelphym*, which measures 15 cm long.

b. Restoration of *Eomaia*, the oldest known placental mammal. It was only 12 or 13 cm long.

Figure 18.16 Paleocene and Eocene Mammals

a. This squirrel-sized animal is the Paleocene genus *Ptilodus*, a member of a group of mammals known as multituberculates. Multituberculates existed from the Early Jurassic to the Early Oligocene.

b. Scene from the Eocene showing the rhinoceros-sized animal known as *Unitatherium*, one of the first giants among mammals. It had three pairs of horns and saber-like upper canine teeth.

and soon became the most abundant land-dwelling vertebrates. Now more than 4,000 species exist, ranging from tiny shrews to elephants and whales.

Numerous groups of mammals evolved during the Cenozoic, and some, such as camels and horses and their relatives, have excellent fossil records. Camels evolved from small, four-toed ancestors and were particularly abundant in North America, where most of their evolutionary history is recorded. They became

Figure 18.17 Miocene Mammals of Nebraska

This scene depicts some of the mammals that lived in Nebraska during the Miocene Epoch, about 21 million years ago. Although a few of these mammals are unfamiliar, most of them look much like mammals living today. Shown in this image are relatives of today's elephants (center), a clawed horselike animal called a chalicothere (right center), hoofed animals called oreodonts (in the water), and small camels (left front) being chased by a bear dog. A saber-toothed cat in the tree feasts on its prey, and cranes circle overhead.

Chase Studio/Photo Researchers, Inc.

primate Any of the mammals that belong to the order Primates; characteristics include large brain, stereoscopic vision, and grasping hands.

extinct in North America during the Pleistocene, but not before some species migrated to South America and Asia. Horses and their living relatives, rhinoceroses and tapirs, also evolved from small Early Cenozoic ancestors (Figure 18.18). Horses and rhinoceroses were common in North America, and like camels, they died out here but survived in the Old World.

Whales provide an excellent example of how our knowledge of life history is constantly improving. Fossil whales have always been common, but until several years ago, fossils linking fully aquatic animals with land-dwelling ancestors were largely absent. It turns out that this transition took place in a part of the world where the fossil record was poorly known. Now, however, a number of fossils are available, showing that whales evolved from land-dwelling ancestors during the Eocene.

Pleistocene Faunas

One remarkable aspect of the history of mammals is that so many very large species existed during the Pleistocene Epoch. In North America, there were mastodons and mammoths, giant bison, huge ground sloths, giant cam-

els, and beavers nearly 2 m long (Figure 18.19). Australia and Europe also had giant mammals. In addition to mammals, giant birds up to 3.5 m tall and weighing 585 kg existed in New Zealand, Madagascar, and Australia.

Many smaller mammals were also present, but the major evolutionary trend in mammals was toward large body size. Perhaps this was an adaptation to the cooler temperatures of the Pleistocene. Large animals have proportionately less surface area compared to their volume, and therefore retain heat more effectively than do small animals.

At the end of the Pleistocene, almost all of the large terrestrial mammals of North America, South America, and Australia became extinct. Extinctions also occurred on the other continents, but they had considerably less impact. These Pleistocene extinctions were modest by comparison to earlier ones, but they were unusual in that they affected mostly large terrestrial mammals. The debate over the cause of this extinction continues between those who think the large mammals could not adapt to the rapid climatic changes at the end of the Ice Age and those who think these mammals were killed off by human hunters, a hypothesis known as *prehistoric overkill*.

Primate Evolution

Several evolutionary trends in regard to **primates** help to define the order, some of which relate to their *arboreal*,

Figure 18.18 Evolution of Horses

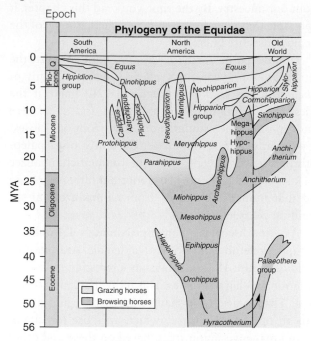

a. Summary chart showing the relationships among the genera of horses. During the Oligocene two lines emerged, one leading to three-toed browsers and the other to one-toed grazers, including the present-day horse, *Equus*.

Equus
(Pleistocene)

Pliohippus
(Pliocene)

Merychippus
(Miocene)

Mesohippus
(Oligocene)

Hyracotherium
(Eocene)

b. Simplified diagram showing some trends in horse evolution. Trends include a size increase, a lengthening of the limbs, a reduction of the number of toes, and the development of high-crowned teeth with complex chewing surfaces.

or tree-dwelling, ancestry. These include changes in the skeleton and mode of locomotion; an increase in brain size; a shift toward smaller, fewer, and less specialized teeth; and the evolution of stereoscopic vision and a grasping hand with opposable thumb. Not all of these trends took place in every primate group, nor did they evolve at the same rate in each group.

The Primate order is divided into two suborders. The *prosimians*, include the lemurs, lorises, tarsiers, and tree shrews and are the oldest primate lineage, with a fossil record extending back to the Paleocene. Sometime during the Late Eocene, the *anthropoids*, primates that include monkeys, apes, and humans, evolved from a prosimian lineage, and by the Oligocene, the anthropoids were a well-established group. The *hominoids*, the group containing apes and humans, diverged from Old World monkeys sometime before the Miocene, but exactly when is still being debated. It is, however, generally accepted that the hominoids evolved in Africa from an ancestral anthropoid group.

During the Miocene, Africa collided with Europe, which changed Africa's climate and allowed animals to migrate between the two landmasses.

Two apelike groups evolved during this time that ultimately gave rise to present-day hominoids. Members of one of these groups was most likely the ancestor of all later hominoids.

The **hominids** (family Hominidae), the primate family that includes present-day humans and their extinct ancestors, have a fossil record extending back almost 7 million years (Figure 18.20). Several features distinguish them from other hominoids. For one, hominids are bipedal; that is, they have an upright posture and walk on two legs rather

hominid Abbreviated form of Hominidae, the family to which humans belong; bipedal primates include *Australopithecus* and *Homo*.

Figure 18.19 Pleistocene Mammals

Restoration of a mammoth trapped in the sticky tar (asphalt) as a present-day oil seep at the La Brea Tar Pits in Los Angeles, California.

Sue Monroe

than four. In addition, they show a trend toward a large and internally reorganized, complex brain. Other features include a reduced face and small canine teeth, omnivorous feeding, increased manual dexterity, and the use of sophisticated tools.

At present, there is no clear consensus on the evolutionary history of the hominid lineage. This lack of agreement is due, in part, to the incomplete fossil record of hominids, as well as new discoveries. In fact, new discoveries of fossil hominids, as well as new techniques

for scientific analysis, are leading to new hypotheses about our ancestry. By the time you read this chapter, it is possible that new discoveries may change some of the conclusions stated here.

Discovered in northern Chad's Djurab Desert, the nearly 7-million-year-old skull and dental remains of *Sahelanthropus tchadensi* (Figure 18.20) make it the oldest known hominid yet unearthed and at or very near the time when humans diverged from our closest living relative, the chimpanzee. Currently, most paleoanthropologists accept that the human-chimpanzee stock separated from gorillas about 8 million years ago and that humans separated from chimpanzees about 5 million years ago.

A recent discovery of an approximately 4.4-million-year-old skeleton and skull of a female *Ardipithecus ramidus* (Figure 18.20), shows an interesting mosaic of evolutionary characteristics. This individual, nicknamed "Ardi," had a dexterous hand for grasping and a foot with an opposable big toe, which apes use for climbing and maneuvering in trees. Based on these and other features, scientists have concluded that Ardi was able to walk upright on the ground while retaining the ability to climb and maneuver in trees.

Australopithecine is a collective term for all members of the genus *Australopithecus*, which currently includes five species (Figure 18.20). Many paleontologists accept

Figure 18.20 Fossil Record of Hominids

The geologic ranges for the commonly accepted species of hominids.

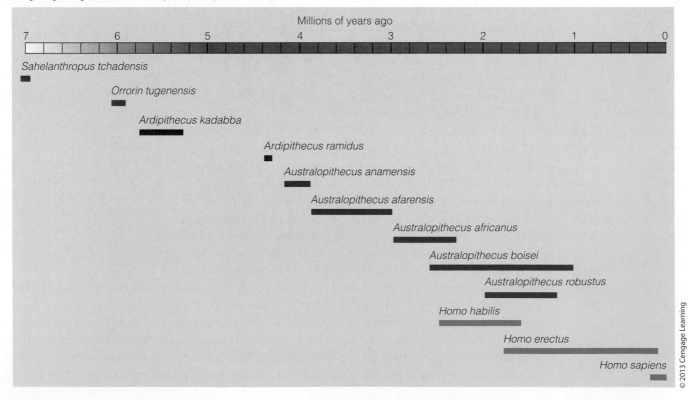

© 2013 Cengage Learning

the evolutionary scheme in which *Australopithecus anamensis*, the oldest known australopithecine, is ancestral to *A. afarensis* (Figure 18.21), who in turn is ancestral to *A. africanus* and the genus *Homo*, as well as the side branch of australopithecines represented by *A. robustus* and *A. boisei*.

The earliest member of our own genus *Homo* is *Homo habilis*, who lived 2.5 to 1.6 million years ago (Figure 18.20). Evolving from *H. habilis* between 1.8 and 1.6 million years ago, was *H. erectus*, a widely distributed species that migrated from Africa during the Pleistocene Epoch. The archaeological record indicates that *H. erectus* made tools, used fire, and lived in caves, an advantage for those living in more northerly climates.

Debate still surrounds the transition from *H. erectus* to our own species, *Homo sapiens* (Figure 18.20). Paleoanthropologists are split into two camps. One camp supports the "out of Africa" hypothesis; that is, the idea that early modern humans evolved from a single woman in Africa, whose offspring then migrated from Africa—perhaps as recently as 100,000 years ago—to populate Europe and Asia, driving the earlier hominid population to extinction.

The alternative explanation, the "multiregional" hypothesis, maintains that early modern humans did not have an isolated origin in Africa, but rather that they established separate populations throughout Europe and Asia. Occasional contact and interbreeding between these populations enabled our species to maintain its overall cohesiveness while still preserving the regional differences in people that we see today. At this time, not enough evidence points to which, if either, theory is correct, but our species, *H. sapiens*, most certainly evolved from *H. erectus*.

Perhaps the most famous of all fossil humans are the *Neanderthals*, who inhabited Europe and the Near East from about 200,000 to 30,000 years ago, and according to the best estimates, never exceeded 15,000 individuals in western Europe. Based on specimens from more than 100 sites, we now know that Neanderthals were not much different from us, only more robust. Europe's Neanderthals were the first humans to move into truly cold climates, enduring miserably long winters and short summers as they pushed north into tundra country. Their remains are found chiefly in caves and hutlike rock shelters, which also contain a variety of specialized stone tools and weapons.

About 30,000 years ago, humans closely resembling modern Europeans moved into the region inhabited by the Neanderthals and completely replaced them. *Cro-Magnons*, the name given to successors of Neanderthals

Figure 18.21 *Australopithecus afarensis*
Recreation of a Pliocene landscape showing *Australopithecus afarensis*, an early hominid species, gathering and eating fruits and seeds.

© 2013 Cengage Learning

in France, lived from about 35,000 to 10,000 years ago; during this period, the development of art and technology far exceeded anything the world had seen before.

With the appearance of Cro-Magnons, human evolution has become almost entirely cultural rather than biological. This means that although we are becoming rapidly more sophisticated in terms of the technology and knowledge we can bring to bear on our problems, we are not necessarily changing physically as a species.

The largest single cultural change for our species occurred not with the computer revolution, but about 10,000 to 15,000 years ago when our forefathers learned to farm rather than to rely strictly on a hunter-gatherer lifestyle. The agricultural revolution occurred at several different places around the world at roughly the same time, as the last cold period of the Pleistocene was ending. It is what made settled life in towns possible—and eventually civilization as we know it.

We are still expanding the area of the planet that we have under the plow. This fact is just one aspect of our increasing impact on the globe. We now consume fossil fuels originally formed in the Paleozoic and Mesozoic eras, and we mine mineral resources that date back all the way to the Proterozoic.

Many species in Earth history have changed the planet, from the humble stromatolites forward in time, but no species is changing the planet as rapidly as we humans are now doing. On the bright side, no species up until our own has been able to understand the changes it imposes on Earth—still our only home.

Index

C

Calcareous tufa, 267
Calcite, 63, 64, 65, 66, 67, 67, 68, 113, 121, 125, 125t
Calcium, 59
Calcium carbonate, 63, 123, 127
  hot springs/geysers and, 267, 269, 269
  in ooze, 310
Calderas, 95–96, 96, 101
Caledonian orogeny, 362
Caliche, 118
California
  earthquakes in, 156, 157, 161, 161, 167, 169–170, 175
  mass wasting in, 206, 209, 211, 211, 212, 213, 220
  San Andreas fault in (See San Andreas fault)
Calving, 272, 274, 275
"Cambrian explosion," 391
Cambrian Period, 332, 343, 343, 361, 363, 366
  history of life in, 391–392, 392t
  sandstone of, 354
  Sauk sequence in, 365
  vertebrates in, 394
Canadian shield, 279, 357, 358, 365, 366
Canyons, 241, 243, 243, 245, 245, 385
Cap rock, 137
Capillary fringe, 249, 250
Capitol Reef National Park, 132, 342, 343
Carbon, 59, 59, 329
Carbon-14 dating, 350–351, 351
Carbonate minerals, 63, 64, 67, 68, 372, 385
Carbonate radical, 60, 61, 61t, 63, 125
Carbonate rocks, 125, 125t, 256, 257, 358
Carbon dioxide, 113–114, 393, 394
Carbon dioxide cycle, 9
Carbonic acid, 113
Carboniferous Period, 362, 365, 367, 393, 396t
Carnotite, 137
Cascade Range, 89, 102, 106, 191, 384
  andesite in, 82
  circum-Pacific belt and, 101
Castle Geyser, 269, 269
Catskill Delta, 372
Cave of Crystals, 246, 256
Caves/cave deposits, 256–258, 257, 258, 352–353, 353
Cementation, 122–123, 124, 124
Cenozoic Era, 332, 344, 376, 377, 378, 381, 381–383

Cordilleran orogeny in, 378, 379
  eastern North America in, 385
  Gulf Coast in, 385
  history of life in, 397, 403, 404–409
  lava flows in, 385, 403
  mammals in, 404–406
  natural resources and, 48–49
  Ouachita mobile belt in, 372
Chalcopyrite, 152
Chalk, 125
Challenger Deep, 31
Channel flow, 227
Chemical sediment, 121, 123
Chemical sedimentary rocks, 124, 125–127, 125t, 127, 249
Chemical weathering, 112–115, 115, 120, 121, 209
Chert, 125t, 126, 128, 137, 358
China (Paleozoic), 362, 363, 364
Chinese earthquakes, 169, 172, 173
Chinle Formation, 380
Chloride, 308
Chlorine, 56, 56, 59, 327, 393
Chlorite, 143, 144, 145t, 149
Chromium, 69
Chrysotile, 152
Chugach Mountains, 387
Cinder cones, 97, 97–98, 387
Circum-Pacific belt, 101, 101, 103, 160, 161, 161, 172, 194
  in Cenozoic, 382
  in North American Cordillera, 383, 383
Cirques, 278, 279, 280
Claron Formation, 108, 385
Clastic texture, 124, 125, 125t
Clastic wedge, 372
Clasts, 124, 126
Clay, 121, 136, 205, 207, 211, 310
Clay minerals, 62, 68, 68
  chemical weathering and, 112, 114
  clay vs., 121
  metamorphism and, 143, 144, 150
  soil and, 116, 117
  uses of, 69
Claystone, 124, 124
Cleavage, 64–65, 66
Climate, 9–10, 113, 115, 362. See also Global warming/cooling
  and breakup of Pangaea, 374
  deserts and, 289, 298, 299
  and distribution of life, 50
  geologic time and, 352–353, 353
  glaciers and, 269, 284, 287

mass wasting and, 205, 215, 217
  soil and, 118, 119
Coal, 125t, 126, 127, 128, 136
Coal swamps, 362, 368, 369, 381, 396, 396, 397
Coast Mountains, 270, 283, 286
Coast Range, 378, 381
Coasts, 323–327, 325. See also Beaches; Shorelines
  emergent vs. submergent, 324–325, 325
  Mesozoic, 376–377
  shorelines vs., 311, 311
Cobalt, 69, 329
Coke, 136
Colorado earthquakes, 173–174
Colorado Plateau, 344, 345, 380, 383, 383
Columbia River basalt, 100, 100
Columnar joints, 93, 93
Columns, 133, 258
Comets, 13
Compaction, 122–123, 124, 124
Complex movement, 210t, 217–218, 219
Composite volcanoes, 98, 98, 102, 103
Compounds, 55–57, 56, 57
Compression (of rock layers), 182, 182, 187, 191, 192
  faults and, 190
  joints and, 188
Concordant plutons, 84, 86
Cone of ascension, 259, 260
Cone of depression, 251, 252, 260
Conglomerate, 19, 124, 124, 126
Constancy of interfacial angles, 58
Contact metamorphism, 142–143, 143, 149, 151, 152
Continent(s), 356
  breakup of, 39 (See also Pangaea)
  Cambrian, 362, 363
  deformation and, 181
  and distribution of life, 50–51, 51
  Earth's crust and, 198–200, 198–200
  floating, 198, 199
  igneous rocks and, 71
  sedimentary rocks and, 110
Continental accretion, 181, 197
Continental–continental boundaries, 41, 42, 43, 196–197, 197
Continental crust, 15, 15, 81, 176t, 179, 199
  continental margins and, 29
  evolution of, 356, 357
  mountains/continents and, 198

Continental deposition, 122, 123
Continental divide, 237
Continental drift, 25–28, 28, 34–35, 35
Continental fit, 25, 25
Continental glaciers, 270–271, 271, 275, 279
  glacial surge in, 276
  sediment in, 277, 281
  stratified drift and, 283
Continental interior, 385
Continental margins, 29, 29, 32–33, 33, 196, 383
  active vs. passive, 32–33, 33, 43
  deformation and, 181
  oceanic circulation and, 310
  passive, 32–33, 33, 43, 358, 360, 376, 385
Continental rifting, 39, 41, 41
Continental rise, 29, 29, 30, 32, 33, 39
Continental shelf, 29, 29, 32, 33, 39, 322, 324
Continental slope, 29, 29, 32, 33, 39
Convection cells, 14, 15, 47, 47–48, 48
Convergent boundaries, 41–44, 42, 181
  batholiths and, 86
  continental–continental, 41, 42, 43
  deformation and, 181, 187
  granitic rocks at, 82
  igneous rocks and, 71
  metallic ores and, 49–50, 50
  metamorphism and, 144, 150–151, 151
  mountain building and, 181, 190, 191, 192, 193
  oceanic–continental, 41, 42, 43, 372
  oceanic–oceanic, 41–43, 42
  recognizing ancient, 44
  volcanoes and, 103, 181
Copper, 49–50, 50, 53, 63, 329
  batholiths and, 86
  metallic bonding in, 57
  metamorphism and, 152
  plutons and, 71
Coquina, 125, 127
Coral reefs, 310, 310, 404
Cordilleran batholiths, 378, 381
Cordilleran mobile belt, 372, 377, 381
Cordilleran orogeny, 362, 365, 378–387, 381
Core (of Earth), 14, 14, 15, 175
  density/composition of, 176t, 177–178
  as Earth subsystem, 3, 4
Core-mantle boundary, 176, 179

Mineral grains, 78–79
Miocene Epoch, *332*, 382, 404, *406*, 407
Mississippian Period, *332*, *338*, 343, *343*
  Absaroka sequence in, 367
  Cordilleran mobile belt in, 372
  history of life in, 395, 397
  Kaskaskia sequence in, 365
  Ouachita mobile belt in, 372
  vertebrates in, 395
Mississippi River, 227, *235*, 237, *237*
"Mixed rocks," 147, *148*
Mobile belts, 371–374
Modified Mercalli Intensity Scale, 165, 166t
Mohorovičić discontinuity (Moho), 178, *178*
Mohs, Friedrich, 66
Moine Thrust Zone, 144
Mojave Desert, *303*
Molecules, 57
Monocline, 185, *185*
Monotremes, 402, 404
Montana earthquake, 171, *171*
Monument Valley, *86*, *291*, *305*
Moon, *312*
Moraines, 281–282, *283*, *284*, *284*, *285*
Morrison Formation, 381, *382*
Mountain building. *See* Orogeny/mountain building
Mountain glaciers, 270. *See also* Valley glaciers
Mountain ranges, 18, 181, 191
  continental fit and, *25*, *25–26*, *26*
  convergent boundaries and, *42*, 181
  in desert regions, 304
  glaciers and, 271
Mount Rushmore, *70*
Mud, 121, 122–123, *124*
Mud cracks, 130
Mudflows, 205, 210t, 213–214, *215*
Mudrock, 124, *124*
Mudslides, 203, 207
Mudstone, 124, *124*
Muscovite, 63, *68*, 69
Mylonites, 143–144, *144*

N

Native elements, 59
Natural Bridges National Monument, *131*
Natural gas, 68, 109, 136–137, *137*
  from anticlines, 187
  geologic structures and, 181, 187

Natural glass, 79
Natural levees, 232–233
Natural resources, 6, 7, 21, 53, 68–69, 119, 120
  energy, 68, 267, 329
  erosion and, 109
  groundwater as, 247, 258
  metamorphism and, 152
  from oceans, 327–329
  plate tectonics and, 24, 48–50, *50*
  in sedimentary rocks, 136–137
Natural selection, 18
Navajo Sandstone, *133*, 136, *136*, 242, *242*, 243, *243*, 380
Nazca plate, 43
Neanderthals, 409
Nearshore currents, 316, 320
Nearshore sediment budget, 323, *324*
Nebraskan glaciation, 385, 386
Neogene Period, *332*, *381*, 382, 396t
Neon, 56
Neoproterozoic Eon, 360
Neptune, 90
Neutron, 54, *55*
Nevadan orogeny, 378, 379, 380, *381*
Nevado del Ruiz, 98, 104
New Orleans hurricane, 326, 327
Newton, Isaac, 198
New Zealand earthquake, *154*
Nickel, 176t, 177, 178, 329
Nile River, *224*, 225–226, *235*, 269
Nitrates, 310
Noble gases, 56, 57
Nonconformity, 337, *340*, 341
Nonferromagnesian silicate, 63, 64, 66, 68, 81
  in Bowen's reaction series, 75, *75*
  and classifying igneous rocks, 81, 82
Nonfoliated metamorphic rock, 18, *19*, 144, 145t, 148–149, *149*
Nonfoliated texture, 147–148, *148*
Nonmetallic resources, 68
Normal fault, 190, *192*, 193
North America, evolution of, 365–381, *378*, 385, 386, 387
North American Cordillera, *383*, 383–387
North American craton, 357
No-till planting, 120
Novarupta lava dome, 74
Novarupta volcano, 104
Nuclear forces, 6, 12
Nucleus (of atom), 54, *55*

Nuée ardente, 99–100, 104
Numerical ages, 20

O

Oblique-slip fault, 191, *192*
Obsidian, 79, *80*, 83, *85*
Ocean(s), 307, *308*, 308–310, 327–329
Ocean basins, 30, 37, 50, 310, 326
Ocean currents, 50, 307, *308*, 309–310
Oceanic circulation, 308, *308*, 309–310
Oceanic–continental boundaries, 41, *42*, 43, 372, 377–378
  metamorphic facies and, 151, *151*
  orogenies at, 194–196, *197*, 383
Oceanic crust, 15, *15*, 31, 81, 176t, 179, 199
  continental margins and, 29
  igneous rocks and, 71
  magnetism and, 37
  mountains/continents and, 198
  plate boundaries and, 39
  volcanoes and, 89, 103
Oceanic–oceanic boundaries, 41–43, *42*, 193–194, *195*
Oceanic ridges, 29, 30, 32, 39, *40*, 44, 199
Oceanic trenches, 29, 31, 44
Oil, 136, 137, *137*, 267. *See also* Petroleum
Oil spills, *328*
Old Faithful geyser, 266, *266*, 268–269, *268–269*
Oldham, R. D., 176
Oligocene Epoch, *332*, 382, 404, *405*, 407
Olivine, 59, 60, 61, *63*, 64, *68*, 82
  Bowen's reaction series and, 74
  crystal settling and, 77, *78*
  plate boundaries and, 44–45
Olympic National Park, *320*
On the Origin of Species, 18
Ooids, 125, *127*, 135–136
Oolitic limestone, 125, *127*
Ooze, 310
Ophiolites, 44, *44*
Ordovician Period, *332*, 343, 344, 362, *363*, 371
  plants in, 396, 396t
  Sauk sequence in, 365
  Tippecanoe sequence in, 365, 392

Organic evolution, theory of, 18
  fossils and, 131 (*See also* Fossil record; Fossils)
  of horses, *406*, 407
  of humans, 407–408, *408*
  of land plants, *396*
  "out of Africa" *vs.* "multiregional," 409
  plate tectonics and, 50–51, *51*
  of primates, 406–409
Original horizontality, 184, 334, *334*, 337, 342
Origin of Continents and Oceans, The, 24
Ornithischia, 398, *399*. *See also* Bird-hipped dinosaurs
Orogens, 360
Orogeny/mountain building, 16, 191–198
  batholiths and, 86
  at continental–continental boundaries, 196–197, *197*
  at convergent boundaries, *42*
  deformation and, 181, 191–198
  Earth's crust and, 198–200, *198–200*
  Mesozoic, 376–381
  metamorphism and, 142
  North American Cordillera and, *383*, 383–387
  at oceanic–continental boundaries, 194–196, *197*
  at oceanic–oceanic boundaries, 193–194, *195*
  Paleozoic, 362, 371–374
  plate tectonics and, 38, 192–193
  Precambrian, 357
  Proterozoic, 360
Orthoclase, 60, 67, *68*, 114
Ostracoderms, *394*
Ouachita mobile belt, 372, 374
Ouachita orogeny, 263, 362, 366
Outgassing, 308
Outlet glaciers, 275
Outwash plains, 283
Overloading, 207
Overpopulation, 8–9
Owen, Richard, 398
Oxbow lakes, 232, *233*
Oxidation, 114
Oxide, 61t, 114
Oxygen, 59, 60, *60*, 176t, 177
  and history of life, 395
  weathering by, 112, 113, 114
Ozone, 393

PERIODIC TABLE OF THE ELEMENTS

Note: Atomic masses are 1993 IUPAC values (up to four decimal places). Numbers in parentheses are atomic masses or mass numbers of the most stable isotope of an element.

Lanthanides

Actinides

Chapter Summary

LO1　Define geology. Geology, the study of Earth, is divided into two broad areas: Physical geology is the study of Earth materials, as well as the processes that operate within Earth and on its surface; historical geology examines the origin and evolution of Earth, its continents, oceans, atmosphere, and life.

LO2　Understand the impact that the formulation of theories has had on the study of geology. The scientific method is an orderly, logical approach that involves gathering and analyzing facts about a particular phenomenon, formulating hypotheses to explain the phenomenon, testing the hypotheses, and finally, proposing a theory. A theory is a testable explanation for some natural phenomenon that has a large body of supporting evidence.

LO3　Explain how geology relates to the human experience. Geology is part of the human experience. We can find examples of it in art, music, and literature. Geology has also played an important role in history, as empires have risen and fallen—often as a result of the distribution and exploitation of natural resources.

LO4　Explain how geology affects our everyday lives. Geology affects our lives most obviously through natural disasters, but also through connections with economic, social, and political issues. Natural disasters usually cannot be prevented, but we can learn more about how to predict and mitigate their severity. Geology influences political and economic events as nations exchange and compete for Earth's limited resources.

LO5　Describe global geologic and environmental issues facing humankind. Overpopulation, which many scientists would argue is the greatest problem facing the world today, is actually a geologic issue. People will have to be fed, clothed, housed, and generally supported by Earth's resources. Overpopulation in poorer areas can rapidly deplete both nonrenewable and renewable natural resources. Industrialized countries, on the other hand, must be concerned with pollution. A prime example of the interconnection of Earth's subsystems is the relationship between the greenhouse effect and global warming.

LO6　Describe the origin of the universe and solar system, and Earth's place in them. The universe is thought to have originated approximately 14 billion years ago in what is known as the Big Bang. Astronomers have deduced this age by observing that celestial objects are moving away from each other in an ever-expanding universe. Furthermore, the universe has a background radiation of 2.7 K above absolute zero (2.7 K = –270.3°C), which is thought to be the faint afterglow of the Big Bang.

About 4.6 billion years ago, our solar system formed from a rotating cloud of interstellar matter. As this cloud condensed, it eventually collapsed under the influence of gravity and flattened into a counterclockwise-rotating disk. Within this rotating disk, the Sun, planets, and moons formed from the turbulent eddies of nebular gases and solids.

Earth formed from a swirling eddy of nebular material 4.6 billion years ago, accreting as a solid body, and soon thereafter differentiated into a layered planet during a period of internal heating.

Key Terms

asthenosphere The part of the mantle that lies below the lithosphere; it behaves plastically and flows slowly.

Big Bang A model for the evolution of the universe in which a dense, hot state was followed by expansion, cooling, and a less-dense state.

core The interior part of Earth beginning at a depth of 2,900 km that probably consists mostly of iron and nickel.

crust Earth's outermost layer, and the upper part of the lithosphere; consists of continental and oceanic crust.

fossils The remains or traces of once-living organisms.

geologic time scale A chart arranged so that the designation for the earliest part of geologic time appears at the bottom, followed upward by progressively younger time designations.

geology The science concerned with the study of Earth materials (minerals and rocks), surface and internal processes, and Earth history.

hypothesis A provisional explanation for observations that is subject to continual testing. If well-supported by evidence, a hypothesis may be called a theory.

igneous rock Any rock formed by cooling and crystallization of magma or lava or by the consolidation of pyroclastic materials.

Jovian planets Any of the four planets (Jupiter, Saturn, Uranus, and Neptune) that resemble Jupiter. All are large and have low mean densities, indicating that they are composed mostly of lightweight gases, such as hydrogen and helium, and frozen compounds, such as ammonia and methane.

lithosphere Earth's outer, rigid part, consisting of the upper mantle, oceanic crust, and continental crust.

magma Molten rock material generated within Earth.

mantle The thick layer between Earth's crust and core.

metamorphic rock Any rock that has been changed from its original condition by heat, pressure, and the chemical activity of fluids, as in marble and slate.

mineral A naturally occurring, inorganic, crystalline solid that has characteristic physical properties and a narrowly defined chemical composition.

organic evolution The theory holding that all living things are related and that they descended with modification from organisms that lived during the past.

plate An individual segment of the lithosphere that moves over the asthenosphere.

plate tectonic theory The theory holding that large segments of Earth's outer part (lithospheric plates) move relative to one another.

principle of uniformitarianism A principle holding that we can interpret past events by understanding present-day processes, based on the idea that natural processes have always operated in the same way.

rock A solid aggregate of one or more minerals, as in limestone and granite, or a consolidated aggregate of rock fragments, as in conglomerate, or masses of rocklike materials, such as coal and obsidian.

rock cycle A group of processes through which Earth materials may pass as they are transformed from one major rock type to another.

scientific method A logical, orderly approach that involves gathering data, formulating and testing hypotheses, and proposing theories.

sedimentary rock Any rock composed of sediment, such as limestone and sandstone.

solar nebula theory A theory for the evolution of the solar system from a rotating cloud of gas.

system A combination of related parts that interact in an organized fashion; Earth systems include the atmosphere, hydrosphere, biosphere, and solid Earth.

terrestrial planets Any of the four innermost planets (Mercury, Venus, Earth, and Mars). They are all small and have high mean densities, indicating that they are composed of rock and metallic elements.

theory An explanation for some natural phenomenon that has a large body of supporting evidence. To be scientific, a theory must be testable (e.g., plate tectonic theory).

LO7 Explain why Earth is a dynamic and evolving planet. Earth's outermost layer is the crust, which is divided into continental and oceanic portions. The crust and underlying solid upper mantle, which together form the lithosphere, overlie the asthenosphere, a zone that behaves plastically and flows slowly. Underlying the asthenosphere is the solid lower mantle. Earth's core consists of an outer liquid portion and an inner solid portion.

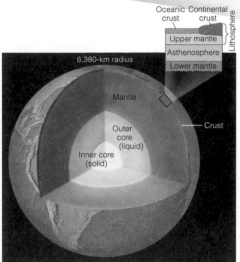

The lithosphere is broken into a series of plates that diverge, converge, and slide sideways past one another. Plate tectonic theory provides a unifying explanation for many geologic features and events.

LO8 Describe the rock cycle. The three major rock groups are igneous, sedimentary, and metamorphic. Igneous rocks result from the crystallization of magma or lava or the consolidation of volcanic ejecta. Sedimentary rocks are typically formed by the consolidation of rock fragments, precipitation of mineral matter from solution, or the compaction of plant or animal remains. Metamorphic rocks result from the alteration of other rocks, usually beneath Earth's surface, by heat, pressure, and chemically active fluids. The rock cycle illustrates the interactions between Earth's internal and surface processes and how the three rock groups are interrelated.

LO9 Define organic evolution and its role in the history of life. The central thesis of the theory of organic evolution is that all living organisms have evolved (descended with modifications) from organisms that existed in the past.

LO10 Describe geologic time and uniformitarianism. Time sets geology apart from the other sciences, except astronomy, and an appreciation of the immensity of geologic time is central to understanding Earth's evolution. The geologic time scale is the calendar geologists use to date past events. The principle of uniformitarianism is basic to the interpretation of Earth history. This principle holds that the laws of nature have been constant through time and that the same processes operating today have operated in the past, although not necessarily at the same rates.

LO11 Explain how the study of geology benefits us. Geology is an integral part of our lives. Our standard of living depends directly on our consumption of natural resources, resources that formed millions and billions of years ago.

Review Questions

1. Explain how the three rock groups are related and how each rock group can be converted into a different rock group.

2. Why is viewing Earth as a system a good way to study it? Are humans a part of the Earth system? If so, what role, if any, do we play in Earth's evolution?

3. Why is plate tectonic theory so important to geology? How does it fit into the systems approach to the study of Earth?

Chapter Summary

LO1 **Review early ideas about continental drift.** The concept of continental movement is not new. Alfred Wegener is generally credited with developing the hypothesis of continental drift. He provided abundant geologic, paleontologic, and paleoclimatic evidence to show that the continents were once united into one supercontinent that he named Pangaea. Unfortunately, he could not explain how the continents moved.

LO2 **Explain the evidence for continental drift.** Evidence used to support the hypothesis of continental drift includes the fit of the shorelines of continents, the appearance of the same rock sequences and mountain ranges of the same age on continents now widely separated, the matching of glacial deposits and paleoclimatic zones, and the similarities of many extinct plant and animal groups whose fossil remains are found today on widely separated continents.

LO3 **Recognize and describe the features of the seafloor.** Many of the topographic features found on the seafloor and along continental margins are manifestations of Earth's internal processes taking place along plate margins such as oceanic ridges, oceanic trenches, active continental margins, and passive continental margins.

LO4 **Describe Earth's magnetic field.** The strength of Earth's magnetic field is not constant, but varies—being strongest at the poles and weakest at the equator. The magnetic poles do not coincide with the geographic (rotational) poles, and there presently is an 11.5° angle difference between them.

LO5 **Explain paleomagnetism and polar wandering.** The hypothesis of continental drift was revised during the 1950s when paleomagnetic studies of rocks indicated the presence of multiple magnetic north poles instead of just one, as there is today. This paradox was resolved by constructing a map on which the continents could be moved into different positions such that the magnetic data would then be consistent with a single magnetic north pole.

LO6 **Explain magnetic reversals and seafloor spreading.** Seafloor spreading was confirmed by the discovery of magnetic anomalies—reversals of magnetic polarity—in the ocean crust that were both parallel to and symmetric around the oceanic ridges.

LO7 **Explain plate tectonic theory as a unifying theory.** Plate tectonic theory became widely accepted by the 1970s because the evidence overwhelmingly supports it and because it provides geologists with a powerful theory for explaining such phenomena as volcanism, earthquake activity, mountain building, global climate change, the distribution of the world's biota, and the distribution of mineral resources.

LO8 **Identify and describe the three types of plate boundaries.** Divergent boundaries are where two plates move away from each other; convergent boundaries are where two plates collide; and transform boundaries are where two plates slide past each other.

Key Terms

abyssal plain Vast flat area on the seafloor adjacent to the continental rise of a passive continental margin.

active continental margin A continental margin with volcanism and seismicity at the leading edge of a continental plate where oceanic lithosphere is subducted.

aseismic ridge A ridge, or broad area, rising above the seafloor that lacks seismic activity.

continental–continental plate boundary A convergent plate boundary along which two continental lithospheric plates collide.

continental drift The theory that the continents were joined into a single landmass that broke apart, with the various fragments (continents) moving with respect to one another.

continental margin The area separating the part of a continent above sea level from the deep seafloor.

continental rise The gently sloping part of the continental margin between the continental slope and the abyssal plain.

continental shelf The very gently sloping part of the continental margin between the shoreline and the continental slope.

continental slope The relatively steeply inclined part of the continental margin between the continental shelf and the continental rise or between the continental shelf and an oceanic trench.

convergent plate boundary The boundary between two plates that move toward each other.

Curie point The temperature at which iron-bearing minerals in cooling magma or lava attain their magnetism.

divergent plate boundary The boundary between two plates that are moving apart.

***Glossopteris* flora** A Late Paleozoic association of plants found only on the Southern Hemisphere continents and India; named for its best-known genus, *Glossopteris*.

guyot A flat-topped seamount of volcanic origin rising more than 1 km above the seafloor.

hot spot A localized zone of melting below the lithosphere that probably overlies a mantle plume; detected by volcanism at the surface.

magnetic anomaly Any deviation, such as a change in average strength, in Earth's magnetic field.

magnetic field The area in which magnetic substances are affected by lines of magnetic force emanating from Earth.

magnetic reversal The phenomenon involving the complete reversal of the north and south magnetic poles.

magnetism A physical phenomenon resulting from moving electricity and the spin of electrons in some solids in which magnetic substances are attracted toward one another.

oceanic–continental plate boundary A convergent plate boundary along which oceanic lithosphere is subducted beneath continental lithosphere.

oceanic–oceanic plate boundary A convergent plate boundary along which two oceanic plates collide and one is subducted beneath the other.

oceanic ridge A mostly submarine mountain system composed of basalt found in all ocean basins.

oceanic trench A long, narrow feature restricted to active continental margins and along which subduction occurs.

paleomagnetism Residual magnetism in rocks, studied to determine the intensity and direction of Earth's past magnetic field.

Pangaea The name Alfred Wegener proposed for a supercontinent consisting of all Earth's landmasses at the end of the Paleozoic Era.

passive continental margin A continental margin within a tectonic plate, as in the East Coast of North America, where little seismic activity and no volcanism occur; characterized by a broad continental shelf and a continental slope and rise.

plate tectonic theory The theory holding that large segments of Earth's outer part (lithospheric plates) move relative to one another.

seafloor spreading The theory that the seafloor moves away from spreading ridges and is eventually consumed at subduction zones.

seamount A submarine volcanic mountain rising at least 1 km above the seafloor.

submarine hydrothermal vent A crack, or fissure, in the seafloor through which superheated water issues.

thermal convection cell A type of circulation of material involving only the asthenosphere or the entire mantle during which hot material rises, moves laterally, cools, and sinks, then is reheated and continues the cycle.

Ancient plate boundaries can be recognized by their associated rock assemblages and geologic structures.

LO9   Describe hot spots and mantle plumes. A mantle plume is a stationary column of magma within the mantle that slowly rises to Earth's surface and forms a volcano (hot spot). If mantle plumes remain stationary within the mantle while the plates move over them, the resulting hot spots leave a trail of extinct and progressively older volcanoes called aseismic ridges that record the movement of the plate.

LO10   Explain plate movement and motion. The average rate of movement and relative motion of the plates can be calculated in several ways. The results of these different methods all agree and indicate that the plates move at different average velocities.

LO11   Explain the driving mechanism of plate tectonics. Geologists think that some type of convective heat system is the major driving force for plate movement, but the extent to which other mechanisms, such as slab-pull and ridge-push, are involved is still unresolved.

© 2013 Cengage Learning

LO12   Recognize the role of plate tectonics in the distribution of natural resources. A close relationship exists between the formation of some mineral deposits, petroleum, and plate boundaries.

LO13   Recognize the role of plate tectonics in the distribution of life. The distribution of plants and animals is not random, but is controlled mostly by climate and geographic barriers, which are controlled to a great extent by the movement of plates.

Review Questions

1. In addition to the volcanic eruptions and earthquakes associated with convergent and divergent plate boundaries, why are these boundaries also associated with the formation and accumulation of various metallic ore deposits?

2. Plate tectonic theory builds on the continental drift hypothesis and the theory of seafloor spreading. As such, it is a unifying theory of geology. Explain why it is a unifying theory.

transform fault A fault along which one type of motion is transformed into another; commonly displaces oceanic ridges; on land, recognized as a strike-slip fault, such as the San Andreas fault.

transform plate boundary A plate boundary along which plates slide past one another and crust is neither produced nor destroyed.

Chapter Summary

LO1 **Define matter.** Matter is composed of chemical elements, each of which consists of atoms. Protons and neutrons are present in an atom's nucleus, and electrons orbit around the nucleus in electron shells.

The number of protons in an atom's nucleus determines its atomic number. The atomic mass number is the number of protons plus neutrons in the nucleus.

Bonding results when atoms join with other atoms; different elements bond to form a compound. With few exceptions, minerals are compounds.

Ionic and covalent bonds are most common in minerals, but metallic and van der Waals bonds are found in some.

^{12}C (Carbon 12) ^{13}C (Carbon 13) ^{14}C (Carbon 14)

© 2013 Cengage Learning

LO2 **Explore the world of minerals.** Minerals are crystalline solids, which means that they possess an ordered internal arrangement of atoms. Mineral composition is indicated by a chemical formula, such as SiO_2 for quartz. Some minerals have a range of compositions because different elements substitute for one another if their atoms are about the same size and have the same electrical charge.

TABLE 3.1
Mineral groups recognized by geologists

MINERAL GROUP	NEGATIVELY CHARGED ION OR RADICAL	EXAMPLES	COMPOSITION
Carbonate	$(CO_3)^{-2}$	Calcite	$CaCO_3$
		Dolomite	$CaMg(CO_3)^2$
Halide	Cl^{-1}, F^{-1}	Halite	$NaCl$
		Fluorite	CaF_2
Hydroxide	$(OH)^{-1}$	Brucite	$Mg(OH)_2$
Native element	—	Gold	Au
		Silver	Ag^*
		Diamond	C
Phosphate	$(PO_4)^{-3}$	Apatite	$Ca_5(PO_4)_3(F,Cl)$
Oxide	O^{-2}	Hematite	Fe_2O_3
		Magnetite	Fe_3O_4
Silicate	$(SiO_4)^{-4}$	Quartz	SiO_2
		Potassium feldspar	$KAlSi_3O_8$
		Olivine	$(Mg,Fe)_2SiO_4$
Sulfate	$(SO_4)^{-2}$	Anhydrite	$CaSO_4$
		Gypsum	$CaSO_4 \cdot 2H_2O$
Sulfide	S^{-2}	Galena	PbS
		Pyrite	FeS_2
		Argentite	Ag_2S^*

© 2013 Cengage Learning

Note that silver is found as both a native element and a sulfide mineral.

Key Terms

atom The smallest unit of matter that retains the characteristics of an element.

atomic mass number The number of protons plus neutrons in the nucleus of an atom.

atomic number The number of protons in the nucleus of an atom.

bonding The process whereby atoms join to other atoms.

carbonate mineral A mineral with the carbonate radical $(CO_3)^{-2}$, as in calcite $(CaCO_3)$ and dolomite $[CaMg(CO_3)_2]$.

cleavage Breakage along internal planes of weakness in mineral crystals.

compound Any substance resulting from the bonding of two or more different elements (e.g., water, H_2O, and quartz, SiO_2).

covalent bond A chemical bond formed by the sharing of electrons between atoms.

crystal A naturally occurring solid of an element or compound with a specific internal structure that is manifested externally by planar faces, sharp corners, and straight edges.

crystalline solid A solid in which the constituent atoms are arranged in a regular, three-dimensional framework.

density The mass of an object per unit volume; usually expressed in grams per cubic centimeter (g/cm^3).

electron A negatively charged particle of very little mass that encircles the nucleus of an atom.

electron shell Electrons orbit an atom's nucleus at specific distances in electron shells.

element A substance composed of atoms that all have the same properties; atoms of one element can change to atoms of another element by radioactive decay, but otherwise they cannot be changed by ordinary chemical means.

ferromagnesian silicate Any silicate mineral that contains iron, magnesium, or both.

hardness A term used to express the resistance of a mineral to abrasion.

ion An electrically charged atom produced by adding or removing electrons from the outermost electron shell.

ionic bond A chemical bond resulting from the attraction between positively and negatively charged ions.

luster The appearance of a mineral in reflected light. Luster is metallic or nonmetallic, although the latter has several subcategories.

mineral A naturally occurring, inorganic, crystalline solid that has characteristic physical properties and a narrowly defined chemical composition.

neutron An electrically neutral particle found in the nucleus of an atom.

nonferromagnesian silicate A silicate mineral that has no iron or magnesium.

nucleus The central part of an atom, consisting of protons and neutrons.

proton A positively charged particle found in the nucleus of an atom.

reserve The part of the resource base that can be extracted economically.

resource A concentration of naturally occurring solid, liquid, or gaseous material in, or on, Earth's crust in such form and amount that economic extraction of a commodity from the concentration is currently or potentially feasible.

rock A solid aggregate of one or more minerals, as in limestone and granite, or a consolidated aggregate of rock fragments, as in conglomerate, or masses of rocklike materials, such as coal and obsidian.

rock-forming mineral Any mineral common in rocks that is important in their identification and classification.

silica A compound of silicon and oxygen.

silica tetrahedron The basic building block of all silicate minerals; consists of one silicon atom and four oxygen atoms.

silicate A mineral that contains silica, such as quartz (SiO_2).

specific gravity The ratio of a substance's weight, especially of a mineral, to an equal volume of water at 4°C.

LO3   Identify mineral groups recognized by geologists. More than 3,500 minerals are known, and most of them are silicates. The two types of silicates are ferromagnesian and nonferromagnesian. In addition to silicates, geologists recognize carbonates, native elements, hydroxides, oxides, phosphates, halides, sulfates, and sulfides.

LO4   Identify physical properties of minerals. Structure and composition control the physical properties of minerals, such as luster, crystal form, hardness, color, cleavage, fracture, and specific gravity.

LO5   Recognize rock-forming minerals. A few minerals, designated rock-forming minerals, are common enough in rocks to be essential to their identification and classification. Most rock-forming minerals are silicates, but some carbonates are also common.

LO6   Explain how minerals form. Several processes account for the origin of minerals, including cooling magma, weathering, evaporation of seawater, metamorphism, and organisms using dissolved substances in seawater to build their shells.

LO7   Recognize natural resources and reserves. Many resources are concentrations of minerals or rocks of economic importance. They are further characterized as metallic resources, nonmetallic resources, and energy resources. Reserves are that part of the resource base that can be extracted profitably. Distinguishing a resource from a reserve depends on market price, labor costs, geographic location, and developments in science and technology. The United States must import many resources to maintain its industrial capacity. Canada is more self-reliant, but it too must import some commodities.

Review Questions

1. What is a silica tetrahedron and how may they bond to form silicate minerals?

2. How do rock-forming minerals differ from any other mineral? Give examples from three categories of rock-forming minerals.

3. Give three examples of mineral cleavage and explain how it occurs. Why do some minerals lack cleavage?

4. How would the color and density of a rock made up mostly of ferromagnesian silicates differ from one composed primarily of nonferromagnesian silicates?

Chapter Summary

LO1 Describe the properties and behavior of magma and lava. Magma is the term for molten rock material below Earth's surface, whereas the same material at the surface is called lava. Silica content distinguishes between ultramafic (< 45 percent silica), mafic (45 to 52 percent silica), intermediate (53 to 65 percent silica), and felsic (> 65 percent silica) magmas. Magma and lava viscosity depend mostly on temperature and composition. The more silica, the greater the viscosity.

LO2 Explain how magma originates and changes. According to Bowen's reaction series, cooling mafic magma yields a sequence of minerals, each of which is stable within specific temperature ranges. Only ferromagnesian silicates are found in the discontinuous branch of Bowen's reaction series. The continuous branch of the reaction series yields only plagioclase feldspars that become increasingly enriched with sodium as cooling occurs.

A chemical change in magma may take place as early ferromagnesian silicates form and because of their density, settle in the magma. Compositional changes also take place in magma when it assimilates country rock or one magma mixes with another.

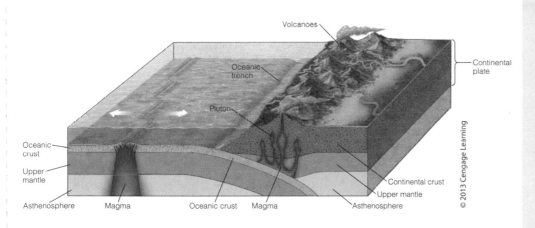

© 2013 Cengage Learning

LO3 Identify and classify igneous rocks by their characteristics. Geologists recognize two broad categories of igneous rocks; volcanic or extrusive, and plutonic or intrusive. Rapid cooling accounts for the aphanitic textures of volcanic rocks, whereas comparatively slow cooling yields the phaneritic textures of plutonic rocks. Igneous rocks with markedly different-sized minerals are porphyritic. Minerals crystallize from magma and lava when small crystal nuclei form and grow.

Texture and composition are the criteria used to classify igneous rocks, although a few are defined only by texture. Igneous rock composition is determined mostly by the composition of the parent magma, but magma composition can change so that the same magma may yield more than one type of igneous rock. Crystallization from water-rich magma results in very large minerals in rocks known as pegmatite. Most pegmatite has an overall composition similar to granite.

Key Terms

aphanitic texture A texture in igneous rocks in which individual mineral grains are too small to be seen without magnification; results from rapid cooling of magma and generally indicates an extrusive origin.

assimilation A process whereby magma changes composition as it reacts with country rock.

batholith An irregularly shaped, discordant pluton with at least 100 km² of surface area.

Bowen's reaction series A series of minerals that form in a specific sequence in cooling magma or lava; originally proposed to explain the origin of intermediate and felsic magma from mafic magma.

concordant pluton Intrusive igneous body whose boundaries parallel the layering in the country rock.

country rock Any preexisting rock that has been intruded by a pluton or altered by metamorphism.

crystal settling The physical separation and concentration of minerals in the lower part of a magma chamber or pluton by crystallization and gravitational settling.

dike A tabular, or sheetlike, discordant pluton.

discordant pluton Pluton with boundaries that cut across the layering in the country rock.

felsic magma Magma with more than 65% silica and considerable sodium, potassium, and aluminum, but little calcium, iron, and magnesium.

hot spot Localized zone of melting below the lithosphere that probably overlies a mantle plume.

igneous rock Any rock formed by cooling and crystallization of magma or lava or the consolidation of pyroclastic materials.

intermediate magma Magma with a silica content between 53% to 65% and an overall composition intermediate between mafic and felsic magma.

laccolith A concordant pluton with a mushroom-like geometry.

lava Magma that reaches Earth's surface.

lava flow A stream of magma flowing over Earth's surface.

mafic magma Magma with between 45% to 52% silica and proportionately more calcium, iron, and magnesium than intermediate and felsic magma.

magma Molten rock material generated within Earth.

magma chamber A reservoir of magma within Earth's upper mantle or lower crust.

magma mixing The process whereby magmas of different composition mix together to yield a modified version of the parent magmas.

mantle plume A cylindrical mass of magma rising from the mantle toward the surface; recognized at the surface by a hot spot, an area such as the Hawaiian Islands where volcanism takes place.

phaneritic texture Igneous rock texture in which minerals are easily visible without magnification.

pluton An intrusive igneous body that forms when magma cools and crystallizes within the crust, such as a batholith or sill.

plutonic (intrusive igneous) rock Igneous rock that formed from magma intruded into or formed in place within the crust.

porphyritic texture An igneous texture with minerals of markedly different sizes.

pyroclastic (fragmental) texture A fragmental texture characteristic of igneous rocks composed of pyroclastic materials.

pyroclastic materials Fragmental substances, such as ash, that are explosively ejected from a volcano.

sill A tabular, or sheetlike, concordant pluton.

stock An irregularly shaped, discordant pluton with a surface area smaller than 100 km².

stoping A process in which rising magma detaches and engulfs pieces of the country rock.

ultramafic magma Magma with a silica content of less than 45%.

vesicle A small hole or cavity formed by gas trapped in cooling lava.

viscosity A fluid's resistance to flow.

volcanic (extrusive igneous) rock An igneous rock formed when magma is extruded onto Earth's surface where it cools and crystallizes, or when pyroclastic materials become consolidated.

volcanic neck An erosional remnant of the material that solidified in a volcanic pipe.

volcanic pipe The conduit connecting the crater of a volcano with an underlying magma chamber.

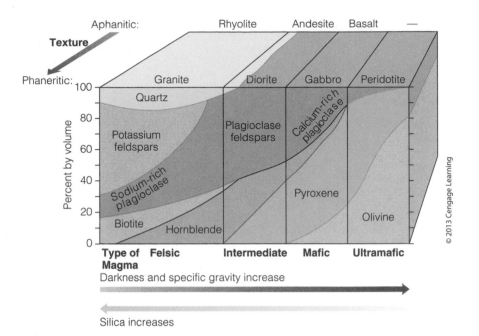

LO4   Recognize intrusive igneous bodies, or plutons. Intrusive igneous bodies known as plutons vary in their geometry and in their relationship to country rock. Some are concordant, whereas others are discordant. The largest plutons, known as batholiths, consist of multiple intrusions of magma during long periods of time. Most plutons, including batholiths, form at or near divergent or convergent plate boundaries.

LO5   Explain how batholiths intrude into Earth's crust. Geologists now generally agree that batholiths were emplaced by forceful injection as magma moved upward. It appears that magma deforms and shoulders aside the country rock, and as it rises farther, some of the country rock fills the space beneath the magma. At shallower depths, where the crust is more rigid, batholiths may move upward by the process called stoping.

Review Questions

1. What factors control the viscosity of magma and lava?

2. What are the similarities and differences among dikes, sills, and laccoliths?

3. Explain what a porphyritic texture is and how one might develop.

4. Use Bowen's reaction series to explain the sequence in which minerals crystallize to form mafic magma.

Chapter Summary

LO1   Understand volcanism and volcanoes. Volcanism encompasses those processes by which magma rises to the surface as lava flows and pyroclastic materials and gases are released into the atmosphere. Gases make up only a few percent by weight of magma. Most is water vapor, but sulfur-gases may have far-reaching climatic effects.

Aa lava flows have surfaces of jagged, angular blocks, whereas the surfaces of pahoehoe flows are smoothly wrinkled. Several other features of lava flows are spatter cones, pressure ridges, lava tubes, and columnar joints. Lava erupted under water typically forms bulbous masses known as pillow lava.

LO2   Identify the types of volcanoes. Volcanoes are found in various shapes and sizes, but all form where lava and pyroclastic materials erupt from a vent. The summits of volcanoes have either a crater or a much larger caldera. Most calderas form following voluminous eruptions, and the volcano peak collapses into a partially drained magma chamber.

Shield volcanoes have low, rounded profiles and are composed mostly of mafic flows that cool and form basalt. Small, steep-sided cinder cones form around a vent where pyroclastic materials erupt and accumulate. Composite volcanoes are made of lava flows and pyroclastic materials of intermediate composition and volcanic mudflows.

Viscous, bulbous masses of lava, generally of felsic composition, form lava domes which are dangerous because they erupt explosively.

LO3   Identify other volcanic landforms. Fluid mafic lava from fissure eruptions spreads over large areas to form a basalt plateau. Pyroclastic sheet deposits result from huge eruptions of ash and other pyroclastic materials, particularly when calderas form.

LO4   Identify the distribution of volcanoes. Approximately 80 percent of all volcanic eruptions take place in the circum-Pacific belt and the Mediterranean belt, mostly at convergent plate boundaries. The majority of other eruptions occur along mid-oceanic ridges or their extensions onto land. The circum-Pacific belt includes North American volcanoes in the Pacific Northwest, as well as those in Alaska. Active volcanoes in North America include some in the Cascade Range in the Pacific Northwest and those in Alaska. Notable volcanoes include Mount Spurr and Novarupta in Alaska, and California's Lassen Peak, the world's largest lava dome. Mount St. Helens in Washington erupted in 1980 and from 2004 to 2005.

LO5   Understand the relationship between plate tectonics, volcanoes, and plutons. Most volcanism and emplacement of plutons takes place at or near divergent and convergent plate boundaries. Excellent examples of divergent plate boundary volcanism are found along the Mid-Atlantic Ridge, as well as the East Pacific Rise and the Indian Ridge and along the East African Rift. Examples of volcanism at convergent plate boundaries include Mount Pinatubo and the Mayon volcano in the Phillippines, and Mount St. Helens in Washington state. Mauna Loa and Kilauea on the island of Hawaii are located far from any divergent or convergent plate boundary.

Key Terms

aa Lava flow with a surface of rough, angular blocks and fragments.

basalt plateau A plateau built up by horizontal or nearly horizontal overlapping lava flows that erupted from fissures.

caldera A large, steep-sided, oval to circular depression, usually formed when a volcano's summit collapses into a partially drained underlying magma chamber.

Cascade Range A volcanic mountain range stretching from Lassen Peak in northern California north through Oregon and Washington to Meager Mountain in British Columbia, Canada.

cinder cone A small, steep-sided volcano made up of pyroclastic materials resembling cinders that accumulate around a vent.

circum-Pacific belt A zone of seismic and volcanic activity and mountain building that nearly encircles the Pacific Ocean basin.

columnar jointing The phenomenon of forming columns bounded by fractures in some igneous rocks as they cooled and contracted.

composite volcano (stratovolcano) A volcano composed of lava flows and pyroclastic layers, typically of intermediate composition, and mudflows.

crater An oval to circular depression at the summit of a volcano resulting from the eruption of lava, pyroclastic materials, and gases.

fissure eruption A volcanic eruption in which lava or pyroclastic materials issue from a long, narrow fissure (crack) or group of fissures.

lahar A mudflow composed of pyroclastic materials such as ash.

lava dome A bulbous, steep-sided mountain formed by viscous magma moving upward through a volcanic conduit.

lava tube A tunnel beneath the solidified surface of a lava flow through which lava moves; also, the hollow space left when the lava within a tube drains away.

Mediterranean belt A zone of seismic and volcanic activity extending through the Mediterranean region of southern Europe and eastward to Indonesia.

nuée ardente A fast-moving, dense cloud of hot pyroclastic materials and gases ejected from a volcano.

pahoehoe A type of lava flow with a smooth, ropy surface.

pillow lava Bulbous masses of basalt, resembling pillows, formed when lava is rapidly chilled under water.

pyroclastic sheet deposit A vast, sheetlike deposit of felsic pyroclastic materials erupted from fissures.

shield volcano A dome-shaped volcano with a low, rounded profile built up mostly by overlapping basalt lava flows.

volcanic ash Pyroclastic materials that measure less than 2 mm.

volcanic explosivity index (VEI) A semi-quantitative scale for determining the size of a volcanic eruption based on evaluation of criteria, such as volume of material explosively erupted and height of eruption cloud.

volcanic tremor Ground motion lasting from minutes to hours, resulting from magma moving beneath the surface, as opposed to the sudden jolts produced by most earthquakes.

volcanism The process whereby magma and its associated gases rise through the crust and are extruded onto the surface or into the atmosphere.

volcano A hill or mountain formed around a vent as a result of the eruption of lava and pyroclastic materials.

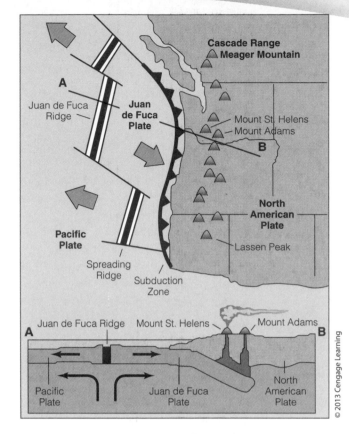

© 2013 Cengage Learning

LO6   Understand volcanic hazards, volcano monitoring, and forecasting eruptions. Geologists have devised a volcanic explosivity index (VEI) to give semiquantitative measure of the size of an eruption. Volume of material erupted and the height of the eruption plume are criteria used to determine the VEI; fatalities and property damage are not considered.

To effectively monitor volcanoes, geologists evaluate several physical and chemical aspects of volcanic regions. Of particular importance in monitoring volcanoes and forecasting eruptions is detecting volcanic tremor and determining the eruptive history of a volcano.

Review Questions

1. Why are eruptions of mafic magma rather quiet, whereas eruptions of felsic magma are commonly explosive?

2. Why do shield volcanoes have such gentle slopes, whereas cinder cones have very steep slopes?

3. What does volcanic tremor indicate, and how does it differ from the shaking caused by most earthquakes?

4. Suppose that you find rocks on land that consist of layers of pillow lava overlain by deep-sea

sedimentary rocks. Where and how did the pillow lava form, and what type of rock would you expect to find beneath the pillow lava?

5. Explain what types of volcanoes form at divergent and convergent plate boundaries. What accounts for the difference(s)?

6. What are lava domes and why are they so dangerous?

Chapter Summary

LO1   Explain how Earth materials are altered. Mechanical and chemical weathering disintegrate and decompose parent material so that it is more nearly in equilibrium with new physical and chemical conditions. The products of weathering include solid particles and substances in solution.

Mechanical weathering includes such processes as frost action, pressure release, salt crystal growth, thermal expansion and contraction, and the activities of organisms. Particles liberated by mechanical weathering retain the chemical composition of the parent material. The chemical weathering processes of solution, oxidation, and hydrolysis bring about chemical changes of the parent material. Clay minerals and substances in solution form during chemical weathering. Mechanical weathering aids chemical weathering by breaking parent material into smaller pieces, thereby exposing more surface area.

LO2   Explain how soil forms and deteriorates. Mechanical and chemical weathering produce regolith, some of which is soil if it consists of solids, air, water, and humus and supports plant growth. Soils are characterized by horizons that are designated, in descending order, as O, A, E, B, and C. Soil horizons differ from one another in texture, structure, composition, and color.

Soils are residual or transported depending on whether they formed in place or their constituents were transported from the weathering site. Laterite is a soil that results from intense chemical weathering in the tropics. Soil erosion, caused mostly by sheet and rill erosion, is a problem in some areas. Human practices such as construction, agriculture, and deforestation can accelerate losses of soil to erosion.

LO3   Know how weathering and resources are related. Intense chemical weathering is responsible for the origin of residual concentrations, many of which contain valuable minerals such as iron, lead, copper, and clay.

LO4   Identify sediment and sedimentary rocks. Sedimentary particles are designated in order of decreasing size as gravel, sand, silt, and clay. Sedimentary particles are rounded and sorted during transport, although the degree of rounding and sorting depends on particle size, transport distance, and depositional process.

Any area in which sediment is deposited is a depositional environment. Major depositional settings are continental, transitional, and marine, each of which includes several specific depositional environments.

Lithification involves compaction and cementation, which convert sediment into sedimentary rock. Silica

Labels: Glacial environment, Stream environment, Lake, Beach, Delta, Sand dunes, Tidal flat, Alluvial fan, Lagoon, Continental shelf, Submarine fan, Barrier island, Deep marine environment

© 2013 Cengage Learning

Key Terms

bed An individual layer of rock, especially sediment or sedimentary rock.

biochemical sedimentary rock Any sedimentary rock produced by the chemical activities of organisms.

carbonate rock Any rock, such as limestone and dolostone, made up mostly of carbonate minerals.

cementation The process whereby minerals crystallize in the pore spaces of sediment and bind the loose particles together.

chemical sedimentary rock Sedimentary rock made up of minerals that were dissolved during chemical weathering and later precipitated from seawater, more rarely lake water, or extracted from solution by organisms.

chemical weathering The decomposition of rocks by chemical alteration of parent material.

compaction Reduction in the volume of a sedimentary deposit that results from its own weight and the weight of any additional sediment deposited on top of it.

cross-bedding A type of bedding in which layers are deposited at an angle to the surface on which they accumulate, as in sand dunes.

depositional environment Any site such as a floodplain or beach where physical, biologic, and chemical processes yield a distinctive kind of sedimentary deposit.

detrital sedimentary rock Sedimentary rock made up of the solid particles (detritus) of preexisting rocks.

differential weathering Weathering that occurs at different rates on rocks, thereby yielding an uneven surface.

erosion The removal of weathered materials from their source area by running water, wind, glaciers, and waves.

evaporite Any sedimentary rock, such as rock salt, formed by inorganic chemical precipitation of minerals from evaporating water.

exfoliation dome A large, rounded dome of rock resulting when concentric layers of rock are stripped from the surface of a rock mass.

fossil The remains or traces of once-living organisms.

frost action The disaggregation of rocks by repeated freezing and thawing of water in cracks and crevasses.

graded bedding Sedimentary layer in which a single bed shows a decrease in grain size from bottom to top.

hydrolysis The chemical reaction between hydrogen (H^+) ions and hydroxyl (OH^-) ions of water and a mineral's ions.

laterite A red soil, rich in iron or aluminum, or both, resulting from intense chemical weathering in the tropics.

lithification The process of converting sediment into sedimentary rock by compaction and cementation.

marine regression The withdrawal of the sea from a continent or coastal area, resulting in the emergence of the land as sea level falls or the land rises with respect to sea level.

marine transgression The invasion of a coastal area or a continent by the sea, resulting from a rise in sea level or subsidence of the land.

mechanical weathering Disaggregation of rocks by physical processes that yields smaller pieces that retain the composition of the parent material.

mud crack A crack in clay-rich sediment that forms in response to drying and shrinkage.

oxidation The reaction of oxygen with other atoms to form oxides or, if water is present, hydroxides.

parent material The material that is chemically and mechanically weathered to yield sediment and soil.

pressure release A mechanical weathering process in which rocks that formed under pressure expand on being exposed at the surface.

regolith The layer of unconsolidated rock and mineral fragments and soil that covers most of the land surface.

ripple mark Wavelike (undulating) structure produced in granular sediment, especially sand, by unidirectional wind and water currents or by oscillating wave currents.

salt crystal growth A mechanical weathering process in which salt crystals growing in cracks and pores disaggregate rocks.

sediment Loose aggregate of solids derived by weathering from preexisting rocks, or solids precipitated from solution by inorganic chemical processes or extracted from solution by organisms.

sedimentary facies Any aspect of a sedimentary rock unit that makes it recognizably different from adjacent sedimentary rocks of the same or approximately the same age.

sedimentary rock Any rock composed of sediment, such as limestone and sandstone.

and calcium carbonate are the most common chemical cements, but iron oxide and iron hydroxide cements are important in some rocks.

LO5 Explain how sedimentary rocks are classified. Detrital sedimentary rocks consist of solid particles derived from preexisting rocks. Chemical sedimentary rocks are derived from substances in solution by inorganic chemical processes or the biochemical activities of organisms. Geologists also recognize a subcategory called biochemical sedimentary rocks.

LO6 Understand sedimentary facies. Sedimentary facies are bodies of sediment or sedimentary rock that are recognizably different from adjacent sediments or rocks. Some sedimentary facies are geographically widespread because they were deposited during marine transgressions or marine regressions.

LO7 Read the story preserved in sedimentary rocks. Sedimentary structures such as bedding, cross-bedding, and ripple marks commonly form in sediments when, or shortly after, they are deposited.

Geologists determine the depositional environments of ancient sedimentary rocks by studying sedimentary textures and structures, examining fossils, and making comparisons with present-day depositional processes.

LO8 Recognize important resources in sedimentary rocks. Many sediments and sedimentary rocks, including sand, gravel, evaporites, coal, and banded iron formations, are important resources. Most oil and natural gas are found in sedimentary rocks.

Review Questions

1. How do climate, particle size, and parent material control the rate of chemical weathering?

2. How do the following pairs of detrital sedimentary rocks compare? Conglomerate and sedimentary breccia; quartz sandstone and arkrose.

3. What are the similarities between the mechanical weathering processes frost wedging and salt crystal growth?

sedimentary structure Any feature in sedimentary rock that formed at or shortly after the time of deposition, such as cross-bedding, animal burrows, and mud cracks.

soil Weathered materials containing water, air, and humus that can support vegetation.

soil degradation Any process leading to a loss of soil productivity; may involve erosion, chemical pollution, or compaction.

soil horizon A distinct soil layer that differs from other soil layers in texture, structure, composition, and color.

solution A reaction in which the ions of a substance become dissociated in a liquid and the solid substance dissolves.

spheroidal weathering A type of chemical weathering in which corners and sharp edges of rocks weather more rapidly than flat surfaces, thus yielding spherical shapes.

strata (singular, **stratum**) Refers to layering in sedimentary rocks.

talus Accumulation of coarse, angular rock fragments at the base of a slope.

thermal expansion and contraction A type of mechanical weathering in which the volume of rocks changes in response to heating and cooling.

weathering The physical breakdown and chemical alteration of rocks and minerals at or near Earth's surface.

Chapter Summary

LO1   Identify the agents of metamorphism. Metamorphic rocks result from the transformation of other rocks, usually beneath Earth's surface, as a consequence of one or a combination of three agents: heat, pressure, and fluid activity. Heat for metamorphism comes from magma, lava flows, or deep burial. Pressure is either lithostatic (uniformly applied stress) or differential (stress unequally applied from different directions). Fluids trapped in sedimentary rocks or emanating from intruding magmas can enhance chemical changes and the formation of new minerals.

LO2   Identify the three types of metamorphism. The three major types of metamorphism are contact, dynamic, and regional. Contact metamorphism takes place when magma or lava alters the surrounding country rock. Dynamic metamorphism is associated with fault zones where rocks are subjected to high differential pressure. Regional metamorphism occurs over a large area and is usually caused by tremendous temperatures, pressures, and deformation within the deeper portions of the crust.

Metamorphic grade generally characterizes the degree to which a rock has undergone metamorphic change. Index minerals—minerals that form only within specific temperature and pressure ranges—allow geologists to recognize low-, intermediate-, and high-grade metamorphism.

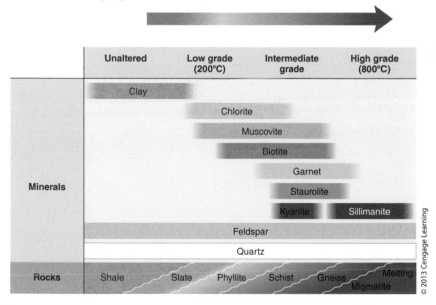

LO3   Explain how metamorphic rocks are classified. Metamorphic rocks are primarily classified according to their texture. In a foliated texture, platy and elongate minerals have a preferred orientation. A nonfoliated texture does not exhibit any discernable preferred orientation of the mineral grains. Foliated metamorphic rocks are arranged in order of increasing grain size, perfection of their foliation, or both. Slate is fine grained, followed by (in increasingly larger grain size) phyllite and schist; gneiss displays segregated bands of minerals. Amphibolite is another fairly common foliated metamorphic rock. Migmatites have both igneous and high-grade metamorphic characteristics. Marble, quartzite, greenstone, hornfels, and anthracite are common nonfoliated metamorphic rocks.

Key Terms

aureole A zone surrounding a pluton in which contact metamorphism took place.

contact (thermal) metamorphism Metamorphism of country rock adjacent to a pluton.

differential pressure Pressure that is not applied equally to all sides of a rock body.

dynamic metamorphism Metamorphism in fault zones where rocks are subjected to high differential pressure.

fluid activity An agent of metamorphism in which water and carbon dioxide promote metamorphism by increasing the rate of chemical reactions.

foliated texture A texture in metamorphic rocks in which platy and elongate minerals are aligned in a parallel fashion.

heat An agent of metamorphism.

index mineral A mineral that forms within specific temperature and pressure ranges during metamorphism.

lithostatic pressure Pressure exerted on rocks by the weight of overlying rocks.

metamorphic facies A group of metamorphic rocks characterized by particular minerals that formed under the same broad temperature and pressure conditions.

metamorphic rock Any rock that has been changed from its original condition by heat, pressure, and the chemical activity of fluids, as in marble and slate.

metamorphic zone The region between lines of equal metamorphic intensity known as isograds.

metamorphism The phenomenon of rocks changing as a result of heat, pressure, and fluids, so that they are in equilibrium with a new set of environmental conditions.

nonfoliated texture A metamorphic texture in which there is no discernable preferred orientation of minerals.

regional metamorphism Metamorphism that occurs over a large area, resulting from high temperatures, tremendous pressure, and the chemical activity of fluids within the crust.

Review Questions

1. Where does contact metamorphism occur, and what types of changes does it produce?

2. Why is metamorphism more widespread along convergent plate boundaries than along any other type of plate boundary?

3. How do metamorphic rocks record the influence of differential pressure in their structures and mineral textures?

4. What specific features about foliated metamorphic rocks would make them unsuitable as foundations for dams? Are there any metamorphic rocks that would make good foundations? Explain your answer.

5. How can aureoles be used to determine the effects of metamorphism?

6. Describe the two types of metamorphic texture, and discuss how they are produced.

7. Do you think metamorphic rocks exist on other planets in our solar system? Why?

8. Why is it important that the average citizen know something about metamorphism, metamorphic rocks, and how they form?

9. Discuss the role each of the three major agents of metamorphism plays in transforming any rock into a metamorphic rock.

10. If plate tectonic movement did not exist, could there be metamorphism?

LO4   Recognize the difference between metamorphic zones and facies.
Metamorphic zones are based on index minerals and are areas of rock that all have similar grades of metamorphism; that is, they have all experienced the same intensity of metamorphism. A metamorphic facies is a group of metamorphic rocks whose minerals all formed under a particular range of temperatures and pressures. Each facies is named after its most characteristic rock or mineral.

LO5   Understand how plate tectonics affects metamorphism.
Metamorphism occurs along all three types of plate boundaries, but is most common at convergent plate margins. Metamorphic rocks formed near Earth's surface along an oceanic–continental convergent plate boundary result from low-temperature, high-pressure conditions. As a subducted oceanic plate descends, it is subjected to increasingly higher temperatures and pressures that result in higher-grade metamorphism.

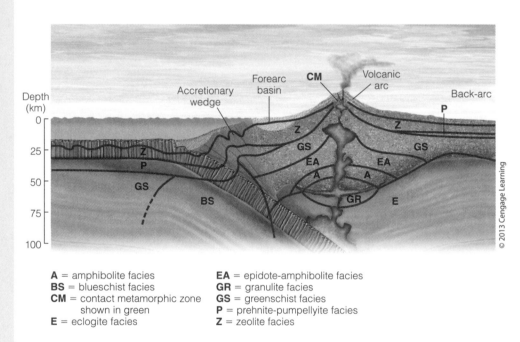

© 2013 Cengage Learning

A = amphibolite facies
BS = blueschist facies
CM = contact metamorphic zone shown in green
E = eclogite facies

EA = epidote-amphibolite facies
GR = granulite facies
GS = greenschist facies
P = prehnite-pumpellyite facies
Z = zeolite facies

LO6   Understand the relationship between metamorphism and natural resources. Many metamorphic rocks and minerals, such as marble, slate, graphite, and talc, are valuable natural resources. In addition, many ore deposits are the result of metamorphism and include copper, tin, tungsten, lead, iron, and zinc.

Chapter Summary

LO1   Explain the elastic rebound theory. Earthquakes are vibrations caused by the sudden release of energy, usually along a fault. The elastic rebound theory is an explanation for how energy is released during earthquakes. As rocks on opposite sides of a fault are subjected to force, they accumulate energy and slowly deform until their internal strength is exceeded. At that time, a sudden movement occurs along the fault, releasing the accumulated energy, and the rocks snap back to their original undeformed shape.

Fault — Fence
Original position
Deformation
Rupture and release of energy
Rocks rebound to original undeformed shape
© 2013 Cengage Learning

LO2   Describe seismology. Seismology is the study of earthquakes. Earthquakes are recorded on seismographs, and the record of an earthquake is a seismogram. An earthquake's focus is the location where rupture within Earth's lithosphere occurs and energy is released. The epicenter is the point on Earth's surface directly above the focus.

LO3   Identify where and how often earthquakes occur. Approximately 80 percent of all earthquakes occur in the circum-Pacific belt, 15 percent within the Mediterranean–Asiatic belt, and the remaining 5 percent mostly in the interior of the plates and along oceanic spreading ridges.

LO4   Identify the different types of seismic waves. The two types of body waves are P-waves (primary waves) and S-waves (secondary waves). P-waves are the fastest seismic waves and travel through all materials, whereas S-waves are somewhat slower and can travel only through solids. P-waves are compressional (expanding and compressing the material through which they travel), whereas S-waves are shear (moving material perpendicular to the direction of travel). Rayleigh (R-waves) and Love waves (L-waves) move along or just below Earth's surface.

LO5   Discuss how earthquakes are located. An earthquake's epicenter is determined using a time-distance graph of the P- and S-waves to calculate how far away a seismic station is from an earthquake. The greater the difference in arrival times between the two waves, the farther away the seismic station is from the earthquake. Three seismographs are needed to locate the epicenter.

LO6   Explain how the strength of an earthquake is measured. Intensity is a subjective or qualitative measure of the kind of damage done by an earthquake. It is expressed in values from I to XII in the Modified Mercalli Intensity Scale. The Richter Magnitude Scale measures an earthquake's magnitude, which is the total amount of energy released by an earthquake at its source. It is an open-ended scale with values beginning at 1. Each increase in magnitude number represents about a 30-fold increase in energy released. The seismic-moment magnitude scale more accurately measures the total energy released by very large earthquakes.

Key Terms

discontinuity A boundary across which seismic wave velocity or direction of travel changes abruptly, such as the mantle–core boundary.

earthquake Vibrations caused by the sudden release of energy, usually as a result of displacement of rocks along faults.

elastic rebound theory An explanation for the sudden release of energy that causes earthquakes when deformed rocks fracture and rebound to their original undeformed condition.

epicenter The point on Earth's surface directly above an earthquake's focus.

focus The site within Earth where an earthquake originates and energy is released.

geothermal gradient Earth's temperature increase with depth; it averages 25°C/km near the surface but varies from area to area.

intensity The subjective measure of the kind of damage done by an earthquake, as well as people's reaction to it.

Love wave (L-wave) A surface wave in which the individual particles of material move only back and forth in a horizontal plane perpendicular to the direction of wave travel.

magnitude The total amount of energy released by an earthquake at its source.

Modified Mercalli Intensity Scale A scale with values from I to XII used to characterize earthquakes based on damage.

Mohorovičić discontinuity (Moho) The boundary between Earth's crust and mantle.

P-wave A compressional, or push-pull, wave; the fastest seismic wave and one that can travel through solids, liquids, and gases; also called a primary wave.

P-wave shadow zone An area between 103 and 143 degrees from an earthquake focus where little P-wave energy is recorded by seismographs.

Rayleigh wave (R-wave) A surface wave in which individual particles of material move in an elliptical path within a vertical plane oriented in the direction of wave movement.

Richter Magnitude Scale An open-ended scale that measures the amount of energy released during an earthquake.

seismógraph An instrument that detects, records, and measures the various waves produced by earthquakes.

seismology The study of earthquakes.

S-wave A shear wave that moves material perpendicular to the direction of travel, thereby producing shear stresses in the material it moves through; also known as a secondary wave; S-waves travel only through solids.

S-wave shadow zone Those areas more than 103 degrees from an earthquake focus where no S-waves are recorded.

tsunami A large sea wave that is usually produced by an earthquake, but can also result from submarine landslides and volcanic eruptions.

National Geophysical Data Center

Review Questions

1. How does the elastic rebound theory account for the energy released during an earthquake?

2. What is the difference between an earthquake's focus and its epicenter? Why is the earthquake's epicenter the location that is usually reported in the news?

3. If Earth were completely solid and homogenous throughout, how would P- and S-waves behave as they travel through the planet? How do they actually behave?

LO7   Describe the destructive effects of earthquakes. The destructive effects of earthquakes include ground shaking, fire, tsunami, landslides, and disruption of vital services. These can be factors in the number of deaths and injuries as well as property damage resulting from an earthquake.

LO8   Discuss earthquake prediction methods. Earthquake precursors are changes preceding an earthquake and include seismic gaps, changes in surface elevations, and fluctuations of water levels in wells. A variety of earthquake research programs are underway in various earthquake-prone countries. Studies indicate that most people would probably not heed a short-term earthquake warning.

LO9   Discuss earthquake control methods. Because of the tremendous energy involved, it seems unlikely that humans will ever be able to prevent earthquakes. However, it might be possible to release small amounts of the energy stored in rocks and thus avoid a large earthquake and the extensive damage that typically results.

LO10   Describe Earth's interior. Earth has an outer layer of oceanic and continental crust, below which lies a rocky mantle and an iron-rich core with a liquid outer part and a solid inner part. Studies of P- and S-waves, laboratory experiments, comparisons with meteorites, and studies of inclusions in volcanic rocks provide evidence for Earth's internal structure and composition.

LO11   Explain the features of Earth's core. Geologists use the behavior of P- and S-waves and the presence of the P- and S-wave shadow zones to estimate the density and composition of Earth's interior, as well as to estimate the size and depth of the core and mantle. Earth's inner core is probably made up of iron and nickel, whereas the outer core is mostly iron with 10 to 20 percent other substances.

LO12   Describe the structure, density, and composition of Earth's mantle. Peridotite, an igneous rock composed mostly of ferromagnesian silicates, is the most likely rock making up Earth's mantle.

LO13   Explain the source of Earth's internal heat. Much of Earth's internal heat is generated by radioactive decay, especially the decay of isotopes of uranium and thorium. The geothermal gradient, which is Earth's temperature increase with depth, is 25°C/km. However, it cannot continue at that rate to great depths because, despite increasing pressure, the temperature would soon be high enough to melt all known rocks.

LO14   Describe the features of Earth's crust. Oceanic crust is composed of basalt and gabbro, whereas continental crust has an overall composition similar to granite. The Moho is the boundary between the crust and the mantle.

Chapter Summary

LO1 Explain how rock deformation occurs. Folded and fractured rocks have been deformed or strained by applied stresses. Stress is compression, tension, or shear. Elastic strain is not permanent, but plastic strain and fracture are, meaning that rocks do not return to their original shape or volume when the deforming forces are removed.

LO2 Understand strike and dip—the orientation of deformed rock layers. Strike and dip are used to define the orientation of deformed rock layers. This same concept applies to other planar features such as fault planes.

LO3 Identify the types of deformation and geologic structures. Anticlines and synclines are up- and down-arched folds, respectively. They are identified by strike and dip of the folded rocks and the relative ages of rocks in these folds. Domes and basins are the circular to oval equivalents of anticlines and synclines, but they are commonly much larger structures.

The two structures that result from fracture are joints and faults. Joints may open up but they show no movement parallel with the fracture surface, whereas faults do show movement parallel with the fracture surface. Joints are very common and form in response to compression, tension, and shear.

On dip-slip faults, all movement is up or down the dip of the fault. If the hanging wall moves relatively down, it is a normal fault; but if the hanging wall moves up relative to the footwall, it is a reverse fault. Normal faults result from tension; reverse faults result from compression. In strike-slip faults, all movement is along the strike of the fault. These faults are either right-lateral or left-lateral, depending on the apparent direction of offset of one block relative to the other. Oblique-slip faults show components of both dip-slip and strike-slip movement.

LO4 Understand deformation and the origin of mountains. A variety of processes account for the origin of mountains. Some involve little or no deformation, but the large mountain systems on the continents resulted from deformation at convergent plate boundaries. A volcanic island arc, deformation, igneous activity, and metamorphism characterize orogenies at oceanic–oceanic plate boundaries,

Key Terms

anticline A convex upward fold in which the oldest exposed rocks coincide with the fold axis and all strata dip away from the axis.

basin A circular fold in which all strata dip inward toward a central point and the youngest exposed strata are in the center.

compression Stress resulting when materials are squeezed by external forces directed toward one another.

continental accretion Orogenics along convergent plate boundaries that result in adding material to a continent.

deformation A general term for any change in shape or volume, or both, of rocks in response to stress; involves folding and fracturing.

dip A measure of the maximum angular deviation of an inclined plane from horizontal.

dip-slip fault A fault on which all movement is parallel with the dip of the fault plane.

dome A rather circular geologic structure in which all rock layers dip away from a central point and the oldest exposed rocks are in the dome's center.

elastic strain A type of deformation in which the material returns to its original shape when stress is relaxed.

fault A fracture along which rocks on opposite sides of the fracture have moved parallel with the fracture surface.

fault plane A fault surface that is more or less planar.

fold A type of geologic structure in which planar features in rock layers such as bedding and foliation have been bent.

footwall block The block of rock that lies beneath a fault plane.

fracture A break in rock resulting from intense applied pressure.

geologic structure Any feature in rocks that results from deformation, such as folds, joints, and faults.

hanging wall block The block of rock that overlies a fault plane.

isostatic rebound The phenomenon in which unloading of the crust causes it to rise until it attains equilibrium.

joint A fracture along which no movement has occurred or where movement is perpendicular to the fracture surface.

monocline A bend or flexure in otherwise horizontal or uniformly dipping rock layers.

normal fault A dip-slip fault on which the hanging wall block has moved downward relative to the footwall block.

oblique-slip fault A fault showing both dip-slip and strike-slip movement.

orogeny An episode of mountain building involving deformation, usually accompanied by igneous activity, and crustal thickening.

plastic strain Permanent deformation of a solid with no failure by fracturing.

principle of isostasy The concept of Earth's crust "floating" on a dense underlying layer.

reverse fault A dip-slip fault on which the hanging wall block has moved upward relative to the footwall block.

shear stress The result of forces acting parallel to one another but in opposite directions; results in deformation by displacement of adjacent layers along closely spaced planes.

strain Deformation caused by stress.

stress The force per unit area applied to a material such as rock.

strike The direction of a line formed by the intersection of an inclined plane and a horizontal plane.

strike-slip fault A fault involving horizontal movement of blocks of rock on opposite sides of a fault plane.

syncline A down-arched fold in which the youngest exposed rocks coincide with the fold axis and all strata dip toward the axis.

tension A type of stress in which forces act in opposite directions but along the same line, thus tending to stretch an object.

terrane Fragments of seamounts, island arcs, and small pieces of continents that were carried on oceanic plates that collided with continental plates.

thrust fault A type of reverse fault in which a fault plane dips less than 45 degrees.

whereas orogeny at an oceanic–continental plate boundary is a result of subduction of an oceanic plate beneath a continental plate. Some mountain systems are within continents far from a present-day plate boundary. These mountains formed when two continental plates collided and became sutured. Geologists now realize that orogenies also involve collisions of terranes with continents.

LO5   Describe Earth's continental crust. Continental crust is characterized as granitic, and it is much thicker and less dense than oceanic crust that is composed of basalt and gabbro. According to the principle of isostasy, Earth's crust floats in equilibrium in the denser mantle below. Continental crust stands higher than oceanic crust because it is thicker and less dense.

Review Questions

1. What are the similarities and differences between a basin and a syncline?

2. Describe how time, rock type, pressure, and temperature influence rock deformation.

3. Suppose that rocks were displaced 200 km along a strike-slip fault during a period of 5 million years. What was the average rate of movement per year? Is the average likely to represent the actual rate of displacement on this fault? Explain.

4. Is there any connection between the principle of isostasy and mountain building? If so, what?

5. What is meant by the elastic limit of rocks, and what happens when rocks are strained beyond their elastic limit?

6. How does a dip-slip fault differ from a strike-slip fault?

7. What is the concept of isostatic rebound? Give an example of how it may take place.

8. What kinds of evidence would indicate that mountain building once took place in an area where mountains are no longer present?

Chapter Summary

LO1   List the factors that influence mass wasting. Mass wasting is the downslope movement of material under the direct influence of gravity. It frequently results in loss of life, as well as millions of dollars in damage annually. Mass wasting occurs when the gravitational force acting on a slope exceeds the slope's shear strength (the resisting forces that help maintain slope stability).

The major factors causing mass wasting include slope angle, weathering and climate, water content, overloading, and removal of vegetation. It is usually a combination of several of these factors that result in slope failure.

LO2   Describe the types of mass wasting. Mass movements are generally classified on the basis of their rate of movement (rapid versus slow), type of movement (falling, sliding, or flowing), and type of material (rock, soil, or debris). Rockfalls are a common mass movement in which rocks free-fall. They are common along steep canyons, cliffs, and road cuts. Two types of slides are generally recognized. Slumps are rotational slides that involve movement along a curved surface and are most common in poorly consolidated or unconsolidated material. Rock slides occur when movement takes place along a more or less planar surface, and they usually involve solid pieces of rock.

Key Terms

complex movement A combination of different types of mass movements in which no single type is dominant; usually involves sliding and flowing.

creep A widespread type of mass wasting in which soil or rock moves slowly downslope.

debris flow A type of mass wasting that involves a viscous mass of soil, rock fragments, and water that moves downslope; debris flows have larger particles than mudflows and contain less water.

earthflow A mass wasting process involving the downslope movement of water-saturated soil.

mass wasting The downslope movement of Earth materials under the influence of gravity.

mudflow A flow consisting mostly of clay- and silt-sized particles and up to 30% water that moves downslope under the influence of gravity.

permafrost Ground that remains permanently frozen.

quick clay A clay deposit that spontaneously liquefies and flows like water when disturbed.

rapid mass movement Any kind of mass wasting that involves a visible downslope displacement of material.

rockfall A type of extremely fast mass wasting in which rocks fall through the air.

rock slide Rapid mass wasting in which rocks move downslope along a more or less planar surface.

shear strength The resisting forces that help maintain a slope's stability.

slide Mass wasting involving movement of material along one or more surfaces of failure.

slow mass movement Mass movement that advances at an imperceptible rate and is usually detectable only by the effects of its movement.

slump Mass wasting that takes place along a curved surface of failure and results in the backward rotation of the slump mass.

solifluction Mass wasting involving the slow downslope movement of water-saturated surface materials; occurs especially at high elevations or high latitudes where the flow is underlain by frozen soil.

Review Questions

1. **What are some of the ways slope stability can be maintained in order to reduce the likelihood of mass movements?**

2. **What potential value would a slope stability map be to a person seeking to purchase new home property?**

3. **How do topography and the underlying geology contribute to slope failure?**

4. **Why are slumps such a problem along highways and railroad tracks in areas with relief?**

5. **How would removing preexisting vegetation tend to affect most slopes in humid regions with respect to mass wasting?**

6. **How are the different factors that affect mass wasting connected?**

7. **How could mass wasting be recognized on other planets or moons? What would that tell us about the geology, and perhaps the atmosphere, of the planet or moon on which it occurred?**

8. **If an area has a documented history of mass wasting that has endangered or taken human life, how should people and governments prevent such events from happening again? Are most large mass wasting events preventable or predictable?**

9. **Why is it important to know about the different types of mass wasting?**

Several types of flows are recognized on the basis of their rate of movement (rapid versus slow), type of material (rock, sediment, or soil), and amount of water. Mudflows consist of mostly clay- and silt-sized particles and contain up to 30 percent water. They are most common in semiarid and arid environments and generally follow preexisting channels. Debris flows are composed of larger particles and contain less water than mudflows. Earthflows move more slowly than either debris flows or mudflows, and move downslope as thick, viscous, tongue-shaped masses of wet regolith. Quick clays are clays that spontaneously liquefy and flow like water when they are disturbed. Creep, the slowest type of flow, is the imperceptible downslope movement of soil or rock. It is the most widespread of all types of mass wasting. Complex movements are combinations of different types of mass movements in which no single type is dominant. Most complex movements involve sliding and flowing.

LO3   Understand how to recognize and minimize the effects of mass wasting. The most important factor in reducing or eliminating the damaging effects of mass wasting is a thorough geologic investigation to outline areas susceptible to mass movements. Although mass movement cannot be eliminated, its effects can be minimized by building retaining walls, draining excess water, regrading slopes, and planting vegetation.

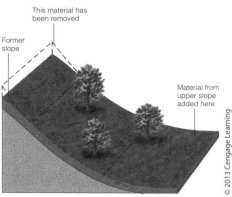

Chapter Summary

LO1   Identify sources of water on Earth. Water continuously evaporates from the oceans, rises as water vapor, condenses, and falls as precipitation. Approximately 20 percent falls on land and eventually returns to the oceans, mostly by surface runoff.

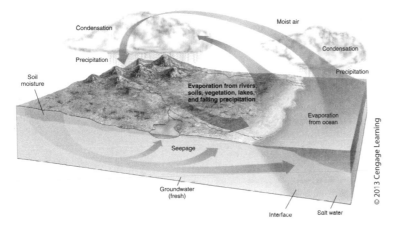

LO2   Describe the role of running water. Runoff takes place by sheet flow, a thin, more or less continuous sheet of water, and by channel flow, confined to long, troughlike stream and river channels. The vertical drop in a given distance, or the gradient, for a channel varies from steep in its upper reaches to more gentle in its lower reaches. Flow velocity and discharge are related, so that if either changes, the other changes as well.

LO3   Explain how running water erodes and transports sediment. Erosion by running water takes place by hydraulic action, abrasion, and solution. The bed load in channels is made up of sand and gravel, whereas suspended load consists of silt- and clay-sized particles. Running water also transports a dissolved load.

LO4   Describe deposition by running water. Braided waterways have a complex of dividing and rejoining channels, and their deposits are mostly sheets of sand and gravel. A single sinuous channel is typical of meandering streams that deposit mostly mud, with subordinate point-bar deposits of sand or more rarely, gravel. Broad, flat, floodplains adjacent to channels are the sites of oxbow lakes, which are simply abandoned meanders. An alluvial deposit at a river's mouth is a delta. Some deltas conform to the three-part division of bottomset, foreset, and topset beds, but large marine deltas are much more complex and are characterized as stream-, wave-, or tide-dominated. Alluvial fans are fan-shaped deposits of sand and gravel on land that form best in semiarid regions. They form mostly by deposition from running water, but debris flows are also important.

LO5   Question whether floods can be predicted and controlled. People have tried to control floods for thousands of years, most commonly by building dams that impound reservoirs and building levees along stream banks. A floodway is a channel constructed to divert part of the excess water in a stream around populated areas or areas of economic importance. Some communities build floodwalls to protect them from floods.

Key Terms

abrasion The process whereby rock is worn smooth by the impact of sediment transported by running water, glaciers, waves, or wind.

alluvial fan A cone-shaped accumulation of mostly sand and gravel deposited where a stream flows from a mountain valley onto an adjacent lowland.

alluvium A collective term for all detrital sediment transported and deposited by running water.

base level The level below which a stream or river cannot erode; sea level is ultimate base level.

bed load The part of a stream's sediment load, mostly sand and gravel, transported along its bed.

braided stream A stream with multiple dividing and rejoining channels.

delta A deposit of sediment where a stream or river enters a lake or the ocean.

discharge The volume of water in a stream or river moving past a specific point in a given interval of time; expressed in cubic meters per second (m^3/sec) or cubic feet per second (ft^3/sec).

dissolved load The part of a stream's load consisting of ions in solution.

divide A topographically high area that separates adjacent drainage basins.

drainage basin The surface area drained by a stream or river and its tributaries.

drainage pattern The regional arrangement of channels in a drainage system.

floodplain A low-lying, flat area adjacent to a channel that is partly or completely water-covered when a stream or river overflows its banks.

graded stream A stream that has an equilibrium profile in which a delicate balance exists among gradient, discharge, flow velocity, channel characteristics, and sediment load so that neither significant deposition nor erosion takes place within its channel.

gradient The slope over which a stream or river flows, expressed in m/km or ft/mi.

hydraulic action The removal of loose particles by the power of moving water.

hydrologic cycle The continuous recycling of water from the oceans, through the atmosphere, to the continents, and back to the oceans, or from the oceans, through the atmosphere, and back to the oceans.

incised meander A deep, meandering canyon cut into bedrock by a stream or river.

infiltration capacity The maximum rate at which soil or sediment absorbs water.

meandering stream A stream that has a single, sinuous channel with broadly looping curves.

natural levee A ridge of sandy alluvium deposited along the margins of a channel during floods.

oxbow lake A cutoff meander filled with water.

point bar The sediment body deposited on the gently sloping side of a meander loop.

runoff The surface flow in streams and rivers.

stream terrace An erosional remnant of a floodplain that formed when a stream was flowing at a higher level.

superposed stream A stream that once flowed on a higher surface and eroded downward into resistant rocks while maintaining its course.

suspended load The smallest particles (silt and clay) carried by running water, which are kept suspended by fluid turbulence.

valley A linear depression bounded by higher areas such as ridges or mountains.

velocity A measure of distance traveled per unit of time, as in the flow velocity in a stream or river.

© 2013 Cengage Learning

LO6 **Understand drainage systems.** Thousands of waterways, which are parts of larger drainage systems, flow directly or indirectly into the oceans. Rivers and streams carry runoff from their drainage basins, which are separated from one another by divides. The most common drainage pattern is dendritic drainage, which consists of a network of channels resembling tree branching.

LO7 **Recognize the significance of base level.** Sea level is ultimate base level, the lowest level to which streams or rivers can erode. Local base levels may be lakes or where streams or rivers flow across resistant rocks. Graded streams tend to eliminate irregularities in their channels, so they develop a smooth, concave profile of equilibrium.

LO8 **Understand the evolution of valleys.** A combination of processes, including downcutting, lateral erosion, sheetwash, mass wasting, and headward erosion, are responsible for the origin and evolution of valleys. Stream terraces and incised meanders usually form when a stream or river that was formerly in equilibrium begins a new episode of downcutting.

© 2013 Cengage Learning

Review Questions

1. Calculate the daily discharge of a river 148 m wide and 2.6 m deep, with a flow velocity of 0.3 m/sec.

2. Diagram and describe how an ideal small delta forms.

3. Explain how the hydrologic cycle works.

4. Describe the processes whereby running water erodes.

5. The discharge of most rivers and streams increases downstream, but in a few cases it actually decreases and they eventually disappear. Why? Give an example.

6. Explain how stream terraces form and what they indicate about the dynamics of a stream or river system.

7. Explain how a superposed stream develops. Also describe the landforms that result from superposition.

Chapter Summary

LO1 **Describe groundwater and its role in the hydrologic cycle.**
Groundwater is part of the hydrologic cycle and an important natural resource. It consists of all subsurface water trapped in the pores and other open spaces in rocks, sediment, and soil.

LO2 **Define porosity and permeability.** Porosity is the percentage of a materials' total volume that is pore space. Permeability is the capacity to transmit fluids. Permeability is dependent on porosity, but also on the size of the pores or fractures and their interconnections.

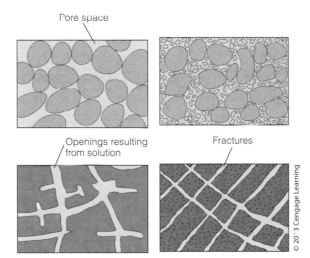

Pore space

Openings resulting from solution

Fractures

© 2013 Cengage Learning

LO3 **Describe the water table.** The water table is the surface separating the zone of saturation (in which pores are filled with water) from the overlying zone of aeration (in which pores are filled with air and water). The water table is a subdued replica of the overlying land surface in most places.

LO4 **Understand patterns of groundwater movement.** Groundwater moves slowly downward under the influence of gravity through the pore spaces in the zone of aeration to the zone of saturation. Some groundwater moves along the surface of the water table, and the rest moves from areas of high pressure to areas of low pressure.

LO5 **Identify the differences between springs, water wells, and artesian systems.** Springs are found wherever the water table intersects the surface. Some springs are the result of a perched water table—that is, a localized aquiclude within an aquifer and above the regional water table. Water wells are openings made by digging or drilling down into the zone of saturation. When water is pumped from a well, the water table in the area around the well is lowered, forming a cone of depression. In an artesian system, confined groundwater builds up high hydrostatic pressure. For an artesian system to develop, an aquifer must be confined above and below by aquicludes with the rock units usually tilted so as to build up hydrostatic pressure, and the aquifer must be exposed at the surface so it can be recharged.

Key Terms

artesian system A confined groundwater system with high hydrostatic pressure that causes water to rise above the level of the aquifer.

cave A natural subsurface opening generally connected to the surface and large enough for a person to enter.

cone of depression A cone-shaped depression around a well where water is pumped from an aquifer faster than it can be replaced.

geothermal energy Energy that comes from steam and hot water trapped within Earth's crust.

geyser A hot spring that periodically ejects hot water and steam.

groundwater Underground water stored in the pore spaces of soil, sediment, and rock.

hot spring A spring in which the water temperature is warmer than the temperature of the human body (37°C).

hydrothermal A term referring to hot water, as in hot springs and geysers.

karst topography Landscape consisting of numerous caves, sinkholes, and solution valleys formed by groundwater solution of rocks such as limestone and dolostone.

permeability A material's capacity to transmit fluids.

porosity The percentage of a material's total volume that is pore space.

sinkhole A depression in the ground that forms by the solution of the underlying carbonate rocks or by the collapse of a cave roof.

spring A place where groundwater flows or seeps out of the ground.

water table The surface that separates the zone of aeration from the underlying zone of saturation.

water well A well made by digging or drilling into the zone of saturation.

zone of aeration The zone above the water table that contains both air and water within the pore spaces of soil, sediment, or rock.

zone of saturation The area below the water table in which all pore spaces are filled with water.

Review Questions

1. What is the difference between an artesian system and a water well? Is there any difference between the water obtained from an artesian system and the water from a water well?

2. Describe the configuration of the water table beneath a humid area and beneath an arid region. Why are the configurations different?

3. Explain how some Earth materials can be porous yet not permeable. Give an example.

4. Explain how groundwater weathers and erodes Earth materials.

5. Why does groundwater move so much slower than surface water?

6. Explain how saltwater incursion takes place and why it is a problem in coastal areas.

7. Discuss the role of groundwater in the hydrologic cycle.

8. Why should we be concerned about how fast the groundwater supply is being depleted in some areas?

9. Why isn't geothermal energy a virtually unlimited source of energy?

LO6   Discuss erosion and deposition caused by groundwater. Karst topography largely develops by groundwater erosion in many areas underlain by soluble rocks and is characterized by sinkholes, caves, solution valleys, and disappearing streams. Caves form when groundwater in the zone of saturation weathers and erodes soluble rock such as limestone. Cave deposits, called dripstone, result from the precipitation of calcite.

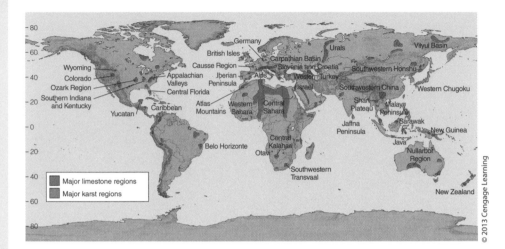

© 2013 Cengage Learning

LO7   Describe the effects of modifications to the groundwater system. Modifications of the groundwater system can cause serious problems such as lowering of the water table, saltwater incursion, subsidence, and contamination. Groundwater contamination by humans is becoming a serious problem and can result from landfills, septic systems, toxic waste sites, and industrial effluents, all of which affect the quality of the groundwater.

LO8   Describe hydrothermal activity. Hydrothermal refers to hot water, typically heated by magma, but also resulting from Earth's geothermal gradient as it circulates deeply beneath the surface. Manifestations of hydrothermal activity include fumaroles, hot springs, and geysers. Geothermal energy is energy produced from Earth's internal heat and comes from the steam and hot water trapped within Earth's crust. It is a relatively nonpolluting form of energy that is used as a source of heat and to generate electricity.

10. One concern that geologists have about burying nuclear waste in present-day arid regions such as Nevada is that the climate may change during the next several thousand years and become more humid, thus allowing more water to percolate through the zone of aeration. Why is this a concern? What would the average rate of groundwater movement have to be during the next 5,000 years to reach buried canisters containing radioactive waste buried at a depth of 400 m?

Chapter 13 IN REVIEW

Glaciers and Glaciation

Chapter Summary

LO1 Identify the different kinds of glaciers. Glaciers currently cover about 10 percent of the land surface. Valley glaciers are confined to mountain valleys and flow from higher to lower elevations, whereas continental glaciers cover vast areas and flow outward in all directions from a zone of accumulation.

James S. Monroe

LO2 Recognize that glaciers are moving bodies of ice on land. A glacier forms when winter snowfall exceeds summer melt and therefore accumulates year after year. Snow is compacted and converted to glacial ice, and when the ice is about 40 m thick, pressure causes it to flow. Glaciers move by plastic flow and basal slip.

LO3 Understand the glacial budget of accumulation and wastage. The behavior of a glacier depends on its budget, which is the relationship between accumulation and wastage. If a glacier has a balanced budget, its terminus remains stationary; a positive or negative budget results in the advance or retreat of the terminus, respectively. Glaciers move at varying rates depending on slope, discharge, and season. Valley glaciers tend to flow more rapidly than continental glaciers.

LO4 Identify features resulting from erosion and transport by glaciers. Glaciers effectively erode and transport because they are solids in motion. They are particularly effective at eroding soil and unconsolidated sediment, and they can transport any size sediment supplied to them. Erosion of mountains by valley glaciers yields several sharp, angular landforms including cirques, arêtes, and horns. U-shaped glacial troughs, fiords, and hanging valleys are also products of valley glaciation. Continental glaciers abrade and bevel high areas, producing a smooth, rounded landscape known as an ice-scoured plain.

Key Terms

abrasion The process whereby rock is worn smooth by the impact of sediment transported by running water, glaciers, waves, or wind.

arête A narrow, serrated ridge between two glacial valleys or adjacent cirques.

basal slip Movement involving a glacier sliding over its underlying surface.

cirque A steep-walled, bowl-shaped depression on a mountainside at the upper end of a glacial valley.

continental glacier A glacier that covers a vast area (at least 50,000 km²) and is not confined by topography; also called an ice sheet.

drumlin An elongate hill of till formed by the movement of a continental glacier or by floods.

end moraine A pile or ridge of rubble deposited at the terminus of a glacier.

esker A long, sinuous ridge of stratified drift deposited by running water in a tunnel beneath stagnant ice.

fiord An arm of the sea extending into a glacial trough eroded below sea level.

firn Granular snow formed by partial melting and refreezing of snow; transitional material between snow and glacial ice.

glacial budget The balance between expansion and contraction of a glacier in response to accumulation versus wastage.

glacial drift A collective term for all sediment deposited directly by glacial ice (till) and by meltwater streams (outwash).

glacial erratic A rock fragment carried some distance from its source by a glacier and usually deposited on bedrock of a different composition.

glacial ice Water in the solid state within a glacier, forms as snow partially melts and refreezes and compacts so that it is transformed first to firn and then to glacial ice.

glacial polish A smooth, glistening rock surface formed by the movement of sediment-laden ice over bedrock.

glacial striation A straight scratch rarely more than a few millimeters deep on a rock caused by the movement of sediment-laden glacial ice.

glacial surge A time of greatly accelerated flow in a glacier. Commonly results in displacement of the glacier's terminus by several kilometers.

© 2013 Cengage Learning

© 2013 Cengage Learning

glaciation Refers to all aspects of glaciers, including their origin, expansion, and retreat, and their impact on Earth's surface.

glacier A mass of ice on land that moves by plastic flow and basal slip.

ground moraine The layer of sediment released from melting ice as a glacier's terminus retreats.

hanging valley A tributary glacial valley whose floor is at a higher level than that of the main glacial valley.

horn A steep-walled, pyramid-shaped peak formed by the headward erosion of at least three cirques.

ice cap A dome-shaped mass of glacial ice that covers less than $50{,}000$ km^2.

kame Conical hill of stratified drift originally deposited in a depression on a glacier's surface.

lateral moraine Ridge of sediment deposited along the margin of a valley glacier.

Little Ice Age An interval from about 1500 to the mid- to late-1800s during which glaciers expanded to their greatest historic extent.

medial moraine A moraine carried on the central surface of a glacier; formed where two lateral moraines merge.

Milankovitch theory An explanation for the cyclic variations in climate and the onset of ice ages as a result of irregularities in Earth's rotation and orbit.

outwash plain The sediment deposited by meltwater discharging from a continental glacier's terminus.

plastic flow The flow that takes place in response to pressure and causes deformation with no fracturing.

recessional moraine An end moraine that forms when a glacier's terminus retreats, then stabilizes, and a ridge or mound of till is deposited.

stratified drift Glacial deposits that show both stratification and sorting.

terminal moraine An end moraine consisting of a ridge or mound of rubble marking the farthest extent of a glacier.

till All sediment deposited directly by glacial ice.

U-shaped glacial trough A valley with steep or vertical walls and a broad, rather flat floor formed by the movement of a glacier through a stream valley.

valley glacier A glacier confined to a mountain valley or an interconnected system of mountain valleys.

valley train A long, narrow deposit of stratified drift confined within a glacial valley.

LO5 Identify the types of glacial deposits. Depositional landforms include moraines, which are ridge-like accumulations of till. The several types of moraines are terminal, recessional, lateral, and medial. Drumlins are composed of till that was apparently reshaped into streamlined hills by continental glaciers or floods. Stratified drift in outwash plains and valley trains consists of sand and gravel deposited by meltwater streams issuing from glaciers. Ridges, known as eskers, and conical hills called kames are also composed of stratified drift.

LO6 Explain what scientists think caused the Ice Age. Major glacial intervals separated by tens or hundreds of millions of years probably occur as a result of the changing positions of tectonic plates, which in turn cause changes in oceanic and atmospheric circulation patterns. Currently, the Milankovitch theory is widely accepted as the explanation for glacial-interglacial intervals. The reasons for short-term climatic changes, such as the Little Ice Age, are not understood. Two proposed causes are changes in the amount of solar energy received by Earth and volcanism.

Review Questions

1. How is it possible for glacial ice, which is solid, to flow?

2. Draw a side view of a drumlin, indicate the direction of flow, and explain how drumlins form.

3. A valley glacier has a cross sectional area of $400{,}000$ m^2 and a flow velocity of 2 m per day. How long will it take for 1 km^3 of ice to move past a given point?

4. Explain how a glacier can erode below sea level but streams and rivers cannot.

5. How does the Milankovitch theory explain the onset of Pleistocene episodes of glaciation?

6. What kinds of evidence would indicate that an ice-free area was once covered by glacial ice?

7. What are glacial surges and what might cause one to occur?

8. Explain in terms of a glacial budget how a once-active glacier becomes stagnant.

9. How do valley glaciers, continental glaciers, and ice caps differ?

10. Where are the only two areas where continental glaciers are present today?

zone of accumulation The part of a glacier where additions exceed losses and the glacier's surface is perennially covered with snow. Also refers to horizon B in soil where soluble material leached from horizon A accumulates as irregular masses.

zone of wastage The part of a glacier where losses from melting, sublimation, and calving of icebergs exceed the rate of accumulation.

Chapter Summary

LO1   Discuss the role wind plays in transporting sediment. Wind transports sediment in suspension or as bed load. Suspended load is the material that can be carried in suspension by water or wind. Silt- and clay-sized particles constitute most of a wind's suspended load. Bed load is the material that is too large or heavy to be carried in suspension by water or wind. Particles are moved along the surface by saltation, rolling, or sliding.

LO2   Explain the two processes of wind erosion. Wind erodes material by either abrasion or deflation. Abrasion is the impact of saltating sand grains on an object. Ventifacts are common products of wind abrasion. Deflation is the removal of loose surface material by wind. Deflation hollows resulting from differential erosion of surface material are common features of many deserts, as is desert pavement, which effectively protects the underlying surface from additional deflation.

LO3   Identify the different types of wind deposits. The two major wind deposits are dunes and loess. Dunes are mounds or ridges of wind-deposited sand that form when wind flows over and around an obstruction, resulting in the deposition of sand grains, which accumulate and build up a deposit of sand. The four major dune types are barchan, longitudinal, transverse, and parabolic. The amount of sand available, prevailing wind direction, wind velocity, and the amount of vegetation determine which type of dune will form. Loess is wind-blown silt and clay deposits composed of angular quartz grains, feldspar, micas, and calcite. It is derived from deserts, Pleistocene glacial outwash deposits, and river floodplains in semiarid regions. Loess covers approximately 10 percent of Earth's land surface and weathers to a rich, productive soil.

Key Terms

abrasion The process whereby rock is worn smooth by the impact of sediment transported by running water, glaciers, waves, or wind.

alluvial fan A cone-shaped accumulation of mostly sand and gravel deposited where a stream flows from a mountain valley onto an adjacent lowland.

barchan dune A crescent-shaped sand dune with its tips pointing downwind.

butte An isolated, steep-sided, pinnacle-like hill formed when resistant cap rock is breached, allowing erosion of less resistant underlying rocks.

Coriolis effect The apparent deflection of a moving object from its anticipated course because of Earth's rotation. Winds and oceanic currents are deflected clockwise in the Northern Hemisphere and counterclockwise in the Southern Hemisphere.

deflation The removal of loose surface sediment by wind.

desert Any area that receives less than 25 cm of rain per year and that has a high evaporation rate.

desert pavement A surface mosaic of close-fitting pebbles, cobbles, and boulders found in many dry regions; it results from wind erosion of sand and smaller particles.

desertification The expansion of deserts into formerly productive lands.

dune A mound or ridge of wind-deposited sand.

loess Wind-blown deposits of silt and clay.

longitudinal dune A long ridge of sand generally parallel to the direction of the prevailing wind.

mesa A broad, flat-topped erosional remnant bounded on all sides by steep slopes.

parabolic dune A crescent-shaped dune with its tips pointing upwind.

pediment An erosion surface of low relief gently sloping away from the base of a mountain range.

playa A dry lake bed found in deserts.

transverse dune A ridge of sand with its long axis perpendicular to the wind direction.

ventifact A stone with a surface polished, pitted, grooved, or faceted by wind abrasion.

Review Questions

1. Is it possible for Mars to have the same type of sand dunes as Earth? What does that tell us about the climate and geology of Mars?

2. Using what you now know about deserts, their location, how they form, and the various landforms found in them, how can you determine where deserts may have existed in the past?

3. Why are low-latitude deserts so common?

4. If deserts are dry regions in which mechanical weathering predominates, why are so many of their distinctive landforms the result of running water and not of wind?

5. Why is desert pavement important in a desert environment?

6. How do dunes form and migrate? Why is dune migration a problem in some areas?

7. Much of recent desertification has been greatly accelerated by human activity. Can you think of anything that can be done to slow the process?

8. What is loess, and why is it important?

LO4 Describe air-pressure belts and global wind patterns. The winds of the major air-pressure belts, oriented east-west, result from the rising and cooling of air. The winds are deflected clockwise in the Northern Hemisphere and counterclockwise in the Southern Hemisphere by the Coriolis effect to produce Earth's global wind patterns.

LO5 Describe the distribution of deserts. Dry climates, located in the low and middle latitudes, where the potential loss of water by evaporation exceeds the yearly precipitation, cover 30 percent of Earth's land surface and are subdivided into semiarid and arid regions. Semiarid regions receive more precipitation than arid regions, yet are moderately dry. Arid regions, generally described as deserts, are dry and receive less than 25 cm of rain per year. The majority of the world's deserts are in the dry climates of the low and middle latitudes. The remaining dry climates of the world are found in the middle and high latitudes, mostly within continental interiors in the Northern Hemisphere.

LO6 Identify the various characteristics of deserts. Deserts are characterized by high temperatures, little precipitation, and sparse plant cover. Rainfall is unpredictable and when it does occur, tends to be intense and of short duration. Mechanical weathering is the dominant form of weathering in deserts, and coupled with slow rates of chemical weathering, results in poorly developed soils. Running water is the major agent of erosion in deserts. Wind is an erosional agent in deserts and is very effective in transporting and depositing unconsolidated fine-grained sediments.

LO7 Describe the different types of desert landforms. Important desert landforms include playas, which are dry lakebeds; when temporarily filled with water, they form playa lakes. Alluvial fans are cone-shaped sedimentary deposits. Pediments are erosional bedrock surfaces of low relief that slope gently away from mountain bases. A mesa is a broad, flat-topped erosional remnant bounded on all sides by steep slopes. Continued weathering and stream erosion form pillar-like structures known as buttes.

Copyright and Photograph by Dr. Parvinder S. Sethi

Chapter Summary

LO1   Understand the composition and circulation of seawater and describe the types of sediment on the seafloor. The oceanic waters contain many dissolved chemicals, but the most common ones are sodium and chlorine, which give seawater its salinity. The upper layer of ocean water, called the photic zone, is where organisms live that practice photosynthesis. Water in the oceans circulates horizontally in surface currents driven by wind and deflected by the Coriolis effect. Deep water circulation also occurs because of density differences in adjacent masses of water. Vertical circulation in the oceans takes place in areas of downwelling (transfer of warm surface water to depth), and, more importantly, in areas of upwelling (transfer of cold water from depth to the surface). Sediments on the deep seafloor consist mostly of pelagic clay and ooze. Reefs composed of animal skeletons are found in warm, shallow seawater and are characterized as fringing reefs, barrier reefs, and atolls

LO2   Describe the behavior of tides, waves, and nearshore currents. Shorelines are continuously modified by the energy of waves and longshore currents and, to a limited degree, by tidal currents. The surface of the oceans rises and falls twice daily in response to the gravitational attraction of the Moon and Sun in most shoreline areas.

Waves are oscillations on water surfaces that transmit energy in the direction of wave movement. Surface waves affect the water and seafloor only to wave base, which is equal to half the wavelength. Little or no net forward motion of water occurs in waves in the open sea. When waves enter shallow water, they are transformed into waves in which water moves in the direction of wave advance.

James S. Monroe

Low and high tide at Turnagain Arm in Alaska.

Wind-generated waves, especially storm waves, are responsible for most geologic work on shorelines, but waves can also be generated by faulting, volcanic explosions, and rockfalls. Breakers form where waves enter shallow water and disrupt the orbital motion of water particles. The waves become oversteepened and plunge forward or spill onto the shoreline, thus expending their kinetic energy. Waves approaching a shoreline at an angle generate a longshore current. These currents are capable of considerable erosion, transport, and deposition. Narrow surface currents called rip currents carry water from the nearshore zone seaward through the breaker zone.

Key Terms

aphotic zone The depth in the ocean below which sunlight does not penetrate.

barrier island A long, narrow island of sand parallel to a shoreline but separated from the mainland by a lagoon.

baymouth bar A spit that has grown until it closes off a bay from the open sea or lake.

beach Any deposit of sediment extending landward from low tide to a change in topography or where permanent vegetation begins.

breaker A wave that steepens as it enters shallow water until its crest plunges forward.

Coriolis effect The apparent deflection of a moving object from its anticipated course because of Earth's rotation. Winds and oceanic currents are deflected clockwise in the Northern Hemisphere, and counterclockwise in the Southern Hemisphere.

downwelling The slow transfer of ocean surface water to depth.

emergent coast A coast where the land has risen with respect to sea level.

fetch The distance the wind blows over a continuous water surface.

gyre A system of ocean currents rotating clockwise in the Northern Hemisphere, and counterclockwise in the Southern Hemisphere.

headland Part of a shoreline commonly bounded by cliffs that extends out into the sea or a lake.

longshore current A current resulting from wave refraction found between the breaker zone and a beach that flows parallel to the shoreline.

marine terrace A wave-cut platform now above sea level.

nearshore sediment budget The balance between additions and losses of sediment in the nearshore zone.

ooze Deep-sea sediment composed mostly of shells of marine animals and plants.

pelagic clay Brown or red deep-sea sediment composed of clay-sized particles.

photic zone The sunlit layer in the oceans where plants photosynthesize.

reef A mound-like, wave-resistant structure composed of the skeletons of organisms.

rip current A narrow surface current that flows out to sea through the breaker zone.

salinity A measure of the dissolved solids in seawater, usually expressed in parts per thousand.

shoreline The area between mean low tide and the highest level on land affected by storm waves.

spit A fingerlike projection of a beach into a body of water such as a bay.

storm surge The surge of water onto a shoreline as a result of a bulge in the ocean's surface beneath the eye of a hurricane and wind-driven waves.

submergent coast A coast along which sea level rises with respect to the land or the land subsides.

tide The regular fluctuation of the sea's surface in response to the gravitational attraction of the Moon and Sun.

tombolo A type of spit that extends out from the shoreline and connects the mainland with an island.

upwelling The slow circulation of ocean water from depth to the surface.

wave An undulation on the surface of a body of water, resulting in the water surface rising and falling.

wave base The depth corresponding to about one-half wavelength, below which water is unaffected by surface waves.

wave-cut platform A beveled surface that slopes gently seaward; formed by the erosion and retreat of a sea cliff.

wave refraction The bending of waves so that they move nearly parallel to the shoreline.

Review Questions

1. **Explain what wave base is and how it affects waves as they enter shallow water.**

2. **How are waves responsible for transporting sediment long distances along shorelines?**

3. **Why does an observer at a shoreline location experience two high and two low tides daily?**

4. **How do rising sea level and coastal development, including developments on barrier islands, complicate efforts to control shoreline erosion?**

LO3   Understand the processes of shoreline erosion and deposition, and the concept of the nearshore sediment budget. Many shorelines are characterized by erosion. Such shorelines have sea cliffs ands wave-cut platforms. Other features commonly present include sea caves, sea arches, and sea stacks. Depositional coasts are characterized by long, sandy beaches, deltas, and barrier islands. Beaches, the most common shoreline depositional feature, are continuously modified by nearshore processes, and their profiles generally exhibit seasonal changes. Spits, baymouth bars, and tombolos all form and grow as a result of longshore current transport and deposition. Barrier islands are nearshore sediment deposits of uncertain origin. They parallel the mainland but are separated from it by a lagoon. The volume of sediment, or nearshore sediment budget, in a nearshore system remains rather constant unless the system is somehow disrupted, as when dams are built across streams that supply sand to the system.

LO4   Identify the different types of coasts and their features. Depositional coasts, such as the U.S. Gulf Coast, are characterized by abundant detrital sediment and depositional landforms. Erosional coasts are steep, irregular, and generally lacking in well-developed beaches. Many along the West Coast of North America fall into this category. Submergent and emergent coasts are defined on the basis of their relationship to changes in sea level. Coasts that form as a result of plate tectonics tend to be straighter.

LO5   Discuss the effects of storm waves and coastal flooding, and explain how coasts are managed as sea level rises. Coastal flooding during storms by waves and storm surge is an ongoing problem in many areas and is often responsible for the greatest damage and number of fatalities during hurricanes. The most significant disaster to hit the U.S. Gulf Coast in recent decades was Hurricane Katrina in August 2005, followed shortly by Hurricane Rita in September of the same year.

During the last century, sea level rose about 12 cm worldwide. The absolute sea-level rise in a shoreline region depends on the volume of water in the ocean basins and the rate of uplift of a coastal area. Many of the nearly 300 barrier islands along the East and Gulf Coasts of the United States are migrating landward as sea level rises. U.S. beaches and coastal areas are also threatened by beach erosion and sediment compaction. Seawalls and riprap are used in some areas to protect shorelines. Some states also regulate the development of coastal property. Other states have tried replacing eroded sand with new sand, but this can be expensive.

LO6   Recognize the types of resources found in the oceans. Table salt, as well as several other chemical compounds, is derived from seawater or deposits that formed from evaporating seawater. In addition, much of the phosphate-rich sedimentary rock that is used for fertilizers and animal feed supplements comes from rocks deposited where upwelling of oceanic waters took place. Seafloor deposits that are economically important include sand, manganese nodules, oil, and natural gas. The United States claims an area extending 200 nautical miles (371 km) offshore as an Exclusive Economic Zone (EEZ).

5. **While driving along North America's West Coast, you see a surface sloping gently toward the sea with several masses of rock rising above it. How would you explain the origin of this feature to your children?**

6. **How do a spit and a baymouth bar form?**

7. **Explain why waves are so small in ponds and lakes but much larger ones develop in the oceans.**

Chapter Summary

LO1   Explain how geologic time is measured. Time is defined by the methods used to measure it. Relative dating places geologic events in sequential order as determined from their positions in the geologic record. Absolute dating provides specific dates for geologic rock units or events, expressed in years before the present.

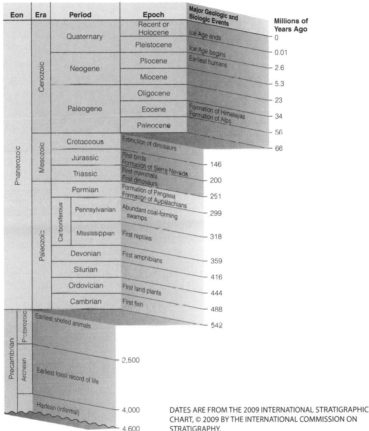

Eon	Era	Period	Epoch	Major Geologic and Biologic Events	Millions of Years Ago
Phanerozoic	Cenozoic	Quaternary	Recent or Holocene	Ice Age ends	0
			Pleistocene	Ice Age begins	0.01
		Neogene	Pliocene	Earliest humans	2.6
			Miocene		5.3
		Paleogene	Oligocene		23
			Eocene	Formation of Himalayas Formation of Alps	34
			Paleocene		56
	Mesozoic	Cretaceous		Extinction of dinosaurs	66
		Jurassic		First birds Formation of Sierra Nevada	146
		Triassic		First mammals First dinosaurs	200
	Paleozoic	Permian		Formation of Pangaea Formation of Appalachians	251
		Carboniferous	Pennsylvanian	Abundant coal-forming swamps	299
			Mississippian	First reptiles	318
		Devonian		First amphibians	359
		Silurian			416
		Ordovician		First land plants	444
		Cambrian		First fish	488
Precambrian	Proterozoic			Earliest shelled animals	542
					2,500
	Archean			Earliest fossil record of life	4,000
		Hadean (informal)			4,600

DATES ARE FROM THE 2009 INTERNATIONAL STRATIGRAPHIC CHART, © 2009 BY THE INTERNATIONAL COMMISSION ON STRATIGRAPHY.

LO2   Review early concepts of geologic time and Earth's age. During the 18th and 19th centuries, attempts were made to determine Earth's age based on scientific evidence rather than revelation. Although some attempts were ingenious, they yielded a variety of ages that now are known to be much too young.

LO3   Understand James Hutton's contribution to the recognition of geologic time. James Hutton, considered by many to be the father of modern geology, thought that present-day processes operating over long periods of time could explain all of the geologic features of Earth. His observations were instrumental in establishing the principle of uniformitarianism and the fact that Earth is much older than earlier scientists thought.

Uniformitarianism, as articulated by Charles Lyell, soon became the guiding principle of geology. It holds that the laws of nature have been constant through time and that the same processes operating today have operated in the past, although not necessarily at the same rates.

Key Terms

absolute dating Uses various radioactive decay dating techniques to assign ages to rocks in years before the present.

angular unconformity An unconformity below which older rocks dip at a different angle (usually steeper) than overlying strata.

carbon-14 dating technique Absolute dating technique relying on the ratio of C^{14} to C^{12} in an organic substance; useful back to about 70,000 years ago.

correlation Demonstration of the physical continuity of rock units or biostratigraphic units, or demonstration of time equivalence as in time-stratigraphic correlation.

disconformity An unconformity in which the rock layers above and below are parallel.

guide fossil Any easily identified fossil with an extensive geographic distribution and short geologic range useful for determining the relative ages of rocks in different areas.

half-life The time necessary for half of the original number of radioactive atoms of an element to decay to a new, and more stable daughter product; for example, the half-life for potassium 40 is 1.3 billion years.

nonconformity An unconformity in which stratified sedimentary rocks overlie an erosion surface cut into igneous or metamorphic rocks.

principle of cross-cutting relationships A principle holding that an igneous intrusion or fault must be younger than the rocks it intrudes or cuts across.

principle of fossil succession A principle holding that fossils, and especially groups or assemblages of fossils, succeed one another through time in a regular and predictable order.

principle of inclusions A principle holding that inclusions or fragments in a rock unit are older than the rock unit itself; for example, granite inclusions in sandstone are older than the sandstone.

principle of lateral continuity A principle holding that rock layers extend outward in all directions until they terminate.

principle of original horizontality According to this principle, sediments are deposited in horizontal or nearly horizontal layers.

principle of superposition A principle holding that in a vertical sequence of undeformed sedimentary rocks, the relative ages of the rocks can be determined by their position in the sequence—oldest at the bottom, followed by successively younger layers.

radioactive decay The spontaneous change of an atom to an atom of a different element by emission of a particle from its nucleus (alpha and beta decay) or by electron capture.

relative dating The process of determining the age of an event as compared to other events; involves placing geologic events in their correct chronological order but does not involve the consideration of when the events occurred in number of years ago.

unconformity A break in the geologic record represented by an erosional surface separating younger strata from older rocks.

Review Questions

1. **What is the difference between relative dating and absolute dating?**

2. **How can the various principles of relative dating be used to reconstruct the geologic history of Mars?**

3. **If you wanted to calculate the absolute age of an intrusive body, what information would you need?**

4. **Given the current debate over global warming and the many short-term consequences for humans, can you visualize how the world might look 10,000 years or even 1 million years from now? Use what you've learned about plate tectonics in terms of how plate movement and global warming will affect ocean currents and weather patterns to make your predictions.**

LO4   Describe relative dating methods and apply them to decipher the geologic history of an area. Besides uniformitarianism, the principles of superposition, original horizontally, lateral continuity, cross-cutting relationships, inclusions, and fossil succession are basic for determining relative geologic ages and for interpreting Earth history.

An unconformity is a surface of erosion, nondeposition, or both, separating younger strata from older strata. These surfaces encompass long periods of geologic time for which there is no geologic record at that location.

Three types of unconformities are recognized. A disconformity separates younger from older sedimentary strata that are parallel to each other. An angular unconformity is an erosional surface on tilted or folded rocks over which younger sedimentary rocks were deposited. A nonconformity is an erosional surface cut into igneous or metamorphic rocks and overlain by younger sedimentary rocks.

LO5   Explain the process of correlating rock units. Correlation is the demonstration of time equivalency of rock units in different areas. Similarity of rock type, position within a rock sequence, key beds, and fossil assemblages are all used to correlate rock units.

LO6   Explain absolute dating methods. Radioactivity was discovered during the late 19th century, and soon afterward, radiometric dating techniques enabled geologists to determine absolute ages for rock units and geologic events.

Absolute dates for rocks are usually obtained by determining how many half-lives of a radioactive parent element have elapsed since the sample originally crystallized. A half-life is the time it takes for one-half of the original unstable radioactive parent element to decay to a new, more stable daughter element.

The most accurate radiometric dates are obtained from long-lived radioactive isotope pairs in igneous rocks. The most reliable dates are those obtained by using at least two different radioactive decay series in the same rock.

Carbon-14 dating can be used only for organic matter such as wood, bones, and shells and is effective back to approximately 70,000 years ago. Unlike the long-lived isotopic pairs, the carbon-14 dating technique determines age by the ratio of radioactive carbon 14 to stable carbon 12.

LO7   Describe the development of the geologic time scale. The geologic time scale was developed primarily during the 19th century through the efforts of many people. It was originally a relative geologic time scale, but with the discovery of radioactivity and the development of radiometric dating methods, absolute age dates were added at the beginning of the 20th century. Since then, refinement of the time-unit boundaries has continued.

LO8   Describe and understand the methods used to determine the chronology of climate change. To reconstruct past climate changes and link them to possible causes, geologists must have a geologic calendar that is as precise and accurate as possible. Thus, they must be able to date geologic events and the onset and duration of climate changes as precisely as possible.

Chapter Summary

LO1   Describe Earth's Precambrian history. Cratons consist of a broad expanse of exposed ancient rocks known as shields, which are bordered by platforms of ancient rocks buried beneath sedimentary rocks and sediments. Each continent possesses at least one shield. In North America, the Canadian shield is made up of rocks as old as 4 billion years. All cratons have a history of accretion along their margins as they evolved to their present size and shape.

Archean rocks are mostly granite-gneiss complexes and greenstone belts, which are deformed, metamorphosed rocks that probably formed in back-arc basins that eventually closed. Many Proterozoic rocks are characterized as quartzite-carbonate-shale assemblages that were deposited on passive continental margins. Continental red beds, glacial deposits, and banded iron formations also formed during the Proterozoic.

LO2   Describe the Paleozoic geography of Earth.

At the beginning of the Paleozoic Era, there were six major continents. Laurentia moved northward, and Gondwana moved toward the South Pole during the Early Paleozoic (Cambrian–Silurian).

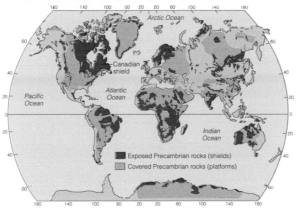

A. M. Goodwin, "The Most Ancient Continental Margins," In The Geology of Continental Margins, Burk and Drake (Eds), 1974, p. 768 (Fig 1), © Springer-Verlag.

Baltica and Laurentia collided to form Laurasia during the Late Paleozoic. Siberia and Kazakhstania collided, then were sutured to Laurasia. Gondwana moved over the South Pole, underwent several glacial-interglacial periods, which produced global sea level changes, and transgressions and regressions along the low-lying craton margins. Beginning in the Carboniferous, Lauasia and Gondwana underwent a series of collisions, resulting in the formation of Pangaea during the Permian.

LO3   Describe the Paleozoic evolution of North America. North America's geologic history is divided in cratonic sequences reflecting widespread transgressions and regressions. The first major transgression of North America was that of the Sauk Sea, which, at its maximum, covered the craton except for part of the Canadian shield and the Transcontinental Arch, a series of large, northeast-southwest-trending islands. An extensive sandstone deposited over the exposed and eroded Sauk landscape marks the beginning of the Tippecanoe sequence, which also includes widespread carbonate deposits. The Kaskaskia sequence was dominated by carbonates and evaporites, except for widespread Upper Devonian and Lower Mississippian black shales. During the Pennsylvanian Period of the Absaroka sequence, repetitive transgressive and regressive sequences of marine and nonmarine sediments produced cyclothems. There was also cratonic mountain building during the

Key Terms

Absaroka sequence A widespread association of sedimentary rocks bounded above and below by unconformities; deposited during a Late Mississippian to Early Jurassic transgressive-regressive cycle of the Absaroka Sea.

Acadian orogeny An episode of Devonian deformation in the northern Appalachian mobile belt resulting from the collision of Baltica with Laurentia.

Canadian shield The exposed part of the North American craton; mostly in Canada but also in Minnesota, Wisconsin, Michigan, and New York.

craton The relatively stable part of a continent; consists of a shield and a buried extension of a shield known as a platform; the ancient nucleus of a continent.

cyclothem A vertical sequence of cyclically repeating sedimentary rocks resulting from alternating periods of marine and nonmarine deposition; commonly containing a coal bed.

greenstone belt A linear or podlike association of igneous and sedimentary rocks; typically synclinal and consists of lower and middle volcanic units and an upper sedimentary rock unit.

Hercynian-Alleghenian orogeny Pennsylvanian to Permian orogenic event during which the Appalachian mobile belt in eastern North America and the Hercynian mobile belt of southern Europe were deformed.

Kaskaskia sequence A widespread sequence of Middle Devonian to Upper Mississippian sedimentary rocks bounded above and below by unconformities; deposited during a transgressive-regressive cycle of the Kaskaskia Sea.

Laramide orogeny A Late Cretaceous to Early Cenozoic episode of deformation in the area of the present-day Rocky Mountains.

Laurentia A Proterozoic continent composed mostly of North America, Greenland, and parts of Scotland and Scandinavia.

mobile belt An elongated area of deformation indicated by folds and faults; generally adjacent to a craton.

Nevadan orogeny Late Jurassic to Cretaceous deformation that strongly affected the western part of North America.

orogen A linear part of Earth's crust that was, or is, being deformed during an orogeny.

Ouachita orogeny A period of mountain building that took place in the Ouachita mobile belt during the Pennsylvanian Period.

platform The broad area extending from a shield, but covered by younger rocks; a platform and shield form a craton.

Sauk sequence A widespread association of sedimentary rocks bounded above and below by unconformities; deposited during a Late Neoproterozoic to Early Ordovician transgressive-regressive cycle of the Sauk Sea.

Sevier orogeny Cretaceous deformation that affected the continental shelf and slope areas of the Cordilleran mobile belt.

shield A vast area of exposed ancient rocks on a continent; the exposed part of a craton.

Taconic orogeny An Ordovician episode of mountain building that resulted in the deformation of the Appalachian mobile belt.

Tippecanoe sequence A widespread body of sedimentary rocks bounded above and below by unconformities; deposited during a Middle Ordovician to Early Devonian transgressive-regressive cycle of the Tippecanoe Sea.

Review Questions

1. **What is the largest exposed area of the North American craton?**

2. **Discuss how plate movement during the Paleozoic Era affected global weather patterns.**

3. **The breakup of Pangea began with initial Triassic rifting between which two continental landmasses?**

4. **What did the westward drift of North America and its collision with the Pacific-Farallon Ridge during the Cenozoic Era result in?**

Pennsylvanian, with thick accumulations of nonmarine detrital rocks and evaporites. The Absaroka Sea overlapped a narrow zone of the south-central craton and had completely retreated by the end of the Permian.

LO4   Understand and explain the history of Paleozoic mobile belts. An oceanic–continental convergent plate boundary formed during Tippecanoe time. This resulted in the Taconic orogeny, the first of several to affect the Appalachian mobile belt. The Taconic Highlands that formed shed sediments into a shallow western sea, producing a clastic wedge called the Queenston Delta. The Acadian orogeny followed the Taconic orogeny and began as an oceanic–continental convergent plate boundary. After Baltica collided with Laurentia, a continental–continental convergent plate boundary formed. The suturing of Laurasia and Gondwana and the formation of Pangaea are marked by the Hercynian-Alleghenian orogeny.

LO5   Explain the role of microplates in Paleozoic geology. The formation of Pangea also involved accretion of microplates, explaining previously anomalous geologic terranes.

LO6   Describe the stages in the breakup of Pangaea. Four stages are recognized in the breakup of Pangaea: 1) The separation of North America from Africa during the Late Triassic, followed by the separation of North America from South America; 2) the separation of Antarctica, India, and Australia from South America and Africa during the Jurassic, while India broke away from the still united Antarctica and the Australia landmass; 3) the separation of South America from Africa, while Europe and Africa began to converge; 4) the separation of Greenland from North America and Europe.

LO7   Explain the Mesozoic history of North America. The initial site of the separation of North America from Africa, which began during the Late Triassic, was the eastern coastal region. As North America rifted from South America during the Jurassic, the Gulf Coastal region was the site of major evaporate accumulation. During the Cretaceous, the north-south trending Cretaceous Interior Seaway divided North America into two landmasses. In the western region of North America, Mesozoic rocks were deposited in a variety of continental and marine environments. The region also was affected by four interrelated orogenies: the Sonoma, Nevadan, Sevier, and Laramide. Batholithic intrusions and eastward thrust faulting and folding were involved in each of these orogenies.

LO8   Explain Cenozoic Earth history. The Alpine-Himalayan orogenic belt and the circum-Pacific orogenic belt—each composed of smaller units called orogens— are the locations of most Cenozoic orogenic activity.

LO9   Describe the geologic history of the North American Cordillera. The complex mountainous region that extends from Alaska to Mexico is called the North American Cordillera. The Cenozoic evolution of the Cordillera involved deformation during the Laramide orogeny, extensional tectonics that formed the Basin and Range structures, intrusion and volcanism, and uplift and erosion. Westward drift of North America resulted in collision with the Pacific-Farallon Ridge. The present topography of the Appalachian Mountains is a result of Cenozoic uplift and erosion. During the Pleistocene, there were several intervals of widespread glacial and interglacial periods in North America, with about 20 recognized warm-cold climatic cycles.

Chapter Summary

LO1 Discuss the evolution of life during the Precambrian. All known life forms during much of the Precambrian were varieties of single-celled bacteria, with multicelled algae appearing at least 1 billion years ago and multicelled animals about 700 million years ago.

LO2 Review life during the Paleozoic Era. During the Cambrian and Ordovician periods, marine invertebrates with durable skeletons appeared in abundance; they diversified throughout the rest of the Paleozoic. Three major groups dominated the Cambrian invertebrate community—the brachiopods, the archeocyathids, and the trilobites. Dominance of the marine invertebrate community by the shelly fauna and the start of large-scale reef building began in the Ordovician. Major extinctions of many of the invertebrate phyla occurred at the end of the Ordovician. Diverse faunas, dominated by reef-building animals, characterize the Silurian and Devonian periods. There was a great decline in invertebrate diversity during the Carboniferous and Permian periods, with mass extinctions at the end of the Permian.

Edaphosaurus

Dimetrodon

© 2013 Cengage Learning

The earliest known vertebrates, the fish, appeared during the Late Cambrian. Their later history included jawless and jawed armored forms (ostracoderms and placoderms), cartilaginous forms, bony forms, and a lobe-finned group that gave rise to the amphibians. The critical factor in the reptiles' ability to colonize all parts of the land was the development of an egg that could be laid on land. The dominant reptile groups during the Permian were the pelycosaurs and therapsids.

The same basic problems faced by animals—desiccation, reproduction, and gravity—had to be overcome by plants in making the transition from water to land. Small, leafless stalks were the earliest terrestrial vascular plants, whose ancestors were probably some type of green algae. Many of the major structural features of today's plants evolved from this simple beginning.

LO3 Discuss life of the Mesozoic Era. The marine invertebrate survivors of the Permian extinction diversified, giving rise to increasingly complex Mesozoic marine invertebrate communities. Seedless plants and gymnosperms formed the Triassic and Jurassic land-plant communities, with angiosperms appearing during the Early Cretaceous, and diversifying rapidly to become the dominant land plant. Dinosaurs, which evolved during the Late Triassic, became most diverse and abundant during the Jurassic and Cretaceous. Two distinct orders, based on hip structure, are recognized: the saurichians (lizard-hipped) and ornithischians (bird-hipped). Evidence exists that supports endothermy in some dinosaurs, but the debate over whether dinosaurs were ectotherms or endotherms, or both, continues. The first flying

Key Terms

angiosperm Any of the vascular plants that have flowers and seeds; the flowering plants.

archosaur One of a group of animals including dinosaurs, flying reptiles (pterosaurs), crocodiles, and birds.

bipedal Walking on two legs as a means of locomotion, as in humans, birds, and some dinosaurs.

cynodont A type of advanced mammal-like reptile; the ancestor of mammals was among the cynodonts.

dinosaur Any of the Mesozoic reptiles that belong to the groups designated as ornithischians and saurischians.

eukaryotic cell A cell with an internal membrane-bounded nucleus that contains chromosomes and other internal structures; possessed by all organisms except bacteria

fossil The remains or traces of once-living organisms.

gymnosperm A flowerless, seed-bearing land plant.

hominid Abbreviated form of Hominidae, the family to which humans belong; bipedal primates include *Australophithecus* and *Homo*.

ichthyosaur Any of the porpoiselike, Mesozoic marine reptiles.

marsupial mammal Any of the pouched mammals such as kangaroos and wombats that give birth to their young in a very immature state; most common in Australia.

monotreme Any of the egg-laying mammals; includes only the platypus and spiny anteater of the Australian region.

Ornithischia One of the two orders of dinosaurs, characterized by a birdlike pelvis, includes ornithopods, stegosaurs, ankylosaurs, pachycephalosaurs, and ceratopsians.

placental mammal Any of the mammals that have a placenta to nourish the developing embryo; most living and fossil mammals.

plesiosaur A type of Mesozoic marine reptile.

primate Any of the mammals that belong to the order Primates; characteristics include large brain, stereoscopic vision, and grasping hands.

prokaryotic cell A cell that lacks a nucleus and organelles such as mitochondria and plastids; the cells of bacteria and cyanobacteria (blue-green algae).

pterosaur Any of the Mesozoic flying reptiles.

quadrupedal Walking on four legs as a means of locomotion.

Saurischia An order of dinosaurs characterized by a lizardlike pelvis; includes theropods, prosauropods, and sauropods.

seedless vascular plant A type of land plant that has specialized tissues for transporting fluids and nutrients throughout the body and that reproduces by spores rather than seeds (e.g., ferns and horsetail rushes).

stromatolite A biogenic sedimentary structure, especially in limestone, produced by the entrapment of sediment grains on sticky mats of photosynthesizing bacteria.

vertebrate Any animal that has a segmented vertebral column; includes fish, amphibians, reptiles, birds, and mammals.

vertebrates were the pterosaurs, at least one species of which was probably endothermic, as indicated by its having had hair.

The aquatic, porpoise-like ichthyosaurs ate fish, and the females probably hatched their eggs within their bodies to bear live young. The long-necked marine plesiosaurs probably laid their eggs ashore. During the Jurassic, birds probably evolved from small carnivorous dinosaurs. Mammals evolved during the Late Triassic, but the several types present during the Mesozoic were small, and their diversity remained low. Mesozoic mass extinctions at the end of the Cretaceous account for the disappearance of the dinosaurs, several other groups of reptiles, and many marine invertebrates. Impact of a large meteorite or comet with Earth may have caused the extinctions, according to one hypothesis.

Archaeopteryx

Tom McHugh/Photo Researchers, Inc.

LO4   Review life during the Cenozoic Era. The marine invertebrate groups surviving the Mesozoic extinctions continued expanding and diversifying during the Cenozoic. Mesozoic holdovers and a number of new groups comprised the Early Cenozoic mammalian fauna, with much diversification occurring. Much of the success of the placental mammals is attributed to their method of reproduction. Primates probably evolved during the Late Cretaceous. Traits that characterize and differentiate hominids from other mammal groups include walking on two legs rather than four, increased brain size, and the ability to manufacture and use sophisticated tools. The hominid fossil record extends back approximately 7 million years. Hominids migrated from Africa between 1 and 2 million years ago. Whether evolution from *Homo erectus*, the most recent link to our own species, *Homo sapiens*, occurred within a single population center, or in multiple population centers, is presently being debated. A remarkable aspect of Cenozoic mammal history is that so many very large species existed during the Pleistocene Epoch. This may have been an adaptation to the cooler temperatures existing then.

Review Questions

1. **What kinds of organisms existed during the Archean?**

2. **Describe the problems that had to be overcome before organisms could inhabit and completely colonize the land.**

3. **The greatest recorded mass extinction to affect Earth occurred at the end of which period?**

4. **What kinds of evidence indicate that dinosaurs were endotherms? Does this evidence apply equally to all dinosaurs?**

5. **What was the most important event in the Mesozoic evolution of land plants?**

6. **Which were the first hominids to migrate out of Africa and from which we evolved?**